Fundamentals of Gas Phase Ion Chemistry

NATO ASI Series

Advanced Science Institutes Series

A Series presenting the results of activities sponsored by the NATO Science Committee, which aims at the dissemination of advanced scientific and technological knowledge, with a view to strengthening links between scientific communities.

The Series is published by an international board of publishers in conjunction with the NATO Scientific Affairs Division

A Life Sciences **B Physics**	Plenum Publishing Corporation London and New York
C Mathematical and Physical Sciences **D Behavioural and Social Sciences** **E Applied Sciences**	Kluwer Academic Publishers Dordrecht, Boston and London
F Computer and Systems Sciences **G Ecological Sciences** **H Cell Biology** **I Global Environmental Change**	Springer-Verlag Berlin, Heidelberg, New York, London, Paris and Tokyo

NATO-PCO-DATA BASE

The electronic index to the NATO ASI Series provides full bibliographical references (with keywords and/or abstracts) to more than 30000 contributions from international scientists published in all sections of the NATO ASI Series.
Access to the NATO-PCO-DATA BASE is possible in two ways:

– via online FILE 128 (NATO-PCO-DATA BASE) hosted by ESRIN, Via Galileo Galilei, I-00044 Frascati, Italy.

– via CD-ROM "NATO-PCO-DATA BASE" with user-friendly retrieval software in English, French and German (© WTV GmbH and DATAWARE Technologies Inc. 1989).

The CD-ROM can be ordered through any member of the Board of Publishers or through NATO-PCO, Overijse, Belgium.

Series C: Mathematical and Physical Sciences - Vol. 347

Fundamentals of Gas Phase Ion Chemistry

edited by

Keith R. Jennings

Department of Chemistry,
University of Warwick,
Coventry, U.K.

Kluwer Academic Publishers

Dordrecht / Boston / London

Published in cooperation with NATO Scientific Affairs Division

Proceedings of the NATO Advanced Study Institute on
Fundamentals of Gas Phase Ion Chemistry
Mont Ste. Odile, France
June 25 – July 6, 1990

Library of Congress Cataloging-in-Publication Data

NATO Advanced Study Institute on Fundamentals of Gas Phase Ion Chemistry (1990 : Sainte-Odile, France)
Fundamentals of gas phase ion chemistry : proceedings of the NATO Advanced Study Institute on Fundamentals of Gas Phase Ion Chemistry, Mont Ste. Odile, France, 25th June-6th July, 1990 / edited by Keith R. Jennings.
p. cm. -- (NATO ASI series. Series C, Mathematical and physical sciences ; vol. 347)
Includes index.
ISBN 0-7923-1423-9
1. Chemical reactions, Conditions and laws of--Congresses.
2. Gases, Ionized--Congresses. I. Jennings, Keith R., 1932- II. Title. III. Series: NATO ASI series. Series C, Mathematical and physical sciences ; no. 347.
QD501.N365 1990
541.3'9--dc20 91-26411

ISBN 0–7923–1423–9

Published by Kluwer Academic Publishers,
P.O. Box 17, 3300 AA Dordrecht, The Netherlands.

Kluwer Academic Publishers incorporates the publishing programmes of
D. Reidel, Martinus Nijhoff, Dr W. Junk and MTP Press.

Sold and distributed in the U.S.A. and Canada
by Kluwer Academic Publishers,
101 Philip Drive, Norwell, MA 02061, U.S.A.

In all other countries, sold and distributed
by Kluwer Academic Publishers Group,
P.O. Box 322, 3300 AH Dordrecht, The Netherlands.

Printed on acid-free paper

Printed in the Netherlands

CONTENTS

PREFACE

This volume presents the proceedings of the 1990 Advanced Study Institute entitled "Fundamentals of Gas Phase Ion Chemistry" held at Mont Ste. Odile , Alsace, France, 25th June - 6th July, 1990. The Institute brought together over 100 physicists, physical and organic chemists working on a wide variety of topics with gas-phase ion chemistry as the common theme. Many different viewpoints, making use of very different experimental and theoretical approaches, were brought to bear on the subject and provided a stimulating and up-to-date account of the subject. Although the Institute was built around the invited lectures, many specific points were addressed in workshops which consisted of informal discussion groups which were organised by participants during the Institute. This volume therefore contains not only chapters based on the lectures but summaries of many of the workshops which adds considerably to the diversity of information presented.

This Advanced Study Institute was the fifth in a series of NATO-sponsored institutes devoted to various aspects of the physics and chemistry of gas phase ions. These meetings have been held every four years since the first, held in Biarritz in 1974, considered "Interactions between Ions and Molecules". The five volumes which comprise the proceedings of these meetings illustrate very clearly the many advances in theory and experiment which have taken place over the last 20 years. As this volume indicates, the increasing use of lasers has allowed the study of state-selected species, cluster ion chemistry is now a very active field and the processes involved in the unimolecular and bimolecular reactions of organic ions are becoming clearer. The increased understanding of gas phase ion chemistry has led to mass spectrometry being used in increasingly diverse ways in industry as described in two chapters in this volume.

Much of the local planning of the meeting at Mont Ste Odile fell upon Rose Marx whose invaluable help I should like to acknowledge. The meeting also benefited considerably from the experience of Pierre Ausloos and Sharon Lias and from the assistance in planning the meeting given by Alzira Almoster-Ferreira and Nico Nibbering.

The Organising Committee would like to acknowledge the assistance given by the staff at Mont Ste. Odile and the support of Dr. Luis V da Cunha, Director of the ASI Programme of the Scientific Affairs Division of NATO.

Keith R. Jennings
University of Warwick,
Coventry,
United Kingdom.
June, 1991

K. R. Jennings (ed.), Fundamentals of Gas Phase Ion Chemistry, ix.

MULTIPHOTON IONISATION: PART I FUNDAMENTAL PRINCIPLES

R J DONOVAN
Department of Chemistry
University of Edinburgh
West Mains Road
Edinburgh EH9 3JJ
Scotland

ABSTRACT: The fundamental processes involved in resonant and non-resonant multiphoton ionisation of atoms and molecules in the gas phase, using pulsed lasers, are described. The mechanisms for ion fragmentation are considered and the change from *ladder switching* to *ladder climbing*, with decreasing laser pulse duration (10^{-9} - 10^{-13} s), is discussed. Multiple ionisation and above threshold ionisation processes are also briefly discussed.

1. Introduction

Multiphoton ionisation techniques have been used extensively for fundamental spectroscopic and photophysical (dynamic) studies of atoms and molecules [1]. Applications to mass spectrometry have until recently been rather slow to materialise but the pace is increasing [2] and commercial instruments are becoming more generally available. Powerful new techniques which combine laser desorption of involatile and fragile (biological) molecules, combined with laser multiphoton ionisation and mass spectrometry, are also being developed [3].

In this chapter we briefly review the fundamental principles which underpin the multiphoton ionisation (MPI) technique and discuss the various ways in which both *resonant* and *non-resonant* ionisation processes can be used to their best advantage. In the following chapter a number of applications of the MPI technique will be outlined. As we shall see, two of the principle advantages of MPI in mass spectrometry are the ability to *selectively* ionise one component in a mixture and to bring about *either soft* or *hard* ionisation, simply by changing the intensity or the frequency of the laser radiation.

The main attraction of the MPI technique for analytical applications is the high sensitivity achievable (single atom detection is possible) [4,5]. We note that multiphoton excitation is operationally very much more convenient than the use of vacuum ultraviolet radiation (i.e. single photon ionisation) as it avoids the problems associated with evacuating the optical path and the use of LiF or MgF_2 optics. The

K. R. Jennings (ed.), Fundamentals of Gas Phase Ion Chemistry, 1–7.

selection rules for MPI are also more *generous* than those for single photon excitation and this allows a wider range of excited states to be accessed [1].

2. Resonant multiphoton ionisation

Multiphoton ionisation involves the absorption of two or more photons by the same molecule, the combined energy being sufficient to cause ionisation (i.e. $nh\upsilon > IE$; $n > 2$). The efficiency of the process is greatly enchanced if a *real* intermediate state is resonant with a simple multiple of the laser frequency and the highest efficiencies are observed when a single photon is sufficiently energetic to reach the intermediate state and a second photon sufficient to bring about ionisation from the intermediate state (see figure 1a). The important point here is that in many cases the intermediate state will have a relatively long lifetime ($> 10^{-8}$ s), unless predissociation is important, and thus the absorption of a second photon during the same laser pulse, by the excited molecule, has a relatively high probability. In this 1+1 *resonance enhanced* multiphoton ionisation (REMPI) process the photons are absorbed *sequentially*, i.e. incoherently. The intensities available with pulsed laser systems are generally high enough to saturate most electronic transitions (i.e. produce a number density in the upper state that is comparable to that in the lower state). The only disadvantage of this 1+1 REMPI scheme is that tunable far ultra-violet radiation (λ200 - 300nm) is required in order to provide sufficient energy to ionise most molecules. This requires frequency doubling and auto-tracking the output of a tunable dye laser which adds another level of sophistication and expense to the techniques involved. Commercial systems for frequency doubling and auto-tracking are now widely available but their optimisation and routine application is still significantly more difficult than the direct use of the fundamental output from a pulsed dye laser.

An alternative approach is to use the non-linear properties of the molecules themselves to bring about simultaneous two-photon absorption (see figure 1b). In this process a *virtual* intermediate state of the molecule is involved. This virtual state can be thought of as a state in which the molecule is "dressed" or perturbed by the photon field (NB a virtual state is *not* an eigen state of the molecule). Thus, if a second photon is incident on the molecule, within a sufficiently short period of time (*ca* 10^{-15} s), both photons are absorbed simultaneously (i.e. coherently). Clearly excitation only occurs when the resonance condition, $2h\upsilon = \Delta E$, is met. The selection rules for this type of process are similar to the Raman selection rules, as Raman scattering is itself a two photon process. Higher order processes involving 3, 4 and more photons have been observed but the cross- sections for such processes fall rapidly [1] and the laser intensities required for such high order processes can cause broadening of spectra and molecular fragmentation. Non-radiative processes (e.g. predissociation) which deplete the intermediate state will lead to a reduction in ionisation efficiency. However, the up-pumping rate even with nanosecond pulses is sufficiently rapid that some ionisation can be generally detected in all but the most extreme cases. The use of pico- and femtosecond pulses clearly allows more efficient ionisation, but at considerably greater expense and technical sophistication.

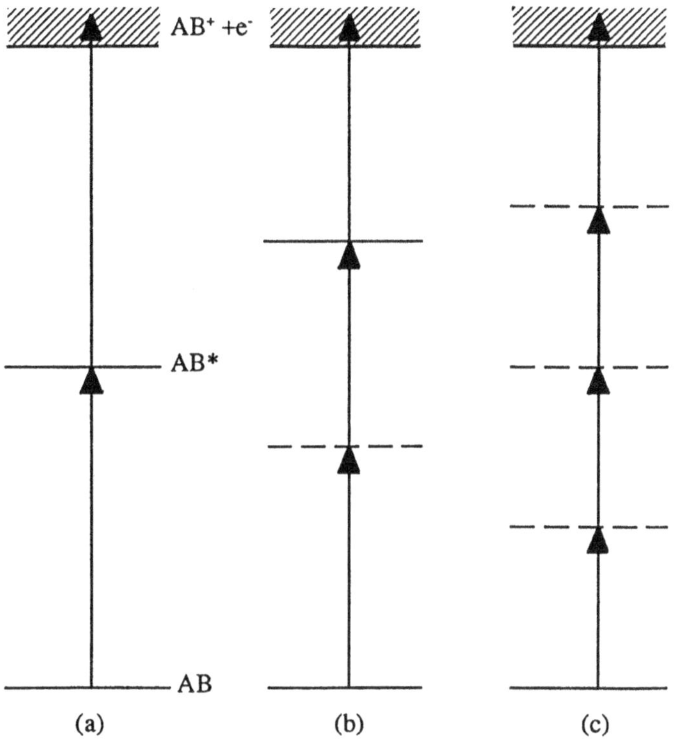

Figure 1. Some examples of excitation schemes for multiphoton ionisation:
(a) 1 + 1 REMPI
(b) 2 + 1 REMPI
(c) non-resonant MPI
The dashed lines represent *virtual* states of the molecule AB (see text).

3 Non-resonant multiphoton ionisation

The discussion at the end of the last section suggests that two or more photons could cause direct ionisation, without the involvement of a real intermediate state (see figure 1c), as the ionised molecule plus electron is also a state of the system (i.e. a continuum state). It will also be clear from the above discussion that such non-resonant ionisation will have a lower probability than processes involving real intermediate states. Notwithstanding, there are some advantages to using non-resonant ionisation, particularly if the ionisation of all components in a mixture is required. As a first step in the analysis of a sample ionisation of all the constituents is generally required and non-resonant MPI can be used in much the same way as electron impact ionisation. In principle MPI is capable of higher efficiency than electron impact ionisation, as there are no space charge limitations on the photon intensity. However, it should be noted that tight focussing of the laser beam is generally required to achieve such high intensities and the volume of gas ionised is thus likely to be small.

4 Fragmentation processes

At low laser intensities the parent ion is generally produced by MPI and this is useful for molecular weight determination. However, fragmentation is required for structural studies and this can be readily achieved by increasing the laser intensity and/or the frequency [2]. Figure 2 illustrates the effect of increasing the laser intensity on the fragmentation of o-xylene [6]. At 10^7 Wcm^{-2} the parent ion dominates the mass spectrum but as the intensity is raised fragmentation becomes increasingly important and at 10^9 Wcm^{-2} fragments down to C^+ are observed.

There has been some controversy over the mechanism that leads to ion fragmentation. The debate has centred around whether *ladder climbing* or *ladder switching* is more important. In the ladder climbing mechanism the molecules continue to absorb photons above the ionisation limit forming a super-excited state with sufficient energy to undergo fragmentation as well as ionisation, i.e. a process analogous to electron impact ionisation.

$$M + nh\upsilon \rightarrow M^* \xrightarrow{mh\upsilon} M^{**} \rightarrow F_1^+ + F_2 + e^- \qquad (1)$$

In contrast the ladder switching mechanism invokes separate steps for the ionisation and fragmentation processes. Thus, the additional photons required to bring about fragmentation are absorbed by the *ions* which are broken down in sequential steps, i.e.

$$M + nh\upsilon \rightarrow M^+ + e^- \qquad (2)$$

$$M^+ + h\upsilon \rightarrow F_1^+ + F_2 \qquad (3)$$

$$F_1^+ + h\upsilon \rightarrow F_3^+ + F_4 \text{ etc.} \qquad (4)$$

The available evidence points strongly towards the ladder switching mechanism being more important with nanosecond laser pulses [3] but ladder climbing is expected to become more important with shorter and higher intensity pulses (see below).

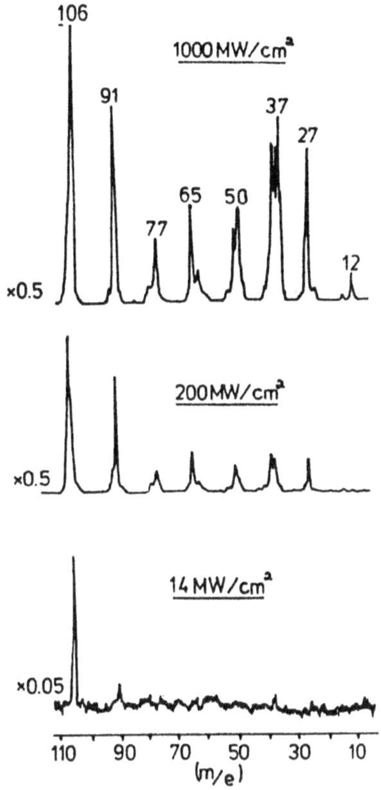

Figure 2. Effect of increasing laser intensity on the 1 + 1 REMPI mass spectrum of o-xylene (λ=268nm). At 14 $MWcm^{-2}$ the parent ion is dominant but at 10^9 Wcm^{-2} extensive fragmentation occurs down to C^+.

5 Multiple Ionisation and Above Threshold Ionisation

Multiple ionisation of atoms has been observed to become increasingly important when short high intensity laser pulses are used. Thus non-resonant multiple ionisation of xenon, up to Xe^{+6} is readily achieved using picosecond pulses with intensities in the region of 10^{14} Wcm^{-2} [7].

Above threshold ionisation has also been observed [8] with femtosecond pulses and a series a equally spaced peaks (ΔE=hυ) are seen in the photoelectron spectrum, corresponding to the absorption of up to eight additional (above threshold) photons, before ejection of the electron takes place.

The mechanism for the formation and fragmentation of multiply charged *molecular* ions has also been studied [9-11]. For diatomic molecules the

Coulomb explosion of the multiply charged parent molecule leads to fragment ions being ejected preferentially along the E–field direction of the polarised laser beam. This occurs because the molecules are more easily ionised when their axis lies along the E-field direction. An interesting new *triple* coincidence technique, termed *covariance mapping* has been used successfully to correlate the ionic products with the various possible fragmentation channels. Larger molecules have also been studied and the importance of the ladder climbing mechanism with very short pulses was confirmed. Extensive fragmentation of n-alkanes has been studied and the kinetic energies of the fragment ions determined. The results are interesting but difficult to interpret in any detail [11].

Clearly much remains to be done with very short pulse excitation but MPI with nanosecond laser pulses is relatively well understood. In Part II we shall use the principles discussed here to explore a number of applications based on the MPI technique.

References

[1] Ashfold, M.N.R. (1986) 'Multiphoton probing of molecular Rydberg states', Mol. Phys. 58, 1-20.

[2] Lubman, D.M. (1988) 'Analytical multiphoton ionization mass spectrometry', Mass Spectrometry Reviews 7, 535-554 and 559-592.

[3] Grotemeyer, J. and Schlag, E.W. (1988) 'Multiphoton-ionisation-mass-spectrometry (MUPI-MS)', Angew. Chem. Int. Edn. 27, 447-459.

[4] Hurst, G.S. (1982) 'Counting the atoms: some applications in chemistry', J. Chem. Educ. 59., 895-899.

[5] Engelke, F. Hahn, J.H., Henke, W. and Zare, R.N. (1987) 'Determination of phenylthiohydantoin - amino acids by two step laser desorption/multiphoton ionization', Anal. Chem. 59, 909-912.

[6] Blease, T.G., Donovan, R.J., Langridge-Smith, P.R.R., Ridley, T. and Wilkinson, J.P.T. (1986) 'Xylene isomer analysis using multiphoton ionisation spectroscopy', Inst. Phys. Conf. Ser. 84, 217-222.

[7] Perry, M.D., Szoke, A., Landen, O.L. and Campbell, E.M. (1988) 'Non-resonant multiphoton ionisation of noble gases: theory and experiment', Phys. Rev. Letters 60, 1270-1273.

[8] Muller, H.G., van Linden van den Heuvell, H.B., Agostini, P., Petite, G., Antonetti, A., Franco, M. and Migus, A., (1988) 'Multiphoton ionisation of xenon with 100-fs laser pulses', Phys. Rev. Letters 60, 565-568.

[9] Codling, K., Frasinski, L.J., Hatherly, P.A. and Barr, J.R.M. (1987), J. Phys. B20, L525.

[10]Frasinski, L.J., Codling, K., Hatherley, P.A., Barr, J.R.M., Ross, I.N. and Toner, W.T. (1987), Phys. Rev. Letters 58, 2424.

[11]Hatherly, P.A., Frasinski, L.J., Codling, K. and Barr, J.R.M. (1988) 'Multiphoton ionisation of n-alkanes using an intense sub-picosecond laser', Chem. Phys. Letters 149, 477-481.

MULTIPHOTON IONISATION: PART II APPLICATIONS

R J DONOVAN
Department of Chemistry
University of Edinburgh
West Mains Road
Edinburgh EH9 3JJ
Scotland

ABSTRACT: Some of the principal applications of the multiphoton ionisation technique are discussed. The following areas are covered; Rydberg state spectroscopy and the determination of ionisation energies, the detection of transient species, the production of state-selected ions, analytical applications and laser desorption of involatile and fragile molecules.

1. Introduction

In this chapter we build on the fundamental aspects of multiphoton ionisation discussed in Part I. Our discussion will be divided into five parts, namely:

(i) Rydberg state spectroscopy and the determination of ionisation energies,
(ii) Detection of transient species,
(iii) Production of state-selected ions,
(iv) Selective ionisation and analytical applications,
(v) Laser desorption and ionisation of involatile and fragile molecules.

Rydberg state spectroscopy is important to several of the other areas to be discussed, notably the production of state selected ions, and it is therefore appropriate that we start with this topic.

2. Rydberg State Spectroscopy and Determination of Ionisation Energies

This area will be illustrated by reference to recent work on molecular bromine [1,2]. Numerous studies of the Rydberg States of Br_2, using multiphoton ionisation techniques, have been reported. However, all of these studies have been limited to rather narrow regions of the spectrum, presumably due to the inconvenience and expense involved in multiple laser dye changes. Indeed, spectroscopic studies using dye

K. R. Jennings (ed.), Fundamentals of Gas Phase Ion Chemistry, 9–19.

lasers over wide spectral regions are very rare. This is unfortunate as assignments become difficult if only narrow regions are observed. In the work outlined below it was deliberate policy to scan the entire region between 50,000 and 80,000 cm^{-1}, using 2+1 resonance enhanced multiphoton ionisation (REMPI) techniques [1,2]. This appears to be the widest range covered to date in a single study and involved six dye changes in combination with the fundamental and doubled outputs from a dye laser. These points are mentioned to illustrate the operational difficulties involved in using dye lasers over a wide frequency range, when compared with synchrotron radiation: the vacuum ultraviolet spectrum of Br_2 (110 - 200 nm) can be scanned in only a few hours at a similar resolution, using synchrotron radiation. However, it should be remembered that spectra recorded with synchrotron radiation and REMPI spectra are complimentary, the former yielding data on the ungerade state manifold whilst the latter yields data on the gerade manifold. Furthermore, higher resolution is more readily achieved, over a limited spectral range, with laser techniques, and lasers are generally more accessible than synchrotrons.

The study of Br_2 Rydberg states was carried out simply by detecting the total ionisation, with charges collected by a pair of parallel plate electrodes biased at *ca* 100V, in a gas cell. Increasingly, *mass resolved* REMPI studies, employing molecular beam sample handling, are being carried out and some of the advantages of this approach will be evident from the discussion which follows.

The lowest Rydberg group based on the 5*s* Ryberg orbital is shown in Figure 1. It should be noted that the energetic limit for 2+1 REMPI of Br_2 lies at 354 nm, although some weak bands are observed below this limit [1]. Some of the weak bands are due to hot-band transitions but others must be due to another process. Ionisation of Br atoms is also observed in this region through 3+1 resonance ionisation, but the cross-section for 3+1 *molecular* ionisation would be too small to account for the observed bands. From separate studies using synchrotron radiation we know that direct ion-pair formation occurs below the ionisation limit for the parent molecule, through coupling between a Rydberg state and the ion-pair state continuum [3].

$$Br_2 + h\nu \rightarrow Br_2^* \text{ (Ry)} \rightarrow Br^+ + Br^- \qquad (1)$$

The weak bands observed above 354 nm may thus be due to a (2+1) excitation process, yielding Br^+ and Br^-.

An alternative explanation which is currently being investigated is that the Br atoms give rise to third harmonic generation and that ionisation occurs through absorption of the third harmonic followed by a further laser photon, taking the excited Br_2 molecule into the ionisation continuum. Clearly, these two mechanisms can be distinguished by direct observation of the ionic species formed.

A further Rydberg group based on the 7*s* orbital is illustrated in Figure 2. Rydberg states as high as 11*s* have been observed but a rising background ionisation makes observations beyond this very difficult [2]. This rising background may well be due to the ion-pair production noted above. Mass resolved studies would clearly eliminate this background.

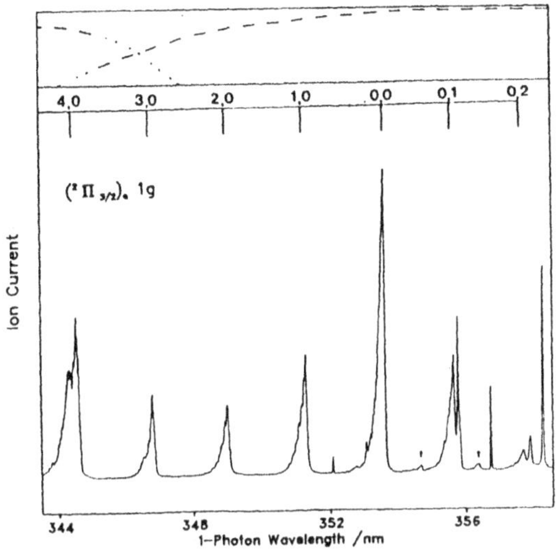

Figure 1. (2+1) REMPI spectrum of Br_2 uncorrected for laser power dependence with wavelength. The laser power is graphically indicated by the dashed curve in the upper portion of the figure. Note that the (4, 0) band was recorded with the shorter wavelength dye and its intensity is therefore enhanced relative to the (3, 0) band; all bands, other than the (4, 0) band, were obtained with the longer wavelength dye. The two peaks identified by (†) are due to impurity IBr. The atomic lines at 352.15, 355.84, 356.82 and 358.31 nm are due to (3+1) REMPI of Br produced via $Br(4s^2 4p^4 4d)\,^2D_{3/2}$, $^4P_{5/2}$, $^4P_{3/2}$, and $^2F_{5/2}$ Rydberg states, respectively.

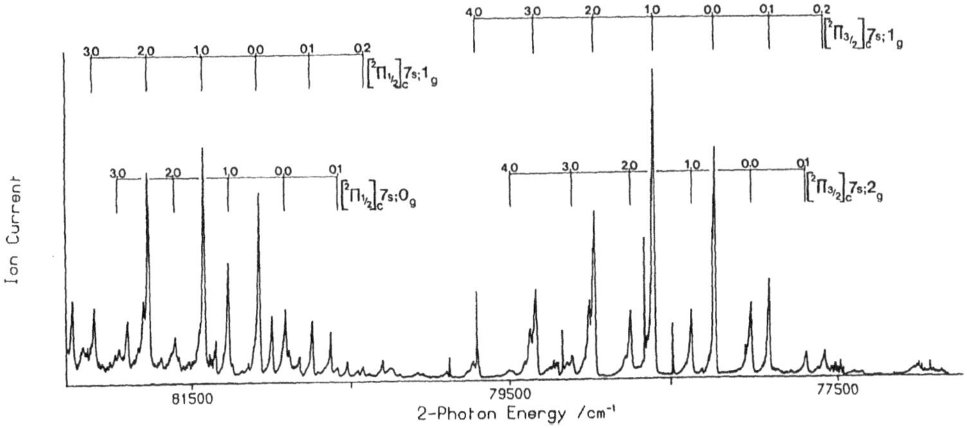

Figure 2. The (2+1) REMPI spectrum of Br_2 for λ_{ex}=243-259 nm normalized to the square of the laser power.

The single photon spectrum of Br_2 was studied at high resolution by Venkateswarlu [4] using conventional spectroscopic techniques, and numerous Rydberg series of ungerade symmetry were identified. Some series were observed to very high values of the principal quantum number ($n \cong 50$), however the ionisation energy derived from these series limits has been shown to be in error by approximately $350 cm^{-1}$ (this is close to one quantum of vibrational energy in either the ion or the ground electronic state, the two being quite close in energy) and the assignments must therefore been questioned. The Rydberg series observed using REMPI are consistent with the now accepted value for the ionisation energy of Br_2 and thus reinforce its validity [2]. The difficulty in obtaining an accurate ionisation energy for Br_2, from convergence limits for Rydberg series, arises from the similar geometries and vibrational frequencies in Br_2 and Br_2^+.

It is interesting, that the gerade Rydberg states Br_2 appear to be less perturbed than those in the ungerade manifold. Thus vibrational progressions and quantum defects are much more regular in the gerade states, compared with the ungerade manifold. It is not obvious why this should be and it therefore cannot be expected to be a general feature of Rydberg state manifolds. Clearly, in the case of Br_2 it greatly assists the assignment of gerade Rydberg states and the determination of the ionisation energy.

More direct and precise ionisation energies can be obtained using optical-optical double resonance techniques. However, this requires the use of *two* tunable lasers, the first to pump the molecule to a selected real intermediate state and the second to pump out of the intermediate state and *scan* across the ionisation threshold. Field ionisation effects have to be carefully taken into account as a field of about 1V can cause a shift of some $10\ cm^{-1}$. By measuring the ionisation energy over a range of known field strengths and extrapolating to zero field, ionisation energies can be obtained with precisions of the order of one wave number.

3 Detection of Transient Species

The most detailed quantitative studies have been carried out with atoms in order to compare experimental cross-sections with theory. A wide range of atoms, generally produced in discharge flow systems, have been studied but we shall concentrate here on oxygen atoms and the work of Bischel and coworkers [5,6] to illustrate the area.

The absolute two-photon cross section was obtained for the transition $3p^3P_J \leftarrow 2p^3P_2$. The integrated cross section, over all upper J states, was obtained as $(1.87 \pm 0.60) \times 10^{-35} cm^4$ by monitoring the induced two-photon fluorescence. This value, together with similar values measured for the n = 4 and 5 levels were in reasonable agreement with *ab initio* calculations. The photo-ionisation cross section out of the $3p^3P$ electronic state was then measured as $(5.3 \pm 2.0) \times 10^{-19} cm^2$. Thus the overall 2+1 multiphoton process has been quantitatively characterised for oxygen atoms excited via the np^3P states (n = 3 and 4). An important aspect of these studies was the attention paid to the mode structure of the laser. A single frequency laser was used, thus eliminating uncertainties over unresolved temporal fluctuations: the results were compared with those obtained with a multi-mode laser [6].

A wide range of simple free radicals have been observed using multiphoton ionisation techniques and a good general review of this area has been given by Hudgens [7]. The reader is referred to this review for further details.

4 Production of State-Selected Ions

Vibrational state selection is relatively straight forward with diatomic and small polyatomic molecules. The important point is that the resonance intermediate state, which is normally a Rydberg state, determines the state of the ion produced as the geometry of the ion will be close to that of the Rydberg state and the Franck-Condon factors will strongly favour transitions with $\Delta v = 0$. Thus, provided the spectroscopy of the intermediate state is reasonably well understood the excitation process to form the ion will be well defined. Numerous studies have examined the energy of the photoeletrons produced by multiphoton ionisation and these confirm that the state of the ion can generally be predicted from a knowledge of the resonant intermediate state. The dynamics of the excitation process have been well reviewed by Kimura [8].

Rotational state selection is considerably more difficult as the rotational state of the resonant intermediate is generally not so easily resolved, due to congestion (particularly for non-hydrides) or due to line broading (e.g. predissociation). Furthermore the resolution of conventional photoelectron techniques (typically 10meV) is not sufficient to allow precise identification of the ionic states produced.

In principle, threshold, or zero kinetic energy photoelectron methods (resolution $\hat{=}$ 1μeV) can be employed to overcome this problem. However, this introduces another level of experimental sophisication and threshold techniques have been less widely used than conventional photoelectron techniques. This situation is changing quite rapidly but it will probably be sometime before threshold techniques are in routine use.

An alternative approach is to use laser induced fluoresence techniques to identify the ro-vibrational state of the ions produced. This technique should be widely applicable provided the spectroscopy of the ion is sufficiently well understood to allow detailed identification of the ionic state. A few pioneering studies have reported the production of ro-vibrational state selected diatomic ions. Xie and Zare [9] have reported studies of ro-vibronic state selected HBr^+, Opitz *et al* [8] results for N_2^+, and Ito *et al* results for CO^+ [11].

5. Selective Ionisation and Analytical Applications

The involvement of a resonant intermediate state in the multiphoton excitation process ensures that *selective* ionisation of specific atoms or molecules occurs, even in the presence of a large excess of other species. We shall illustrate this point by reference to work on the selective ionisation of a single isomer of xylene in a mixture of the three isomers, ortho-, meta- and para-xylene [12]. Before we do this however, we need to consider a problem associated with the complexity of molecular spectra. The high density of rovibronic states associated with molecules makes their spectra far more complex than those of atoms and also dilutes the states accessible for excitation with monochromatic radiation (i.e. the population is spread over a large number of states and only a few states can interact with the monochromatic radiation), under normal laboratory conditions. This problem becomes increasingly severe with increase in molecular size. However, the problem can be overcome to a larger extent by cooling molecules to *ca* 10K using a pulsed supersonic jet. The effect of cooling para-xylene in a supersonic jet is shown in Figure 3.

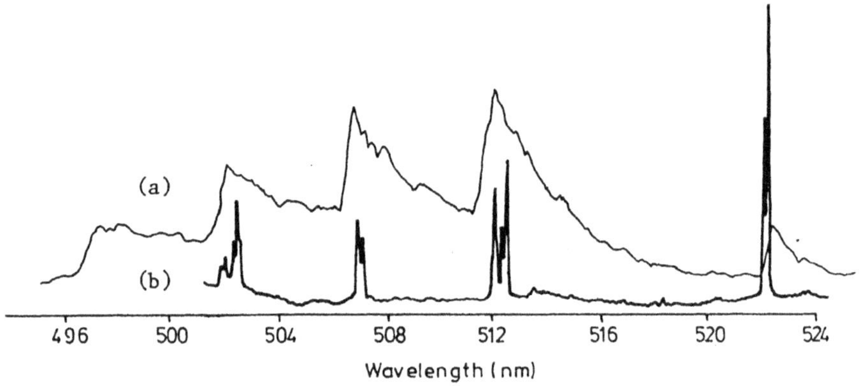

Figure 3. Two-photon resonant four-photon ionisation (2+2) REMPI spectra of para-xylene obtained (a) at room temperature and (b) with a supersonic jet, illustrating the effect of expansion cooling.

The spectrum at room temperature consists of a series of broad overlapping peaks, while at *ca* 10 K (supersonic jet) the peaks are sharp and there is very little overlap [12]. As the spectra of the three isomers all lie in the same region and overlap extensively with each other at 300 K, jet cooling is essential.

Two ionisation schemes were investigated. The first and most sensitive scheme was resonant two-photon ionisation, i.e. (1+1) REMPI:

$$Xy(S_0) \xrightarrow{h\upsilon_1} Xy^*(S_1) \xrightarrow{h\upsilon_1} Xy^+ + e^- \qquad (2)$$

Wavelengths in the region of 270nm are required for this scheme and the doubled output of a dye laser was therefore used. The resonant two-photon ionisation spectra of the three isomers are shown in Figure 4.

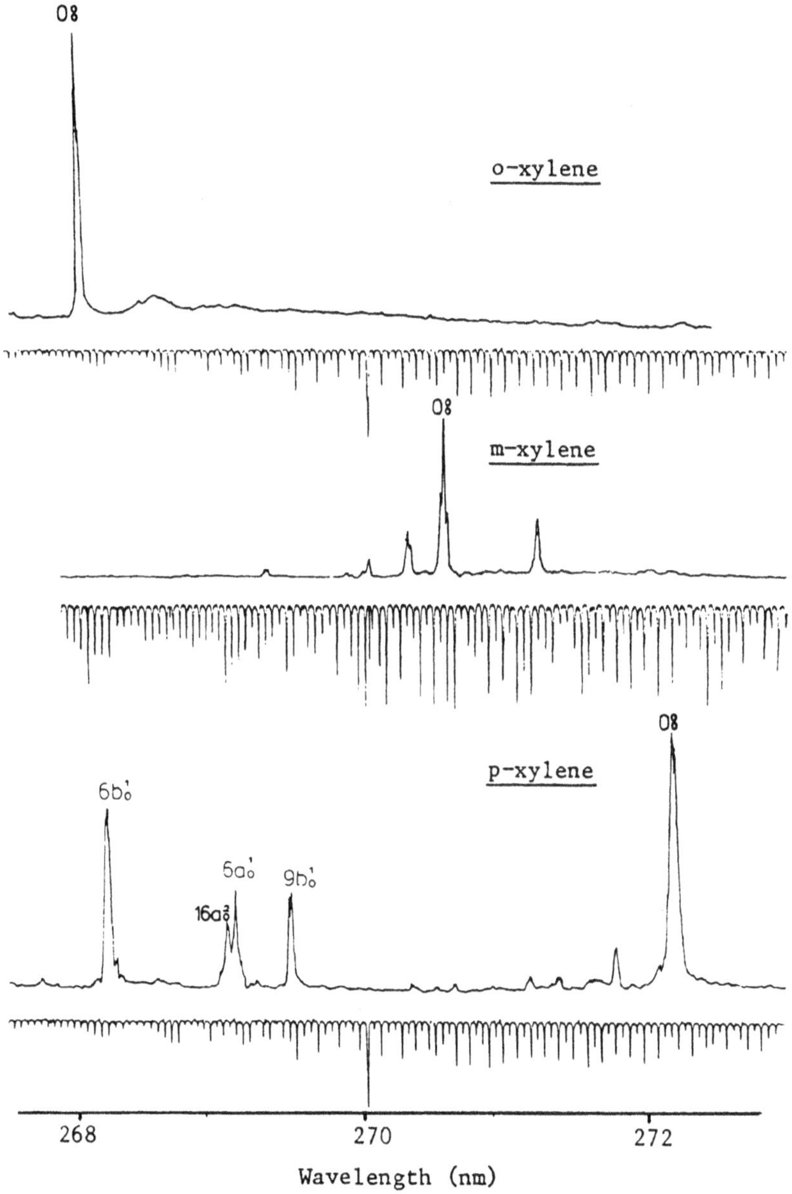

Figure 4. One photon resonant two-photon ionisation (1+1) REMPI spectra of the jet-cooled xylene isomers in the region of the origin bands. The lower traces are a combination of optogalvanic lines and etalon fringes.

The observed resonances correspond with transitions to the first singlet state (S_1) of the xylene isomers [12].

The second ionisation scheme involved two-photon resonant four-photon ionisation, i.e. (2+2) REMPI:

$$Xy(S_0) \xrightarrow{2h\upsilon_2} Xy^*(S_1) \xrightarrow{2h\upsilon_2} Xy^+ + e^- \qquad (3)$$

This scheme has the advantage of directly employing the visible output of the dye laser (*ca* 500nm), but suffers from the disadvantage that cross-sections associated with the two-photon transitions are inherently much smaller than from single photon absorption. The two-photon resoant four-photon ionisation (2+2) REMPI spectra of the xylene isomers are shown in Figure 5.

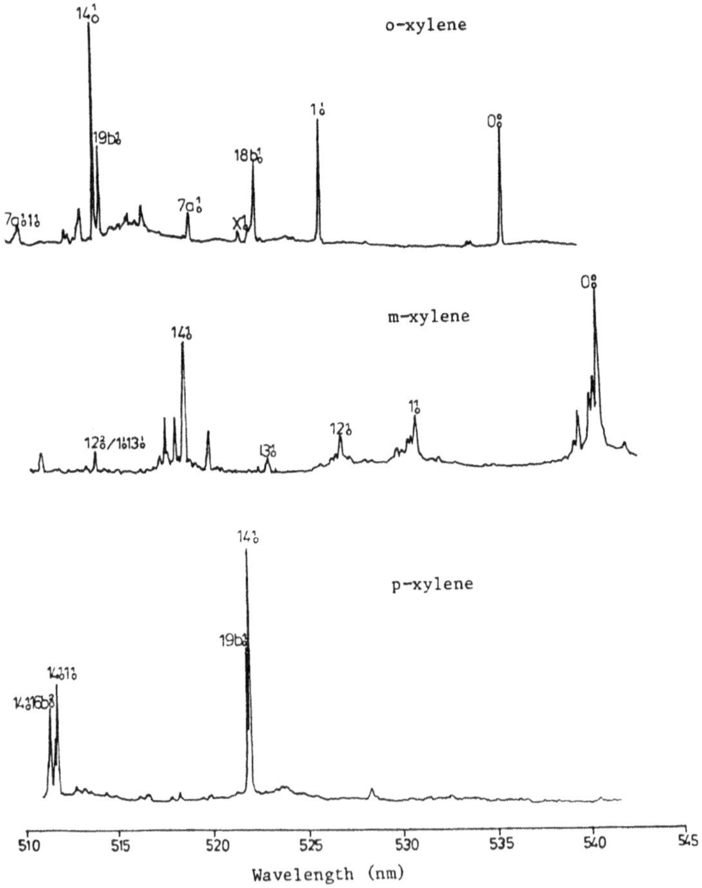

Figure 5. Two-photon resonant four-photon ionisation (2+2) REMPI spectra of the xylene isomers in the region of the origin bands.

It should be noted that the origin band is absent in the two-photon resonant spectrum of para-xylene, as it is symmetry forbidden. Extensive vibronic structure was observed for all three isomers to shorter wavelength [13].

Both ionisation schemes clearly allow the selective ionisation of individual xylene isomers and experiments were carried out to establish the limits for detecting one isomer in the presence of other isomers. The resonant two-photon ionisation scheme was chosen for this, due to the higher signal/noise ratio achieved with this approach. Mixtures of ortho- and para-xylene were examined and a limit for detecting para-xylene in such mixtures was determined as 0.5% (i.e. one part of para-xylene could be detected in the presence of 200 parts of ortho-xylene). This limit could probably be improved with further work. The absolute detection limit for the xylene isomers with the present experimental arrangement was estimated as *ca* 10^{10} xylene molecules cm^{-3}: again it should be possible to improve on this significantly.

6. Laser Desorption and Ionisation of Involatile and Fragile Molecules

Laser desorption techniques can be used to produce large and thermally unstable molecules in the gas phase, without fragmentation*. In most cases a matrix is not required and, if the compound of interest can be thinly deposited on a suitable surface (metal and glass surfaces are often used), it can generally be desorbed intact using a low power laser. This is best achieved by the use of a second laser operating in the infra-red region: UV desorption has also been used but is generally less successful. Selective ionisation of the desorbed neutral molecules is then achieved by using a second UV laser to bring about resonance ionisation.

The pulse duration of most conventional laser systems is typically 10ns and the analysis of the ions produced in a resonance enhanced experiment is therefore very conveniently carried out using time-of-flight mass spectrometry. For large molecules and high resolution studies reflectron ion optics are now commonly employed and a resolution of 10^4 has been achieved.

At low laser intensities the dominant ion produced is the parent molecule ion. However, as the laser intensity is increased fragmentation increases and it is possible to move smoothly from "soft" to "hard" ionisation conditions simply by changing the laser intensity (see Part I).

---

*Direct ionisation, simultaneous with desorption, can be achieved with a single high power laser pulse but this approach leads to extensive fragmentation and is less well controlled compared with neutral desorption referred to here. Furthermore, the yield of ions in this type of process is very low (10^{-4}) compared to the yield of desorbed neutral species.

Increasing emphasis has been placed on the analysis of delicate biological samples and considerable success has been achieved in sequencing bio-polymers [14-19]. High sensitivity has been achieved by Zare and co-workers who claim close to picomole sensitivity [18,19].

The multiphoton ionisation technique has been applied to a large number of other areas but space does not permit their inclusion here. However, one area that must be mentioned is the analysis of solid samples and surface analysis using the *laser microprobe mass spectrometry* technique. Commercial instrumentation for this type of work is available: the historical development and some of the uses of the laser microprobe have been reviewed by Clarke [20].

Although the above discussion is incomplete it is hoped that at least the general flavour of what is possible with multiphoton ionisation techniques has been imparted.

References

[1] Lawley, K.P., Donovan, R.J., Ridley, T., Yencha, A.J. and Ichimura, T. (1990) 'First Rydberg cluster in Br_2 observed via two-photon resonant multiphoton ionisation', Chem. Phys. Letters, 168, 168-172.

[2] Ridley, T., Lawley, K.P., Donovan, R.J. and Yencha, A.J. (1990) 'Resonance enhanced multiphoton ionisation spectrum of Br_2: the lower Rydberg states', Chem. Phys., 148, 315-323.

[3] Yencha, A.J., Kela, D.K., Donovan, R.J., Hopkirk, A. and Kvaran, A. (1990) 'Ion-pair (Br^+ + Br^-) formation from photodissociation of Br_2 near the first ionisation limit', Chem. Phys. Letters, 165, 283-288.

[4] Venkateswarlu, P. (1969) 'The vacuum ultraviolet spectrum of Br_2', Can. J. Phys., 47, 2525-2538.

[5] Bamford, D.J., Saxon, R.P., Jusinski, L.E., Buck, J.D. and Bischel, W.K. (1988) 'Two-photon excitation of atomic oxygen at 200.6, 192.5 and 194.2nm: absolute cross sections and collisional ionization rate constants', Phys. Rev., A37, 3259-3269.

[6] Bamford, D.J., Bischel, W.K., Hickman, A.P. and Dyer, M.J. (1988) ' Measurement of two-photon absorption cross sections in atomic O at 226nm: single frequency versus multimode lasers', SPIE, 912, 139-144.

[7] Hudgens, J.W. (1988) in S.H. Lin (ed.) Advances in multiphoton processes and spectroscopy, World Scientific, Singapore.

[8] Kimura, K. (1987) 'Molecular dynamic photoelectron spectroscopy using resonant multiphoton ionization for photophysics and photochemistry', Int. Rev. Phys. Chem., 6, 195-226.

[9] Xie, Jinchun and Zare, R.N. (1989) 'Rotationally state-selected HBr^+: preparation and characterisation', Chem. Phys. Letters, 159, 399-405.

[10]Opitz, S., Proch, D., Trickl, T. and Kompa, K.L. (1990) 'State-selective ionization of nitrogen by resonance-enhanced three-and four-photon excitation', Chem. Phys., 143, 305-323.

[11]Fugii, A., Ebata, T. and Ito, M. (1989) 'Production of Rotationally state selected ions by resonant enhanced multiphoton ionization of CO in a supersonic free jet', Chem. Phys. Letters, 161, 93-97.

[12]Blease, T.G., Donovan, R.J., Langridge-Smith, P.R.R., Ridley T. and Wilkinson, J.P.T (1986) 'Xylene isomer analysis using multiphoton resonance ionisation spectroscopy', Inst. Phys. Conf. Ser., 84, 217-222.

[13]Blease, T.G., Donovan, R.J., Langridge-Smith, P.R.R. and Ridley, T. (1988) '(2 + 2) Resonance enhanced multiphoton ionization spectra of jet-cooled ortho-, meta- and para- xylene', Laser Chem., 9, 241-263.

[14]Grotemeyer, J. and Schlag, E.W. (1989) 'Biomolecules in the gas phase: multiphoton ionisation mass spectrometry', Acc. Chem. Res., 22, 399-406.

[15]Grotemeyer, J. and Schlag, E.W. (1988) 'Multiphoton-ionisation-mass-spectrometry (MUPI-MS)', Angew. Chem. Int. Edn. 27, 447-459.

[16]Beavis, R.C., Lindner, J. Grotemeyer, J. and Schlag, E.W. (1989) 'Resonance enhanced multiphoton ionization of biological molecules', Ber. Bunsenges. Phys. Chem., 93, 365-370.

[17]Lubman, D.M. (1988) 'Analytical multiphoton ionization mass spectrometry', Mass Spectrometry Reviews 7, 535-554 and 559-592.

[18]Engelke, F., Hahn, J.H., Henke, W. and Zare R.N. (1987) 'Determination of phenylthiohydantoin - amino acids by two step laser desorption/multiphoton ionization', Anal. Chem. 59, 909-912.

[19]Zare, R.N., Hahn, J.H. and Zenobi, R. (1988) 'Mass spectrometry of molecular adsorbates using laser desorption/laser multiphoton ionization', Bull Chem. Soc. Japan, 61, 87-92.

[20]Clarke, N.S. (1989) 'Laser microprobe mass spectrometry', Chem. in Britain, 25, 484-488.

ION DISSOCIATION AND ION-MOLECULE REACTIONS STUDIED WITH STATE-SELECTED IONS

O. DUTUIT
Laboratoire de Physico-chimie des Rayonnements et LURE
Bât.350
Centre Universitaire d'Orsay
91405- Orsay CEDEX
FRANCE

ABSTRACT. The internal energy dependence is examined in ion dissociation and ion-molecule reactions with state selected ions produced by photoionisation with synchrotron radiation and threshold photoelectron-photoion coincidences. The dissociation mechanism of the methane ion and its deuterated compounds stongly varies with internal energy of the parent ion. Whereas the CH_4^+ X ground state dissociation is statistical, the A $(2a_1)^{-1}$ state (22.4-25 eV photon energy) and the doubly excited states (22.4-32 eV) show direct dissociation mechanisms. The preparation mode of state selected ions by autoionisation is shown to have an influence on the ion dissociation. Examples are given for the dissociation of CO_2^+, CF_3I^+ and H_2O^+. In the second part, the ion-molecule reaction mechanisms are also shown to be strongly dependent on the internal energy (electronic, vibrational and spin-orbit) of the reactant ion. A near resonant charge transfer reaction $N_2^+(X,A,v) + Ar \rightarrow Ar^+ + N_2$ and a non resonant charge transfer reaction $O_2^+(X,a,v) + Ar \leftrightarrow Ar^+ + O_2$ are discussed.

Introduction

State selected chemistry contributed significantly to the understanding of molecular dynamics for both unimolecular and more recently bimolecular processes.

The role of electronic energy in chemical reactivity has been known for a long time, but it is only recently that the role of vibrational energy was investigated. Even more recently some data showed the influence of rotational and fine structure energies.

1. Main techniques for state selection of ions

Electron impact techniques to produce ions analysed by mass spectrometry have provided a lot of data with non-selected ions. State selected chemistry was then developed with photoionisation techniques : single photon ionisation (with helium lamps or synchrotron

K. R. Jennings (ed.), Fundamentals of Gas Phase Ion Chemistry, 21–54.

radiation light sources) associated to coincidence methods as well as laser multiphoton ionisation techniques.

The principles for ionic state selection by photoionisation methods were recently reviewed by T.Baer [1] to study ion dissociation and by C.-Y. Ng [2] to study ion-molecule reactions.

1.1. PHOTOELECTRON-PHOTOION COINCIDENCES

Photoelectron-photoion coincidences (PEPICO) have been most successfully used to select vibronic states of ions produced by photoionisation [3]. This technique is very generally applicable to any kind of molecule. In the ionisation process with a vacuum ultraviolet photon

$$M + h\nu \dashrightarrow M^+ + e^-$$

the ion M^+ can be formed in a range of internal energy states from its ground state to an energy $h\nu$ - IP, where IP is the ionisation potential. The excess energy is released as kinetic energy of the ejected photoelectron because of its very light mass compared to the ion M^+. The energy relation is simply given by

$$E_{int}(M^+) = E_{h\nu} - IP - KE_e$$

where $E_{int}(M^+)$ is the internal energy of the ion M^+, $E_{h\nu}$ the excitation photon energy and KE_e the kinetic energy of the ejected photoelectron. In order to select the internal energy of the ion M^+, it is thus necessary to detect ions in coincidence with a fixed photoelectron kinetic energy for a known photon energy. Two different techniques can be used experimentally : a fixed photon energy (usually 21.21 eV from a helium lamp) and a variable photoelectron kinetic energy or a fixed photoelectron kinetic energy (usually near zero energy electrons also called threshold electrons) and a tunable photon energy (synchrotron radiation). The second method called TPEPICO (Threshold PhotoElectron-PhotoIon Coincidences) has several advantages. Firstly the electron detection efficiency is much higher for threshold electrons (near 100%) than for energetic electrons (a few percent for most electron analysers). Secondly, a larger variety of vibronic ionic states can be populated by resonant autoionisation which only occurs with continuously variable photon energy, as will be explained just below.

Figure 1 shows a comparison of the vibrational levels of O_2^+ which are produced in the same energy region with a helium lamp and with synchrotron radiation. Fig.1a shows that with the 21.21 eV helium line, the O_2^+ vibrational levels are populated according to the Franck-Condon factors, which in particular implies the population of only the first five vibrational levels of the $X^2\Pi_g$ ground state. In contrary Fig.1b shows that all vibrational levels up to v=20 of the $X^2\Pi_g$ are populated with tunable synchrotron radiation. In this latter case, in addition to the direct ionisation which populates vibrational levels according to Franck-Condon factors, it is possible to excite neutral Rydberg states located above the ionisation potential which can decay by autoionisation. Among the different decay processes of these Rydberg states, they can loose a near-zero kinetic energy electron to form the energy resonant ionic state. This process is the one occuring in the case of Fig.1b which is a plot of the amount of threshold electrons as a function of photon energy (Threshold PhotoElectron Spectrum or TPES). It is very general (observed in all molecules and in any energy region up to the double ionisation threshold) and is not governed by the Franck-Condon principle. In some very specific cases, such as H_2^+ and $C_2H_2^+$ ground state, this autoionisation process is dominant and produces selected vibronic levels without requiring coincidences. This has been used in the pioneer work of W.Chupka [4] and later in the group of Y.T.Lee [5] to study state selected ion-molecule reactions.

The TPEPICO method requires a continuously tunable photon energy source in the vacuum ultra-violet. Some experiments have been done with lamps producing the helium or the H_2 many line spectrum, but their intensity is lower. VUV lasers are hard to operate above 10 eV and are tunable only in very restricted energy regions. So up to now, synchrotron radiation has been the best source. Because of the difficulties to have access to synchrotron radiation beam lines with such sophisticated experiments as those employing coincidence methods, only a few groups have developed this field [6,7].

1.2. MULTIPHOTON IONISATION WITH LASERS

The second powerful method to produce state selected ions uses multi-photon ionisation with lasers. This method gives a population of ion states which cannot be state selected by usual coincidence methods, because several ions are produced by one laser pulse.

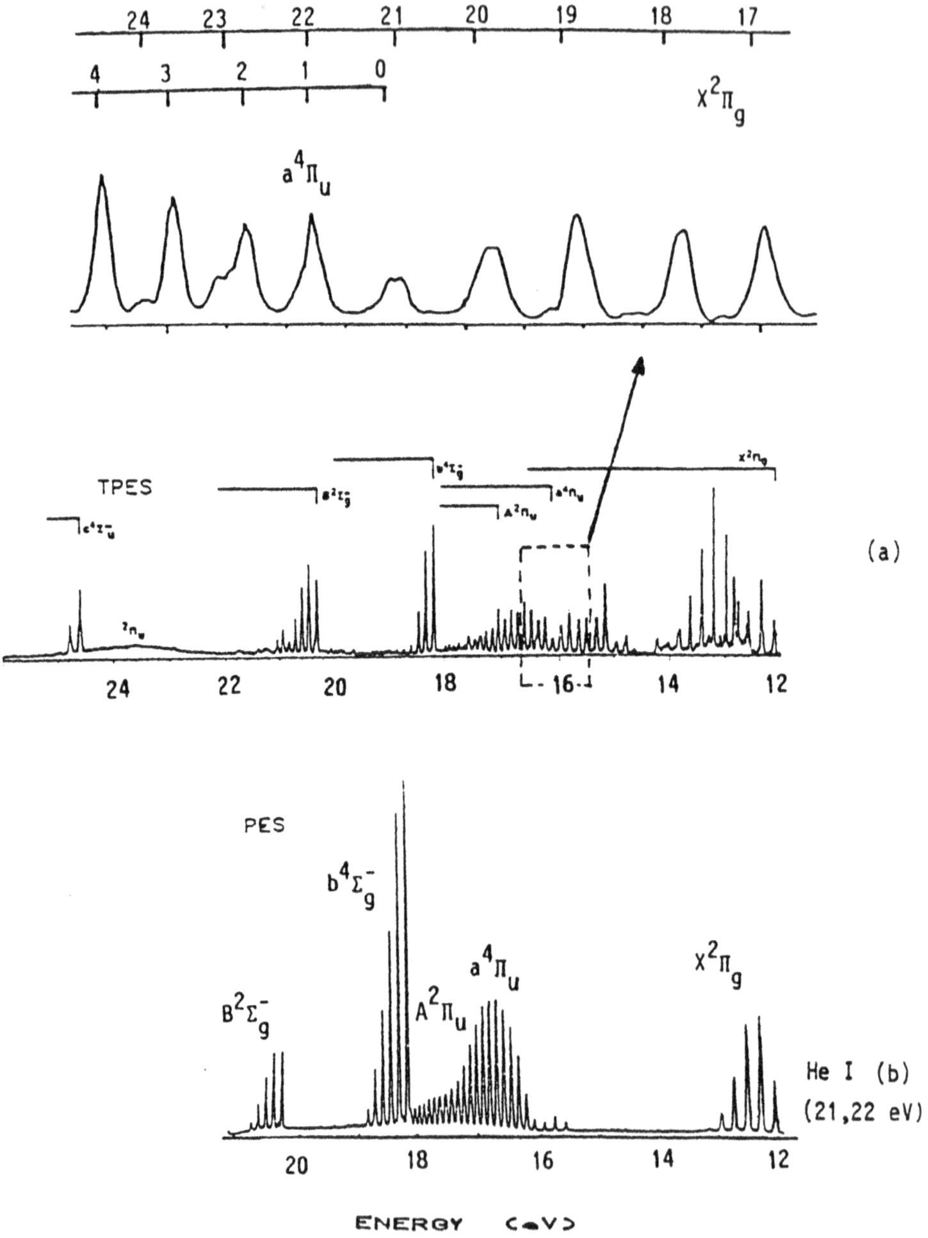

Figure 1. Comparison of the threshold photoelectron spectrum TPES (a) and the photoelectron spectrum PES obtained with the HeI radiation line (b) of O_2. On top of the figure is an enlargement of a part of the TPES.

However if the last photon is absorbed before ionisation by a Rydberg state, it is then possible to produce state selected ions. This method, called Resonant Enhanced Multi-Photon Ionisation (REMPI) consists in the following : the first absorbed photon(s) populate(s) a rovibrational level of a pure neutral Rydberg state converging to an ionic state with a similar geometry, because the removal of a Rydberg electron has little effect on the ion core. The last photon ionises the Rydberg state following a $\Delta v=0$ selection rule, as direct photoionisation is governed by Franck-Condon factors. In principle this method produces a single rovibrational ion state. In practice, it is not always the case for two main reasons. Firstly, Rydberg states are seldom pure and are often mixed with other states such as valence states, for example. Secondly, multiphoton ionisation experiments require high laser intensities and so the ion might absorb several photons. In the case of NO^+, which has been extensively studied, the selectivity is greater than 70% using the E $^2\Sigma^+$ v=0-3 and C $^2\Pi$ v=4-6 intermediate neutral states [8].

TPEPICO experiments with synchrotron radiation and REMPI with lasers are complementary techniques, as the first is very general, but has a rather poor resolution (10 meV), whereas the second strongly depends on the spectroscopy of the molecule and the laser characteristics, but has a high resolution (selection of a specific vibrational mode in $C_2H_2^+$ for example [9]).

2. Ion dissociation

Dissociative photoionisation of small molecules has been extensively studied over the past years (see the review of T.Baer [1]) on the dissociation dynamics of energy-selected ions). The measurement of fragmentation branching ratios, dissociation rates and fragment ion angular distributions as a function of parent ion internal energy gave a lot of informations about the dissociation mechanisms.

Diatomic and triatomic ion dissociation strongly depend on the internal energy of the parent ion. For example in the case of N_2O^+ [10], the high vibrational levels of N_2O^+ (X) ground state dissociate mainly into NO^+ + N except just below the N_2O^+ (A) state where they dissociate mainly into O^+ + N_2. At about the same internal energy, the ground vibrational level (0,0,0) of N_2O^+(A) state is stable and decays by fluorescence, while the excited vibrational levels of N_2O^+(A) dissociate mainly into NO^+ + N. These results illustrate very well the essential role of both the electronic and vibrational energy in the dissociation. It can be rather well understood by dynamical theories involving potential surfaces.

For polyatomic ions made up of at least five or six atoms, the situation is different. Usually the internal energy is rapidly converted to the ground electronic state and redistributed statistically on all vibrational modes, prior to dissociation. It has been shown that the statistical RRKM theory can explain most of the results concerning the breakdown diagram and the dissociation rates when the internal energy of the parent ion is not too high [1].

For intermediate size molecular ions, the situation is more complex and attempts have been made to combine both theoretical approaches [11].

Under what circonstances might dissociations be statistical ? First the redistribution of the available internal energy on the different vibrational modes of the parent ion has to be rapid compared to the dissociation rate. Secondly the dissociation must occur on a single potential surface.

In the first part of this section, we will present the dissociative photoionisation of CH_4 [12], which represents an intermediate case by its size and high degree of symmetry. We will investigate the statistical or non-statistical behaviour of its dissociation as a function of the internal energy of the parent ion.

In the second part, we will show on different examples that for the same internal energy, the dissociation pattern of the ion can depend on the preparation mode of the parent ion (direct photoionisation or autoionisation).

2.1. ROLE OF INTERNAL ENERGY IN THE DISSOCIATION OF A SMALL POLYATOMIC ION : METHANE AND ITS DEUTERATED COMPOUNDS

We studied the dissociative photoionisation of methane as a function of internal energy, from the ionisation potential, i.e. 12.6 eV, up to 30 eV photon energy.

The electronic configuration of methane in the T_d symmetry group is $(1a_1)^2(2a_1)^2(1t_2)^6$ and the two valence states of the ion are the X ground state $(1t_2)^{-1}$ and the A excited state $(2a_1)^{-1}$, which have an ionisation potential of 12.61 eV and 22.39 eV respectively. In fact the X ground state has been shown to be split into three states by the Jahn-Teller effect [13]. Its Franck-Condon region from the neutral CH_4 ground state extends up to about 16 eV [14]. Previous photoionisation mass spectrometry studies have been summarized by J. Berkowitz [15]. The high energy of the first excited electronic state and the unusually large (≅ 6.4 eV) Franck-Condon gap between the X and A states has limited until now the study of the dissociative photoionisation of state selected methane up to 16 eV [16]. But now with the TPEPICO technique used with synchrotron radiation, it is possible to produce parent ions with a continuously variable internal energy, even in the Franck-Condon gap. In this energy range, resonant autoionisation produces threshold electrons. It allows us to continuously follow the variation of the dissociation mechanisms of the methane ion as a function of its internal energy in a very wide range (12.6-30 eV).

We measured the dissociation branching ratio from the integrated fragment ion peaks in the threshold photoelectron photoion coincidence time of flight spectra and thus obtained the breakdown diagram for the following three methane ion isotopes : CH_4^+, CD_4^+ and CD_3H^+. In figure 2a, we present the breakdown diagram for the CH_4^+ dissociation. All possible fragments were observed, but H^+, H_2^+ and C^+ remain minor fragments compared to CH_3^+, CH_2^+ and CH^+ in the whole range up to 28 eV photon energy. Most fragments appear at their thermodynamic threshold [17]. The first dissociation onset at 14.3 eV leads to CH_3^+ + H fragments and the second at 15.2 eV to CH_2^+ + H_2 fragments. The sharp increase of CH_2^+ around 21 eV is due to a secondary dissociation i.e. CH_4^+ —> CH_3^+ + H —> CH_2^+ + H + H. It does not occur at the thermodynamical threshold of this process as will be discussed below. The CH^+ fragments can come either from the direct dissociation of CH_4^+ into CH^+ + H_2 + H or from sequential dissociations of CH_3^+ or CH_2^+.

We performed RRKM calculations using the exact counting method [18] to calculate the state density of the parent ion and of the transition state respectively and assumed a complete internal conversion to the CH_4^+ ground state in the whole energy range up to 28 eV photon energy. We took into account all possible sequential fragmentations. The calculated breakdown curve for CH_3^+ and CH_2^+ is in excellent agreement with the published curves calculated for internal energies up to 10 eV, which did not take into account the CH^+ + H_2 + H dissociation channel in this energy range [19-20]. It is compared to the experimental curve in Figure 2. The overall shape is rather well reproduced at low internal energy. When the energy increases, more and more descrepancies appear. Firstly, the sequential dissociation of CH_3^+ into CH_2^+ + H experimentally appears 1.7 eV above the predicted calculated energy value. Secondly CH^+ fragments are overestimated. Thirdly there are some structures in the region of the A state, whereas the RRKM calculations give smoooth curves. Finally we observe a sharp increase

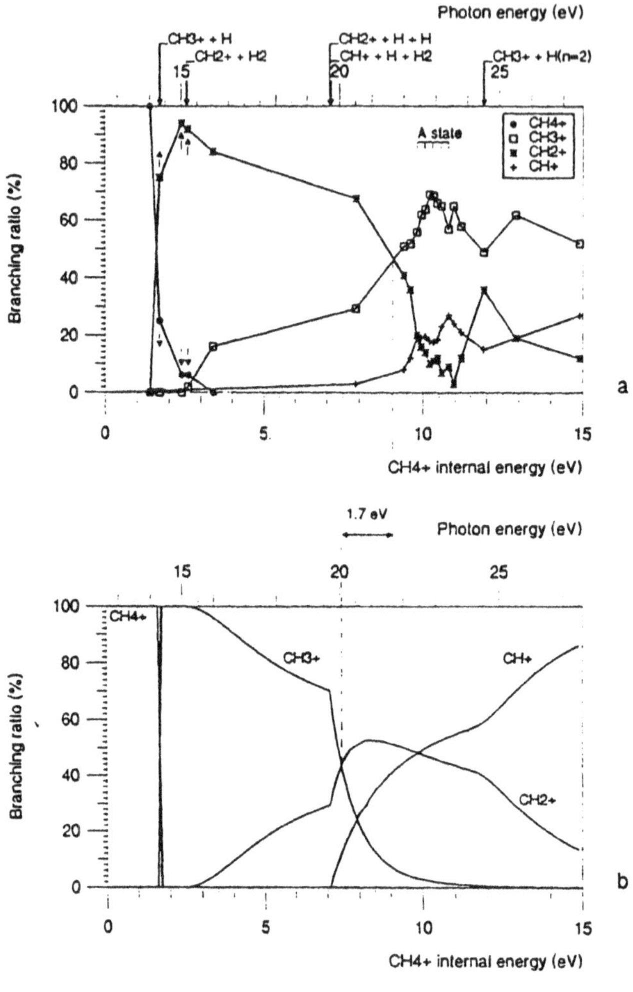

Figure 2. Breakdown diagram of CH_4^+. (a) Experimental results. The upper arrows indicate the main dissociation limits. (b) RRKM calculations (see text).

of the CH_3^+ production in the region of the satellite states at 24.5 eV, whereas this fragment should not be produced according to the statistical theory.

2.1.1. *X state region.* In the region of the X state, previous studies had already shown that the CH_4^+ dissociation can be very well explained with the statistical theory [16]. For example the breakdown curve is well reproduced by the RRKM calculations. Moreover the isotope effects in the dissociation of CH_4^+ near the CH_3^+ + H dissociation threshold are well explained by a tunneling effect through the centrifugal barrier included in a statistical theory [21-22].

2.1.2. *Franck-Condon gap region.* In the Franck-Condon gap, there is a disagreement between the RRKM calculations and the experimental results concerning the onset of the sequential dissociation $CH_4^+ \longrightarrow CH_3^+ + H \longrightarrow CH_2^+ + H + H$. The 1.7 eV energy difference between the calculated and experimental onsets cannot be explained by a wrong set of parameters used in the RRKM calculations or by the fact that we neglected angular momentum conservation, because it has little effect on the breakdown curve as shown by E.Klots in this case [20]. A second hypothesis would question the basics of the statistical theory i.e. that the energy is not randomly distributed on all CH_4^+ degrees of freedom. However this seems very unlikely. Finally W.Chupka [23] observed the $CH_3^+ \longrightarrow CH_2^+ + H$ dissociation at its thermodynamical onset, which excludes the presence of an activation barrier. From all these considerations, we deduce that the CH_3^+ fragment which is formed in this energy region, either has less internal energy than predicted or is in an excited electronic state, which is stable up to 1.7 eV above the dissociation limit of the CH_3^+ ground state.

Let us note that in this photon energy range, the ions are only formed by resonant autoionisation of the $(2a_1)^{-1}(3pt_2)$ and $(2a_1)^{-1}(4pt_2)$ Rydberg states . To explain the production of CH_3^+ in an excited electronic state, we could imagine that the autoionisation process populates exclusively the highest Jahn-Teller CH_4^+ state. According to the correlation diagram in C_{3v} symmetry [24], this state (1^2E) is correlated to the $CH_3^+(^3E)$ excited state, which is at 6.26 eV above the $CH_3^+(^1A_1)$ ground state. The corresponding $CH_3^+(^3E)$ + H dissociation limit thus lies at 20.57 eV, which is resonant with the lowest $(2a_1)^{-1}(3pt_2)$ Rydberg state. However the vibronic coupling between the different Jahn-Teller states is expected to be very efficient with such a high density of states.

We propose an alternative explanation invoking the Rydberg dissociation into CH_3^{**} + H neutral fragments followed by resonant autoionisation of the CH_3^{**} fragment. It is most probable that this process will preferentially produce the resonant $CH_3^+(^3E)$ state. This hypothesis implies a competition between autoionisation and dissociation into neutrals in favor of the latter, which is supported by the observation of the H Lyman a emission arising from the Rydberg dissociation into CH_3 + H(n=2) cited above. Such a process involving the dissociation of neutral states prior to autoionisation in the fragment was first proposed by V. Cermak to explain Penning ionisation of oxygen and nitrogen atoms in collision of metastable helium with O_2 and N_2 respectively [25-26]. This same process was then observed in photodissociation of O_2, CO, Cl_2, HCl and Br_2 into atomic autoionizing states using the HeI radiation [27]. Much later it was observed in the case of HBr, CH_3Br and HI [28] excited in inner valence shells and more recently in H_2 [29] , H_2CO [30] and in O_2 [31] excited in valence shells. Our hypothesis is corroborated by the following observed isotope effect : CD_3^+ in CD_4 is less abundant than CH_3^+ in CH_4 in this photon energy range, which can be interpreted by the less favorable $CD_4 \longrightarrow CD_3 + D$ dissociation rate, due to mass considerations.

2.1.3. *A state region.* In the region of the A state, the main fragment is CH_2^+ and structures are observed in the breakdown diagram of the three isotopes. The presence of these structures clearly indicates that the dissociation of the methane ion in the A state region is not statistical and that the internal energy conversion to the X ground state is slow compared to the ion dissociation. In the case of CD_4^+, these structures appear as regular oscillations (see figure 3). We observe alternative oscillations between the CD_2^+ fragment on one hand and the CD^+ and D^+ fragments on the other hand, as a function of the vibrational èxcitation of the CD_4^+ A state. This could be explained by energy resonance phenomena. Indeed the vibrational frequency of the D_2 molecule (0.39 eV) is almost equal to two quantum of the CD_4^+ A state ($v_1 \approx 0.18$ eV from our TPES). Resonances can thus occur between one out of two vibrational A state levels and the $CD_2^+ + D_2(v)$ vibronic dissociation channels.

2.1.4. *Satellite states region.* In the region of the satellite states, all possible fragments are formed including CH_3^+ which should not be produced according to the RRKM theory. It rises again in the breakdown curve exactly at the onset of the $CH_3^+ + H(n=2)$ dissociation limit. The H Lyman α emission ($H(n=2) \longrightarrow H(n=1)$ transition) was indeed observed in this energy range by Wu and Judge [32]. It indicates that these states undergo a rapid dissociation prior to internal conversion to the X ground state, which is not surprising for doubly excited electronic states.

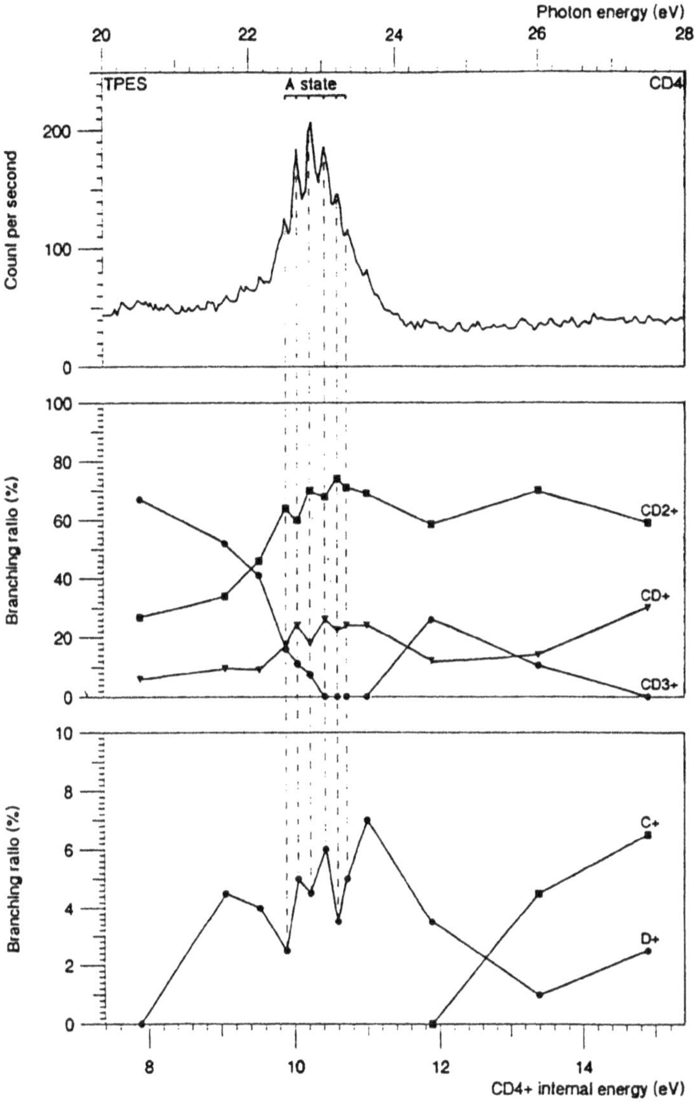

Figure 3. Breakdown diagram of CD_4^+ in the A state region. (a) Threshold photoelectron spectrum. (b) Breakdown diagram.

2.1.5. *Conclusion.* In conclusion, the study of the CH_4^+ dissociation showed that at low internal energies in the Franck-Condon X state zone (12.6-16 eV), the statistical RRKM theory explains the experimental results very well. With a higher internal energy, it is not the case. In the Franck-Condon gap between the X and A states (20.5-22.4 eV), we observe a 1.7 eV shift in the sequential dissociation onset of the CH_3^+ fragment. To explain this, we propose a mechanism involving the dissociation of Rydberg states into neutral fragments followed by autoionisation of the CH_3^{**} fragment towards an excited electronic state of CH_3^+. In the region of the A state (22.4-24 eV), the internal conversion to the ground state is shown to be ineffective. Quantum effects leading to oscillations in the breakdown diagram, which we propose to explain by resonance effects, show that the A state dissociation is a direct process. At even higher internal energy, in the region of the satellite states (26-28 eV), we observe an increase of the CH_3^+ fragment, which is probably due to a direct dissociation process leading to CH_3^+ + H(n=2).

2.2. ROLE OF THE PREPARATION MODE OF STATE SELECTED IONS : DIRECT IONISATION AND AUTOIONISATION.

In dissociative photoionisation studies, ions are produced both by direct ionisation and by autoionisation with proportions which vary according to the wavelengths used in the experiment. We will show several examples where it has an influence on the observed ion dissociation. Autoionisation processes can affect the ion dissociation by different means : the first example shows an influence on the fragment branching ratio in the case of CO_2^+, the second example shows an influence on the angular distribution of the I^+ fragment ions in the CF_3I^+ dissociation and the third example demonstrates the production of fragment ions by an autoionisation process in the fragment following the neutral photodissociation of the molecule.

2.2.1. *Effect in the dissociation branching ratio : example of* CO_2^+ *C* $^2\Sigma_g^+$ *state [33].* Figure 4 shows threshold photoelectron-photoion coincidence spectra obtained for the different vibrational levels of the C $^2\Sigma_g^+$ state of CO_2^+. The parent ions are completely dissociated, the remaining signal (see figure 4) being due to false coincidences which were not perfectly corrected in these old experiments of our group. The first dissociation limit is O^+ + CO and is energetically located below the CO_2^+ C $^2\Sigma_g^+$ state, whereas the second dissociation limit is CO^+ + O and is located between the (0,0,0) and (0,1,0) levels of the

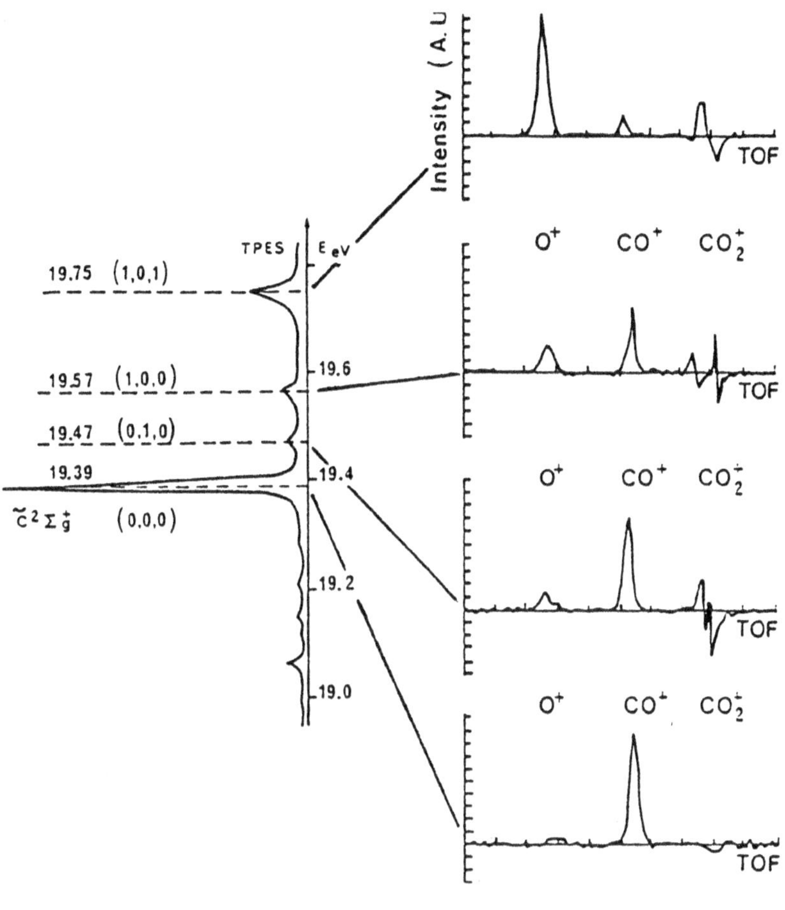

Figure 4. TPEPICO ion time of flight spectra obtained for the different vibrational levels of the CO_2^+ C state. The threshold photoelectron spectrum is displayed in the left part.

C state. The ground vibrational level dissociates almost completely into O^+ + CO. The small amount of CO^+ fragments show that they appear 75 meV below their thermodynamical onset. J.H.D. Eland and J.Berkowitz [34] have already explained this effect by the complete availability of the thermal rotational energy in the CO_2^+ dissociation. The excited vibrational levels, which are located above the CO^+ + O limit dissociate mainly towards this limit. However they still produce some O^+ fragments with a percentage which decreases from 42±4% for the (0,1,0) level to 17±3% for the (1,0,0) level and 6±2% for the (1,0,1) level. In coincidence experiments done at fixed wavelength (21.21 eV) and variable photoelecron energy [35], no O^+ fragments were produced by the dissociation of the same excited vibrational levels of CO_2^+ C $^2\Sigma_g^+$ state. The difference between the two experiments is that in the second case the parent ions are only formed by direct ionisation, whereas in the first case, they are also formed by resonant autoionisation. It is clear from the total O^+ ion yield as a function of wavelength [34],which shows the O^+ fragments appearing as structures in the Franck-Condon gap between the B and C states of CO_2^+, that autoionisation is responsible for the production of O^+ ions in this energy region. So this resonant autoionisation process probably populates very high vibrational levels of the B state which dissociate into O^+ + CO. Above the C state onset, it is thus likely that this process is still present and disappears with the Rydberg series converging to the excited vibrational levels of the C state. This would very well explain our results showing a significant production of O^+ ions, decreasing with vibrational energy of the C state.

2.2.2. *Effect in the fragment ion angular distribution : example of the dissociation of CF_3I^+ [36].* We studied the dissociation of CF_3I^+ ions selected in internal energy in the region of the A^2A_1 state, i.e. 13-15 eV above the ground neutral state. The three principal dissociation products are CF_3^+, I^+ and CF_2I^+.

Figures 5 and 6 show photoelectron-photoion time of flight coincidence spectra obtained for the I^+ fragment ion coming from the dissociation of CF_3I^+. Figure 5 displays threshold photoelectron-photoion coincidence spectra obtained for three different internal energies E_{int} of the parent ion, 13.3, 14.24 and 15.0 eV, equal to the incident photon energy $E_{h\nu}$. In figure 6 the coincidence spectra correspond to the same 13.3 eV internal energy, but for different photon energies, the coincident photoelectron energy being equal to the difference $E_{h\nu}$-E_{int}. In principle the three spectra Fig.5a, Fig.6a and Fig.6b should be identical, as they correspond to the same internal energy of the parent ion. However their shape is very different. The same effect is observed for the CF_2I^+ fragment and less evidently for the CF_3^+ fragment because of the more complicated dissociation mechanism leading to this channel.

The time of flight peak shape reflects the fragment ion velocity and angular distribution with respect to the apparatus axis. For this range of kinetic energy, all fragment ions are collected without any discrimination in our time of flight analyser. The time of flight peak shapes can be analysed by computing a simulation of the kinetic energy release distribution as well as the angular distribution of the fragment ions reflected by the β anisotropy parameter (see ref [36] for further details).

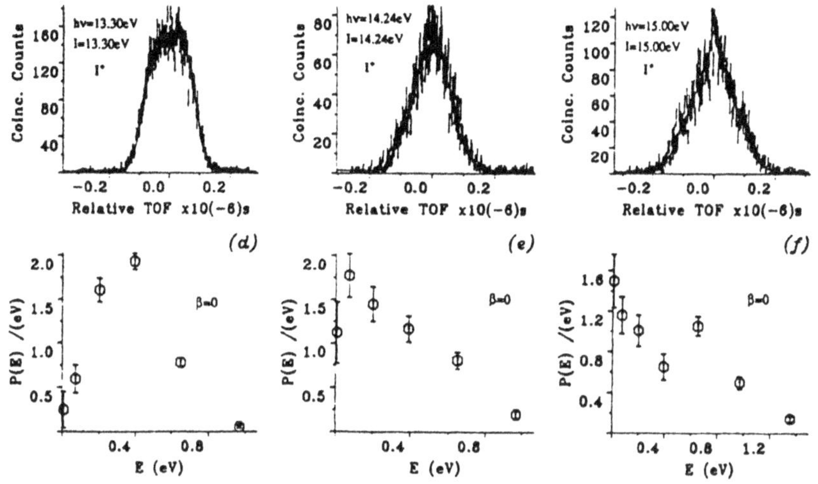

Figure 5. I^+ TPEPICO time of flight spectra recorded at photon energies of (a) 13.3 eV: (b) 14.24 eV; (c) 15.0 eV. The kinetic energy release distributions which are inferred from the simulations, assuming isotropic fragment distribution, are displayed below each time of flight spectrum (d,e,f).

In a idealized case of a single value for the kinetic energy release and an isotropic distribution (β=0), this woud give rise to a square, flat topped time of flight peak (see Figure 7b). The width of the peak is determined by the magnitude of the kinetic energy. Any anisotropy will, however, modify the ion peak shape. In particular a more peaked shape (Fig.7a) is obtained if the ions are ejected predominantly at 90° with respect to the time of flight axis (β<0). Alternitavely a split, but still symmetric peak shape (Fig.7c) is obtained when fragments are ejected in a direction along the time of flight axis (β>0). In a realistic situation where there is a distribution of kinetic energy, the observed time of flight peak shape will be given by a superposition of such idealized shapes. Without further angle resolved data, the information cannot be unambiguously extracted. However, if the observed peak shape is a split type as in Fig.6, it unambiguously indicates some parallel anisotropy.

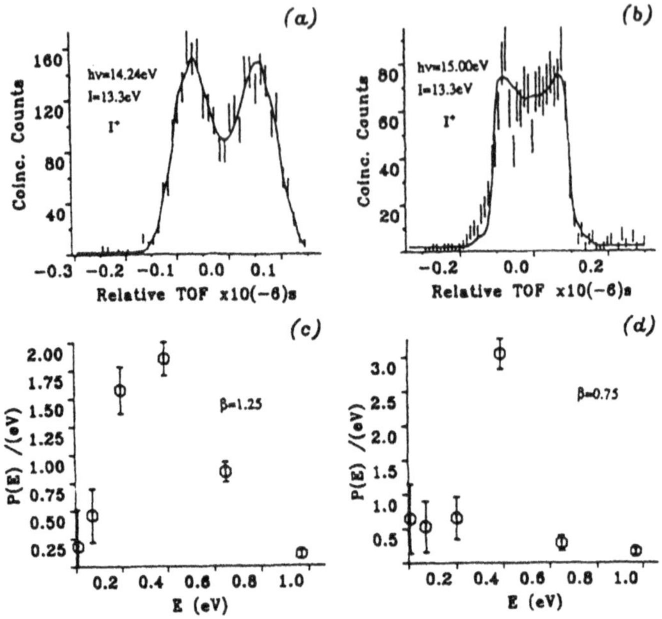

Figure 6. I^+ PEPICO time of flight spectra corresponding to a nominal 13.3 eV internal energy recorded with (a) 14.24 eV, and (b) 15.0 eV photon energies. The kinetic energy release distributions inferred from the simulations are displayed below each time of flight spectrum (c,d). The best fit simulations are obtained for anisotropic fragment distribution with (c) b=1.25 and (d) b=0.75.

The kinetic energy release distribution is roughly the same for the three spectra at 13.3 eV internal energy (Fig.5d, 6c and 6d), but the β parameter varies strongly. We propose to explain this by an alignment of the molecule resulting from autoionisation by the polarised synchrotron radiation source. Around 14 eV, where this phenomenon is the most pronounced, there is an autoionisation peak in the total ion yield showing the existence of Rydberg states at this energy. In order to keep the alignment given by the first step with the photon absorption, the autoionisation and the subsequent parent ion dissociation are required to be rapid compared to the rotation of the molecule. This phenomenon had never been observed before.

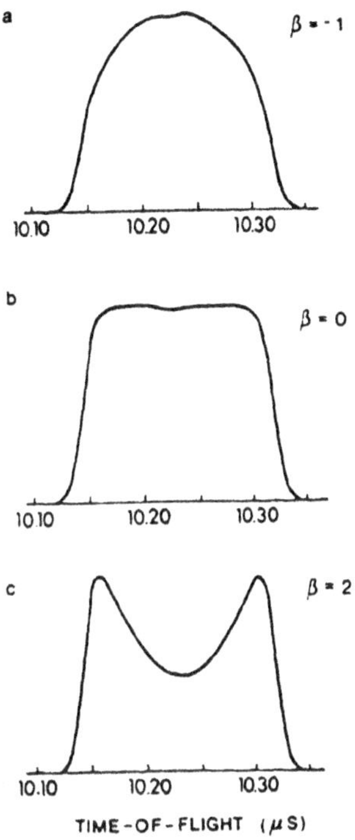

Figure 7. Simulated fragment ion time of flight peak shapes for a single value kinetic energy release. Different peak shapes illustrate the behavior expected for different fragment angular distributions, represented by the indicated b values : (a) b=-1; (b) b=0; (c) b=2.

2.2.3. *Production of fragment ions by autoionisation in the neutral fragment.* Competition between autoionisation and dissociation of Rydberg states into neutral fragments has been studied in a series of molecules [7,37]. One usually implicitely supposes that autoionisation precedes the ion dissociation. However V. Cermak [25-27] had already observed in 1973 in Penning ionisation experiments with metastable helium, a process involving the dissociation of Rydberg states of O_2, N_2, Cl_2 and Br_2 prior to autoionisation in the atomic fragment. It can been summarized the following way :

$$He^* + O_2 \dashrightarrow He + O^+ + O + e^-$$

does not occur through the ion dissociation process

$$He^* + O_2 \rightarrow He + O_2^+ + e^- \rightarrow He + O^+ + O + e^-$$

but through the atomic autoionisation process

$$He^* + O_2 \rightarrow He + O_2^{**} \rightarrow He + O^{**} + O \rightarrow He + O^+ + O + e^-$$

It is only recently that this phenomenon has been re-discovered by several groups [28-31]. For example, A.A.Cafolla et al [31] observed atomic oxygen photoelectron lines, in addition to the expected O_2 molecular photoelectron bands, when molecular oxygen is excited by synchrotron radiation in the range 19 to 22 eV. The atomic autoionisation lines appear at constant kinetic energy, when the photon energy varies, which clearly demonstrates that autoionisation only occurs after dissociation is complete.

In the preceding section, we showed that the breakdown diagram of methane in the 20.5-22.4 eV range could be explained by the same type of mechanisms, involving the dissociation of CH_4^{**} Rydberg states into neutral fragments followed by autoionisation of the CH_3^{**} fragment [12].

Some old results can also be reinterpreted with this mechanism. For example R.Stockbauer [38] observed a step in the threshold photoelectron spectrum of H_2O, appearing exactly at the parent ion first dissociation onset. This observation, which was confirmed by our group in Orsay, was not understood. We can now propose to interpret it by autoionisation in the OH^{**} fragment, after the dissociation of the H_2O^{**} Rydberg states into OH^{**} + H i.e.

$$H_2O + h\nu \rightarrow H_2O^{**} \rightarrow OH^{**} + H \rightarrow OH^+ + H + e^-$$

Below the OH^+ + H thermodynamical onset, the OH^{**} can only decay by fluorescence, as observed by our group [39], whereas above the onset the fragment autoionisation process opens up, giving an increase in the threshold photoelectron yield. This increase, observed in D_2O with the expected isotopic energy shift [38], is relatively less important than in H_2O. It could be due to a mass effect, as the dissociation of D_2O into OD + D is expected to be slower than the one of H_2O into OH + H. So it is expected that the competition between the molecule dissociation and the mass independent molecular autoionisation process is less in favor of the proposed mechanism. Let us note that a difficulty remains in explaining how the dissociation into neutral fragments and the fragment resonant autoionisation process can occur with no kinetic energy release, either in the OH and H fragments or in the ejected threshold photoelectron. This threshold behavior remains unexplained.

3. Ion-molecule reactions. Role of internal energy in charge transfer reactions

3.1. GENERAL

In an ion-molecule reaction, several processes can occur :

$AB^+ + C$	-->	$C^+ + AB$	charge transfer
	-->	$AC^+ + B$	chemical reaction
	-->	$A^+ + B + C$	collision induced dissociation

In order to understand which are the mechanisms involved for the different reaction channels, a big effort has been made in recent years to characterize experimentally the collision conditions. An ideal experiment would be to study a reaction with state selected reactants in a crossed beam apparatus and to characterize completely the products in internal energy, kinetic energy and angular distribution. Such an angle resolved state to state experiment has not been performed yet. In practice, the experimental devices put the effort either on the characterization of the reactants or on the products and crossed beam experiments have been done mainly with ground state reactants. Here we will concentrate on results concerning state selected reactant ions.

Recent developments of coincidence techniques and multiphoton ionisation with lasers provided data, giving new insights into the reaction dynamics. An extensive review has been made by C.-Y. Ng [2] on state selected and state to state ion-molecule reaction dynamics by photoionization methods.

3.1.1. *Role of internal energy.* Experiments are now ablc to investigate the role of different kinds of internal energy : electronic energy, vibrational nergy, rotational energy and spin-orbit energy. The pioneer work of W.Chupka and co-workers [4] showed strong vibrational energy dependence in the reactivities of the reactions $H_2^+ + H_2$ [40], H_2^+ + He (Ne or Ar) [41] and $NH_3^+ + NH_3$ [42]. These first results showed for example that vibrational excitation of H_2^+ is much more efficient than collisional energy to overcome the endothermicity in the reaction with He to produce HeH^+. These authors used the specificity of resonant autoionisation in some molecules to produce state selected ions without using coincidences. Later the same method was used in the group of Y.T.Lee [5] to study the reactivity of state selected H_2^+ and $C_2H_2^+$. However the coincidence techniques developed in the groups of T.Baer [43], I.Koyano [44] and P.M.Guyon [45] allowed a more general study of these types of reactions.

3.1.2. *Role of collision energy.* Ion-molecule reactions also depend strongly on collisional energy. At low collision energy, typically below 1 eV in the center of mass frame, reactive channels dominate, whereas at higher collision energy charge transfer and collision induced dissociation become the dominant channels. The different mechanisms depend also on the impact parameter and on the orientation of the molecules during the collision. However it is extremely difficult to specify them experimentally. No such experiments have been performed for ion-molecule reactions yet. In most experiments the translational energy of the reactant ion is well controlled and the target molecule has a near zero energy (often thermal energy at 300K) in the laboratory frame.

$$\begin{array}{cccc} AB^+ \;+ & C & \longrightarrow & \text{products} \\ v_{lab} & v \cong 0 & & \end{array}$$

Then in the laboratory frame

$$E_{lab} \cong 1/2\,(m_{AB^+})\,(v_{lab})^2$$

and in the center of mass frame

$$E_{CM} \cong E_{lab} \cdot m_C / (m_C + m_{AB^+})$$

where v_{lab} is the velocity in the laboratory frame of the AB^+ reactant ion, and m_{AB^+} and m_C are the masses of the reactant ion and the target molecule respectively.

3.1.3. *Methods*. Several techniques are used to study ion-molecule reactions at thermal energies (flowing afterglows, drift and SIFT tubes, ICR...) i.e. with a thermal energy distribution at temperatures varying from 300K to 20K (see the recent reviews [46,47]). In crossed beam experiments the minimum energy in the laboratory frame is typically a few eV. So to study the reactive channels, it requires the use of other techniques such as merged beam or guided beam techniques (see the recent review on ion beam methods [48]). In radiofrequency octopole ion guides the laboratory energy can be varied from about 20 meV to 10 ev or more, which allows us to bridge the gap between thermal energy and crossed beam experiments.

It is this technique used in our new experiment in Orsay, which combines on one hand the ion state selectivity by photoionisation with synchrotron radiation and threshold photoelectron photoion coincidences and on the other hand the radiofrequency ion guide technique to control the collision energy. This new experiment called CERISES, which is the acronym for "Collisions Et Réactions d'Ions Sélectionnés par des Electrons de Seuil", has been constructed in collaboration with the group of D.Gerlich in Freiburg. Figure 8 presents schematically the experimental set-up.

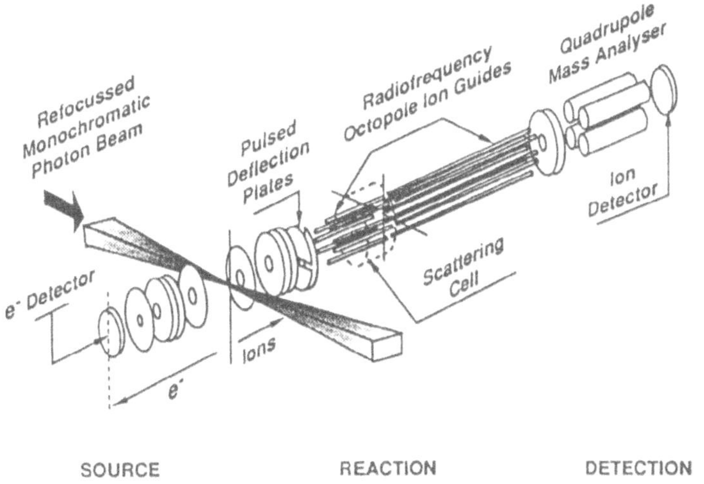

Figure 8. Schematic of the CERISES experimental set-up (see text).

Briefly the apparatus is divided into three parts, which are pumped differentially and which correspond to the source, the reaction and the detection regions. The first part is the source region, where monochromatised synchrotron radiation is refocussed in the center of the source chamber filled with the reactant gas at a pressure of about 10^{-4} to 10^{-3} torr. Electrons and ions are extracted in opposite directions by a small DC field of about 1 V/cm. Threshold electrons are selected by time of flight using the pulsed time structure of synchrotron radiation from the Super-ACO storage ring (at present, pulses of 500 ps with a period of 120 ns). Both angular and time discrimination are used to reject energetic electrons. Ions are gated in coincidence with threshold electrons and focussed into a radiofrequency octopole ion guide in the second zone. There they collide with a target gas in a scattering cell, which is a 4.4 cm long cylinder around the octopoles filled at a pressure of about 10^{-4} torr. The kinetic energy of the reactant ions is defined by the DC potential on the octopole radiofrequency guide, which can be varied between 100 meV to a few eV. Product ions are collected with a 100% efficiency, due to the rf guiding field and mass analysed by a quadrupole mass spectrometer. The octopoles are cut at the exit of the scattering cell, so that ions are accelerated into the second part of the octopole guide and ion time of flight spectra can be measured, which give information on the product ion kinetic energy release. The wide possibilities of the radiofrequency ion guides have been described in detail by Teloy and Gerlich [49a] and more recently by D.Gerlich [49b]. Absolute reaction cross sections and reaction branching ratios are measured from the areas of the coincidence time of flight peaks. False coincidences, which are reduced by an ion "sweeping" pulse in the source region, according to the method of M.Lavollée and G.Henri [50], are corrected from the experimental spectra. Some of the first results obtained with this experimental set-up are given below.

In the following we will discuss the dependence of charge transfer reactions on internal and collisional energy, illustrated by results obtained by different techniques.

Charge transfer reaction mechanisms vary with collision energy. They can be classified as "direct" or "intimate" (collision complex) mechanisms, according to phenomenological characteristics, such as the collision energy dependence of the total cross section or the angular distribution of product ions. An isotropic angular distribution or a forward-backward symmetry in the center of mass frame indicate that there is an intermediate collision complex whose lifetime is long compared to its rotation period. Whereas if an angular peaked distribution is observed, the process is "direct". As soon as the collision energy is high enough (typically above a few hundred meV or 1 eV), charge transfer reactions are "direct", i.e. correspond to a long distance electron jump, and have often been interpreted according to two efficiency criteria : close energy resonance (small energy defect between the reactant and product ions) and favorable Franck-Condon factors between the reactant and product states. We will present below two cases : an example of a near resonant charge transfer reaction

N_2^+(X or A,v) + Ar --> $Ar^+ + N_2$ $\Delta E = +0.18$ eV

and an example of a non-resonant charge transfer

O_2^+ (X or a,v) + Ar <--> $Ar^+ + O_2$ $\Delta E = -3.70$ eV

3.2. EXAMPLE OF A NEAR RESONANT CHARGE TRANSFER REACTION : N_2^+ + Ar

This example is taken from a review of P.M.Guyon and E.E.Gislason [51] on four near resonant charge transfer reactions studied with state selected ions. This reaction has also been studied in the group of C.-Y.Ng, who performed state-to state reaction experiments, by using a secondary probe reaction between the Ar^+ product ions and H_2 [2].

3.2.1. *Experimental results.* Charge transfer cross sections were measured by our group for state selected N_2^+ ions colliding with argon at three relative energies (E_{CM}= 8, 14, and 20 eV) with the first ion-molecule reaction experiment built in Orsay in 1982 [45]. The N_2^+ ions are selected in different vibrational levels of both the gound X state and the A excited electronic state. The experiment gives relative cross sections, but they are normalized to the Ar^+ + Ar cross section. It is also necessary to correct the measured cross sections for the N_2^+(A,v) states for the finite lifetime of these radiative states. Details can be found in ref [45a]. The absolute cross sections are shown in figure 9 for the 20 eV collision energy.

Certain features of the results were quite surprising when they were obtained, because they seemed to be in contradiction with the simple previsions made from the criteria cited above. The cross sections show a strong dependence on the vibrational levels for both reactant electronic states. If the Franck-Condon principle were valid, then the

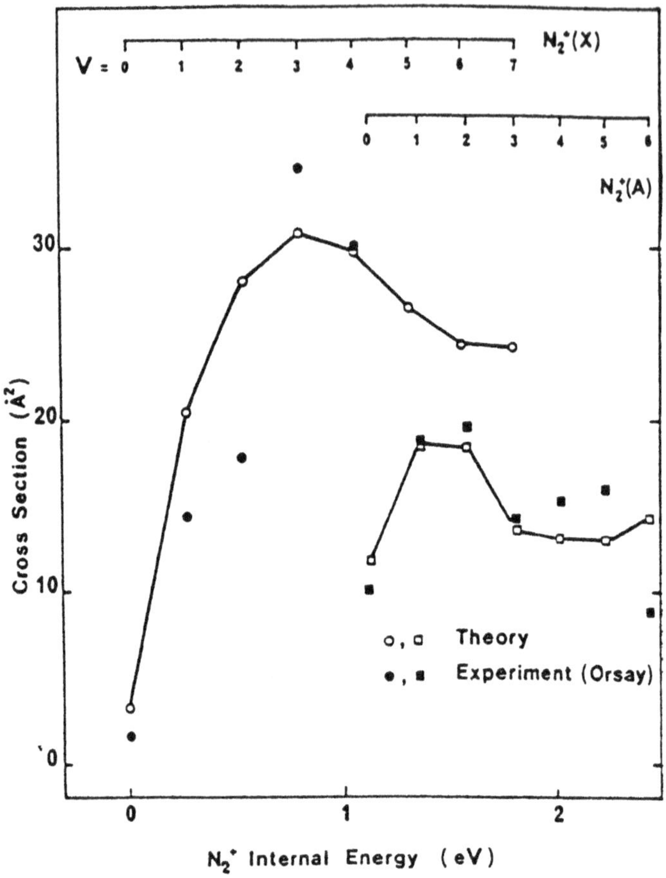

Figure 9. Charge transfer cross sections for $N_2^+(X,A,v)$ + Ar at 20 eV plotted against the internal energy of the N_2^+ ion. (The zero of energy corresponds to $N_2^+(X,v=0)$ + Ar). Theoretical results [56] for v=0 to 7 of the X state and v=0 to 6 of the A state are shown as open circles and squares respectively. Experimental results [45] are shown as solid circles and squares. The lines drawn through the theoretical values are only a guide for the eye.

charge transfer cross section should be independent of the reactant vibrational level [52]. It is well known that the Franck-Condon principle breaks down badly at low collision energy for most charge transfer systems. However the energy is here sufficiently high, so that we could expect a long distance electron jump. Moreover the most remarkable result is the very small cross section for the N_2^+(X v=0) state at all three collision energies (1.6 Å^2 at 20 eV). This occurs even though the Franck-Condon factor is the most favorable (0.92 for the transition N_2^+(X, v=0)-->N_2(X,v=0)) and the energy defect with $Ar^{+2}P_{3/2}$ is only 0.18 eV, which is very close to resonance. Thus this reaction should be strongly "allowed". A third notable feature of the results is the smaller reactivity of the N_2^+(A) state. For example the vibronic states N_2^+(A,v=0) and N_2^+(X,v=4) differ in energy by only 0.06 eV, but their charge transfer cross sections with Ar at 20 eV differ by a factor of two.

3.2.2. *Theoretical calculations*. These results as well as those obtained by other groups [53-54] have inspired a considerable amount of theoretical work on this system [55-58]. Archirel and Levy [57] have calculated a set of potential energy surfaces for the states N_2^+(X) + Ar, N_2^+(A) + Ar and N_2 + Ar^+(^{2}P) as well as the couplings between these surfaces using a novel computational technique. They assumed that transitions could occur when two curves crossed and calculated the cross sections using either the Landau-Zener or Demkov formula, as appropriate. Good agreement was obtained with the experimental values in most cases. Nikitin et al [58] have taken a somewhat similar approach. They estimated the adiabatic vibronic interaction curves for this system, instead of couplings between diabatic curves, and assumed that transitions could occur when any two curves have an avoided crossing. Their calculated cross sections using the Landau-Zener formula for N_2^+(X, v=0-3) also agree with experimental results. The most detailed calculations on this system have been carried out by Spalburg et al [55] and Parlant and Gislason [56]. The latter work used the potential energy surfaces of Archirel and Levy [57] in a classical path treatment of the charge transfer process. The translational motion is treated classically and the vibronic degrees of freedom are treated quantum mechanically using the time dependent Schrödinger equation. This type of calculation is expected to be quite accurate at collision energies above 5 eV. Figure 9 shows at 20 eV that the agreement with experimental results is excellent for both reactant electronic states.

3.2.3. *Interpretation of the data*. The theoretical models discussed here can give explanations of the various experimental features observed. For example the strong dependence of the cross sections on the vibrational level of N_2^+ is no more surprising once the vibronic curves are examined. In particular, no curve crossing is obtained between the N_2^+(X,v=0) + Ar curve and any other exit channel curve. This explains why the cross section for this level is so low in this collision energy range. The smaller reactivity of the $N_2^+(A)$ state is well reproduced by the calculations of Parlant and Gislason [56] and explained by two reasons. Firstly, the electronic coupling between the curves $N_2^+(A)$ + Ar and N_2 + Ar^+ is considerably smaller than for the curves $N_2^+(X)$ + Ar and N_2 + Ar^+. In addition one third of all $N_2^+(A)$ + Ar collisions take place on the Ω =3/2 electronic surface. In this case charge transfer can only occur to the Ω=3/2 state of $Ar^+(^2P_{3/2})$ + N_2 and not to the $Ar^+(^2P_{1/2})$ + N_2 exit channel. This significantly reduces the cross sections for the Ω=3/2 reactants. The theoretical calculations also correctly predict that Ar^+ ions are formed predominantly (>80%) in the $^2P_{3/2}$ state as observed [2].

In conclusion, energy resonance and vibronic couplings are the important factors in this charge transfer process occuring at rather large internuclear distances and Franck-Condon factors only play a minor role.

3.3. EXAMPLE OF A NON RESONANT CHARGE TRANSFER REACTION : O_2^+ + Ar <--> Ar^+ + O_2

3.3.1. *$Ar^+(^2P_{3/2}$ or $^2P_{1/2})$ + O_2 --> O_2^+ + Ar reaction.* For low collision energies the two exit channels of this reaction are :

$Ar^+(^2P_{3/2}$ or $^2P_{1/2})$ + O_2 --> $O_2^+(X,v)$ + Ar ΔE=-3.7 eV exothermic
--> $O_2^+(a,v)$ + Ar ΔE=+0.345 eV endothermic

The first reactive channel is exothermic by 3.7 eV, while the second is endothermic by 0.345 eV. The $O_2^+(a,v)$ states satisfy both the energy resonance and Franck-Condon criteria. The cross sections to produce these states are thus expected to have high values, comparable to the N_2^+ + Ar system, at collision energies higher than the endothermicity. On the other hand the situation is completely different for the $O_2^+(X)$ states. The transition to the resonant $O_2^+(X,v=19$ and $20)$ states has near zero Franck-Condon factors, whereas the transition to the most favored $O_2^+(X,v=0)$ state corresponds to a 3.7 eV energy defect.

Dotan and Lindinger [59] measured the variation of the rate constant of the Ar^+ + O_2 reaction as a function of the collision energy. Their results, which were later confirmed by Scherbarth and Gerlich [60], show that it decreases with energy and then sharply increases at the thermodynamic onset of the $O_2^+(a)$ state production. In these experiments the two spin orbit states of Ar^+(J=3/2 and J=1/2) are populated. Our recent results obtained in Orsay show that this energy dependence is different for each spin orbit state, as can be seen in figure 10 which plots the variation of the cross section σ for both $^2P_{3/2}$ and $^2P_{1/2}$ states. The $\sigma_{1/2}/\sigma_{3/2}$ ratio reaches a maximum of about 3 at 1.8 ev collision energy in the laboratory frame (0.8 eV in the center of mass frame) (see figure 11).

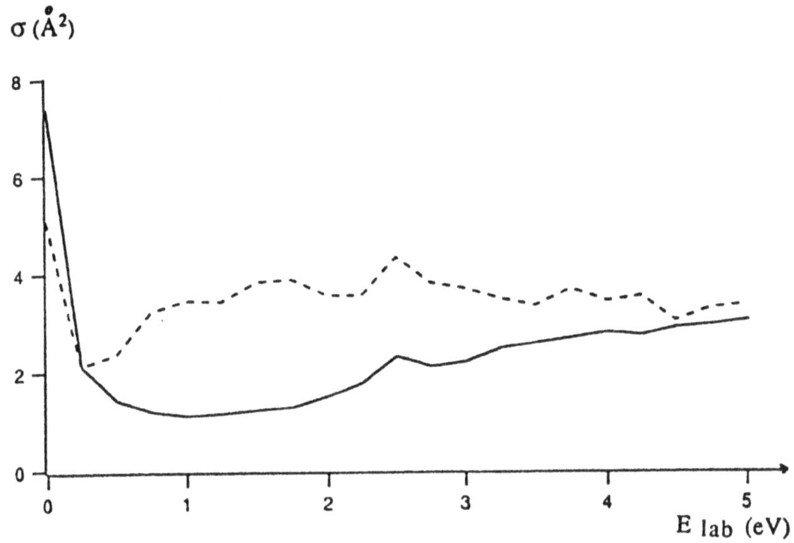

Figure 10. Charge transfer cross sections of the $Ar^+(^2P_J) + O_2$ reaction plotted against the laboratory frame collision energy. The solid line is for the $Ar^+(^2P_{3/2})$ and the dotted line for the $Ar^+(^2P_{1/2})$ state.

The results can be qualitatively interpreted as followed. Below 0.345 eV, the only accessible channel is the formation of $O_2^+(X)$ ground state ions by a collision complex mechanism [61], which breaks down the Franck-Condon principle. The decrease with energy can be explained by the decrease of the intermediate complex lifetime. The sharp increase at a few tenths of an eV is interpreted to be due to the opening of the $O_2^+(a)$ state formation by a "direct" mechanism. The endothermicity for this channel is shifted by 0.18 meV to lower energy for the $Ar^+(^2P_{1/2})$ state compared to the $Ar^+(^2P_{3/2})$ state. This can explain the strong variation of the $\sigma_{1/2}/\sigma_{3/2}$ ratio in this energy region, as it can be rather well reproduced assuming the same but translated σ variation shape for both states (see figure 11).

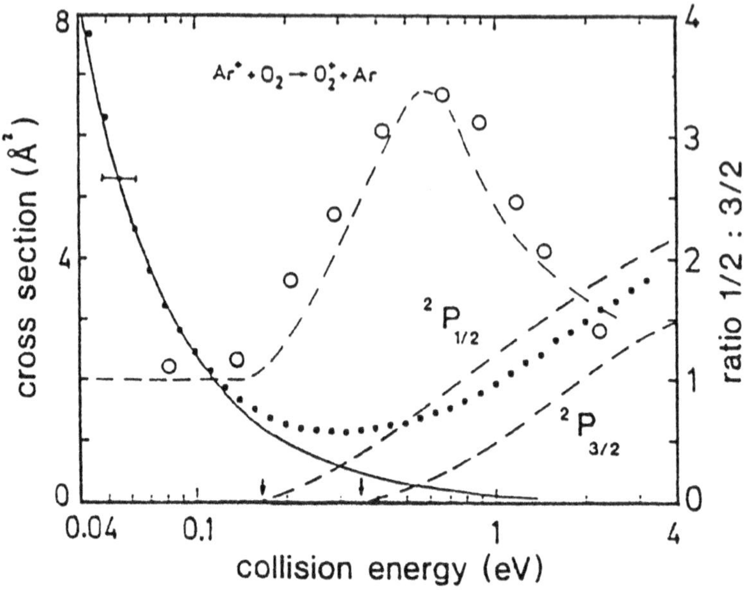

Figure 11. Charge transfer cross sections of the $Ar^+ + O_2$ reaction plotted against the center of mass frame collision energy. The points are the experimental values obtained for the mixture of the two spin-orbit states of Ar^+ [60]. The dashed lines are the supposed cross sections for each spin-orbit state (arrows indicate the O_2^+(a) state thermodynamic threshold for each state) and their ratio, which is compared to our experimental ratio.

Scherbarth and Gerlich [60] obtained kinetic energy distributions of O_2^+ product ions, which are also indicative of a significant change of the reaction mechanism between 0.05 and 0.5 eV collision energy. Experimental evidence was found by Flesch et al [62] supporting the dominating O_2^+(a) state production from its thermodynamic onset, by using a secondary reaction with Ar to probe the product ion electronic state. The nascent rovibrational state distributions of the O_2^+(a) state was measured by laser photofragmentation [60]. The results show that the v=0 is mainly populated with a high rotational excitation (up to J=60). This is in contradiction with a long distance electron jump mechanism. To explain these results, Scherbarth and Gerlich [60] propose a qualitative model based on the DIM (Diatomics In Molecules) potential energy surfaces calculated by Sizun and Kuntz [63]. However better surfaces are required to allow detailed dynamical calculations.

The presence of two different mechanisms in this system is also indicated by the angular distribution of O_2^+ product ions analysed by Futrell et al [64]. At 1.1 eV collision energy, a sideways peaked component is significant of an intimate mechanism and a backward peaked component indicates a "direct" mechanism. The second mechanism becomes dominant when the collision energy increases.

3.3.2. *O_2^+ (X or a,v) + Ar --> Ar^+ + O_2 reverse reaction.* The reverse reaction has been recently studied in our group by threshold photoelectron photoion coincidences as a function of the vibrational and electronic energy of O_2^+ for v=17,18,19,20,23 and 24 of the X state and for v=0 to 6 of the a state. In the first section of this paper we showed in figure 1 that the very high vibrational states of O_2^+(X) state can be populated with synchrotron radiation because of resonant autoionisation. It is particularly interesting to compare the reactivity of the (X,v=23) and (X,v=24) states to the one of the (a,v=2),(a,v=3) and (a,v=4) states, which have about the same total internal energy. Figure 12 shows the results obtained at 0.48 eV collision energy in the center of mass frame (0.87 eV in the laboratory frame). Firstly, a strong vibrational energy dependence is observed within the a state. It is very similar to the variation observed by Kato [65] at a higher collision energy of 1.4 eV, who measured relative cross sections for this reaction. The large value of our measured absolute cross sections agree well with a long distance electron jump mechanism. A maximum is observed for v=5 as for the Franck-Condon distribution. The (X,v=23) and (X,v=24) have a much lower cross section, which might still be somewhat overestimated because of the influence of false coincidences and of the overlap with the nearby (a,v=2 to 4) levels. The low vibrational levels (v<18) of the X state are found non reactive at any collision energy.

3.3.3. *Conclusion.* In conclusion of the study of this non resonant $[Ar,O2]^+$ system it is clear that two different charge transfer mechanisms are involved. At low collision energy, the reaction undergoes through an intermediate collision complex. More experimental data are still required to have a better understanding of this mechanism forming the O_2^+(X) state. In particular it would be very interesting to have more information about the role of the spin orbit energy in this collision energy range to check whether all the available energy is statistically redistributed or not. At higher collision energy the charge transfer mechanism looks "direct" to produce the near resonant states. However the rovibrational energy of the O_2^+(a) state obtained in the Ar^+ + O_2 reaction is not as expected, breaking

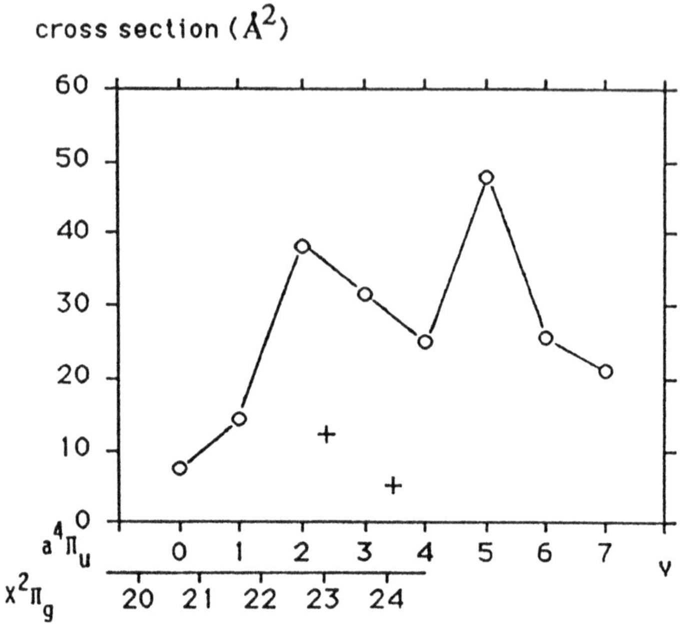

Figure 12. Experimental charge transfer cross sections for the $O_2^+(X,a,v)$ + Ar reaction plotted against the vibrational level at a collision energy of 0.48 eV in the center of mass frame. The crosses correspond to the $O_2^+(X)$ state vibrational levels and the open circles to the $O_2^+(a)$ state.

down the Franck-Condon principle. In this energy range the spin orbit dependence of this reaction can be roughly interpreted with thermodynamical considerations. More work needs to be done concerning the reverse O_2^+ + Ar reaction. The complexity of this system, which remains a challenge for theoreticians, shows that such a "simple" non resonant charge transfer reaction is still not well understood.

ACKNOWLEDGEMENT

The author wants to gratefully acknowledge all her co-workers for the work presented in this paper, in particular P.M.Guyon, M.Richard-Viard, M.Aït-Kaci, J.Lemaire, I.Powis, M.Lavollée, C.Métayer, J.Hepburn, T.Weng, J.B.Ozenne, D.Gerlich and E.A.Gislason. She is particularly grateful to R.Gash who made useful suggestions to the manuscript. We also aknowledge the support provided by the GRECO "Dynamique réactionnelle" and the European Community within the Science Stimulation Program. The staff of LURE is also thanked for their help and support of these experiments, as well as for operating the Storage Ring.

REFERENCES

[1] T.Baer, Advances in Chemical Physics 64, 111 (1986)

[2] C.-Y.Ng, in "Techniques for the study of ion-molecule reactions", Vol.XX, ed.J.M.Farrar and W.H.Saunders,Jr., Wiley, New York, p.417 (1988)

[3] T.Baer, in "Gas phase ion chemistry", Vol.1, ed. M.T.Bowers, Academic, New York, Ch.5, p.153 (1979)

[4] W.Chupka, (a) in "Ion-molecule reactions", ed. J.L.Franklin, Plenum, New York, p.33 (1972); (b) "Chemical spectroscopy and photochemistry in the vacuum ultraviolet", NATO-Advanced Study Institutes Series C, ed. C.Sandorfy, P.J.Ausloos and M.B.Robin, Reidel, Boston, vol.8, p.433 (1973)

[5] (a) F.A.Houle, S.L.Anderson, D.Gerlich, T.Turner and Y.T.Lee, Chem.Phys. Lett. 82, 392 (1981); (b) S.L.Anderson, T.Turner, B.H.Mahan and Y.T.Lee, J.Chem.Phys. 77, 1842 (1982); (c) T.Turner, O.Dutuit and Y.T.Lee, J.Chem Phys. 81, 3475 (1984); (d) T.Turner and Y.T.Lee, J.Chem.Phys. 81, 5638 (1984)

[6] (a) P.M.Guyon, T.Baer, L.F.A.Ferreira, I.Nenner, A.Tabché-Fouhailé, R.Botter and T.R.Govers, J.Phys.B 11, L141 (1978); M.Richard-Viard, O.Dutuit, M.Lavollée, T.Govers, P.M.Guyon and J.Durup, J.Chem.Phys. 82, 4054 (1985)

[7] I.Nenner and J.A.Beswick, Handbook on Synchrotron Radiation, vol.2, ed.G.V.Marr, Elsevier, Ch.6, p.355 (1987)

[8] T.Ebata and R.N.Zare, Chem.Phys.Lett. 130, 467 (1986)

[9] T.M.Orlando, B.Yang and S.L.Anderson, J.Chem.Phys. 90, 1577 (1989)

[10] M.Richard-Viard, O.Atabek, O.Dutuit and P.M.Guyon, J.Chem.Phys. to be published

[11] J.C.Lorquet, C.Barbier and B.Leyh-Nihant, Advances in Mass Spectrometry, ed.J.F.J.Todd, Wiley, New York, p.71 (1985)

[12] O.Dutuit, M.Aït-Kaci, J.Lemaire and M.Richard-Viard, Phys. Scripta T31, 223 (1990)

[13] M.N.Paddow-Row, D.J.Fox, A.J.Pople, K.N.Houk and D.W.Pratt, J.Amer.Chem.Soc. 107, 7696 (1985)

[14] C.R.Brundle, M.B.Robin and H.J.Basch, J.Chem.Phys. 53, 2196 (1970)

[15] J.Berkowitz, in "Photoabsorption, photoionisation and photoelectron spectroscopy", Academic Press (1979)

[16] R.Stockbauer, J.Chem.Phys. 58, 3800 (1973)

[17] P.L.Kronebusch and J.Berkowitz, Int.J.Mass Spec. Ion Phys. 22, 283 (1976)

[18] S.E.Stein and B.S.Rabinovitch, J.Chem.Phys.58, 2483 (1973)

[19] M.Vestal and G.Lerner, in "Fundamental studies relating to the radiation chemistry of small organic molecules", Aerospace Res. Lab. Rept. no 67-0114 (1967)

[20] C.E.Klots, J.Phys.Chem. 75, 1526 (1971)

[21] A.J.Illies, M.F.Jarrold and M.T.Bowers, J.Amer. Chem.Soc.104, 3587 (1982)

[22] I.Powis, J.Chem.Soc.Far.Trans. II 75, 1294 (1979)

[23] W.Chupka, J.Chem.Phys. 48, 1109 (1968)

[24] E.F.Van Dishoeck, W.J.Van Der Hart and M.Van Hemert, Chem.Phys. 50, 45 (1980)

[25] V.Cermak and J.Sramek, J.Electron Spec.Rel.Phenom. 2, 97 (1973)

[26] V.Cermak, J.Electron Spec.Rel.Phenom. 3, 329 (1974)

[27] V.Cermak, J.Electron Spec.Rel.Phenom. 6, 135 (1975)

[28] P.Morin and I.Nenner, Phys.Scripta T17, 171 (1987)

[29] A.P.Hickman, Phys.Rev.Lett. 59, 1553 (1987)

[30] J.R.Appling and M.White, Brookhaven National Lab. Rep., p.181 (1986)

[31] A.A.Cafolla, T.Reddish and J.Comer

[32] C.Y.R.Wu and D.L.Judge, J.Chem.Phys. 75, 172 (1981)

[33] M.Richard-Viard, O.Dutuit, A.Amarkhodja and P.M.Guyon, in "Photophysics and photochemistry above 6 eV", ed.F.Lahmani, Elsevier, p.153 (1985)

[34] J.H.D.Eland and J.Berkowitz, J.Chem.Phys. 67, 2782 (1977)

[35] R.Bombach, J.Dannacher, J.P.Stadelmann and J.C.Lorquet, J.Chem.Phys. 79, 4214 (1983)

[36] I.Powis, O.Dutuit, M.Richard-Viard and P.M.Guyon, J.Chem.Phys. 92, 1643 (1990)

[37] (a) P.M.Guyon, T.Baer and I.Nenner, J.Chem.Phys. 78, 3665 (1983); (b) M.J.Hubin-Frankin, J.Delwiche and P.M.Guyon, Z.Phys.D, 5, 203 (1987)

[38] R.Stockbauer, J.Chem.Phys. 72, 5277 (1980)

[39] O.Dutuit, A.Tabché-Fouhailé, I.Nenner, H.Frohlich and P.M.Guyon, J.Chem.Phys. 83, 584 (1985)
[40] W.A.Chupka, M.E.Russell and K.Refaey, J.Chem.Phys. 48, 1518 (1968)
[41] W.A.Chupka and M.E.Russell, J.Chem.Phys. 49, 5426 (1968)
[42] W.A.Chupka and M.E.Russell, J.Chem.Phys. 48, 1527 (1968)
[43] (a) T.Baer, in "Gas Phase Ion Chemistry", ed.M.T.Bowers, Academic Press, New York, vol.1, p.153 (1979); (b) T.Baer, L.Squires and A.S.Werner, Chem.Phys. 6, 325 (1974); (c) L.Squires and T.Baer, J.Chem.Phys. 65, 4001 (1976)
[44] I.Koyano and K.Tanaka, J.Chem.Phys. 72, 4858 (1980)
[45] (a) T.R.Govers, P.M.Guyon, T.Baer, K.Cole, H.Frohlich and M.Lavollée, Chem.Phys. 87, 373 (1984); (b) P.M.Guyon, T.R.Govers and T.Baer, Z.Phys.D, 4, 89 (1986)
[46] N.G.Adams and D.Smith, in "Techniques for the study of ion-molecule reactions", Vol.XX, ed.J.M.Farrar and W.H.Saunders,Jr., Wiley, New York, p.165 (1988)
[47] P.R.Kemper and M.T.Bowers, in "Techniques for the study of ion-molecule reactions", Vol.XX, ed.J.M.Farrar and W.H.Saunders,Jr., Wiley, New York, p.1 (1988)
[48] J.M.Farrar, in "Techniques for the study of ion-molecule reactions", Vol.XX, ed.J.M.Farrar and W.H.Saunders,Jr., Wiley, New York, p.325 (1988)
[49] (a) E.Teloy and D.Gerlich, Chem.Phys. 4, 417 (1974); (b) D.Gerlich, in "Advances in Chemical Physics", ed. C.-Y.Ng and M.Baer, to be published
[50] M.Lavollée and G.Henri, J.Phys.B, 22, 2019 (1989)
[51] P.M.Guyon and E.E. Gislason, in "Topics in Current Chemistry", vol.151, Springer-Verlag, Berlin, p.161 (1989)
[52] E.E.Gislason and G.Parlant, Comments At. Mol. Phys. 19,157 (1987)
[53] T.Kato, K.Tanaka and I.Koyano, J.Chem.Phys. 77, 834 (1982)
[54] (a) C.L.Liao, R.Xu and C.-Y.Ng, J.Chem.Phys. 85, 7136 (1986); (b) J.D.Shao, Y.G.Li, G.D.Flesch and C.-Y.Ng, J.Chem.Phys. 86, 170 (1987)
[55] (a) M.R.Spalburg, J.Los and E.A.Gislason, Chem Phys. 94, 327 (1985); (b) M.R.Spalburg and E.A.Gislason, Chem.Phys. 94, 339 (1985)
[56] G.Parlant and E.A.Gislason, Chem.Phys. 101, 227 (1986)
[57] P.Archirel and B.Lévy, Chem.Phys. 106, 51 (1986)
[58] E.E.Nikitin, M.Y.Ovchinnikova and D.V.Shalashilin, Chem.Phys. 111, 313 (1987)
[59] I.Dotan and W.Lindinger, J.Chem.Phys. 76, 4972 (1982)

[60] S.Scherbarth and D.Gerlich, J.Chem.Phys. 90, 1610 (1989)

[61] N.Kobayashi, Y.Kaneko, F.Koike and T.Watanabe, J.Phys.Soc.Japan 42, 1701 (1977)

[62] G.D.Flesch, S.Nourbakhsh and C.-Y.Ng, J.Chem.Phys. 92, 3590 (1990)

[63] M.Sizun and P.J.Kuntz, in "Abstracts of contributed papers", II ECAMP, ed. A.E. de Vries and M.J.Van der Wiel, EPS, Amsterdam (1985)

[64] J.Futrell, private communication

[65] T.Kato, J.Chem.Phys. 80, 6105 (1984)

GENERATION AND REACTIVITY OF STATE SELECTED ATOMIC AND SIZE SELECTED TRANSITION METAL CLUSTER IONS

MICHAEL T. BOWERS, PAUL R. KEMPER, GERT von HELDEN and MING-TEH HSU
Department of Chemistry
University of California
Santa Barbara, CA 93106

ABSTRACT.

In this chapter, two distinct studies in the general field of gas phase transition metal ion chemistry are presented. In the first, atomic metal ions of the first row transition series are generated by electron impact on various volatile organometallic reagents. These ions are then mass selected and injected at low energies into a pressure and temperature variable drift reactor. After drifting through the reactor, filled with He buffer gas, the ions exit, are mass analyzed by a quadrupole and detected. By pulsing the ion beam as it enters the drift reactor, arrival time distributions (ATD) at the detector can be obtained. Many of the ATD's were found to be bimodal or trimodal, indicating the atomic metal ion beams contained multiple components that interacted significantly differently with the He buffer gas. Analysis of the data showed the largest splittings in the ATD's were due to ion electronic states with different electronic configurations (either $3d^n$ or $4s3d^{n-1}$). In favorable cases electronic states within the $4s3d^{n-1}$ configuration could be resolved. Clustering of state selected M^+ with He was studied. It was found that only M^+ ions with $3d^n$ configurations formed the M^+–He adduct. Equilibrium constant measurements yielded values of ΔH between 1 and 4 kcal/mol and ΔS between 13 and 18 cal/mol°K. Deactivation of excited states was also observed only for metal ions with $3d^n$ configuration excited states and $4s3d^{n-1}$ ground states. Deactivation rate constants were measured for two systems and a mechanism proposed.

In the second study, size selected Nb_n^+ ions ($1 \leq n \leq 6$) were injected into the drift cell and reacted with O_2, NO, C_2H_6, C_3H_8 and C_6H_6 along with deuterated analogues. Several reactions of NbO^+ are also reported. Both total rate constants and product branching ratios were measured. Examples were chosen to highlight the size selective behavior of the clusters and to point out that truly surprising reactions are observed. Thermodynamic and mechanistic conclusions are drawn where appropriate.

K. R. Jennings (ed.), Fundamentals of Gas Phase Ion Chemistry, 55–85.

1. Introduction

This report will discuss two separate aspects of our work in the chemistry of transition metal ions. The first will be concerned with state selection of atomic transition metal ions and our first results on the reactivity of these species. The second will be concerned with the reactivity of size selected niobium cluster ions with small hydrocarbons. Both areas offer exciting prospects for future study, and brief remarks will be made on directions we intend to pursue.

2. Electronic State Chromatography of Transition Metal Ions

2.1 BACKGROUND

One of the most active areas of research in ion chemistry is the study of how atomic transition metal ions interact and react with organic molecules.[1] The d-orbitals available to these ions open up reactive possibilities that simply do not exist for main group atoms. That is the good news. The bad news is transition metals are involatile and consequently, are not easily transported into the gas phase. The principal method used, until recently, to generate the atomic ions was high energy electron impact on the limited number of volatile compounds available for each metal. Often these compounds did yield substantial M^+ signals under 70 to 100 eV EI conditions. The problem was, and is, that the distribution of excited states was not known. What is known is that each of the metals has metastable excited electronic states whose lifetimes far exceed typical experimental time scales.[2] And, as we will see shortly, generation of these states in high abundance by EI is a common occurrence for metals in the first transition series.

Other methods of generation of atomic transition metal ions have been used. Surface ionization[1b,3] is a potentially useful technique because it presumably generates a Boltzmann distribution of states appropriate for the temperature of the filament. For metals with no low lying excited states, such a process forms essentially pure ground state. In other cases, useful amounts of low lying excited states can be prepared and in favorable cases their reactivity studied.[1b,3] What has not yet been shown is that SI does indeed produce a thermal distribution of states. The new experimental technique of Electronic State Chromatography discussed here will allow such a test to be conducted.

Laser desorption[4] is a universal technique for all metals, and shows great promise for the future. Metal ions can be formed *in situ* in ICR cells or high pressure pulsed sources and reaction chemistry readily carried out. Again, however, the problem of excited state distribution arises. We hope to be able to say something definitive on this issue in the near future.

One method that is unambiguous is resonant multiphoton ionization of neutral metal atoms with concurrent photoelectron spectroscopy of the ejected electron. This elegant method has been applied to a number of systems[5] and has been shown to be able to generate useable signals for at least some metals. The drawback here is the substantial investment that must be made just to generate the ions (two pump lasers, one dye laser, a PES attached and a complex tandem flow reactor). As a consequence, this method will be limited to a narrow band of applications where such things as the effect of total system electronic angular momentum on reactivity can be explored.

Up to now the presence of excited states in atomic metal ion populations has been inferred primarily from reactivity studies[1b,6]. One of two methods is utilized: either a reaction is observed that is endothermic for ground state ions, or thresholds of endoergic reactions are shifted to lower energies than expected from known thermochemistry. This latter method allows estimates of percentages of excited state populations if certain assumptions are made about the shapes and magnitudes of the cross sections of the various electronic states. (More will be said about this later).

Translational energy spectroscopy[7] has also been applied to metal ions.[8] In these experiments, the excited state is observed as a collision generated superelastic peak, separated in energy from the main beam by the energy separation of the ground and excited states. If very careful experiments are performed, estimates of the population of the excited state can be obtained.[9] In general, however, only qualitative information results from these experiments.

The method to be discussed here takes advantage of a remarkable difference in mobilities in helium gas for different electron configurations of the transition metal atomic ions. Differences in mobilities in helium of electronic states of main group atoms have been observed previously for C^+[10] and O^+[11], but these differences were relatively small (of the order of a few percent to perhaps as much as 20%). However, as will be shown, we routinely observe differences in mobilities of 50% for transition metal ions. It is this large difference in mobilities between states, and the richness of metastable electronic states for transition metal ions, that makes the "chromatography" of these states possible. That fact, coupled with the enormous interest in their chemistry, makes these studies both very interesting and of immediate practical importance in the field of gas phase organometallic transition metal ion chemistry.

2.2 EXPERIMENTAL METHODS

A first report on the experimental method, and some preliminary results have been published[12] and a much more detailed paper has just been submitted for publication.[13] In addition, a very detailed description of the instrument recently appeared in the literature.[14] Consequently, this section will be quite brief and will give only enough of an overview that the reader can fully appreciate the experimental results.

2.2.1 *The Instrument.* The atomic metal ions are formed by EI from appropriate volatile organometallic compounds. The pressure in the source is kept low enough that the fraction of ions undergoing a collision in the source is less than 5%. The ions exit the source, are accelerated to 5 kV, and are mass selected by a home-built high resolution reverse geometry mass spectrometer. The ions are then decelerated to 1 to 3 eV and focused on the entrance hole of a variable pressure (0 - 7 torr) and variable temperature (80 - 600K) drift reactor. For the experiments discussed here, typical pressures of helium buffer gas in the drift cell are 2 - 5 torr and typical temperatures 140 to 300K. The entrance and exit holes are 0.5 mm, and the drift length is 4.00 cm. Eight drift guard rings insure the ions experience a very uniform drift field throughout the cell. Ions exiting the cell are mass analyzed by a quadrupole mass filter and detected using an electron multiplier.

2.2.2. *The Mobility Experiment.* The mobility is a measure of how fast an ion will drift through a buffer gas when subjected to a uniform electric field. The greater the mobility, the faster the ions move through the gas. Consequently, ions entering the cell at the same time, but having different mobilities, will soon be both spatially and temporally segregated. It is this property that allows us to perform the Electronic State Chromatography experiment.

In order for temporal separation to occur at the cell exit plate, all ions must start at the entrance plate at the same time. Consequently, the ions must be pulsed into the cell to begin the experiment. The "width" of the start pulse depends on two factors. First, it should be much narrower than the *spread* in ion transit time occurring in the cell due to collisions of the ion with the buffer gas. Second, it has to be broad enough so that sufficient ions are introduced, such that an actual experiment can be done. We use pulse widths of 1 to 3 μs, which satisfies both requirements for most of our systems.

The "start" pulse begins a clock that ends when an ion reaches the detector. If a time-to pulse-height-converter (THPC) is used, care must be taken to limit the collection rate to <5% of the input pulse frequency (typically 1 - 5 kHz). In this mode we have the best resolution (~0.4 μs/channel) in the arrival time distributions, but the experiments take a long time to complete because count rates are so low. If we use a multichannel scaler detection system (MCS), on the other hand, count rates of $3 \times 10^6\ s^{-1}$ can be achieved, but resolution falls off to ≥ 2 μs/channel. The particular experiment dictates which detection method is used.

Details of how we determine the actual mobilities from the data are given elsewhere.[13] What is important here are the factors that govern the arrival time distribution (ATD) at the detector, especially the width of that distribution. It is straightforward to show[13] that the time resolution, R, of this distribution is

$$R = \frac{t_d}{\Delta t_d} = \left(\frac{qV}{8kT}\right)^{1/2} \tag{1}$$

where t_d is the average drift time in the cell (i.e. the center of the ATD), Δt_d is the time width of the ATD, q is the ion charge, V is the drift voltage across the cell, T is the temperature and k is Boltzmann's Constant. This is a very important relationship, and it guides our choice of the cell operating conditions.

At first glance, it appears we want to work at maximum V and minimum T in order to maximize the resolution R. The payoff could be complete base line resolution of all electronic states in the metal ion beam, since each state must interact at least slightly differently with helium. In practice, two restraints are imposed. First, simply raising V increases the effective translational temperature T_{eff}, of the ion:

$$T_{eff} = T + \frac{v_d^2 M_{He}}{3k} \tag{2}$$

where M_{He} is the mass of helium and v_d is the drift velocity, which depends on the drift voltage V.[15] Consequently, the resulting non thermal M^+ ions develop a broad temperature distribution, resulting in a broadened ATD. What must be done is to increase the helium concentration in the cell such that the absolute transit time through the cell does not change as V is increased; i.e. v_d remains constant. Given our current pumping system, we are limited to about 7 torr of He at 300K.

The second problem is that all mobilities approach the Langevin mobility, K_L as $T \rightarrow 0$.[15] Consequently, temperature cannot simply be lowered as far as possible because the mobilities of different electronic states eventually begin approaching K_L faster than their resolution is improving.

Hence, while the ATD's get narrower, they begin to overlap and information is lost. We have found that the best compromise occurs at 140 to 170K for the first row transition metals.

2.2.3. *Clustering of M^+ with He.* The transition metal ions can cluster with He as they traverse the drift reactor:

$$M^+ + 2He \rightleftharpoons M^+{\cdot}He + He. \quad (3)$$

This reacting system can come to equilibrium. Two tests for equilibrium were conducted. First, the drift voltages were decreased until the $M^+{\cdot}He/M^+$ ratio became constant. Second, the arrival time distributions for $M^+{\cdot}He$ and M^+ had to be identical. Both criteria were met in all instances where equilibrium is claimed to have occurred.

The drift field translationally heats the ions. However, in all instances our equilibrium measurements are conducted in the low field limit,[15] with $E/N \leq 2.5$ Td. Under these conditions $T_{eff} \leq T + 0.5K$, which is acceptable over our temperature range of 140 to 150K.

2.2.4. *Quenching of Excited States.* Collisional quenching of the metastable electronic states can be studied under favorable circumstances; that is when the ATD's of the excited and ground state are well resolved and when the quenching rate constant is large enough: $k \geq 1 \times 10^{-16}\ cm^3 s^{-1}$. Evidence for quenching is a "filling in" of the ATD spectrum between the excited state and the ground state. Ions arriving at the detector at these intermediate times spent part of their time in the drift cell as an excited state and part as the ground state. The most accurate way to obtain quenching rate constants from these peaks is to fit the total ATD profile for a given set of experimental conditions.[15] We are currently finishing development of this very non-trivial procedure for application to our data. A simpler, more approximate method of measuring peak height changes with He pressure in the drift cell is used here. This allows reasonably accurate preliminary numbers to be obtained.

2.3 RESULTS AND DISCUSSION

Data has been obtained on all atomic metal ions in the first row transition series except Sc^+. The detailed results are presented elsewhere.[13] Here we will pick several example systems and give a summary of the results.

2.3.1. *The Determination Of The Electronic Ground State.* In most of the systems studied, multiple peaks were observed in the ATD spectra. An example is given for vanadium ions in Figure 1. Clearly three peaks are present, indicating at least three metastable electronic states are formed by 50 eV electron impact on $VOCl_3$. The relative intensity of these three states dramatically changes with electron energy, however, as shown in Figure 2. As electron energy is lowered, peak III begins to dominate, probably heading for 100% of the total V^+ ions at the appearance energy of V^+ from $VOCl_3$. Hence, peak III is almost certainly the $a\,^5D\ 3d^4$ ground state of V^+.

It also appears from the data in Figure 2 that peak I is the highest energy state (or set of states). Hence, one gets essentially no information about state ordering simply by doing an experiment at one

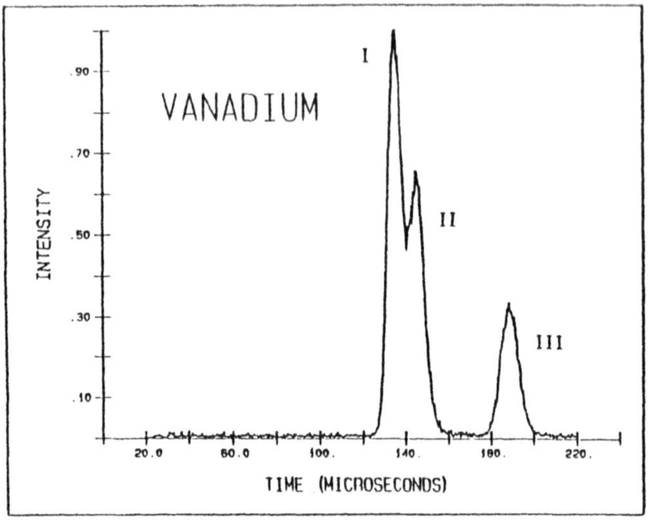

Figure 1. Arrival Time Distribution for V^+ ions formed from $VOCl_3$ by 50 eV electron impact. The cell temperature was 176K.

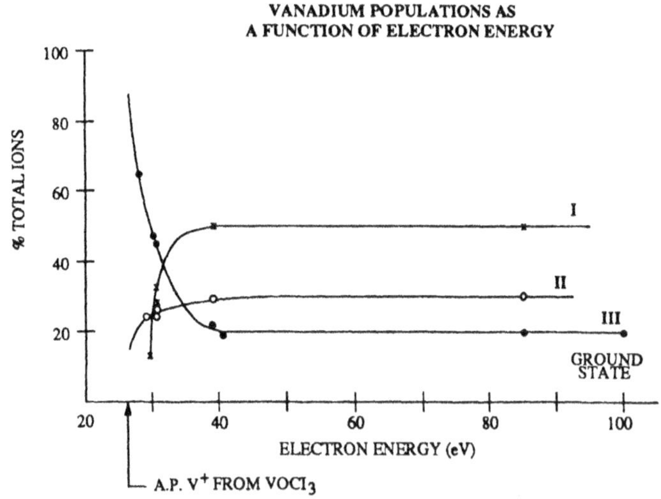

Figure 2. A plot of percent of total ions versus ionizing electron energy for the three V^+ peaks shown in Figure 1. The appearance potential of V^+ from $VOCl_3$ is shown by the arrow.

electron energy, especially if that energy is significantly above threshold.

Similar experiments were done on all systems to sort our state ordering. Assignments of these states, and their apparent relative abundances, will be dealt with shortly.

2.3.2. *Assignment Of Electronic Configurations For Different Peaks In The Arrival Time Distribution.* In Figure 1, three peaks were evident. Peaks I and II partially overlapped and "arrived" at the detector about 50 μs before peak III. The implication is that peaks I and II interact with He in a similar way, and peak III interacts in a substantially different way. A second example is given for Co^+ ions in Figure 3. In this case there are only two peaks, but clearly peak I arrives at the detector

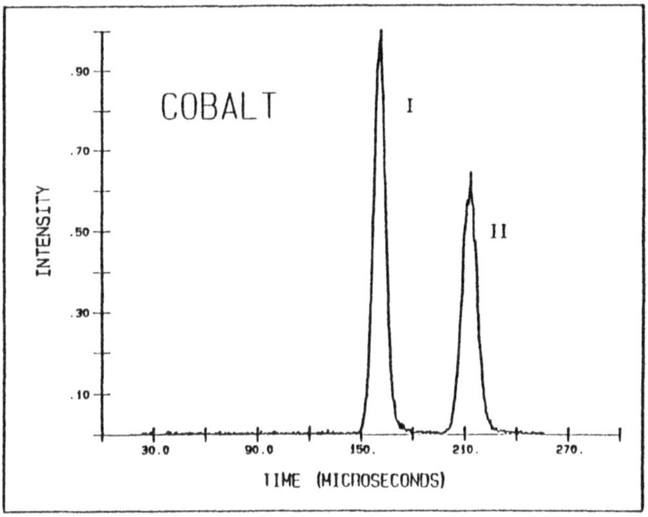

Figure 3. Arrival Time Distribution for Co^+ ions formed from $Co(CO)_3NO$ by 50 eV electron impact. The cell temperature was 152K.

about 50 μs before peak II. These arrival time distributions can be converted to mobilities.[13] Figure 4 is a plot of the mobilities of the various peaks in an ATD spectrum versus the number of valence electrons on the metal ion.

First, consider those ions with six or more valence electrons, that is the region from Mn^+ to Zn^+. It is obvious two widely separated sets of mobilities are observed; one set at about 24 $cm^2V^{-1}s^{-1}$ and one set at about 16 $cm^2V^{-1}s^{-1}$. The ground states are designated as the filled dots within the open circles, ⊙. For Co^+, Ni^+ and Cu^+ the ground states are known to have d^8, d^9 and d^{10} electronic configurations, respectively. Each of these ground states has a mobility of about 16 $cm^2V^{-1}s^{-1}$. On the other hand, Mn^+, Fe^+ and Zn^+ are known to have ground state configurations of $4s^13d^5$, $4s^13d^6$ and $4s^13d^{10}$ respectively. Each of these ground states have mobilities of about 24 $cm^2V^{-1}s^{-1}$.

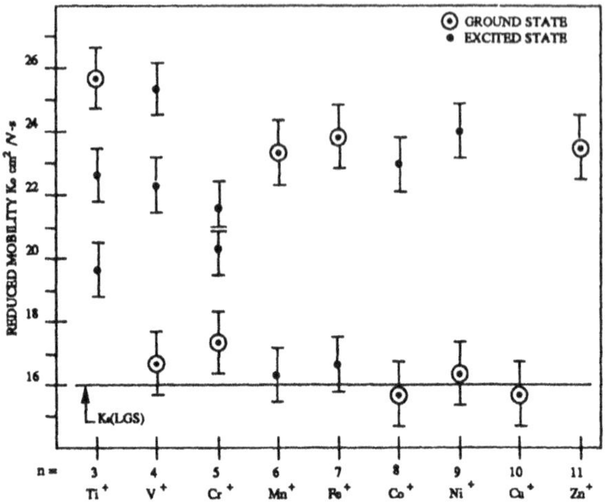

Figure 4. A plot of the reduced mobilities for the various peaks in the ATD spectra versus n, the total number of valance electrons (4s and 3d). The prediction of the Langevin Theory for the reduced mobility is shown by the straight line. Experimentally determined ground state mobilities are designated by ⨀, while mobilities of all other states are designated by •.

Hence, it appears unambiguous that the "low" mobility ions (slow ions) have $3d^n$ electronic configurations and the "high" mobility ions (fast ions) have $4s^1 3d^{n-1}$ configurations. Consequently, a clue to the identity of the electron state designation is given by the relative mobility of the ATD peak. These clues will be of great help in assigning the excited states and their abundances when data from other experiments in the literature are considered.

There are other, more subtle points apparent in Figure 4. For example, the "slow" ions have mobilities near the Langevin mobility. This is not surprising since these are all $3d^n$ configurations and the $3d^n$ orbital radii[16] are much smaller than the Langevin capture radius.[17] Also observed, however, is an odd/even alteration that is reproducible within the experimental error of the measurements: The M^+ ions with an odd number of d electrons clearly have about 5% higher mobility than the M^+ ions with an even number of d electrons. The reason for this is not known.

There is also a slight trend toward increasing mobility as the total number of d electrons decreases. This is almost certainly due to an increase in d orbital size,[16] resulting in a small but discernible repulsive interaction with He.

The "fast" ions are all of $4s3d^{n-1}$ configuration. In this case, capture collisions almost certainly do not occur, in the Langevin sense, due to the very large size of the 4s orbital. It is apparent that substantially different interactions begin to occur for different electronic states with $4s3d^{n-1}$ configurations when the number of valence electrons is five or less. Detailed quantum mechanical calculations will be required to understand fully what is going on. At this state we simply hope to assign electronic states to the various ATD peaks and approximate abundances for a particular set for experimental conditions.

2.3.3. *Assignment Of Electronic States And Relative Abundances To The Atd Spectra.* Determining unambiguous assignments of specific electronic states to particular ATD peaks is not a simple matter. From the mobility data, we know the electron configuration ($4s3d^{n-1}$ or $3d^n$), and from the variation of electron energy studies we know which peak contains the ground state, and can usually infer the energy ordering of the other peaks. Finally, only long lived metastable states are involved. Other very useful information can be obtained from reactivity studies, such as those performed by Armentrout's group[1b,6] or by Ridge and coworkers.[18] Here I will give one example of the kind of reasoning we have gone through to obtain the estimates presented in Table 1.

Figure 5 shows a relatively low resolution ATD spectrum of Ti^+. Electron energy variation studies unambiguously show that peak I contains the ground state. These same studies indicate peak III is highest in energy. The mobility data indicate peaks I and II are definitely $4s3d^2$ configuration, and that peak III is very probably $4s3d^2$. Armentrout and coworkers[1b,6e,19] detect reactivity from a series of states. The principal reactive states have approximate energies of 0.0 eV (ground state and possibly some $b^4F\ 3d^3$ at 0.13 eV), 0.59 eV and one or more states between 1.1 and 1.24 eV. Consideration of our data and those of Armentrout and coworkers leads to the following assignment for the ATD spectrum in Figure 5; peak I is the ground state; peak II is the $a^2F(4s3d^2)$ state at 0.57 eV; and peak III is the $a^2D(4s3d^2)$ or $b^4D(4s3d^2)$ or both. These assignments are given in Table 1 along with the relative intensities taken from data like that of Figure 5. Similar rationals were used to obtain all of the suggested assignments in Table 1. Detailed arguments are presented elsewhere.[13]

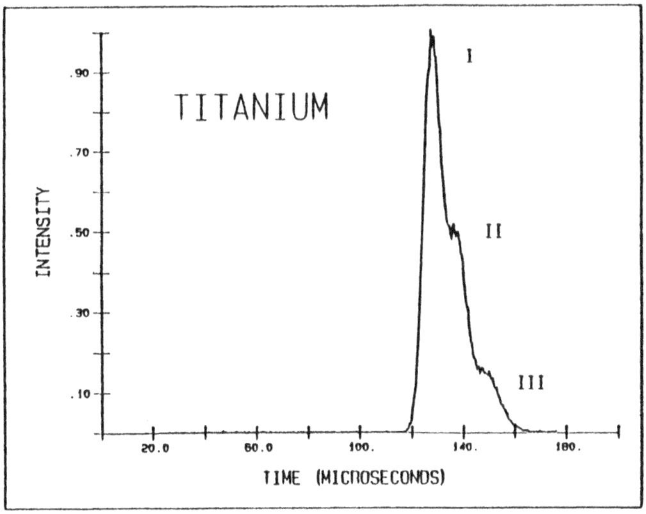

Figure 5. Arrival Time Distribution for Ti^+ formed from $TiCl_4$ by 50 eV electron impact. The temperature of the cell was 300K.

TABLE 1. Populations of metastable electronic states of first row transition metal ions formed by 50 eV electron impact of selected precursor neutral molecules.

M^+	State	Energy, eV[a]	%[b]
Ti^{+c}	$a^4F(4s3d^2)$	0.0	51
	$a^2F(4s3d^2)$	0.57	35
	$a^4D(4s3d^2)$	1.08	14
	$b^4D(4s3d^2)$	1.16	
V^{+d}	$a^5D(3d^4)$	0.0	~20
	$a^5F(4s3d^3)$	0.32	~30
	High Energy States		
	$b^3P(4s3d^3?)$	(1.69)	~50
Cr^{+e}	$a^6S(3d^5)$	0.0	46
	$a^6D(4s3d^4)$	1.48	47
	$a^4D(4s3d^4)$	2.42	7
Mn^{+f}	$a^7S(4s3d^5)$	0.0	88
	$a^5S(4s3d^5)$	1.17	
	$a^5D(3d^6)$	1.08	12
Fe^{+g}	$a^6D(4s3d^6)$	0.0	76
	$a^4D(4s3d^6)$	0.99	
	$a^4F(3d^7)$	0.23	24
	$a^4P(3d^7)$	1.67	
Co^{+h}	$a^3F(3d^8)$	0.0	36
	$b^3F(4s3d^7)$	0.42	64
Co^{+i}	$a^3F(3d^8)$	0.0	83
	$b^4F(4s3d^7)$	0.42	17
Ni^{+j}	$a^2D(3d^9)$	0.0	17
	$a^4F(4s3d^8)$	1.16	83
	$a^2F(4s3d^8)$	1.76	
	$b^2D(4s3d^8)$	2.90	
Cu^{+k}	$a^1S(3d^{10})$	0.0	100
Zn^{+l}	$a^2S(4s3d^{10})$	0.0	100

a) This is the energy of the lowest spin orbit state.
b) These are approximate percentages. If a more detailed account of other work in the literature is considered, minor contributions from other states occur and approximate fractions can be estimated for pairs of states we have grouped together.
c) From $TiCl_4$ d) From $VOCl_3$ e) From CrO_2Cl_2 f) From $CH_3C_5H_4Mn(CO)_3$
g) From $Fe(CO)_5$ h) From $Co(CO)_3NO$ i) From $C_5H_5Co(CO)_2$ j) From $Ni(CO)_4$
k) From Cu(I)Acetate or Cu(II)(Acetate)$_2$ l) From $Zn(C_2H_6)_2$

Several points are worth mentioning about the data in Table 1. First, the ground state is the major state formed about half the time. Second, the actual compound used to generate M^+ can have a strong influence on the state distribution (see Co^+ for an example). Third, deactivation of excited states by collisions with the He buffer gas can make determination of accurate nascent distributions difficult. This latter point will be addressed further later in this paper. Fourth, the ability to assign state distributions opens up the possibility of studying the dynamics of M^+ formation, state selective reactivity, and excited state quenching processes.

2.3.4. *Clustering Of M^+ With He.* We have obtained some preliminary data on the clustering of M^+ ions with helium. Using the relationships

$$\Delta G^\circ = -RT \ln K_p \tag{4}$$

$$\Delta G^\circ = \Delta H^\circ - T\Delta S^\circ \tag{5}$$

and a plot of ΔG° vs T, values of ΔH° and ΔS° can be readily obtained. An example of such a plot is given in figure 6, and the data summarized in Table 2. Note that the four systems that form adducts

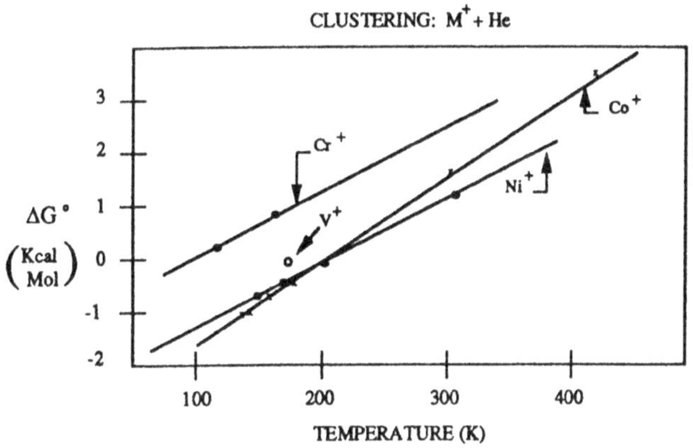

Figure 6. A plot of ΔG° in kcal/mol versus temperature in degrees Kelvin for clustering equilibria $M^+ + 2He \rightleftharpoons M^+{\cdot}He + He$ where $M^+ = Co^+$, Cr^+ V^+ and Ni^+. Only one temperature has been done to date for V^+.

all have $3d^n$ ground state electronic configurations. No clustering was observed for ions with $4s3d^{n-1}$ ground states, although the $a^4F(3d^7)$ excited state of Fe^+ did form an $Fe^+{\cdot}He$ adduct. Consequently, M^+ $4s3d^{n-1}$ states interact primarily repulsively with He and probably do not form bound ground state adducts.

The values of ΔH° vary rather substantially from 1.33 kcal/mol for Cr^+ to 3.68 kcal/mol for Co^+. Clearly there is more than a simple electrostatic interaction determining these binding energies. The Cr^+ ion has a half filled $3d^5$ ground state and appears to be less willing to accommodate the two

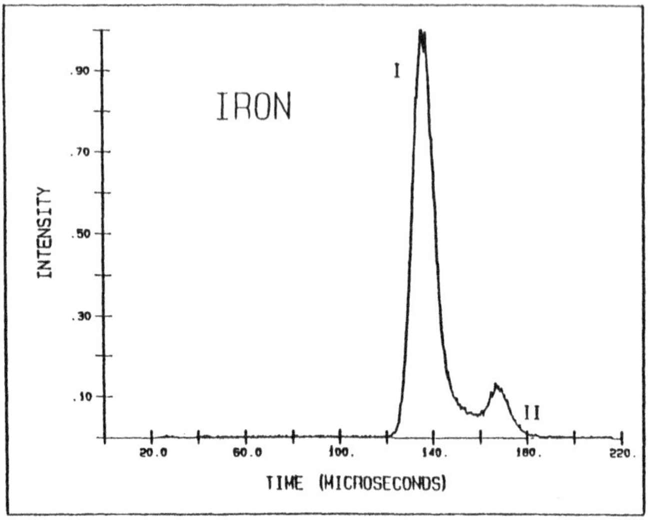

Figure 7. Arrival Time Distribution for Fe^+ from $Fe(CO)_5$ by 50 eV electron impact. The cell temperature was 176K. Note the filling in between the two peaks, characteristic of collision deactivation.

surfaces to interact at energies near or below the M^{+*}/He asymptote and thermal energy deactivation can occur. If, on the other hand, M^+ has a $3d^n$ ground state and a $4s3d^{n-1}$ excited state, then the two M^+–He potential surfaces will not cross until much higher energy and thermal energy deactivation becomes immeasurably slow. A qualitative test of this latter case is for V^+ deactivation. Thermal energy deactivation is immeasurably slow. However, if the drift field in the cell is raised, the V^+ ions gain kinetic energy, and deactivation may become possible. A plot of % V^+ ground state ($a^5D\ 3d^4$) versus drift voltage is given in Figure 8.
Clearly the fraction of ground state dramatically increases as drift voltage increases, a result fully in accordance with the mechanism outlined above.

Experiments are underway with other deactivating neutrals; initially Ne, Ar and possibly N_2. In each case the system will become more attractive at long range, but it is not possible to predict deactivation rates since these will depend on where the curves cross.

3. Reactions of Size Selected Niobium Cluster Ions

3.1 BACKGROUND

This section initiates an essentially new topic, the reaction of size selected niobium cluster ions. Metallic clusters are particles that contain between two and perhaps thousands of atoms whose physical and chemical properties bridge the gap between the bare atom and the bulk. These clusters typically have cohesive energies between 1 and 5 eV, unlike van der Waals or many "ion-molecule" electrons from He than either of the three ions with partially filled 3d subshell. A thorough understanding of the details of the interactions awaits high quality *ab initio* theoretical calculations.

TABLE 2. Thermodynamic data derived from state selected equilibrium constant measurements.

$$M^+ + 2He \rightleftharpoons M^+{\bullet}He + He$$

M^+		ΔH° (kcal/mol)	ΔS° (cal/mol °K)
V^+	($a^5D\ 3d^4$)	-2.6^a	–
Cr^+	($a^6S\ 3d^5$)	−1.33±0.2	−13.1±1.5
Co^+	($a^3F\ 3d^8$)	−3.68±0.2	−17.5±0.5
Ni^+	($a^2D\ 3d^9$)	−3.12±0.1	−14.2±0.5

a) Only one temperature. ΔH° was calculated, assuming ΔS° = −14.5 cal/mol°K.

One interesting point is the accuracy of the ΔH° and ΔS° numbers. We feel it will be possible to obtain ΔS° values to 0.1 or 0.2 cal/mol °K. Since we are dealing with diatomic M^+–He molecules, we should be able to test theoretical predictions of the rotational constant and vibrational frequency of these adducts.

Finally, measurements are underway both to refine these data and to measure ΔH° and ΔS° values using Ne, Ar and other simple reactants in place of He.

2.3.5. *Collisional Quenching Of Excited States.* Quenching of one electron configuration to another shows up in our experiments as a "filling in " between the two peaks in the ATD spectrum. We have observed this phenomena in only two systems, Mn^+ and Fe^+, in spite of the fact deactivation rate constants larger than 1 x 10^{-16} should be detectable in our experiment. An example is given in Figure 7 for Fe^+. It is obvious that substantial "filling in" has occurred between the two peaks. By varying the drift time and He pressure it is possible to measure deactivation rate constants. We find k = 5 ± 3 x 10^{-14} cm^3/s (Fe^+) and k = 4 ± 2 x 10^{-14} cm^3/s (Mn^+). These systems deactivate relatively inefficiently, requiring ~3 x 10^4 collisions with He for 95% deactivation. The Fe^+ result is consistent with the finding of Loh et. al.[4b] that ~10^5 collisions with He deactivate 97% of excited Fe^+.

It is curious why other systems do not deactivate. A possible answer was suggested by Loh et. al.[4b] If the excited state has a $3d^n$ configuration, then it undergoes an attractive interaction with He at long range. If the ground state is $4s3d^{n-1}$ then it undergoes a primarily repulsive interaction with He at long range. This combination of electronic configurations in M^+ allows the two potential

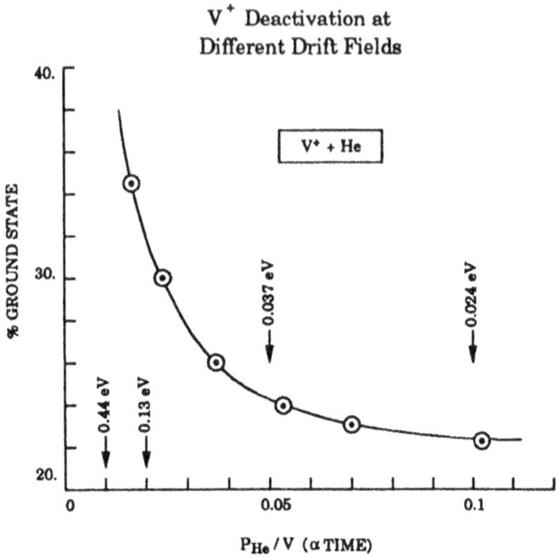

Figure 8. A plot of % V^+ ground state versus P_{He}/V in torr $volt^{-1}$. Several average translational energies of the V^+ ions are shown.

clusters. They have been the focus of intense investigation, particularly within the past 5 years, and a number of excellent compilations exist.[20] In this paper we will be concerned only with a small subset of the many interesting problems metallic clusters present; namely, the generation and reactivity of size selected transition metal ions at thermal energy. Even more specifically, we will focus exclusively on niobium cluster ions.

The only report of thermal energy reactions of size selected niobium cluster ions is the work of Smalley and coworkers using an externally injected FT-ICR machine.[21] These workers studied the adsorption of H_2 onto Nb_n^+ for $2 \leq n \leq 25$. Large variations in reactivity were observed as a function of size, and two reactive forms were found for n = 11 and n = 19. The authors concluded that H_2 dissociatively chemisorbs on the clusters.

Other work has appeared on size selected transition metal cluster ions, but it has been restricted to ions that can be formed by electron impact on volatile organometallic compounds (eg. Mn_2^+ from Mn_2CO_{10}) or by simple synthetic methods in an ICR cell. Pioneers on work of this kind have been in the groups of Ridge[22] and Freiser.[23]

Fast flow reactors coupled to time-of-flight mass spectrometers have been applied to transition metal reactivity studies. Specifically, several studies of niobium cluster ions have been carried out using this method. For example, Zakin, et. al.[24] have looked at the reaction of Nb_n^+, Nb_n and Nb_n^- with D_2 for $1 \leq n \leq 27$. The positive ion results showed qualitative agreement with the ICR work of Smalley's group. They also showed that the reactivity was only weakly dependent on the charge state of the niobium cluster. A similar conclusion had been reached earlier by Brucat, et.

al.[25] from reactions of Nb_n^+ (and Co_n^+) with N_2 and CO.

Perhaps the most interesting work has been the reaction of niobium clusters with benzene. St. Pierre and El-Sayed[26] first observed that neutral niobium clusters could completely dehydrogenate benzene. The probability for doing so was strongly size dependent, with a threshold at n = 5 or n = 6. These results were later confirmed by Zakin, et. al.[27] for C_6H_6. These authors found, however, that C_6D_6 did not lose its deuterium at low detector laser fluence. At higher fluences, all six D's could be removed from the adsorbed benzene for n = 5 and 6. Similar "photochemical" effects were simultaneously reported by El-Sayed's group.[28] These latter results point to a serious problem with laser detection of neutral species in fast flow reactors; namely, it is hard to sort out the native reactivity of the cluster and any photochemical induced reactivity. More will be said about the reactivity of Nb_n^+ with C_6H_6 and C_6D_6 in the following sections.

This brief introduction is intended to give the reader a flavor of the kind of work currently existing in the literature. It is not meant to be exhaustive, and much work that has been done has not been mentioned. Hopefully, it will help put in perspective the work going on in our laboratory, which we will briefly describe next.

3.2 THE METAL CLUSTER INSTRUMENT

The basic instrument is the same as that described in Section 2 for the atomic metal ions. The major difference is the ion source. Clusters are created in a rather standard Smalley type[29] laser desorption source that is designed to utilize either a translating/rotating metal rod or metal disk. Ions are formed directly in the laser plasma and cluster in a 2 mm diameter channel of variable length before exiting via a 30° included angle conical nozzle. A 2 mm diameter beam dynamics skimmer is located 1 - 10 cm downstream. The main chamber is pumped by a 10" diffusion pump, while 6" and 4" diffusion pumps follow the skimmer and are located before and after the mass spectrometer entrance slit, respectively. The main difference between our laser desorption source and others described in the literature is that ours floats at 5 kV, to facilitate mass selection by the mass spectrometer which operates at ground potential. The acceleration of the cluster ions occurs a few centimeters after the skimmer. Great care, and many design changes, were required to keep the system from discharging during operation, both due to the presence of the He carrier gas and due to metallic buildup on insulating surfaces. The source normally operates at 20 Hz, using the mildly focused output of an excimer laser operating at 308 nm(XeCl) with output powers between 100 and 400 mJ/pulse.

The mass selection, refocusing, deceleration, injection into the drift reaction cell, and ion detection are similar to those described in Section 2 and in Reference 14. Under optimal conditions, signal levels are between 2000 and 4000 counts/sec but typically we operate at count levels between 500 and 1500 counts/sec. This is sufficient to quantitatively study even slow reactions ($\sim 10^{-12}$ cm^3/s) and allow unit mass resolution of the products if required.

3.3 REACTION SYSTEMS

In this section we will present results on a number of reaction systems. These data are, for the most part, new and consequently are currently being both extended and repeated. The interpretation of the data is thus incomplete, and of necessity cautious, but a number of obvious and important conclusions can and will be made at this time.

3.3.1 Cluster Distribution. A typical cluster distribution is shown in Figure 9, up to and including Nb_{10}^{+}. This distribution continues beyond Nb_{15}^{+} with only modest decrease in intensity.

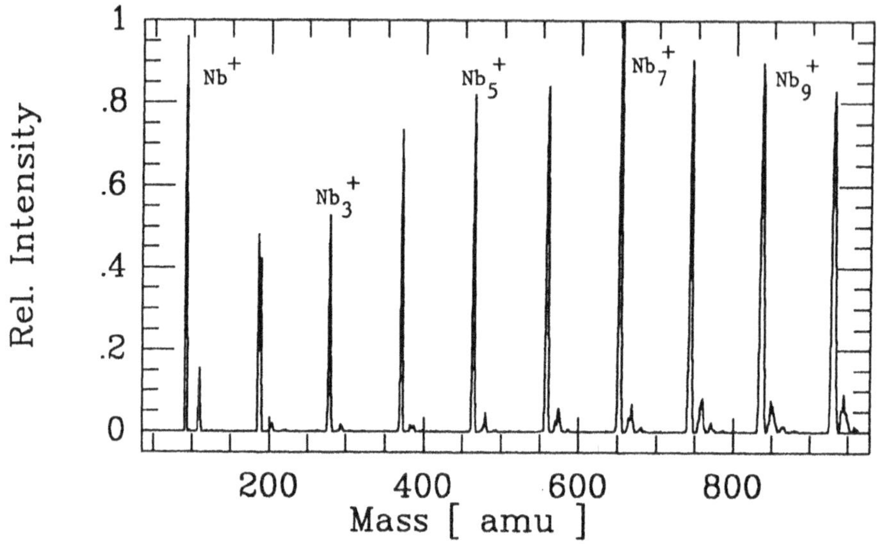

Figure 9. Mass spectrum of Nb_n^{+} cluster ions detected after the drift reaction cell. The distribution continues above n = 15 with little falloff in intensity. The small satellite peaks are oxides.

The important point is that this distribution is detected after the drift cell with the quadrupole mass filter simply turned off and the weakest Nb_n^{+} peak shown is still in excess of 1000 counts/sec. In this paper we will only be able to work with $n \leq 5$ or $n \leq 6$ since the quadrupole in place has an upper mass limit of ~650 Daltons. Funds for a system to extend the mass range to 2000 Daltons have been requested.

3.3.2. Reactions of Nb_n^{+} with O_2. Reactions with O_2 are fast for n = 1 to 6, occurring between 33 and 49% of the collision rate. The reactions are similar for all values of $3 \leq n \leq 6$, and are typified by the reactions of Nb_4^{+}:

Reaction	%
$Nb_4^{+} + O_2 \rightarrow Nb_3O_2^{+} + Nb$	9
$Nb_4^{+} + O_2 \rightarrow Nb_3O^{+} + NbO$	83
$Nb_4^{+} + O_2 \rightarrow Nb_3^{+} + NbO_2$	8

Loss of NbO from the collision complex dominates, presumably due to the strength of the Nb–O bond (~8.12 eV)[30]. While there is not sufficient thermodynamic data to determine the heats of reaction of all the channels, it does appear that thermodynamics rather than detailed kinetics plays a strong role in determining the product distributions in reactions of Nb_n^+ with O_2.

Loh, et. al.[31] have also looked at the reactions of size selected Nb_n^+ with O_2, for n = 1, 2 and 3, in a guided ion beam machine. They find that all three species react at about the Langevin limit at lowest energies, or roughly a factor of two faster than we observe. There are also some minor differences in the product distributions, probably due to the fact Loh et. al. did not correct for multiple collision effects. A more detailed discussion will be published elsewhere.[32]

3.3.3. Reactions of Nb_n^+ with C_3H_8 and C_3D_8. In the reactions with O_2, the Nb_n^+ cluster was broken up, presumably by the formation of the strongly bound NbO molecule. Similar behavior is not to be expected for much more inert alkane molecules, however, as typical M^+–H or M^+–CH_n bond energies are 2 to 3 eV[33] and Nb_n^+–Nb bond energies are 5 ± 1 eV.[34] Substantial interest is present in the scientific community on transition metal mediated C–H or C–C bond activation of alkanes, however.[35] Consequently, it is of interest to see how simple alkanes react with transition metal clusters as the size of the cluster varies.

A summary of our data for n = 1 to 5 is given in Table 3. A number of things are evident. First, the only chemical reaction is dehydrogenation. The number of H_2 molecules lost increases with the cluster size; loss of one H_2 dominates for Nb^+ while three and four losses of H_2 dominate for Nb_5^+ and Nb_6^+. In no instance are particles containing carbon lost from the cluster.

The variation of reactivity with temperature is interesting. As temperature decreases from 300K to 170K, adduct formation becomes very important for Nb_4^+ and Nb_5^+, but no adducts at all are observed for Nb^+ or Nb_3^+. (The curious Nb_2^+ ion, which primarily forms only adducts at all temperatures, will be discussed shortly). In addition, the rate constants for dehydrogenation for Nb_4^+ and Nb_5^+ show a strong negative temperature dependence. Such a dependence is strongly suggestive that the rate determining transition state has an energy just below the Nb_n^+ + C_3H_8 asymptotic energy. Modeling this temperature dependence with statistical phase space theory should give an estimate of the transition state energies, and possibly some information on their structures.[35,36] In any case, it is interesting that Nb_5^+ and Nb^+ can completely dehydrogenate propane, probably resulting in destruction of the carbon skeleton and chemisorption of individual carbon atoms on the Nb_n^+ clusters. The gradual increase in dehydrogenation with cluster size is also interesting, and the mechanism will be investigated in the future, using selectively deuterated propanes.

The reactivity of the dimer is anomalous. Figure 10 shows typical spectra for reactions of Nb_2^+ with C_3H_8 and C_3D_8 at 300K. There is a small H_2–loss peak in reaction with C_3H_8(Figure 10a) but clearly, adduct formation is the overwhelming product. With C_3D_8, however, there is a relatively large D_2–loss peak (~25%) (Figure 10b). This is a very surprising result and indicates a relatively strong inverse isotope effect is occurring. If the mechanism for H_2 (or D_2) loss is initiated by a chemisorption step that dissociates a C–H (or C–D) bond, then the implication is that the transition state for C–D bond breaking is lower in energy than for C–H bond breaking. Smaller, but still discernible, inverse isotope effects are also evident in reactions of Nb_4^+, Nb_5^+ and Nb_6^+ where more pairs of D_2 molecules are lost than H_2 molecules. At this time we have no explanation for this interesting but unexpected observation.

Table 3. Absolute Rate Constants and Branching Ratios for the Reaction

$$Nb_n^+ + C_3H_8 \longrightarrow \text{Products}$$

n	k^a_{TOT} 170K	k^a_{TOT} 300K	$k = aT^{-m}$ m	Number of H_2 Lost[b]	%
1	58	58	~0	1	89
				2	11
2	35	7	3.7	0	100[c]
3	28	28	~0	1	≤2
				2	92
				3	6
4	18	3	–	0[d]	46
			3.5[e]	2	54
5	23	5.5	–	0[d]	38
			3.0[e]	2,3,4[f]	62
6	–	1.9	–	3	60[g]
				4	40[g]

a) In units of 10^{-11} cm^3/s. The error in the absolute numbers is on the order of ± 20%. The relative values are much more accurate, since all were done under the same experimental conditions.
b) The branching ratios for loss of 1 to 4 H_2 units change slightly going from 170 to 300K.
c) There is a small loss of one H_2 molecule at 300K.
d) Clustering is not observed for n = 4 and 5 at 300K.
e) These are the values of m for loss of H_2 units only, not for adduct formation.
f) Approximately equal amounts of 2,3 and 4 H_2 molecules are lost.
g) Data pertains to 300K only.

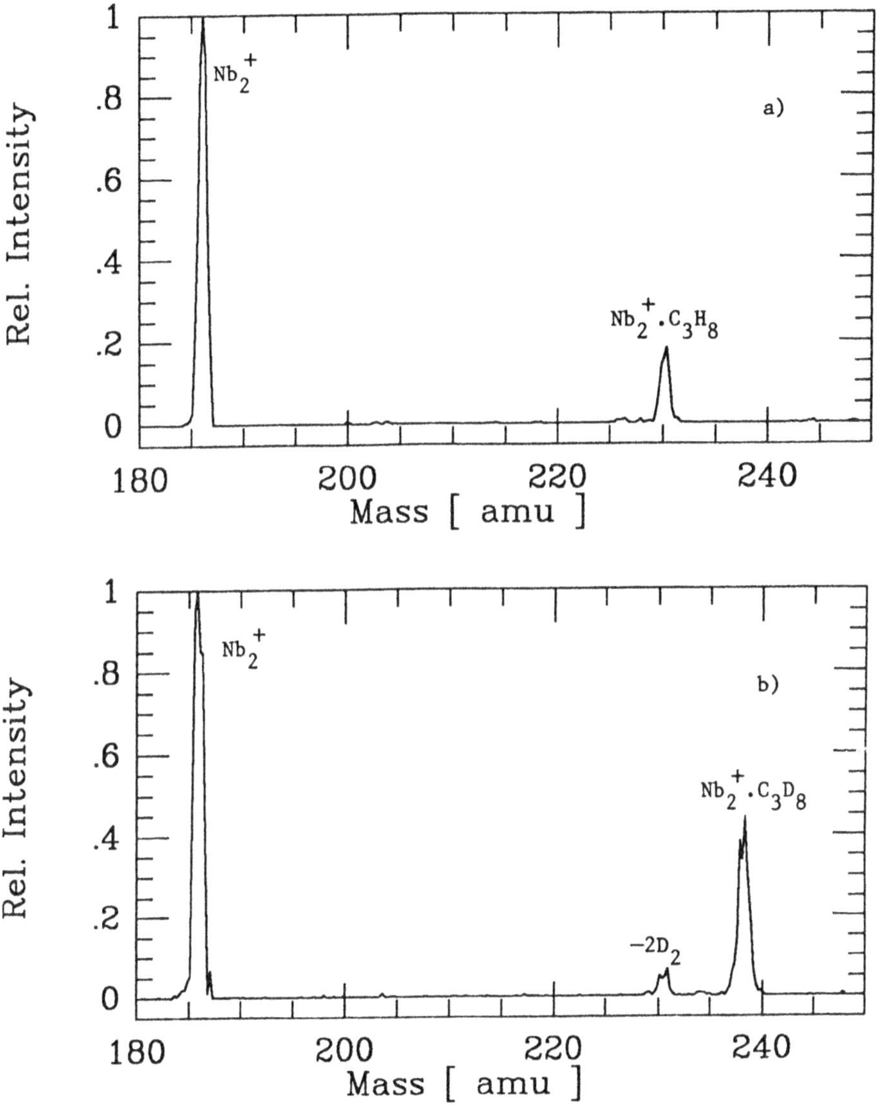

Figure 10. Reaction of Nb_2^+ with a) C_3H_8 and b) C_3D_8. Adduct formation dominates in both cases, but note the large loss of two D_2 units in b).

When the total reaction efficiencies are plotted versus cluster size, an interesting alternation is observed (see Figure 11). The odd clusters react substantially faster than their even neighbors. This alternation is also observed in the Nb_{n-1}^{+}–Nb bond strength. Simplistically, the stronger bonded clusters are less able to react with adsorbed molecules, presumably because their electrons are tied up in intra-cluster bonding.

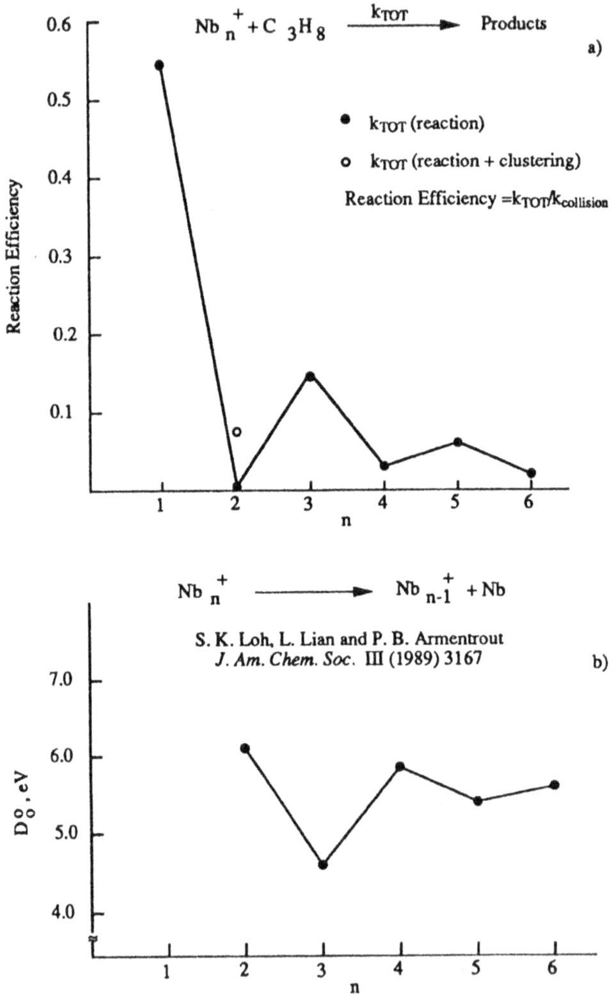

Figure 11. a) Reaction efficiency versus cluster size. b) Experimental binding energies versus cluster size.

3.3.4. Reaction of Nb_n^+ with C_2H_6 and C_2D_6. No details will be given here. However, the reactivity of ethane with Nb_n^+ is essentially identical to propane, including the reaction rates alternation shown in Figure 11. The rate constants are smaller than for propane by about a factor of 2 or 3 for $n \geq 3$, but otherwise this system shows no new chemistry.

3.3.5. Reaction of NbO^+ with C_3H_8 and C_3D_8. We have investigated the reactivity of a series of $Nb_nO_m^+$ clusters with propane, but in this paper we will only comment on NbO^+. The Nb_n^+ clusters only dehydrogenate propane (Table 3). For example, niobium atomic ions primarily singly

			%
$Nb^+ + C_3H_8$	$\longrightarrow$	$NbC_3H_6^+ + H_2$	89
	$\longrightarrow$	$NbC_3H_4^+ + 2H_2$	11

dehydrogenate propane. The low mass spectra for NbO^+ reacting with C_3H_8 and C_3D_8 are given in Figure 12. (Sequential reactions of the primary products also occur but will not be discussed here). The striking result is that demethanation effectively competes with dehydrogenation for this system:

			%
$NbO^+ + C_3H_8$	$\longrightarrow$	$NbOC_3H_6^+ + H_2$	80
	$\longrightarrow$	$NbOC_2H_4^+ + CH_4$	20

One question of interest is whether or not NbO^+ inserts directly into a C–C bond or whether both reaction channels occur via initial C–H insertion. It seems most probable that Nb^+ reacts exclusively via C–H insertion, followed by a β–hydrogen shift and elimination of H_2. Apparently, the β–methyl shift necessary for CH_4 elimination cannot effectively compete. This result is suggestive that initial C–C insertion may be occurring with NbO^+, but at this point nothing more definitive can be concluded. The important, and surprising, observation is that the presence of oxygen on the niobium ion promotes C–C bond activation in propane, where this channel is not observed for pure niobium clusters of any size.

3.3.6. Reactions of Nb_n^+ with C_6H_6 and C_6D_6. As pointed out in the introduction, neutral Nb_n clusters react with benzene and, under certain circumstances, completly dehydrogenate it. However, concerns over the influence of the "photochemical" effects of the laser used to ionize the neutral clusters before detection make any conclusions drawn from this work ambiguous..[27,28] Consequently, Zakin, et. al. investigated reactions of Nb_n^+ with C_6H_6 and C_6D_6 in their flow reactor where the ions were formed in the laser vaporization plasma directly. These authors found that for $n \geq 4$, complete dehydrogenation of benzene could occur, and that a normal isotope effect reducing reaction efficiencies by a factor of ~2 or 3 occurred when C_6D_6 was used. They also indicated that

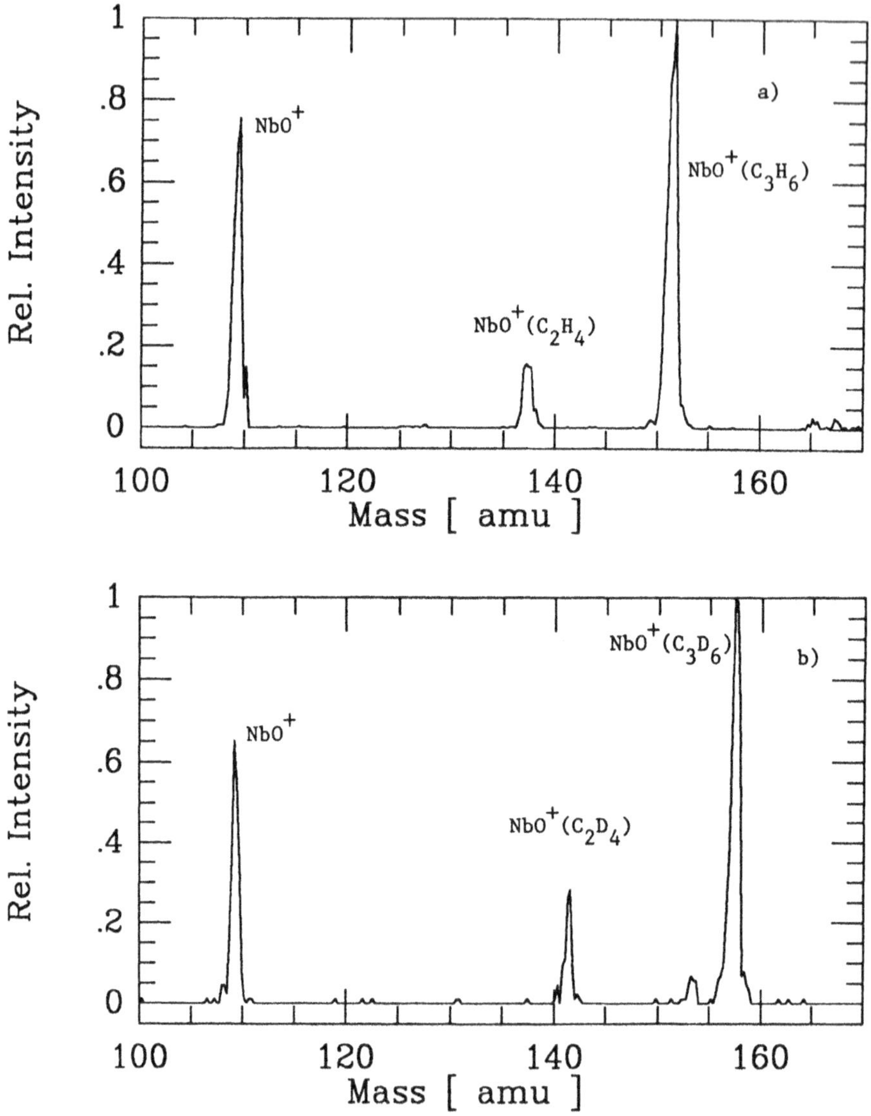

Figure 12. Reaction of NbO^+ with a) C_3H_8 and b) C_3D_8. Note the large loss of methane in both systems.

n = 1, 2 and 3 partially dehydrogenated the adsorbed benzene, producing primarily loss of a single H_2 molecule, but no product distributions nor raw data were presented.

Our results for n = 1 to 5 agree with the data of Zakin, et. al. only in the broadest qualitative sense. For Nb^+, the overwhelming product is adduct formation, with perhaps 1% loss of one H_2. For Nb_2^+, adduct formation is about 70% and loss of one H_2 30%. For Nb_3^+, there is no adduct formation, loss of one H_2 is about 95% and two H_2 molecules 5%. The branching ratio for Nb_4^+ is given below and shown in Figure 13 for C_6D_6. An essentially identical branching ratio is observed for C_6H_6.

			%
$Nb_4^+ + C_6D_6$	$\longrightarrow$	$Nb_4C_6D_4^+ + D_2$	7
	$\longrightarrow$	$Nb_4C_6D_2^+ + 2D_2$	75
	$\longrightarrow$	$Nb_4C_5D^+ + D_2 + CD_3$	18

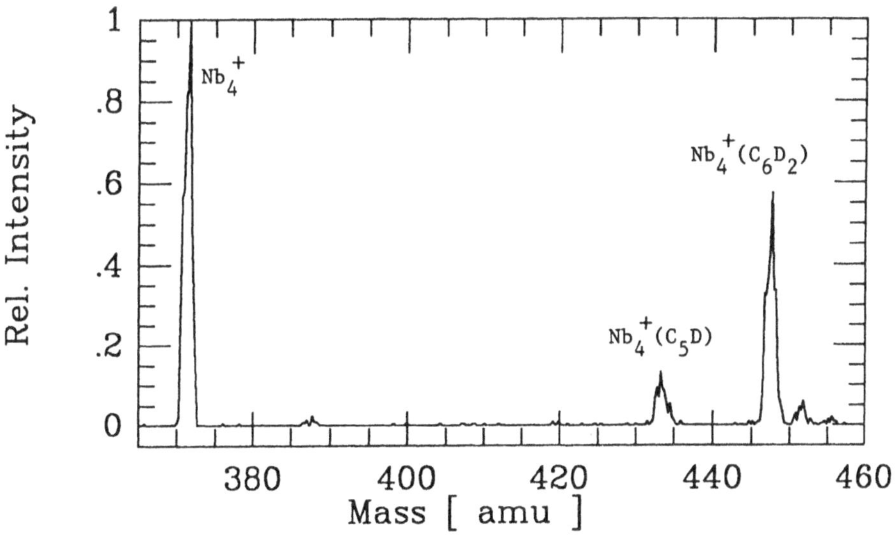

Figure 13. Reaction of Nb_4^+ with benzene-d_6. Note the large peak for loss of $D_2 + CD_3$.

Two observations are noteworthy. First, we see no formation of $Nb_4C_6^+$, and essentially no formation of the adduct $Nb_4C_6D_6^+$. Zakin, et. al. indicate these products dominate for Nb_4^+. The second observation is startling. We see an 18% contribution from $Nb_4C_5D^+$, a product that can only arise from loss of D_2 and CD_3 (or less likely D and CD_4). The chemical composition of this product is confirmed unambiguously by using C_6H_6 and noting the product ion mass is shifted to lower mass by one mass unit. Clearly, C_6H_6 is strongly chemisorbed onto Nb_4^+ with complete structural rearrangement of the hydrogens <u>and</u> the carbon skeleton.

Figure 14 shows the spectra for Nb_5^+ reacting with C_6H_6 and C_6D_6. The only two products observed are loss of two H_2 (D_2) molecules or three H_2(D_2) molecules. There is no sign of disruption of the carbon skeleton. There clearly is a greater loss of three H_2 than there is three D_2, leading to a normal isotope effect of about 1.7.

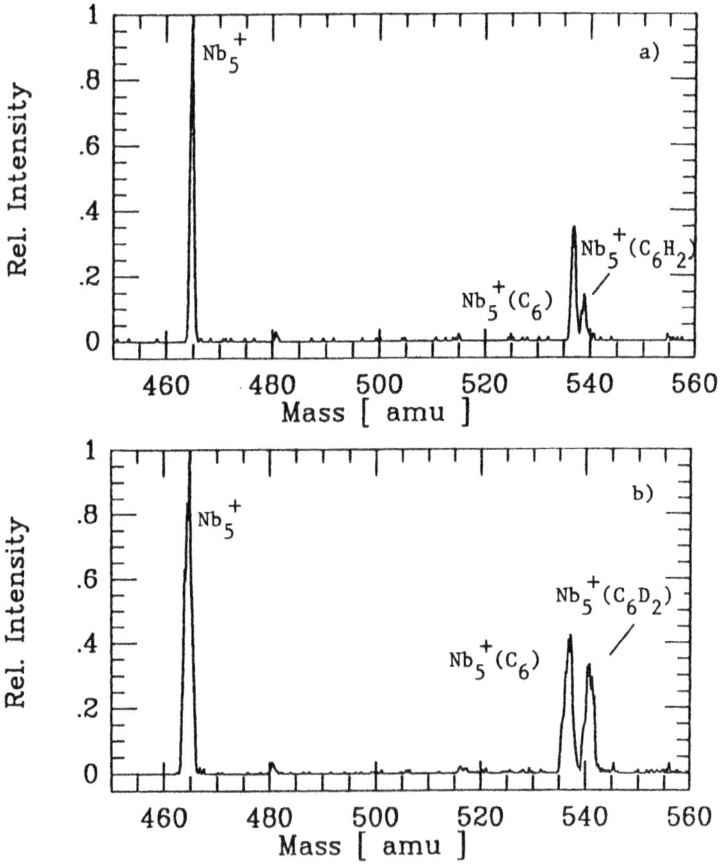

Figure 14. Reaction of Nb_5^+ with a) C_6H_6 and b) C_6D_6.

There is no adduct formation, nor is any loss of a single H_2 molecule evident. Zakin, et. al.[37] report adduct formation about the same intensity as loss of three D_2 molecules using C_6D_6 and about 60% as large as loss of three H_2 molecules using C_6H_6. They do not mention a sizeable loss of two D_2 molecules or two H_2 molecules, and show no raw data.

3.3.7. NbO^+ Reacting with Benzene. The NbO^+ ion reacts only by clustering with benzene. A fairly high conversion spectrum is shown in Figure 15. the interesting point in this spectrum is that two, and only two, C_6H_6 molecules add to NbO^+. A similar observation is made for Nb^+ reacting with benzene. It appears reasonable to deduce that a sandwich type structure exists for these two species, where vacant d-type orbitals on Nb^+ and NbO^+ interact with the π-cloud of the benzene ring.

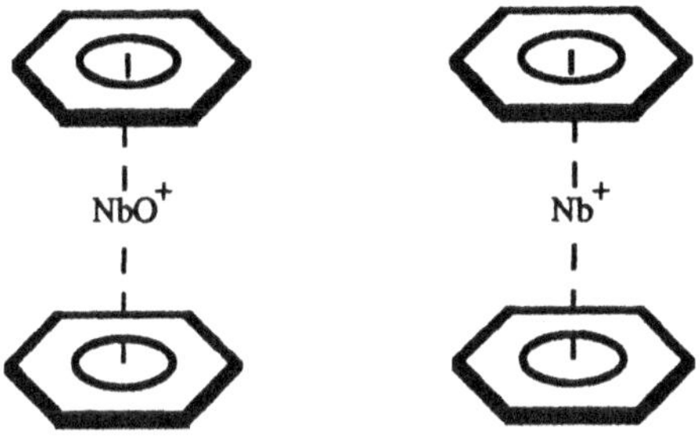

A third benzene ring would be very much more weakly bound than the first two, and consequently is not observed under our experimental conditions.

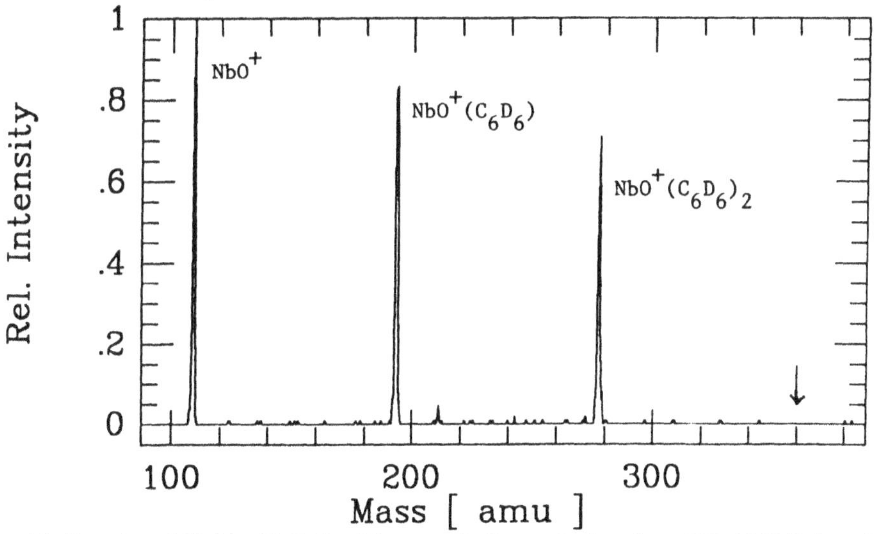

Figure 15. Reaction of NbO^+ with C_6D_6. The arrow indicates the location of $NbO^+(C_6D_6)_3$, which is totally absent.

3.3.8. Reactions of Nb_n^+ with NO. Reactions of O_2 with Nb_n^+ disintegrated the clusters through fast sequential reactions, primarily by loss of NbO neutral molecules. The principal driving force for this process appears to be the strong NbO and NbO^+ bond strengths (8.1 and 7.1 eV) relative to the O_2 bond strength (5.2 eV). The NO molecule is relatively more strongly bound than O_2 (D^o (N–O) = 6.57 eV) and little is known about the strength of NbN or NbN^+ bonds. Consequently, it is of interest to see if NO efficiently dismantles Nb_n^+ clusters similar to O_2 and to see if some thermochemical information can be obtained regarding niobium–nitrogen bonds.

Atomic Nb^+ ions do react with NO

		%
$Nb^+ + NO \longrightarrow$	$NbO^+ + N$	100
$\longrightarrow$	$NbN^+ + O$	0

and form exclusively the NbO^+ product. The formation of NbO^+ is exothermic by 0.53 eV. Since the NbN^+ product is not observed, this reaction channel is probably endoergic. This implies that

$$\Delta H_f(NbN^+) \geq \Delta H_f(Nb^+) + \Delta H_f(NO) - \Delta H_f(O)$$

or

$$\Delta H_f(NbN^+) \geq 12.7 \text{ eV}.$$

From this value an upper limit on the bond strength of NbN^+ can be obtained of $D^o(Nb^+–N) \leq 6.56$ eV. This compares with $D^o(Nb^+–O) = 7.09$ eV.

The NbO^+ ion does not react at all with NO. From known thermochemical data (see Table 4), reaction to form NbO_2^+ + N products is endothermic by 0.92 eV. since $NbNO^+$ + O products are also not observed, a lower limit on the heat of formation of $NbNO^+$ can be obtained, $\Delta H_f(NbNO^+) \geq 8.20$ eV. Interestingly, no adduct formation between NbO^+ and NO is observed. This is surprising since there are no exoergic product channels and indicates either a surprisingly weakly bound NbO^+–NO electrostatic complex (of a few kcal/mol or less) or there is a barrier in the entrance channel due to symmetry or other restrictions.

The branching ratio of Nb_2^+ reacting with NO is given below and is shown in Figure 16.

		%
$Nb_2^+ + NO \longrightarrow$	$NbO^+ + NbN$	50
$\longrightarrow$	$NbN^+ + NbO$	50

Here both products are formed in equal abundance. Since the only difference in the products is the location of the charge, this result strongly implies that IP(NbN) ≅ IP(NbO) = 7.91 eV, where the ionization potential of NbO has been reported by Dyke, et. al.[38] It follows directly that $\Delta H_f(NbN) = \Delta H_f(NbN^+) + 1.03$ eV, using the estimates presented earlier, $\Delta H_f(NbN) \geq 13.73$ eV and $D^o(Nb–N) \leq 5.53$ eV.

Table 4. Thermochemical Data

Species	D^o_{298} (eV)	IP (eV)
Nb	7.8[a]	6.88[b]
Nb^+	14.35[c]	—
Nb–O	8.12[b]	7.91[d]
Nb^+–O	7.09[e]	—
Nb–N	≤5.53[f]	7.91[g]
Nb^+–N	≤6.56[h]	—
O_2	5.17[b]	12.07[b]
NO	6.57[b]	9.25[b]

a) This is the ΔH_f, 298 of atomic niobium: Weast, R.C. (Ed.) (1979), "Handbook of Chemical Physics", CRC Press, Boca Raton, FL.
b) Reference 30.
c) This is the ΔH_f, 298 of atomic niobium ion calculated from ΔH_f, 298 (Nb) and IP(Nb).
d) Reference 38.
e) Calculated from D^o_{298} (Nb^+–O) = D^o_{298} (Nb–O) + IP(Nb) – IP(NbO).
f) See text.
g) Assumed to be the same as IP(NbO) as deduced from the data in Figure 16 (see text).
h) See text.

For Nb_3^+, Nb_4^+ and Nb_5^+ the primary reactions are losses of NbO and NbN from each system. Rapid sequential reactions with NO occur in all cases. The absolute rates have not been accurately determined, but the data in hand indicate that k_{TOT} increases with n. This result is in contrast to reactions with O_2 where no correlation of k_{TOT} with n was apparent.

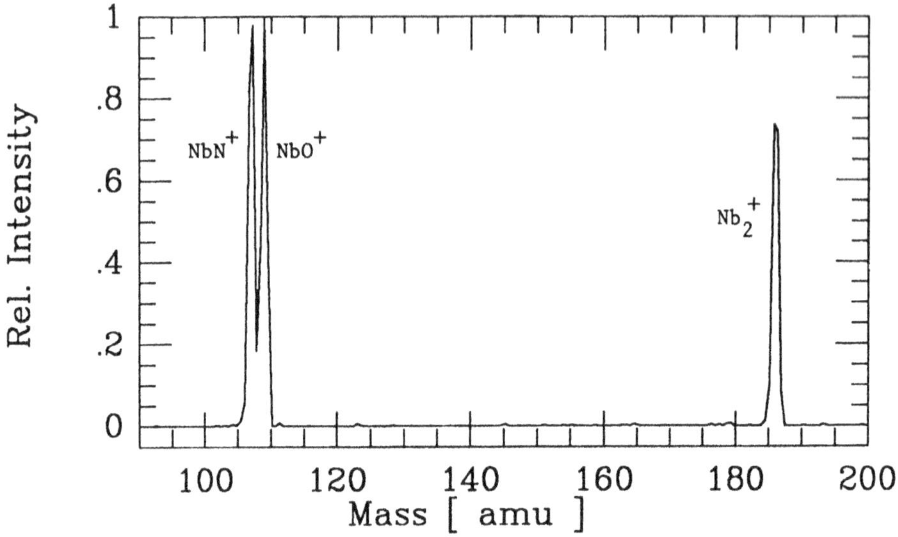

Figure 16. Reaction of Nb_2^+ with NO.

4. Summary and Conclusions

Two rather different new research areas are described here. The intent of the chapter was to introduce the techniques, give some preliminary results, and hopefully, show the promise these methods have for investigating new aspects of the chemistry of transition metal ions and their clusters. The approach of our group is always quantitative, focusing on measurements that yield absolute numbers that can be used to develop and improve theoretical models. For the atomic ions our emphasis to date has been to thoroughly understand the origins of the multiple peaks in the arrival time distributions, and once that was done, to begin to use this information to investigate a variety of topics. In the future we will be measuring binding energies of transition metals to simple neutral particles as well as deactivation rates of electronic excited states by collision. We also intend to investigate state selected chemistry and to inquire into the mechanism of formation of M^+ from various precursor materials.

The cluster chemistry has been limited to niobium at this early stage, but clearly the method is general and can be applied to any metal that clusters relatively easily and whose isotope pattern is relatively simple. Reactions of Nb_n^+ with a number of neutral substrates were discussed with the emphasis on a description of what was observed rather than detailed mechanistic studies. We hope the point was made that useful, interesting, and often surprising results are obtained and that both size selection and truly thermal conditions are crucial in extracting unambiguous information. These studies will be expanded to other metals, and neutrals, in the future.

5. Acknowledgements

The support of the National Science Foundation under Grant CHE88-17201 and the Air Force Office of Scientific Research under Grant AFOSR89-0102 are gratefully acknowledged.

6. References

1. See, for example, a) Russell, D.H. (Ed) (1989), "Gas Phase Inorganic Chemistry", Plenum, New York; b) Armentrout, P.B. (1990), "Ann. Rev. Phys. Chem", Strauss, H. (Ed.); c) Armentrout, P.B. and Beauchamp, J.L. (1989), *Acct. Chem. Res.* 22, 315; d) Marks, T.J. (Ed.) (1990), "Bonding Energetics in Organometallic Compounds", *ACS Symposium Series 428*, Washington D.C.

2. Radiative transition between $3d^n$ and $4s3d^{n-1}$ electronic configurations are parity forbidden, leading to very long lifetimes. See Garstang, R.H. (1962) *Mont. Not. R. Astron. Soc.* 124, 321.

3. a) Sunderlin, L.S. and Armentrout, P.B. (1988), *J. Phys. Chem.* 92, 1209; b) Halle, L.F., Armentrout, P.B. and Beauchamp, J.L. (1981), *J. Am. Chem. Soc.* 103, 962.

4. a) Cody, R.B., Burnier, R.C., Reents, W.D., Carlin, T.J., McCrery, D.A., Lergel, R.K. and Freiser, B.S. (1980), *Int. J. Mass. Spectrom. Ion Phy.* 33, 37; b) Loh, S.K., Fisher, E.R., Lian, L., Shultz, R.H. and Armentrout, P.B. (1989), *J. Phys. Chem.* 92, 1209.

5. Sanders, L., Sappy, A.D. and Weisshaar, J.C. (1986), *J. Phys. Chem.* 85, 6952; ibid (1987), 91, 5145; Hanton, S., Sanders, L. and Weisshaar, J.C. (1989) *J. Phys. Chem.* 93, 1963.

6. See, for example, a) Elkind, J.L. and Armentrout, P.B. (1985), *J. Phys. Chem.* 89, 5626; b) ibid., (1986) *J. Chem. Phys.* 84, 4862; c) ibid. (1986) *J. Phys. Chem.* 90, 5736; 6526; d) ibid. (1987), *J. Chem. Phys.* 86, 1868; e) ibid. (1988), *Int. J. Mass. Spectrom. Ion Proc.* 83, 259.

7. Illies, A.J. and Bowers, M.T. (1982) *Chem. Phys.* 65, 28.

8. Hanratty, M.A., Carter, E.A., Beauchamp, J.C., Goddard III, W.A., Illies, A.J. and Bowers, M.T. (1986) *Chem. Phys. Lett.* 123, 239; Hanratty, M.A., Beauchamp, J.L., Illies, A.J., van Koppen, P.A.M. and Bowers, M.T. (1988), *J. Am. Chem. Soc.* 110, 1.

9. Rincon, M., Kirchner, N. and Bowers, M.T. (1988), *Int. J. Mass. Spectrom. Ion Proc.* 86, 369.

10. Twiddy, N.D., Mohebati, A. and Tichy, M. (1986), *Int. J. Mass Spectrom. Ion Proc.* 74, 251; Grice, S.T., Harland, P.W., Maclagan, R. and Simpson, R.W. (1989), *Int. J. Mass Spectrom. Ion Proc.* 87, 181.

11. Rowe, B.R., Fahey, D.W., Fehsenfeld, F.C. and Albritton, D.L. (1980), *J. Chem. Phys.* 73, 194.

12. Kemper, P.R. and Bowers, M.T. (1990), *J. Am. Chem. Soc.* 112, 3231.

13. Kemper, P.R. and Bowers, M.T. (1991), *J. Am. Chem. Soc.* 113, xxxx.

14. Kemper, P.R. and Bowers, M.T. (1990), *J. Am. Soc. Mass Spectrom.* 1, 197.

15. McDaniel, E.W. and Mason, E.A. (1973), "The Mobility and Diffusion of Ions in Gases," John Wiley and Sons, New York.

16. Bauschlicher, C.W. and Langhoff, S.R. (1988), *Chem. Phys. Lett.* 149, 10.

17. Su, T. and Bowers, M.T. (1979), in "Gas Phase Ion Chemistry," Bowers, M.T. (Ed.), Academic Press, N.Y. pp. 84-119.

18. See, for example, Reents Jr., W.D., Stobel, R., Freas III, R.B., Wronka, J. and Ridge, D.P. (1985), *J. Phys. Chem.* 89, 5666; Strobel, F. and Ridge, D.P. (1989), *J. Phys. Chem.* 93, 3635.

19. Sunderlin, L.S., and Armentrout, P.B. (1988), *J. Phys. Chem.* 92, 1209.

20. See, for example, Gates, B.C., Gucz, L. and Krozinger, H. (Eds.) (1986), "Metal Clusters in Catalysis," Elsevier, Amsterdam; Trager, F. and zu Putlitz, G. (Eds.) (1986), "Metal Clusters," Springer, Berlin; Jena, P., Kao, B.K. and Khanna, S.N. (Eds.) (1987), "Physics and Chemistry of Small Clusters," Plenum, New York.

21. a) Geusic, M.E., Morse, M.D. and Smalley, R.E. (1984), *J. Chem. Phys.* 82, 590; b) Elkind, J.L., Weiss, F.D., Alford, J.M. Laaksonen, R.T. and Smalley, R.E. (1988), *J. Chem. Phys.* 88, 5215.

22) See, for example, Freas, R.B. and Ridge, D.P. (1984), *J. Am. Chem. Soc.* 106, 825.

23. See, for example, Jacobson, D.B. and Freiser, B.S. (1984), *J. Am. Chem. Soc.* 106, 4623; 106, 5351; Buckner, S.W. and Freiser, B.S. (1988) in "Gas Phase Inorganic Chemistry," D.H. Russel (Ed.) Plenum, New York.

24. Zakin, M.R., Brickman, R.O., Cox, D.M. and Kaldor, A. (1988), *J. Chem. Phys.* 88, 3555.

25. Brucat, P.J., Pelleitte, C.L., Yang, S., Zheng, L.-S., Craycraft, M.J. and Smalley, R.E. (1986), *J. Chem. Phys.* 85, 4747.

26. St. Pierre, R.J. and El-Sayed, M.A. (1987), *J. Phys. Chem.* 91, 763.

27. Zakin, M.R., Cox, D.M. and Kaldor, A. (1987), *J. Phys. Chem.* 91, 5224.

28. St. Pierre, R.J., Chronister, E.L. Song, L. and El-Sayed, M.A. (1987), *J. Phys. Chem.* 91, 4648.

29. Powers, D.E., Hansen, S.G., Geusie, M.E., Puiu, A.C., Hopkins, J.B., Dietz, T.G., Duncan, M.E., Langridge-Smith, P.R.R. and Smalley, R.E. (1982), *J. Phys. Chem.* 86, 2556.

30. Chase, M.W., Davies, C.A., Downey, J.R., Frunip, D.J., McDonald, R.A. and Syverced, A.N. (1985) *J. Phys. Chem. Ref. Data* 14, Supple. 1.

31. Loh, S.K., Lian, L. and Armentrout, P.B. (1989), *J. Chem. Phys.* 91, 6148.

32. Radi, P.P., von Helden, G., Hsu, M.-T., Kemper, P.R. and Bowers, M.T. (to be published).

33. See, for example, Armentrout, P.B. and Georgeadis, R. (1988), *Polyhedron* 7, 1573.

34. Loh. S.K., Lian, L. and Armentrout, P.B. (1989), *J. Am. Chem. Soc.* 111, 3167.

35. See, for example, van Koppen, P.A.M., Brodbelt-Lustig, J., Bowers, M.T., Dearden, D., Beauchamp, J.L., Fisher, E.R. and Armentrout, P.B. (1991), *J. Am. Chem. Soc.* 113, xxxx, and references therein.

36. Bass, L.M., Cates, R.D., Jarrold, M.F., Kirchner, N. and Bowers, M.T. (1983) *J. Am. Chem. Soc.* 105, 7124.

37. Zakin, M.R., Brickman, R.O., Cox, D.M. and Kaldor, A. (1988), *J. Chem. Phys.* 88, 5943.

38. Dyke, J.M., Ellis, A.M., Feher, M., Morris, A., Paul, A.J. and Stevens, J.C.H. (1987), *J. Chem. Soc. Faraday Trans. 2* 83, 1555.

TECHNIQUES FOR PREPARING AND STUDYING THE PROPERTIES OF LARGE CLUSTER IONS BY HIGH RESOLUTION MASS SPECTROMETRY.

A.J. STACE
School of Molecular Sciences,
University of Sussex,
Falmer, Brighton BN1 9QJ,
U.K.

ABSTRACT. A brief survey is given of techniques that have been developed both for generating clusters and for studying the properties of their ions by mass spectrometry. Particular consideration is given to the study of metastable fragmentation and the implementation of cluster ion experiments on a tandam mass spectrometer.

1. INTRODUCTION

During the past ten years there has been considerable research activity directed towards qualitative experiments on neutral and ionic clusters [1,2]. Typical of this type of experiment has been the search for "magic" numbers in cluster ions composed of the inert gases [3], and the technique most frequently used to detect and identify these clusters is mass spectrometry. Even clusters as fragile as those formed from the inert gases, e.g. Ar_n, will produce stable ions as a result of ionization by electrons with energies in excess of 70 eV. Such behaviour does not necessarily imply that electron impact is a "soft" ionization technique, but is more a reflection of those processes which lead to the formation and stabilization of cluster ions (see below).

For the experimental measurement of cluster ion intensities, i.e. identification of "magic" numbers etc. resolution requirements are normally not paramount. Therefore, such experiments can often be performed on time-of-flight instruments [3]. In contrast, the study of cluster ion reaction mechanisms frequently requires access to an instrument which is capable of resolving to within ± 1 amu over a wide mass range. Detailed quantitative studies of fragmentation routes via, for example, the observation of unimolecular, CID, or laser-induced metastable decay, can only be done at high masses with a tandem instrument. Only in the past 3 -4 years has serious consideration been given to high resolution mass spectrometric measurements of cluster ions and their fragmentation products [1,2]. The purpose of this article is review the advantages and limitations of using a high resolution, double focusing mass spectrometer to study cluster ions. At the same time, attention will be drawn to the techniques used to generate and ionize clusters.

K. R. Jennings (ed.), Fundamentals of Gas Phase Ion Chemistry, 87–103.

2. EXPERIMENTAL CONSIDERATIONS

The techinques available for studying the properties of cluster ions have developed in parallel with other areas of mass spectrometry. For example, the improvements in magnet technology, designed primarily to facilitate the study of large molecules of biological interests, are clearly of considerable benefit to those interested in high-mass cluster ions. Similarly, improvements in time-of-flight techniquies and, most notably the development of the reflectron [4], have all been to the benefit of cluster studies.

For the most part, the methods used by the various research groups to generate, ionize and detect ion clusters are very similar. In detail, differences in technique lie in the use of conical rather than pin-hole nozzles, or in the use of laser vaporization [5] rather than ovens [6] to generate clusters from refractory materials. Similarly, research groups may choose to use photo-ionization [7] rather than electron impact [8], or alternatively, to generate cluster ions via nucleation about an atomic or molecular ion core [9].

2.1. Cluster formation by adiabatic expansion

Neutral clusters can be generated by the adiabatic expansion of a gas through a small orifice (nozzle) into a vacuum. In terms of the conditions which may prevail within the gas reservior, it is possible to identify two limiting situations, each of which can be classified in terms of the "Knudsen" number [10], K_n, where

$$K_n = \frac{\text{Mean free path in the reservior}}{\text{Diameter of the nozzle}} \qquad (1)$$

i) $K_n > 1$; the atoms or molecules pass through the nozzle without undergoing collisions, and there is little or no redistibution of energy (see below) between the translational and any internal degrees of freedom. Such behaviour is frequently referred to as "effusive" flow.

ii) $K_n < 1$; the atoms or molecules experience a very large number of collisions as they pass through the nozzle. There is an extensive redistribution of energy, with the forward translational energy of the beam increasing at the expense of any local or internal molecular energy. Such behaviour is characteristic of "hydrodynamic" flow [10]. Channelling the relative kinetic energy between atoms or molecules in the beam into the flow velocity has two effects: (1) the most probable forward velocity increases beyond that appropriate for a Maxwellian velocity distribution; (2) the velocity spread in the beam is considerably reduced. Because we identify temperature with the width of a velocity distribution, it is possible to equate the reduction in velocity spread during expansion, with a drop in the local temperature. Obviously, if the passage through the collision region is accompanied by an extensive reduction in the relative kinetic energy (temperature) between atoms or molecules, then dimer and cluster formation can occur. The initial step in any condensation process will involve, at the least, a three-body collision and, therefore, must proceed at an early stage in the expansion. Once a cluster starts to develop, i.e. contains four or more atoms or molecules, further growth can take place via sequences of two-body collisions. Some collisions adding atoms, others removing the excess energy.

For the isentropic ($\Delta S = 0$) expansion of an ideal gas, the temperature reduction from T_i to T_f associated with an expansion is given by [10]

$$T_f = T_i[1+(\gamma-1)/2\ M^2]^{-1} \tag{2}$$

where γ is the heat capacity ratio, C_p/C_v and M is the Mach number, which is the ratio of the final flow velocity to the local velocity of sound. Because the velocity of sound varies as $T^{½}$, Mach numbers > 1 are typical; hence, the technique is often referred to as "supersonic" expansion.

A useful approximation to equation (2) is the expression [11]

$$T_f = T_i[1 + 5896(pD)^{0.8}]^{-1} \tag{3}$$

where p is the reservior pressure in atmospheres and D is the nozzle diameter in cm; this equation has been derived for the expansion of pure argon. Typical expansion conditions (see below) might be p = 5 atm, D = 200 μm, and T_i = 300 K, which gives $T_f \approx$ 2 K. In the case of clusters, two factors could serve to increase the temperature: (1) cluster formation is an exothermic process, and as such the excess energy has to be removed; (2) the extent to which the various degrees of freedom within a molecule experience a temperature reduction, depends upon their relaxation efficiencies. For an isolated species, such as a benzene molecule, in an expansion of argon or helium, it might be expected that $T_{vib} > T_{rot} > T_{trans}$. Once a molecule enters the collision-free region of an expansion (a distance of approximately 50D downstream of the nozzle), any inequalities in the temperatures of the various degrees of freedom will be "frozen in". For a cluster, the situation is slightly different. The translational temperature (velocity spread) between clusters will remain low. However, if a molecule in a cluster retains all or part of the vibrational energy it had in the reservior prior to expansion, that energy will dissipate into the cluster. Further cooling will then proceed via the evaporation of molecules. Similarly, any rotational energy the molecules possess will also equilibriate; the main restriction being conservation of angular momentum. However, Coriolis interactions will allow for a mixing of energy between rotational and molecular bending degrees of freedom. Thus, in a cluster we might expect $T_{vib} \approx T_{rot} > T_{trans}$. In molecular systems where relaxation of the vibrational degrees of freedom is particularly inefficient, cluster could appear liquid-like [12], rather than the solid-like phase expected given the low temperatures normally associated with supersonic nozzle expansion.

2.2. Cluster ionization

The technique outlined above to generate clusters will always produce a distribution, over which the experimentalist has only nominal control. Equation (3) summerises the important variables in a typical experiment.

and variation of, for example, the temperature of the gas reservior will affect the observed cluster distribution, shifting the mean to either smaller or larger clusters, depending on whether the temperature is increased or reduced, respectively.

Many early experiments on clusters assumed that a mass spectrum reflected the distribution of neutral species in a beam, and little or no attention was paid to the possibility of fragmentation [3]. However, it is now appreciated that, in terms of a fragile cluster, electron or photon impact ionization is quite a violent process and that the energy it imparts, either directly or indirectly (see below), can induce quite extensive fragmentation in the resultant cluster ion [1,2]. Consider the case of an argon cluster: Harberland [13] has identified a sequence of events which could follow from the initial ionization step through to the observation of a (stable?) inert gas cluster ion approximately 10^{-5} s later. Some 10^{-12} s after ionization it is thought that the positive charge becomes self-trapped. The driving force being the stability of the $^2\Sigma^+_u$ state of the argon dimer ion, Ar_2^+. Because of a very poor Franck-Condon overlap between the ground vibrational levels of a neutral cluster and the corresponding ion, it has been proposed [14] that ionization at low energies (< 15.75 eV) proceeds through the initial formation of a Rydberg state . Autoionization and self trapping would then be synchronous event. However, autoionization structure which is present in the photoelectron spectra of Ar_2 is not observed in similar experiments on Ar_3 and higher clusters [15]. The exact ionization mechanism for a large cluster at low electron or photon impact energies is still therefore the subject of some uncertainty. At higher energies (> 15.75 eV) ionization can proceed through the initial formation of Ar^+, probably on the surface of the cluster.

Once self-trapping has taken place, energy comparable in magnitude to the dissociation energy of Ar_2^+ ($D_e \approx 1.3$ eV) will be released into the clusters. Clearly, their internal temperatures will increase and, because atoms close to the surface have binding energies of ≈ 0.05 eV, extensive fragmentation will occur. The observation time window on a mass spectrometer is typically in the range 10^{-6} - 10^{-5} s, and several experimental studies [8,16] have shown that on this time-scale the dominant fragmentation route is the loss of one or more monomer units, i.e. argon atoms in the case of Ar_n^+ clusters or CO_2 molecules in the case of $(CO_2)_n^+$ clusters. Recent studies [14,17] have extended the time-scale to $\approx 10^{-4}$ s and shown that over an observation period of $\approx 5\text{x}10^{-5}$ s, a large ion cluster, e.g. Ar_{250}^+, might lose five or six atoms during its passage through a field-free-region. In some respects, the loss of atoms from these large clusters could be viewed as evaporative cooling. Given that the cluster ions are comparatively "cold" by the time they leave the ion source ($\approx 10^{-6}$ s after their formation) it is clear that extensive fragmentation must take place before they have a chance to reach one of the observation time windows. Therefore, the possibility that the cluster ions will retain any memory of their size distribution prior to ionization is very remote.

3. PRACTICAL CONSIDERATIONS

Figure 1 shows a schematic diagram of a typical cluster beam apparatus. As might be anticipated from equation (2), the higher the reservior pressure and/or the larger the nozzle diameter, the greater the degree of cooling. Although the exact dependences on p and D may vary slightly, similar considerations are directly applicable to cluster formation. From a practical viewpoint, the high flow rate associated with a large diameter nozzle or high reservoir pressure can present problems. The fragile nature of weakly-bound inert gas clusters means that the background pressures in both the expansion and collimation chambers must be kept as low as possible; therefore, high capacity pumping is required. Two factors in the design of an experiment can reduce the vacuum requirements. The first is differential pumping [10]; the coldest component in an expansion is the central core, and by placing a skimmer downstream from the nozzle this core can be extracted and passed into the collimation chamber. Thus, the pumping requirements in the collimation chamber need only be sufficient to avoid scattering the beam during its passage from the skimmer to the mass spectrometer. The position of the skimmer with respect to the nozzle is important; the gas expanding from the nozzle creates a shock wave when it interacts with the ambiant background; to be effective, the skimmer must be situated inside the shockwave boundary. The second technical development that has had a considerable impact on supersonic jet experiments, is the pulsed nozzle.

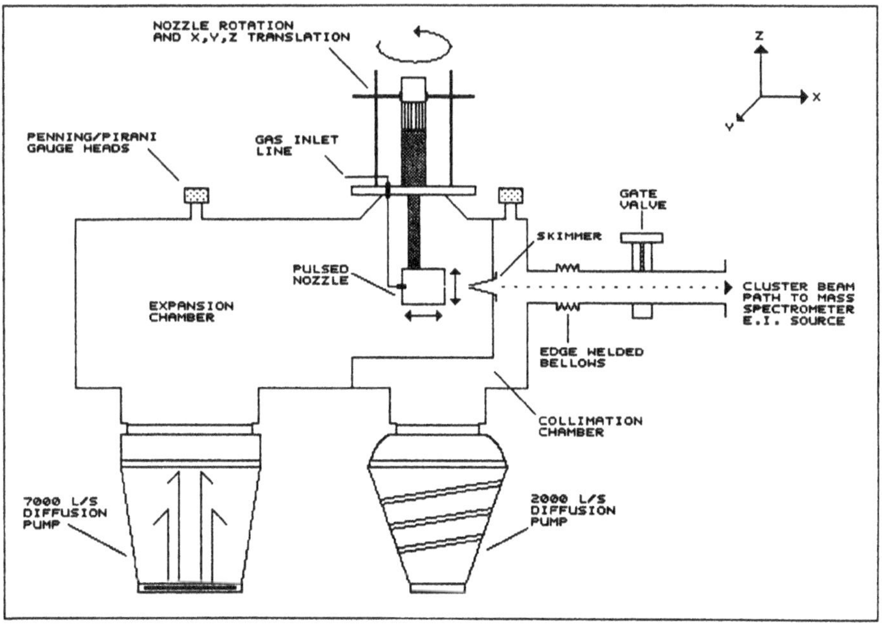

Figure 1. Schematic diagram of supersonic nozzle with differential pumping.

By modulating the gas flow it is possible to conduct experiments at very high reservior pressures, whilst still maintaining a modest pumping requirement. A further advantage is that phase-sensitive detection techniques can be used in conjunction with the pulsed valves. Devices used to modulate the gas range from fast piezoelectric valves giving pulse widths of the order of 100 μs [18], through to modified car fuel injection valves with pulse widths of ≈ 10 ms [19]. The latter provide a cheap and reliable means of constructing a pulsed supersonic nozzle. With a pulsed nozzle system, care has to taken to avoid scattering through the reflection of gas pulses from the walls of the vacuum chamber.

Having identified the essential features of the cluster formation process, there now follows a description of the particular apparatus which we have constructed to study the properties and reactions of cluster ions. Most of our earlier work on ion cluster reactivity was undertaken on a single focusing magnetic sector mass spectrometer which had unit mass resolution out to approximately 1000 amu. Over the past three years however, we have been developing a new experiment which utilizes a high resolution, double focusing instrument. The new experiment consists of a supersonic nozzle system coupled to a VG ZAB-E mass spectrometer [14,17], and a schematic diagram of the complete apparatus is shown in figure 2.

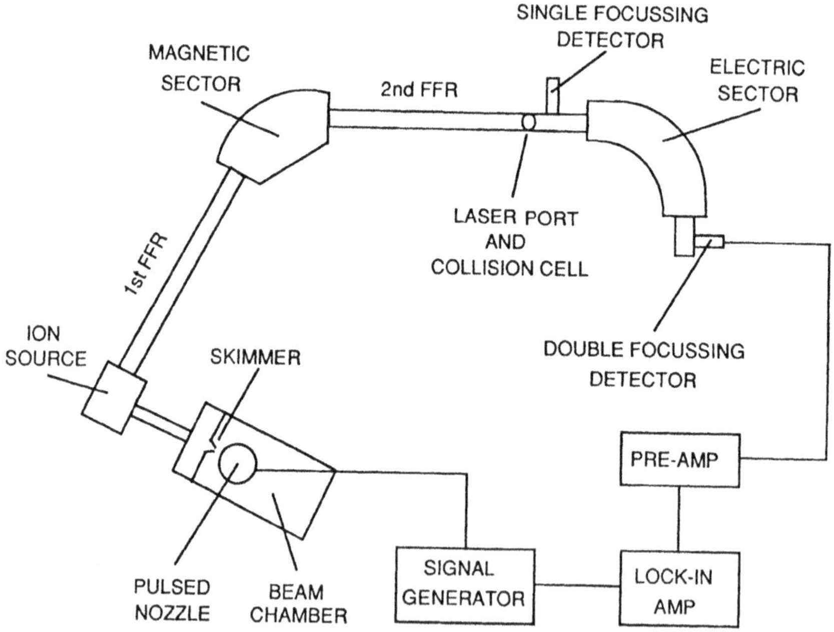

Figure 2. Schematic diagram of molecular beam - mass spectrometer system.

Clusters are generated by the adiabatic expansion of a gas through a pulsed nozzle. Modulation of the gas flow is achieved through the use of a Bosch fuel injection valve (Serial No. 028 150 045), with a 3 mm diameter stainless steel cap forced on to the injector pin and covered with a disc of KALREZ. The valve can operate at frequencies up to 150 Hz; it is our experience however, that the best pulse shapes are obtained in the frequency range 1 - 40 Hz. To produce large clusters, use has been made of a conical (Laval) nozzle, with an initial opening of 200 μm, and a cone angle of approximately 30^0 extending over a distance of 5 mm. In this environment cluster growth is promoted by the very large number of collisions the gas experiences within the cone. The gas expands into a chamber which is evacuated by a 8000 l s^{-1} diffusion pump, backed by a mechanical booster and a two-stage rotary pump. The expansion chamber is separated from a collimation chamber by a 1 mm diameter skimmer positioned approximately 3 cm from the nozzle. The latter chamber is pumped by a 2000 l s^{-1} Edwards Diffstak diffusion pump backed by a two-stage rotary pump. The cluster beam system is connected to the mass spectrometer via a length of edge-welded flexible bellows which allows independent movement of the two pieces of apparatus. Such a decoupling of movement is particularly important if the high resolution of the mass spectrometer is to be maintained. The VG ZAB-E mass spectrometer is a reverse geometry, double focising instrument, which combines high mass range (12,000 amu at a source potential of 8 kV) with high resolution (> 100,000 at 10% valley definition). The ion source has been modified to allow the unimpeded entrance and exit of the cluster beam, and the filament has been upgraded so that the source can operated with a trap current of 1000 μA at an electron energy of 70 eV. Under normal operating conditions and, with an argon pressure of approximately 5 bar behind the nozzle, the pressure in the expansion chamber is 4x10^{-4} mbar, in the collimation chamber it is 2x10^{-6} mbar, and in the ion source the background pressure is below 1x10^{-6} mbar. The low value of the latter enables us to disregard the possibility that ion-molecule reactions between isolated molecules and/or ion clusters may be responsible for any of the reactions we observe.

Within the flight tube of the mass spectrometer a background pressure of less than 6x10^{-9} mbar is maintained by three 700 l s^{-1} Edwards Diffstak diffusion pumps. There is no perceptible increase in flight tube pressure when the cluster apparatus is in operation. These conditions ensure that collision-induced fragmentation during the passage of ions from the source to the detector can be ignored.

The detection system consists of a modified Daly detector [17]. In conjuction with phase-sensitive detection (psd), ion currents of $\geqslant 10^{-18}$ A can be measured. Using a fast photomultiplier tube and photon counting techniques the detection range can be extended to < 1 ion s^{-1}. The combination of psd and photon counting gives a dynamic detection range of 10^{12}. The distance from the nozzle to the ion source is 75 cm, and the total flight path of the clusters from the nozzle to the detector is approximately 5 m.

A particular useful feature of the mass spectrometer is the presence of a mass marker which operates from a Hall probe. Calibrated from the known masses of CsI cluster ions, the marker provides mass identification to within ± 1 amu out as far as 11 000 amu. Although the mass spectrometer is capable of achieving very high resolution, a compromise has to be struck between sensitivity and resolution. Total ion currents are typically of the order of 10^{-8} A; however, the intensities of individual ion clusters, particularly at high masses are often very low. Therefore, a balance has to be found between signal strength and resolution and for most of the results presented here the latter was typically 5000. Figure 3 shows an example of a mass spectrum recorded for argon cluster ions in the region of Ar_{250}^+. The ^{36}Ar isotope peaks are very well resolved and the ^{38}Ar isotope peaks are also evident. It is interesting to note that at these high masses, the cumulative effect of a slight mass deficiency in ^{40}Ar (39.9624 on the ^{12}C scale) becomes significant.

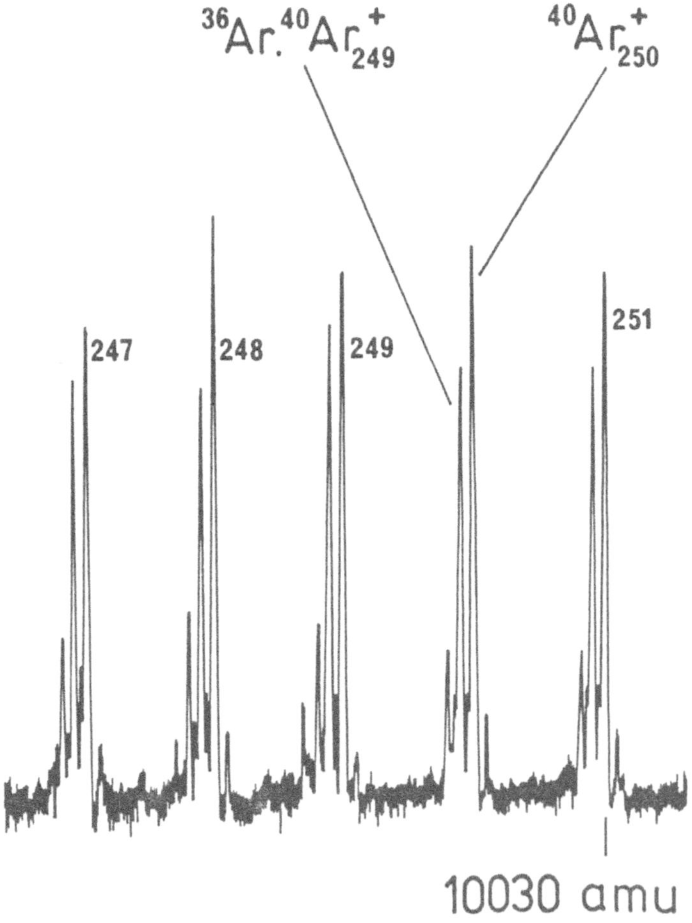

Figure 3. Example of an argon cluster ion mass spectrum. Clusters containing one and two ^{36}Ar atoms are clearly resolved. Taken from ref. [8].

4. MIKE/LINKED SCAN EXPERIMENTS

The combination of a magnetic and an electric sector configured in either a forward or reverse gemometry, enables a double focusing mass spectrometer to be used in one or more linked scanning modes [20]. Of particular value in the case of cluster ions (see below), is the fact that fragmentation in either the first or second field-free-regions (first and second FFR in figure 2) can be studied. Therefore, the different time-windows represented by the two regions can be used to monitor the temporal evolution of stable cluster ion structures [8,17].

4.1. MIKE (mass-analysed ion kinetic energy) spectra

Of the scanning methods available, the MIKE technique is the most straightforward in concept, and can be used to study the fragmentation of cluster ions during their passage through the second FFR. The magnet is set to transmit cluster ions of a single mass, and the electric sector voltage is then scanned towards zero while the accelarating voltage and magnetic field remain constant. Equation (4) provides a means of identifying fragment ions from the magnitude of the electric sector voltage necessary to transmit them. It is common practice to consider the electric sector as a kinetic energy analyser and, therefore, to discuss fragmentation in terms of changes in ion kinetic energy [21].

$$E^* = \frac{m_2 z_1}{m_1 z_2} E_o \qquad (4)$$

E_o is the initial parent ion kinetic energy (8000 eV is a typical value), E^* is the kinetic energy after fragmentation, m_1 and z_1 are, respectively, the mass and charge on the parent ion, and m_2 and z_2 are the mass and charge on the fragment ion.

For an elementary cluster ion reaction of the type

$$Ar_n^+ \rightarrow Ar_{n-m}^+ + mAr \qquad (5)$$

equation (4) reduces to

$$E_{n-m} = \frac{(n-m)}{n} E_o \qquad (6)$$

Thus, m can easily be identified from an accurate measurment of E_{n-m}. Figure (4) shows a section of a mass spectrum obtained for krypton cluster ions with E_0 at 3 kV. In figure (5) the ion Kr_{309}^+ has been mass-selected and a MIKE spectrum taken. The figure shows that during its passage through the second FFR the ion Kr_{309}^+ loses up to six krypton atoms.

4.2. Linked scans

The analog circuitry of the ZAB-E provides the facility for two types of linked scan, B/E and B^2/E. Both scans involve fragmentation in the first

FFR between the ion source and the magnet. In a B/E scan the magnet current and the electric sector voltage are varied in such a way that the ratio B/E remains constant. For reaction (5) the conditions necessary to transmit the parent ion and its fragments are

$$\frac{M_{n-m}}{M_n} = \frac{B_{n-m}\ E_{n-m}}{B_n\ E_n} \qquad (7)$$

As with the MIKE scan, equation (7) can be used to identify m. From a B^2/E scan it is possible to identify all those parent Ar_n^+ ions which contribute to the intensity of a preselected Ar_{n-m}^+ fragment ion generated in the first FFR. In terms of cluster ion mass spectra, where the interest is in reactions of the type given by (5) above, B^2/E scans provide no new information beyond that available from a B/E scan. Such a redundancy arises because a cluster ion mass spectrum contains as many parent ions as there are different-sized clusters, and each can contribute to the formation of more than one fragment.

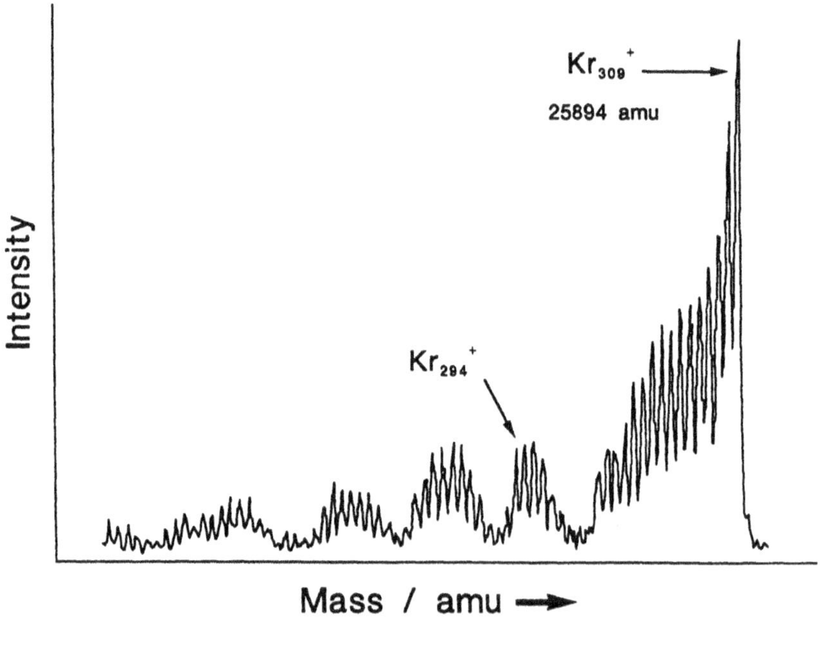

Figure 4. A typical mass spectrum of the Kr_{309}^+ region of a krypton cluster ion mass spectrum. The accelerating voltage is 3 kV. The nominal mass of Kr_{309}^+ is 25,894 amu. Taken from ref. [17].

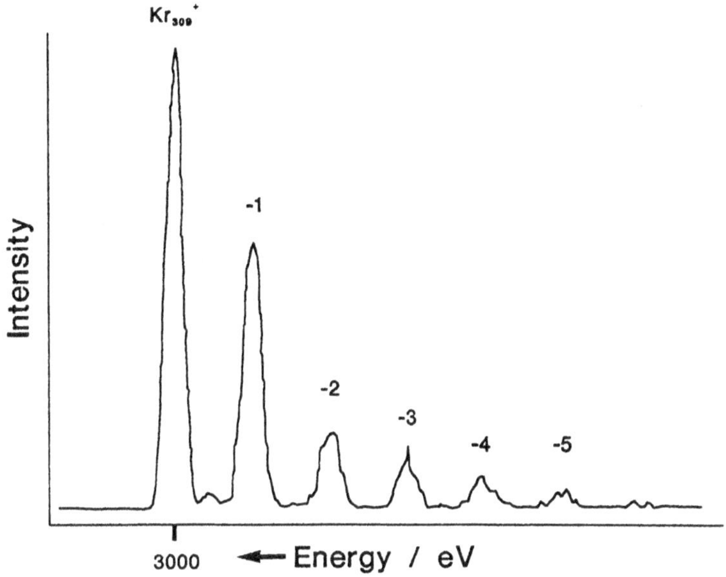

Figure 5. A MIKE scan of Kr_{309}^+ at 3 kV, showing unimolecular fragmentation in the second FFR with the loss of up to 6 krypton atoms. Taken from ref. [17].

Of the scanning procedures discussed above, MIKE scans have the lowest resolution. The release of kinetic energy during the decomposition process, means that fragment ion peak profiles are broader than those of the parent ion. However, because the fragment ions from both the B/E and B^2/E scans pass through both the magnetic and electric sectors, they are better resolved than those from a MIKE scan which pass through the electric sector only. Fortunately, the magnitude of the kinetic energy release in most cluster ion fragmentation processes is very low and, therefore, small changes in mass can easily be resolved in a MIKE scan.

Each of the scanning procedures discussed above can lead to the observation of ion peaks that are not necessarily associated with the fragmentation process under study. These peaks are commonly referred to as artifact peaks [20]. With the increasing use of double focusing mass spectrometers in cluster experiments, it is important that possible interference from the presence of artifact peaks be taken into consideration [14]. It would appear that MIKE scans in particular give rise to a very specific series of artifact peaks, and the following example serves to illustrate how they occur. A typical MIKE scan of the following reaction

$$Ar_{23}^+ \rightarrow Ar_{22}^+ + Ar \qquad (8)$$

would start with the Ar_{23}^+ ion of mass 920 amu being transmitted through the magnet with a kinetic energy of 8000 eV. Following decomposition in the second FFR, Ar_{22}^+ would be detected at the reduced kinetic energy of 7652 eV as given by equation (4). As we have already mention, a cluster ion mass spectrum contains as many "parent" ions as there are clusters and, during the course of an experiment, all these "parent" ions are fragmenting both in the ion source and first FFR flight tube. Hence, at the same time that a MIKE scan of Ar_{23}^+ is being recorded, Ar_{25}^+ ions, for example, are fragmenting in the first FFR to give Ar_{24}^+. Therefore, by analogy with a single focusing experiment, the fragmentation of Ar_{25}^+ will produce a metastable ion with a nominal mass given by M_2^2/M_1 (for $M_1^+ \rightarrow M_2^+ + M_3$) of 921.6 amu. If the single focusing resolution of the magnetic sector is poor (< 600), then both Ar_{23}^+ and the metastable ion will be transmitted through to the second FFR at the same time. However, because the Ar_{24}^+ fragment is generated outside the ion source, its kinetic energy is reduced to a value given by equation (4); 7680 eV to be precise. Hence, a MIKE scan of reaction (8) will contain one peak at 7652 eV corresponding to the true fragment ion and a second peak at 7680 eV which is the artifact ion. Improving the single focusing resolution can eliminate artifact peaks at low masses, but because the ion currents associated with individual clusters are low, such improvements cannot be made idefinitely.

In general, the fragment ions from a MIKE scan of Ar_n^+ will experience interference from artifact ions in the following manner:

$$Ar_n^+ \rightarrow Ar_{n-1}^+ + Ar \text{ (in the 2}^{nd}\text{ FFR)} \qquad (9)$$

will coincide with

$$Ar_{n+2}^+ \rightarrow Ar_{n+1}^+ + Ar \text{ (in the 1}^{st}\text{ FFR)} \qquad (10)$$

Under circumstances where two atoms are lost, the analogous situation is

$$Ar_n^+ \rightarrow Ar_{n-2}^+ + 2Ar \text{ (in the 2}^{nd}\text{ FFR)} \qquad (11)$$

will coincide with

$$Ar_{n+4}^+ \rightarrow Ar_{n+2}^+ + 2Ar \text{ (in the 1}^{st}\text{ FFR)} \qquad (12)$$

Thus, in the case of $Ar_{50}^+ \rightarrow Ar_{48}^+ + 2Ar$, we have a parent ion of mass 2000 amu and the fragment ion appearing at an electric sector voltage of 7680 eV. The artifact ion, Ar_{52}^+, has a nominal mass of 2002.9 amu and appears at an electric sector voltage of 7703 eV. The low kinetic energy release associated with the fragmentation of argon cluster ions means that all these examples can be readily identified [14].

In the B/E scan the main artifact contribution appears to come from the presence of isotope peaks. Although the fragment ion resolution is better than that in, for example, a MIKE scan, the parent ion resolution does not appear to be as good. For this reason, B/E scans of high-mass clusters ions were susceptible to interference from neighbouring peaks containing ^{36}Ar and ^{38}Ar [14]. A second source of interference comes through the fragmentation of cluster ions in the 2 nd FFR. However, because the probability of decay varies exponentially with time, the genuine B/E fragments are always very much more intense that the 2 nd FFR artifacts.

5. CLUSTER ION FRAGMENTATION PROCESSES

Over the past five years, a large number of studies have been undertaken where cluster ion metastable peak intensities have been monitored [1,2]. As illustrations of the type of information that can be extracted from the unimolecular decay pattern, we shall briefly consider two examples:

5.1. Unimolecular decay of X_n^+ clusters

In terms of detecting a metastable peak, by far the most frequently observed unimolecular decay process in both atomic and molecular systems is

$$X_n^+ \rightarrow X_{(n-1)}^+ + X \qquad (13)$$

An example of such behaviour is given in figure 5 for the unimolecular decay of Kr_{309}^+. Reaction (13) is most likely to be the process with the lowest energy, and the presence of a competitive shift will probably prevent other less-favourable decay routes from producing metastable peaks of comparable intensity. Where the loss of more than one atom or molecule is shown to be taking place as, for example, in figure 5, this is most likely to be as a result of the sequential loss of single species during the time period of the observations. Scheier and Mark [22] have recently shown that small argon ion clusters exhibit sequential fragmentation, and the law of microscopic reversibility must certainly exclude a cluster from losing three or more individual atoms in a single step.

Earlier experiments on cluster ion fragmentation [23] showed there to be strong correlation between fragmentation pattern and the number of atoms (or molecules) necessary to adopt certain stable structures. Thus the process

$$Ar_{20}^+ \rightarrow Ar_{19}^+ + Ar \qquad (14)$$

produces a metastable peak which is very much more intense than those generated by its immediate neighbours [16,23]. This result has been interpreted in terms of the formation of a stable Ar_{19}^+ ion, consisting of a central Ar_2^+ surrounded by 17 argon atoms in an icosahedral arrangement [16]. These experiment have recently been extended to cluster ions containing up to 300 argon or krypton atoms [14,17], and the correlation between reactivity and structure appears to persist; certainly out as far as Ar_{148}^+ or Kr_{148}^+. Figure 6 shows the results of a series of MIKE scans on the unimolecular fragmentation of krypton cluster ions, Kr_n^+. Each of the intense fragment ions correlates with the formation of a stable species [24].

5.2. Unimolecular decay of mixed cluster ions.

A simple analytical argument [25] shows that when two decay routes are in competition, the route with the lowest critical energy will produce the more intense metastable peak (exceptions to this rule are to be found in decay processes involving rearrangement steps). By using the idea that a competitive shift can have a strong influence on fragmentation pattern, it

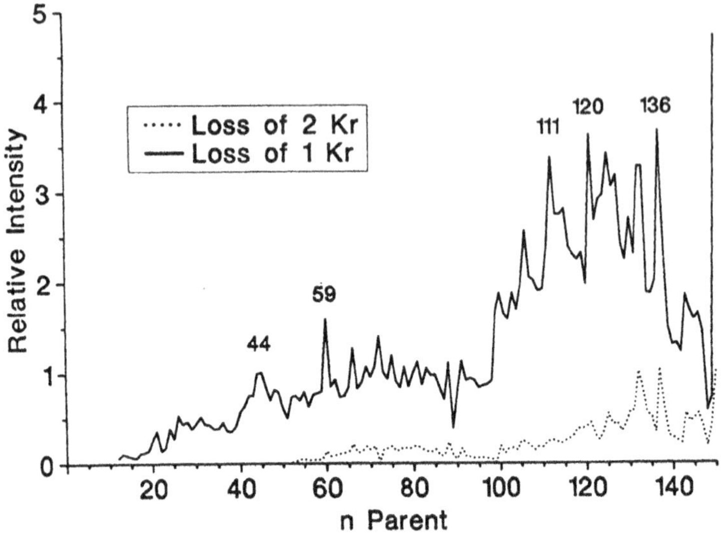

Figure 6. The results of a series of MIKE scans on the unimolecular fragmentation of krypton cluster ions. The ratio of fragment to parent ion intensity is plotted against the parent cluster size.

is possible to study such phenomena as the preferential solvation of ions in mixed clusters.

The electron or photon impact ionization of mixed ammonia/water clusters results in the formation of protonated species with the general formula $[(NH_3)_n.(H_2O)_m]H^+$ [26], and following on from the above discussion, one might seek to use competative fragmentation to answer the question " which of the molecules does the proton become preferentially associated with? ". The are two possible metastable decomposition routes

$$[(NH_3)_n.(H_2O)_m]H^+ \rightarrow [(NH_3)_n.(H_2O)_{m-1}]H^+ + H_2O \quad (15)$$

$$\rightarrow [(NH_3)_{n-1}.(H_2O)_m]H^+ + NH_3 \quad (16)$$

Figure 7 shows the observed metastable peak intensities as a function of n for the case when m = 2. For n $\leqslant$ 5, the only reaction path observed is (15); however, once n becomes equal to 6, there is an immediate switch to reaction (16). This transition can be rationalized in terms of the various electrostatic interactions present in the cluster ion.

If it is assumed that the central core consists of NH_4^+, then the cationic nature of the protons will ensure that ammonia molecules with their higher polarizabilities (2.26 A, as opposed to 1.48 A for water) will form the first solvation shell. Hence the short-ranged ion-induced dipole interaction is responsible for forming a $NH_4^+.(NH_3)_4$ unit, the stability of which accounts for the behaviour observed in figure 7 up to n = 5. Beyond the first solvation shell, it would appear that the longer-ranged ion-dipole interaction becomes comparatively more important, because the decay of larger cluster ions involves only the loss of ammonia (μ = 1.48 D) rather than water (μ = 1.85 D). It is concluded, therefore, that NH_4^+ prefers ammonia molecules in its first solvation shell, but thereafter, water is the preferred solvent. Some of the smaller water/ammonia cluster ions have been the subject of equilibrium thermodynamic experiments [27,28], where it is observed that ΔH^0 values for the addition of either water or ammonia differ by just 2 kcal (0.08 eV). That nonequilibrium experiments of the type discussed here respond to such small energy differences, suggests that relative metastable peak intensities may be quite sensitive as a probe to subtle variations in the electrostatic interactions present in cluster ions.

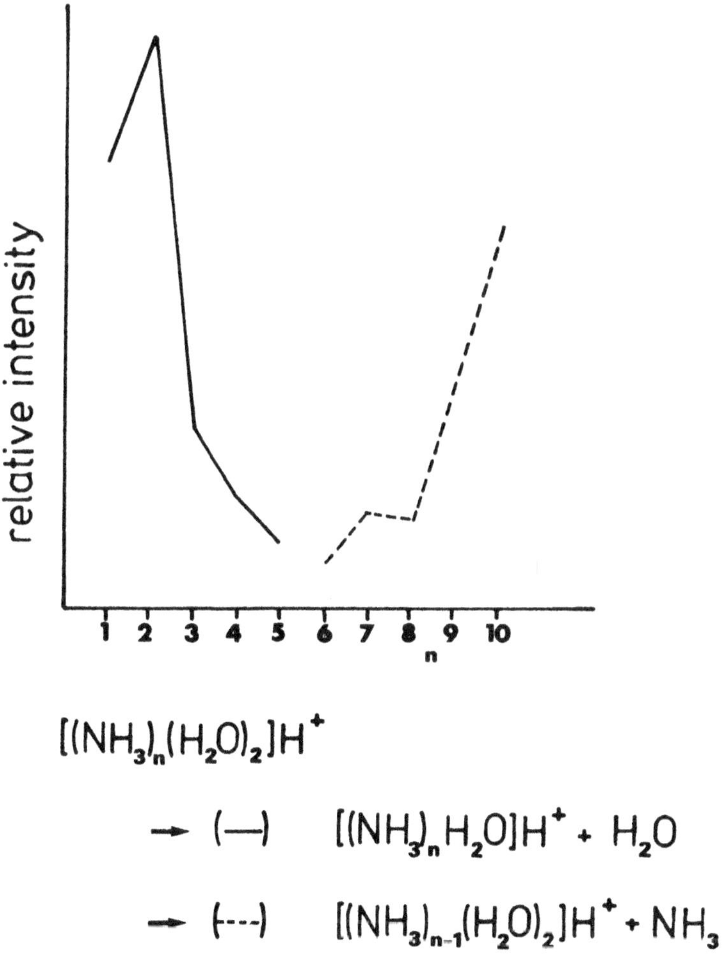

Figure 7. Relative metastable peak intensities for the unimolecular decomposition of $[(NH_3)_n.(H_2O)_2]H^+$ clusters as a function of n.

6. Acknowledgements

The author would like to thank the SERC for the provision of equipment grants which have made these experiments possible. In addition, thanks are due to A.K. Shukla, C. Moore, and P.G. Lethbridge who have undertaken much of the work described.

7. References

[1] T.D. Mark, Int. J. Mass Spectrom. Ion Processes, 79, 1 (1987).
[2] A.J. Stace, in Mass Spectrometry, edited by M.E. Rose (The Royal Society of Chemistry Specialist Periodical Report, London, 1987) Vol. 9, p. 96.
[3] O. Echt, K. Sattler, and E. Recknagel, Phys. Rev. Letts., 47, 1121 (1981).
[4] A.F. Mamyrin, V.I. Karataev, D.V. Shmikk, and V.A. Zagulin, Sov. Phys. JETP, 37, 45 (1973).
[5] M.D. Morse, M.E. Geusic, J.R. Heath, and R.E. Smalley, J. Chem. Phys., 83, 2293 (1985).
[6] C. Brechignac and Ph. Cahuzac, Chem. Phys. Lett., 117, 365 (1985).
[7] W. Kamke, B. Kamke, H.-U. Kiefl, and I.V. Hertel, J. Chem. Phys., 84, 1325 (1986).
[8] A.J. Stace and A.K. Shukla, Int. J. Mass Spectrom. Ion Phys., 36, 119 (1980).
[9] R.J. Beuhler and L. Friedman, J. Chem. Phys., 77, 2549 (1982).
[10] M.A.D. Fluendy and K.P. Lawley, Chemical Applications of Molecular Beam Scattering, (Chapman and Hall, London, 1973).
[11] W.H. Flygare, in Fourier, Hadamard, and Hilbert Transformations in Chemistry, edited by A.G. Marshall (Plenum Press, New York 1982).
[12] A.J. Stace, D.M. Bernard, J.J. Crooks, and K. Reid, Mol. Phys., 60, 671 (1987).
[13] H. Harberland, Surf., Sci. 156, 305 (1985).
[14] P.G. Lethbridge and A.J. Stace, J. Chem. Phys., 89, 4062 (1988).
[15] K. Norwood, J.-H. Guo, and C.Y. Ng, J. Chem. Phys., 90, 2995 (1989).
[16] A.J. Stace and C. Moore, Chem. Phys. Lett., 96, 80 (1983).
[17] P.G. Lethbridge and A.J. Stace, J. Chem. Phys., 91, 7685 (1989).
[18] J.B. Cross and J.J. Valentini, Rev. Sci. Instrum., 53, 38 (1982).
[19] C.E. Otis and P.M. Johnson, Rev. Sci. Instrum., 51, 1128 (1980).
[20] K.R. Jennings and R.S. Mason, in Tandam Mass Spectrometry, edited by F.W. McLafferty (Wiley, New York, 1983).

[21] R.G. Cooks, J.H. Beynon, R.M. Caprioli, and G.R. Lester, Metastable Ions (Elsevier, Amsterdam, 1973).

[22] P. Scheier and T.D. Mark, Phys. Rev. Lett., 59, 1813 (1988).

[23] T.D. Mark, P. Scheier, K. Leiter, W. Ritter, K. Stephan, and A. Stamatovic, Int. J. Mass Spectrom. Ion Processes, 74, 4095 (1986).

[24] P.G. Lethbridge, G. Del Mistro, and A.J. Stace, J. Chem. Phys., 93, 1995 (1990).

[25] A.J. Stace and C. Moore, J. Am. Chem. Soc., 105, 1814 (1983).

[26] A.J. Stace, J. Am. Chem. Soc., 106, 2307 (1984).

[27] J.D. Payzant, A.J. Cunningham, P. Kebarle, Can. J. Chem., 51, 3242 (1973).

[28] I.N. Tang and A.W. Castleman,Jr., J. Chem. Phys., 62, 4576 (1975).

CHEMICAL REACTIONS IN AND ON CLUSTER IONS

A.J. Stace,
School of Molecular Sciences,
University of Sussex,
Falmer, Brighton BN1 9QJ,
U.K.

ABSTRACT. The ionization of molecular clusters frequently leads to the resultant ions undergoing chemical reactivity. Specific examples taken from pure and mixed cluster ions are discussed in order to illustrate those factors, such as time scale and cluster size, which could have an influence on the type of reaction observed.

1. Introduction

Any mass spectrometric analysis of molecular cluster ions will provide evidence of chemical reactivity. Such behaviour falls broadly into two catagories: i) the unimolecular decay of a single species within a cluster (usually the charge carrier); ii) an ion-molecule reaction between two or more of the component molecules [1]. Interest in cluster reactions has developed for two reasons. First, a molecular cluster ion represents a microscopic liquid-like enviroment; therefore, the types of reactions observed might be similar to those initiated in bulk liquids by, for example, radiolysis. Second, there may exist reactions that are unique to small clusters, possibly by virtue of their very high surface-to-volume ratio. An example of the latter may be a sequential three- or four body reaction which might be inhibited in the liquid phase because important intermediates would be free to diffuse apart. Within the confines of a small cluster there are obvious restrictions on how far species can diffuse.

A further class of reaction involves those processes which might be expected to exhibit a gradual change in behaviour as a function of cluster size. Two examples are: i) the cage effect. How large does a cluster have to be before it has sufficient degrees of freedom to absorb the momentum of a dissociating radical pair? Recent experiments by Lineberger and co-workers [2] have shown that such size dependences do exist, and that for the photodissociation reaction

$$I_2^-.(CO_2)_n + h\nu \rightarrow I^-.(CO_2)_m + (n-m)CO_2 + I \quad (1)$$

$$\rightarrow I_2^-.(CO_2)_k + (n-k)CO_2 \quad (2)$$

K. R. Jennings (ed.), Fundamentals of Gas Phase Ion Chemistry, 105–116.

step (2) is only observed when $n > 5$, and becomes the dominant process when $n > 16$; ii) in a unimolecular reaction of the type

$$(CH_3COCH_3)_n^+ \rightarrow (CH_3COCH_3)_{n-1}.CH_3CO^+ + CH_3 \quad (3)$$

how does the size of the cluster influence the extent of reaction? For an isolated acetone ion, reaction (3) is endothermic by approximately 0.7 eV and this energy comes from the ionization process. The same must be true for excitation of the corresponding cluster ion [3]; however, as the size, n, increases so does the number of degrees of freedom. If the entire cluster participates in accommodating the excess energy of an ion (in a sense that decomposition is a statistical/RRKM process), then a gradual decline in fragment ion intensity might be expected as a function of increasing size. However, as figure 1 shows, reaction (3) is observed out to n = 20 and the parent/ fragment ion ratio appears almost constant beyond n = 6. Such behaviour suggests that there exist a finite limit in the extent to which energy is randomized within a cluster. Whether such a limit is imposed by time and/or vibrational coupling remains to be seen. Also shown in figure 1 is the result of an RRKM calculation [4], in which it has been assumed that the energy of the cluster is statistically partitioned among the available degrees of freedom. It is interesting to note the coincidence between the calculated decline in reaction probability and the observed onset of a near-constant fragment ion intensity. This result suggests a localized reaction zone of between 5 and 6 molecules. The fact that reaction (3) has been observed in liquid acetone following excitation by radiolysis [5], is an indication that cluster and liquid could be acting in a similar manner. To an excited ion, the bulk liquid must represent an infinite heat sink and, therefore, the reaction zone must also be very localized.

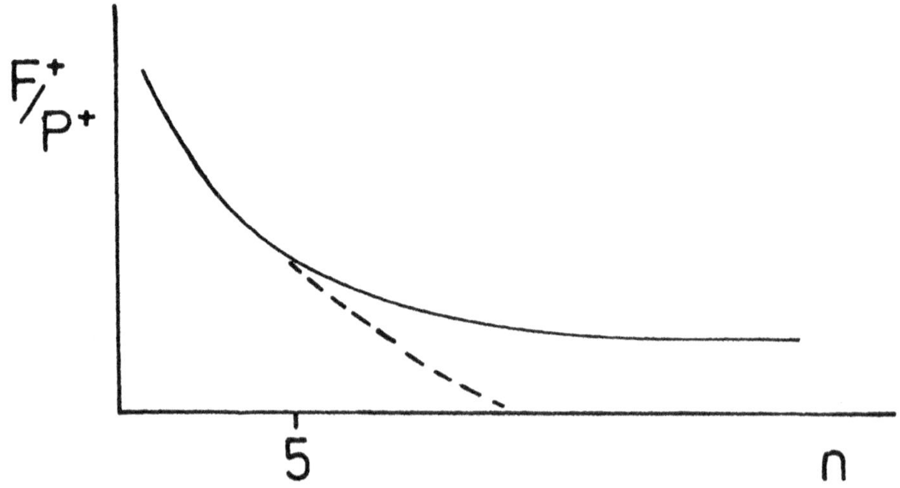

Figure 1. Plot of the ratio $(CH_3COCH_3)_{n-1}.CH_3CO^+/(CH_3COCH_3)_n$, ($F^+/P^+$), as a function of n. Also shown is the ratio predicted from an RRKM calculation [4].

For the most part, molecular clusters can be prepared using the techniques outlined in an earlier chapter. Therefore, we shall restrict our discussion here to an analysis of the results from experiments on chemical reactivity in and on clusters. Two areas will be examined, i) reactions in pure cluster ions and ii) the unimolecular decay of molecular ions in association with inert gas clusters. In each case, the results have predominantly been derived from the observation of mass spectra as opposed to using the MIKE or B/E techniques discussed previously. In most instances, MIKE and B/E scans of molecular cluster ions show evidence of monomer unit loss only. This result implies that the time scale for reactivity is $< 10^{-6}$ s which again would support the notion of a localized reaction site. It is only very recently (see below) that evidence of reaction times $> 10^{-5}$ s has been seen.

2. Reactions in pure cluster ions

Within this catagory, the most frequently observed process is that which leads to the formation of a protonated cluster. In the case of water clusters, for example, ionization leads to the formation of $(H_2O)_nH^+$ via the following sequence of events

$$(H_2O)_n + h\nu(e^-) \rightarrow [(H_2O)_n^+]^\dagger + e^-(2e^-) \quad (4)$$

$$[(H_2O)_n^+]^\dagger \rightarrow (H_2O)_{n-m}H^+ + OH + mH_2O \quad (5)$$

where $[\ldots]^\dagger$ denotes a vibrationally excited state of the ion which could also be metastable [6]. The driving force for step (5) is a cluster analogue of the exothermic gas phase ion/molecule reaction

$$H_2O + H_2O^+ \rightarrow H_3O^+ + OH \ (\Delta H \approx -85 \text{ kJ mol}^{-1}) \quad (6)$$

Similar behaviour is exhibited by other clusters systems consisting of molecules with high proton affinities, e.g. $(CH_3OH)_nH^+$, $(CH_3OCH_3)_nH^+$, $(NH_3)_nH^+$, and also in mixed cluster ions, such as $[(CH_3COCH_3)_n.(H_2O)_m]H^+$ and $[(NH_3)_n.(H_2O)_m]H^+$.

Where the cluster consists of alcohol molecules, i.e. methanol, ethanol etc, a slightly more complex route to protonation has been proposed [7]

$$(ROH)_n + e^- \rightarrow [(ROH)_n^+]^\dagger + 2e^- \quad (7)$$

$$[(ROH)_n^+]^\dagger \rightarrow (ROH)_{n-1}.R'OH^+ + R'' \quad (8)$$

where $R = R' + R''$ and R'' is lost through an α-cleavage reaction which is then followed by the step

$$(ROH)_{n-1}.R'OH^+ \rightarrow (ROH)_{n-1}H^+ + R'O \quad (9)$$

where $R'O$ is either an aldehyde or a ketone. Once protonation has taken place in the larger clusters, the only additional reactions they exhibit, as evidenced by metastable decay for example, are the evaporative loss of monomer units, i.e.

$$(CH_3OH)_nH^+ \rightarrow (CH_3OH)_{n-1}H^+ + CH_3OH \qquad (10)$$

In contrast, the smaller clusters, and in particular the protonated dimer ions, undergo reactions similar to those established for the isolated species in the gas phase. For example, Morgan and Castleman [8] found that the methanol dimer ion undergoes the reaction

$$(CH_3OH)_2H^+ \rightarrow (CH_3)_2OH^+ + H_2O \qquad (11)$$

and Shukla and Stace [7] observed metastable decay corresponding to the following reactions in the 2-propanol dimer ion

$$[(CH_3)_2CHOH]_2H^+ \rightarrow [(CH_3)_2CH]_2OH^+ + H_2O \qquad (12)$$

$$[(CH_3)_2CHOH]_2H^+ \rightarrow (CH_3)_2CHOH_2.H_2O^+ + C_3H_6 \qquad (13)$$

$$[(CH_3)_2CHOH]_2H^+ \rightarrow (CH_3)_2CHOH_2{}^+ + (CH_3)_2CHOH \qquad (14)$$

Reaction (12) is considered to be an important intermediate step in the formation of di-isopropyl ether, $[(CH_3)_2CH]_2O$, during the radiolysis of 2-propanol vapour [9].

Several cluster systems show evidence of quite complex ion/molecule chemistry. In the mass spectrum of carbon dioxide clusters the following ions are observed: $(CO_2)_n{}^+$, $(CO_2)_n.O^+$, $(CO_2)_n.CO^+$, and $(CO_2)_n.O_2{}^+$. On thermodynamic grounds, formation of the latter ion by a direct reaction, i.e.

$$(CO_2)_n{}^+ \rightarrow (CO_2)_{n-1}.O_2{}^+ + C \qquad (15)$$

is very unlikely. Instead, Romanowski and Wanczek [10] proposed that carbon dioxide cluster ions undergo a two-stage ion/molecule reaction

$$(CO_2)_n{}^+ \rightarrow (CO_2)_{n-1}.O^+ + CO \qquad (16)$$

$$(CO_2)_{n-1}.O^+ \rightarrow (CO_2)_{n-2}.O_2{}^+ + CO \qquad (17)$$

The gas phase anologue of reaction (17) having been observed previously in an ICR experiment [11].

Much less work has been undertaken on studying reactivity in negatively charged clusters. However, in common with cations these also exhibit reactions analogous to those observed in gas phase ion-molecule processes. Thus, low-energy electron attachment to neutral carbon dioixde clusters [12], leads to the observation of $(CO_2)_n.O^-$, which is consistent with the gas phase reaction [13]

$$e^- + CO_2 \rightarrow O^- + CO \qquad (18)$$

In a similar manner, electron attachment to nitrous oxide clusters leads to the formation of $(N_2O)_n.O^-$ and $(N_2O)_n.NO^-$. As with the $(CO_2)_n{}^+$ system, formation of $(N_2O)_nNO^-$ [14] can be explained in terms of a sequential ion/molecule reaction

$$e^- + (N_2O)_n \rightarrow (N_2O)_{n-1}.O^- + N_2 \qquad (19)$$

$$(N_2O)_{n-1}.O^- \rightarrow (N_2O)_{n-2}.NO + NO \qquad (20)$$

Both these reactions have been observed for isolated ions in the gas phase [15].

So far, the examples discussed have consisted of monomer units, i.e. CO_2 in the case of carbon dioxide clusters, exhibiting either single decay steps or participating in a single internal ion-molecule reaction. Recent observations on the properties of cluster ions composed of quite complex organic molecules, have revealed the presence of multiple reaction sequences. In the mass spectrum of gas phase n-butanoic acid, the most intense fragment peak comes from a McLafferty rearrangement process

$$CH_3CH_2CH_2COOH^+ \rightarrow CH_2C(OH)_2^+ + C_2H_2 \qquad (21)$$

Following the ionization of n-butanoic acid clusters [16], a series of ions corresponding to the general formula $(CH_3CH_2CH_2COOH)_nH^+$ are observed, together with a second sequence that has the general formula $(CH_3CH_2CH_2COOH)_1H^+.[CH_2C(OH)_2]_m$, for m up to five. The nature of the reaction product suggests that each parent ion, $(CH_3CH_2CH_2COOH)_nH^+$, undergoes a sequence of McLafferty rearrangement reactions. However, this line of reasoning is complicated by the fact that protonated carboxylic acids are known not to undergo McLafferty rearrangement reactions. Our data indicates that the end product of the reaction sequence is $CH_3CH_2CH_2COOH_2^+.[CH_2C(OH)_2]_m$, i.e. the proton is associated with just one molecule which remains unreacted, leaving the additional n-butanoic acid molecules free to participate in reaction (21). We are then left to ask the questions "where is the positive charge?, and what is its role in the reaction sequence?"

A second series of these "chain reactions" has been observed following the ionization of ethyl formate clusters $(CH_3CH_2OC(H)O)_n$ [17]. Two products are observed and they have been distinguished using isotopes:

$$(CH_3CH_2OC(H)O)_n^+ \rightarrow$$

$$(CH_3CH_2OC(H)O)_{n-1}^+.CH_3CH_2OH + CO \qquad (22)$$

followed by

$$(CH_3CH_2OC(H)O)_{n-1}^+.CH_3CH_2OH \rightarrow$$

$$(CH_3CH_2OC(H)O)_{n-1}^+.H_2O + C_2H_4 \qquad (23)$$

Reaction (23) is not observed without (22) having first taken place. For this reason we identify the product of (23) as $(CH_3CH_2OC(H)O)_{n-1}^+.H_2O$ rather than $(CH_3CH_2OC(H)O)_{n-2}^+.CH_3CH_2OH.CO_2H_2$, which would involve consuming a second ethyl formate molecule. Reaction (22) can proceed through two or three steps without (23) being observed. Up to five sequential steps involving reactions (22) and (23) have been identified [17]. We believe that complex reactions of this type in cluster ions are driven, in part, by the thermodynamic stability the products gain through solvation within the cluster.

The ethyl formate cluster reactions have also provided the first example of a decay process occuring on a 10^{-5} s time scale [17]. A MIKE scan involving a parent ion, $(CH_3CH_2OC(H)O)_n^+$, exhibits a very weak signal corresponding to reaction (22) occuring in the 2 nd FFR.

3. The fragmentation of molecular ions in association with inert gas clusters.

A quite different set of circumstances under which reactivity in cluster ions can be observed, arises through the preparation and ionization of mixed clusters of the form $Ar_n.X$ and $(CO_2)_n.X$ [18-22]. X can range from diatomics like I_2 and Cl_2 to complex polyatomics like ethyl formate or diethyl ketone. In each case a mass spectrum of the mixed clusters reveals the presence of parent ions, i.e. $Ar_n.X^+$, and also fragment ions that have arisen from X^+ having decomposed but with the products remaining attached to the argon. Examples of some of the observed reactions are:

$$Ar_n.CH_3COCH_3^+ \rightarrow Ar_n.CH_3CO^+ + CH_3 \quad (24)$$

$$(CO_2)_n.CH_3OH^+ \rightarrow (CO_2)_n.CH_2OH^+ + H \quad (25)$$

$$Ar_n.CH_3OCH_3^+ \rightarrow Ar_n.CH_3OCH_2^+ + H \quad (26)$$

$$Ar_n.CH_3OCH_3^+ \rightarrow Ar_n.CHO^+ + H_2 + CH_3 \quad (27)$$

$$Ar_n.C_2H_4^+ \rightarrow Ar_n.C_2H_3^+ + H \quad (28)$$

$$Ar_n.C_2H_4^+ \rightarrow Ar_n.C_2H_2^+ + H_2 \quad (29)$$

Approximately 50 different examples of reactivity in mixed $Ar_n.X^+$ and $(CO_2)_n.X^+$ clusters has been observed. In many instances, the products of three or more sequential reactions are observed and in each case the charged fragment remains attached to the inert gas component. In some cases the fragmentation pathways of the isolated ion have critical energies of reaction as high as 2.7 eV; by comparison, the binding energy of a single argon atom to the inert gas cluster will be ≈ 0.05 eV. For the purposes of discussion in this section, we shall consider CO_2 to be an inert gas in the sense that there is no chemical interference between it and any of the observed cluster reactions. Using the ZAB-E it has been possible to resolve hydrogen atom loss in reaction (28) from clusters containing up to 130 argon atoms [21]. Figure 2 shows two sections of a mass spectrum recorded for the decomposition of the ethylene ion in association with argon clusters [22]; the reaction steps are given above. Figure 3 shows how the relative intensities of the fragment ions vary as a function of cluster size. Of particular interest, is the fragment

$Ar_{54}.C_2H_3^+$ which appears very much more intense than any of its immediate neighbours. Further discussion of this feature of the results will be presented below.

In the analysis of these reactions, we have attempted to identify three key aspects of their chemistry and physics: i) how does the

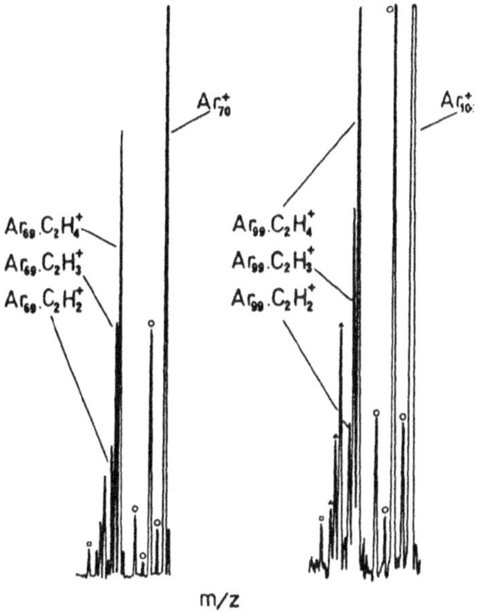

Figure 2. Section from the mass spectrum of ethylene in association with argon clusters. The peaks marked and are isotope peaks arising from the presence of ^{36}Ar and ^{38}Ar. Taken from ref. [22].

molecular ion acquire sufficient energy to react?; ii) if there is sufficient energy to break covalent bonds, i.e. ≈ 2 eV, why does the fragile inert gas cluster remain intact?; iii) can the types of reactions observed, tell us anything about the position of the molecular ion with respect to the remainder of the cluster. Data from three seperate sources has been combined to provide a qualitative picture of the molecular ion/inert gas cluster interaction. These sources consist of the present experiments, experiments on the photoionization efficiencies of small mixed clusters, and observations on excitation transfer in doped matricies.

3.1. Excitation

It was originally proposed that reactive ions were excited by charge transfer following the ionization of inert gas atoms on or close to the surface of a cluster [23]. The molecule (X) then receives as internal excitation the energy difference I.E.(Ar) - I.E.(X), which is sufficient to promote all of the observed reactions. More recently, support for this assumption has come from the photoionization experiments of Ding *et al* [24] and Kamke *el at* [25]. Their experiments have shown that the photoexcitation of an inert gas atom in association with a molecule can ultimately lead to the formation of a molecular ion. The term "intramolecular Penning ionization" has been used to describe the process [25], and it follows the basic form

$$Ar_n.X \;+\; h\nu \;\rightarrow\; Ar^*.Ar_{n-1}.X \qquad (30)$$

$$Ar^*.Ar_{n-1}.X \;\rightarrow\; Ar_{n-m}.X^+ \;+\; mAr \;+\; e^- \qquad (31)$$

where Ar^* denotes an argon atom in an excited electronic state. Evidence of a similar excitation transfer mechanism operating in molecular clusters, has been presented by Dao and Castleman [26]. Further examples, of excitation transfer come from excitation experiments performed on doped inert gas matrices [27]. These suggest that, even within argon, there is a range of excited states all of which can participate in excitation transfer. Obviously, a consideration of processes like (30) and (31) offers the possibility of a more general excitation mechanism than that originally proposed.

3.2. Fragmentation

Based on considerations discussed in the previous section, an upper limit can be estimated as to the amount of internal energy a molecular ion receives following excitation transfer. This figure can, in turn, be used to estimate a lifetime for the ion with respect to energetically accessible decomposition routes. Assuming there is little or no

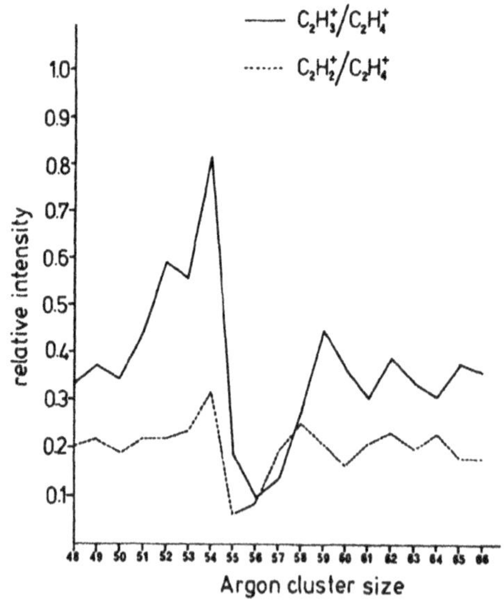

Figure 3. Plot of the relative intensities of the fragments observed following the electron impact ionization of mixed ethylene/argon clusters. Taken from ref. [22].

vibrational coupling between the molecular ion and the inert gas component, unimolecular reaction rate theory (RRKM) gives lifetimes of $< 10^{-12}$ s for the observed reaction pathways [20]. In contrast, the calculated lifetimes for those processes which, although energetically accessible, are not observed, is $> 10^{-10}$ s [20].

Evidence to support the assumption that the inert gas component does not participate in energy randomization prior to fragmentation, is based on the fact that the fragment ions do not exhibit substantial variations in intensity as a function of cluster size. In kinetic terms, it would appear that molecular ions fragment on a time scale which is very much shorter than that appropriate for vibrational predissociation, and that the inert gas component acts as a spectator.

3.3. Position of the molecular ion with respect to the inert gas cluster.

With regard to the structure of molecular ion/inert gas clusters, certain features of the experimental data lead us to make two qualitative observations: (i) many molecular ions appear capable to undergoing quite extensive fragmentation without disrupting the inert gas component; (ii) fragment ions of the type $Ar_n.Y^+$, form intense "magic" number combinations; in particular, $Ar_{54}.Y^+$ appears in many mass spectra as having a high relative intensity [21,22]. An example of such behaviour is shown in figure 3 for the fragments from reactions (28) and (29) [22]. Combining these two features of our results, we would conclude that each *reactive* molecular ion sits on or close to the surface of the inert gas cluster, and that many fragment ions certainly occupy surface sites. The $Ar_{54}.Y^+$ fragment cluster ions are thought to adopt a configuration consisting of a single atom from Y, often an oxygen atom, combining with the inert gas component to form a stable close-packed 55-atom structure [21].

Recent experiments have concentrated on the reactivity of complex organic ions associated with argon clusters [21]. As an example, the ionization of methyl formate/argon clusters, $Ar_n.CH_3OC(H)O$ leads to the following fragments being observed: $Ar_n.CH_3OH^+$, $Ar_n.CH_2OH^+$, $Ar_n.CHOH^+$, $Ar_n.CHO^+$, and $Ar_n.CO^+$. Through the use of isotopes, each fragment can be accounted for in terms of the established gas phase chemistry for the isolated methyl formate ion [21]. Figure 4 shows a short section of the observed mass spectrum. However, in addition to the ions given above, the following fragments are observed: $Ar_n.CH_2OC(H)O^+$ and $Ar_nCHOC(H)O^+$, and these products have no gas phase counterparts. To account for the presence of the latter, it has been proposed that the following steps are taking place.

$$Ar_n.CH_3OC(H)O^+ \rightarrow Ar_n.CH_3OH^+.CO \qquad (32)$$

$$Ar_n.CH_3OH^+.CO \rightarrow Ar_n.CH_2OH^+.CO + H \qquad (33)$$

$$Ar_n.CH_3OH^+.CO \rightarrow Ar_n.CHOH^+.CO + H_2 \qquad (34)$$

The methyl formate ion first undergoes are rearrangement process to give a methanol ion plus neutral carbon dioxide, and both fragments remain

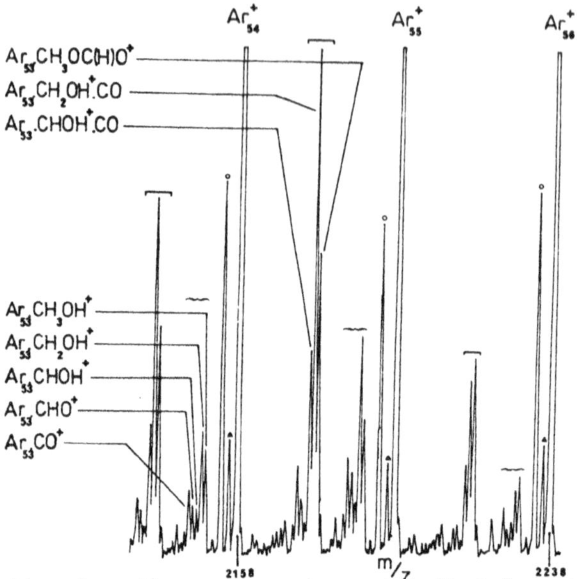

Figure 4. Section from the mass spectrum of methyl formate in association with argon clusters. Similar fragment ions on successive clusters are denoted by brackets and braces. A circle indicates the presence of ^{36}Ar in a cluster and a triangle the presence of ^{38}Ar. Taken from ref. [21].

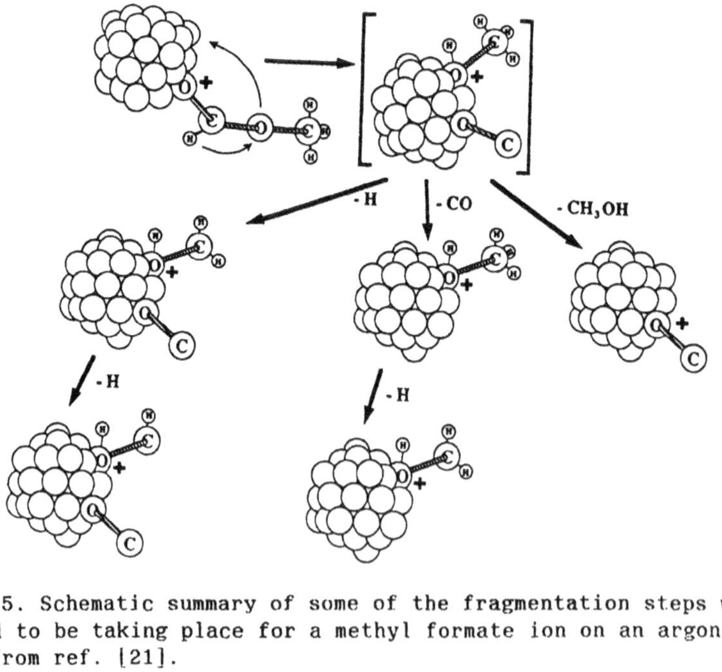

Figure 5. Schematic summary of some of the fragmentation steps which are assumed to be taking place for a methyl formate ion on an argon cluster. Taken from ref. [21].

attached to the inert gas cluster. The methanol ion then proceeds to fragment further via pathways that have been established for the isolated ion, whilst the CO molecule remains associated with the cluster. For this mechanism to operate, the CO molecule cannot have received any internal excitation as a result of the initial rearrangement step. Figure 5 summarises some of the chemical reactions exhibited by mixed methyl formate/argon cluster ions.

The range of chemical pathways observed for complex molecular ions, like methyl formate, in association with inert gas clusters is quite extensive. Qualitative experiments of this type offer numerous opportunities of studying such effects as the influence steric interactions have on ion chemistry. However, the way forward clearly lies in the direction of quantitative experiments, i.e. the mass selection of individual $Ar_n.X^+$ clusters followed by the collision or laser-induced promotion of reactivity.

4. Acknowledgements

The author would like to thank the SERC for the provision of equipment grants which have made these experiments possible. In addition, thanks are due to A.K. Shukla, D. Bernard and N.G. Gotts who have undertaken much of the work described.

5. References

[1] A.J. Stace, in Mass Spectrometry, edited by M.E. Rose (The Royal Society of Chemistry Specialist Periodical Report, London, 1987) Vol. 9, p. 96.

[2] M.L. Alexander, N.E. Levinger, M.A. Johnson, D. Ray, and W.C. Lineberger, J. Chem. Phys., 88, 6200 (1988).

[3] A.J. Stace and A.K. Shukla, J. Phys. Chem., 86, 865 (1982).

[4] A.J. Stace, unpublished results.

[5] L.W. Sieck and P. Ausloos, Radiat Res., 52, 47 (1972).

[6] H. Shinohara, N. Nishi, and N. Washida, J. Chem. Phys., 84, 5561 (1986).

[7] A.K. Shukla and A.J. Stace, J. Phys. Chem., 92, 2579 (1988).

[8] S. Morgan and A.W. Castleman Jr, J. Am. Chem. Soc., 111, 3841 (1989).

[9] H.J. van der Linde and G.R. Freeman, J.Am. Chem. Soc., 92, 4417 (1970).

[10] G. Ronamowski and K.P. Wanczek, Int. J. Mass Spectrom. Ion Proc., 62, 277 (1984).

[11] S. Jaffe and F.S. Klein, Int. J. Mass spectrom. Ion Phys., 14, 458 (1974).

[12] M. Knapp, D. Kreisle, O. Echt, K. Sattler, and E. Recknagel, Surf. Sci., 156, 313 (1985).

[13] D.A. Parkes, J. chem. Soc. Faraday I, 68, 2121 (1972).

[14] M. Knapp. Oecht, D. Kreisle, T.D. Mark, and E. Recknagel, Chem. Phys. Lett., 126, 225 (1986).

[15] D.A. Parkes, J. chem. Soc. Faraday I, 68, 2103 (1972).

[16] D.M. Bernard and A.J. Stace, Int. J. Mass Spectrom. Ion Phys., 84, 215 (1988).
[17] N.G. Gotts and A.J. Stace, unpublished results.
[18] A.J. Stace, J. Am. Chem. Soc., 106, 4380 (1984).
[19] A.J. Stace, J. Am. Chem. Soc., 107, 755 (1985).
[20] A.J. Stace, J. Phys. Chem. 91, 1509 (1987).
[21] D.M. Bernard, N.G. Gotts, and A.J. Stace, Int. J. Mass Spectrum. Ion Proc., 95, 327 (1990).
[22] A.J. Stace and D.M. Bernard, Chem. Phys. Lett., 146, 531 (1988).
[23] A.J. Stace, J. Phys. Chem., 87, 2286 (1983).
[24] A. Ding, J.H. Futrell, R.A. Cassidy, L. Cordis, and J. Hesslich, Surf. Sci., 156, 282 (1985).
[25] B. Kamke, W. Kamke, Z. Wang, E. Ruhl, and B. Brutschy, 86, 2525 (1987).
[26] P.D. Dao and A.W. Castleman Jr., J. Chem. Phys., 84, 1435 (1986).
[27] N. Schwentner, E.E. Koch, and J.J. Jortner, in G. Hohler, editor, Electronic Excitations in Condensed Rare Gases,(Springer, Berlin, 1985) Vol. 107.

SEMICONDUCTOR CLUSTER ION REACTIONS AND ENERGETICS

SCOTT L. ANDERSON
Chemistry Department
State University of New York at Stony Brook
Stony Brook, NY 11794-3400 USA

ABSTRACT. This paper discusses our attempts to use ion beam techniques to unravel the energetics and mechanisms for reactions of boron cluster ions. Results from past work are summarized, and recent results on reactions with CO_2 and N_2O are presented.

1. Introduction

The major problem facing researchers in the area of covalently bound clusters is the nearly complete lack of any information on the geometric or electronic structure of the clusters. For atomic or molecular clusters bound by van der Waals forces, this is a less serious problem because there one is dealing with a collection of well characterized subunits (the atoms or molecules) which interact weakly enough that the clustering can be thought of as a perturbation on the properties of the isolated subunits. For clusters of many metal and semiconductor atoms, the bond energies are much higher (2-8 eV), and the physical and chemical properties of the clusters are not simply related to the properties of the bulk or atomic materials. In addition, since the number of bonds/atom (and thus the strength of the perturbation) changes with cluster size, the properties of the clusters also are strongly dependent on cluster size.

Most of our information on the structure of small molecules has come from various spectroscopies. So far, clusters have proven to be quite resistent to spectroscopic probing. For small clusters (dimers and trimers) there has been high resolution spectroscopy resulting in detailed structures. For larger clusters, to date the only technique giving state resolved information is photodetachment, which gives electronic and some vibrational information. No data with sufficient detail to allow structural determination has been reported. This lack of structural data makes study of cluster reactions especially interesting and frustrating. As will be seen below, it is common to observe oscillations in cluster properties with size and collision energy. These clearly have some relation to the changing (but unknown) structures of the clusters. In a perverse reversal of the usual situation, this means that we have to use the chemical behavior of the clusters to provide clues about their physical properties.

The lack of experimental structural information makes theory especially important in the cluster field, because calculations are really the only source (other than fevered imaginations) for structural information. For this reason, I believe that it is important for some experimental groups to concentrate their efforts on systems where it is reasonable to expect theoretical work to be <u>quantitative</u>. We have been concentrating on

K. R. Jennings (ed.), Fundamentals of Gas Phase Ion Chemistry, 117–130.

cluster ions of aluminum, carbon, and boron, which with only 3 valance electrons and 5 in total, is should be a good candidate for theoretical work. This paper will concentrate entirely on boron cluster ions.

For background it is useful to review some properties of bulk boron and boron rich materials[1]. Solid boron is a hard, brittle, refractory, wide band-gap semiconductor with high strength/weight ratio and interesting thermoelectric properties. Boron is a small "electron deficient" atom, and as a consequence, crystal structures of these materials are quite unusual. The simplest isomorph of boron (α-rhombahedral) for example consists of hollow B_{12} icosahedra, which are packed together to form a 3-d network. Other boron isomorphs, and boron rich materials such as boron carbide and metal borides, are more complex variations on this theme. The 12 atom hollow icosahedron is also known in stable compounds such as $B_{12}H_{12}^{=}$ and the carboranes ($B_{10}C_2H_{12}$), the ortho isomer of which was once produced in ton quantities as a prospective rocket fuel.

These unusual structures are held together by unusual bonds. Unlike two center bonds in which bonding electron density is found along the axis connecting pairs of atoms, in boron it is more common to have three center - two electron bonds. Three atoms forming a triangular face share a pair of electrons, and the bonding electron density is concentrated in middle of the face. Other multi-center arrangements include pentagons and hexagons of B atoms sharing electrons. These network structures are quite rigid, and to allow their formation, the B atom valence electrons are extensively hybridized.

Given this propensity for strong directional bonding, it is hard to predict *a priori* what the structures of the clusters might be like. For this reason we undertook SCF-CI calculations of the structures of B_n^+ (n=2-6)[2]. Due to lack of analytic gradients in our computer code, we did geometry searches by starting for each size cluster with all symmetric structures, optimizing them for symmetric displacements, them doing some local non-symmetric optimization. Especially for the larger clusters, this scheme is inefficient, and increases the probability that the global minimum is not found. Nonetheless, the results are worth examining since they are the only ones available and give a reasonable idea of the type of structures which we are dealing with: B_3^+ - equilateral triangle; B_4^+ - rhombus; B_5^+ - trigonal bipyramid;, B_6^+ - pentagonal prism. These are clearly not closest-packed structures, and have as a theme the presence of a large number of multi-center bonds. Most of these involve 3 atoms making up triangular faces, which is a pattern reminiscent of bulk boron bonding arrangements. We speculate that the structures of the larger cluster ions continue this theme, building bigger networks with high coordination and stability. Fortunately, several theoretical groups have begun to do more elaborate calculations, and presumably accurate structures will result.

2. Experimental Setup.

Most of our work is experimental, and we have two cluster beam instruments. Figure 1 shows the first, which has been described in detail previously[3]. Clusters

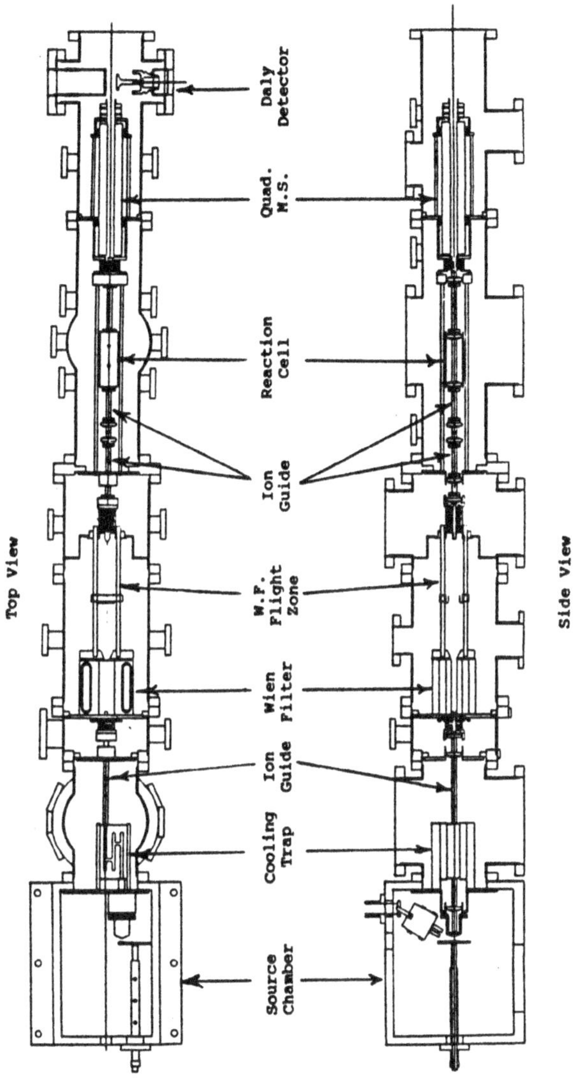

Fig. 1. The Smaller Stony Brook cluster beam machine. Total path length is 2.3 meters.

are made by either fast atom bombardment (for metals) or laser ablation (boron, carbon), inside a radio-frequency ion trap that helps collect the ejected cluster ions. Because we don't want to work with the internally hot nascent clusters, we first cool them off by storage in a special radio-frequency cooling trap. This trap is constructed from a set of stacked 2mm thick stainless steel plates, which have a slot machined through them in the form of a maze. This creates a labyrinthine chamber where ions can be stored. Radio frequency fields applied to the walls and DC potentials on the top and bottom plates, prevent low energy ions from hitting the trap, and once injected, the only way for the clusters to escape the trap is to diffuse through the maze, which is filled with a buffer gas (helium or N_2). The average storage time is about 100 ms, during which time the cluster ions undergo tens of thousands of collisions and equilibrate (we assume!) at the buffer gas temperature, which for these experiments is 350K.

The cooled clusters are passed through a Wien filter which allows us to select a particular size and composition to study, then injected into a set of octapole ion guides. The desired collision energy is set, then the ions are guided through a scattering cell containing a target molecule. The products and unreacted cluster ions are then collected, mass analyzed, and counted. The experiment consists of measuring the integral cross sections for all product ions, as a function of collision energy and reactant cluster size and composition.

3. Previous results.

The first set of experiments we perform on any new cluster system is collision-induced-dissociation (CID). We measure appearance potentials and integral cross sections for all the observed fragment ions, and the results for boron are summarized in figure 2. The line plot gives the stabilities (lowest energy fragment appearance potentials) for the cluster ions as a function of cluster size. The bars give the partial and total cross sections for 10 eV collision energy, which is well above the fragmentation thresholds. This data directly gives semiquantitative information about the cluster stabilities, and indirect evidence regarding the cluster structures and ionization potentials.

As the figure shows, the stabilities of the clusters generally increase with increasing cluster size, though there are fluctuations of almost 2 eV. This increase is consistent with our proposed 3-dimensional network structures for the clusters. If the structures where chains, or rings (ie. topologies where the number of bonds/atom does not increase with size) then we would expect to see little increase in stability past the 5 or 6 atom size range. The fluctuations in stability do not fit any kind of shell model, however, they are consistent with the changes in average coordination number for the icosahedral fragment structures we have proposed[2]. The CID cross sections are more or less constant as cluster size increases, but the fragment ion branching ratios vary considerably. For the small clusters we observe mostly B^+ production, while for the larger ones B^+_{n-1} is the dominant fragment. It appears that the dominant fragmentation pathway is ejection of a single B atom, and the branching simply reflects whether the

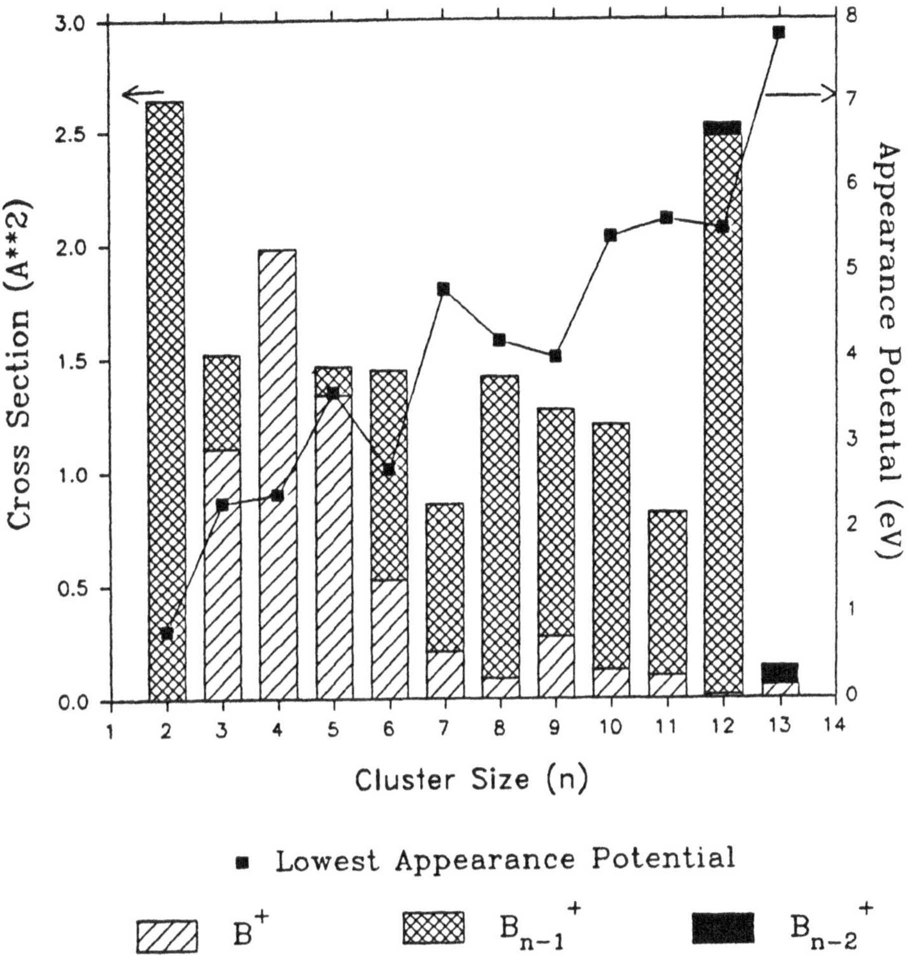

Fig. 2. Appearance Potentials and Fragment Branching Ratios for CID of Boron Cluster Ions.

charge ends up on the B atom or the B_{n-1} fragment. This pattern is also consistent with 3-d network structures. If chain or ring structures were important, we would expect to see more fragments from fission of the parent cluster ions.

A final bit of information we can infer from the both the branching ratios and the thresholds for production of B^+/B^+_{n-1} is that the ionization potentials for B_n vary with n. For 5 atoms and smaller, the IP of B_n is significantly higher that for B (8.298 eV), while for the larger clusters, the IP slowly drops below that of B. This information is useful in interpreting the product branching observed in chemical reactions, as well as providing some evidence regarding the electronic structure of the clusters.

To date we have studied chemistry of boron cluster ions with a number of small molecules[4] with an emphasis on oxidation chemistry. For this paper I will use our recent studies of the reactions with CO_2 and N_2O as examples. Lets begin with CO_2 since the chemistry is simpler. For reference note the following thermochemistry: D_0(OC-O) = 5.45 eV, D_0(C-O) = 11.09 eV, D(B_n^+-O) = 4-9 eV. (The energies for binding O to B_n^+ depend on cluster size and are estimated from our results on various boron cluster ion oxidation reactions).

4. Reaction of B_n^+ with CO_2.

Figure 3 shows cross sections for the major product channels in reaction of B_n^+, n = 6, 8, and 12 with CO_2. B_6^+ is typical of small boron cluster ions, B_{12}^+ is a typical large cluster, and B_8^+ is a transitional cluster. The major low collision energy product for all size reactant cluster ions is production of B_nO^+ via the reaction:

$$B_n^+ + CO_2 \quad \text{-->} \; B_nO^+ + CO.$$

There are also a variety of other channels (not shown) which though minor, are important in that they give some insight into the mechanism. For example we observe the reaction:

$$B_n^+ + CO_2 \quad \text{-->} \; B_{n-1}CO^+ + BO.$$

This is consistent with a mechanism for interaction of B_n^+ with CO_2 involving a complex of the form $[OB_nCO]^+$. The lowest energy decomposition channel for this species should be CO elimination (giving the B_nO^+ product) however BO elimination (giving $B_{n-1}CO^+$) should also be a low energy process, assuming that CO has a reasonable binding energy to the cluster ion. Further discussion of BO elimination is given below.

At high collision energies we observe B_{n-1}^+ production. (This channel is not plotted for B_6^+ because it does not show up when plotted on a 120Å^2 scale). There are two contributions to B_{n-1}^+ production:

$$B_n^+ + CO_2 \quad \text{-->} \; B_nO^+ + CO \quad \text{-->} \; B_{n-1}^+ + BO + CO,$$

and

$$B_n^+ + CO_2 \quad \text{-->} \; B_{n-1}^+ + B + CO_2.$$

As shown below, BO elimination from the B_nO^+ product is relatively facile, and we also know that at high energies simple CID becomes important[2].

Very little evidence is seen for reactions involving complete dissociation of CO_2 when it interacts with the cluster. One reaction of this type is:

$$B_n^+ + CO_2 \quad \text{-->} \; B_{n-2}C^+ + B_2O_2,$$

which can be seen in the plot for B_{12}^+. The reaction is quite inefficient, and is insignificant for reactant clusters which form B_nO^+ + CO efficiently.

Also observed is $B_nO_2^+$ which would be consistent with complete dissociation of CO_2 on the cluster. Pressure dependent studies suggest however that this channel is in fact the result of a sequential reaction:

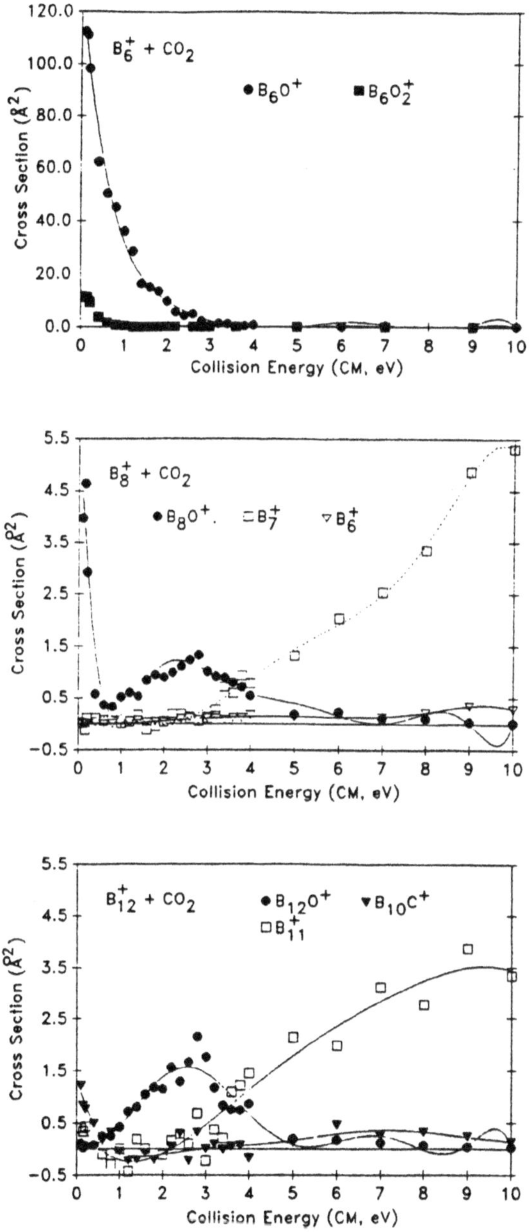

Fig. 3. Cross sections for the major product channels in reaction of a few sizes of B_n^+ with CO_2

$$B_n^+ + CO_2 \quad \text{-->} \; B_nO^+ + CO$$
$$B_nO^+ + CO_2 \quad \text{-->} \; B_nO_2^+ + CO.$$

In fact if we increase the scattering cell pressure to 10^{-3} Torr (20 times our normal pressure) we can drive this sequence of reactions for many steps. For example, for B_6^+ we observe $B_6O_x^+$ species containing up to seven O atoms. This raises two interesting points. The fact that we see such a long sequence of reactions means that each step in the sequence must be exoergic with no activation barriers. This is in spite of the change in composition from pure boron to boron oxide. In addition, the fact that we see very little fragmentation of the cluster during the sequence of oxidation reactions suggests that the energy deposited in the cluster-oxide product ion is less than about 1 eV, otherwise we would begin to see fragmentation (see below). In sum, it appears that for reactant clusters for which first step of the oxidation sequence (B_nO^+ production) is exoergic, subsequent steps are also exoergic, but not very.

The cross section for oxidation changes dramatically with reactant cluster ion size as shown in figure 4. As shown, the major low energy process is B_nO^+ production, which occurs with 30-100% probability for n= 2-7 and 9, but in only a few percent of collisions for n=8, and 10-14. The drop off in reactivity is due to energetics. Evidently the reaction becomes endoergic for the larger cluster ions, as might be expected from the estimated thermochemistry given above. For the larger clusters, thresholds to oxidation are observed. This decrease in reactivity with cluster ion size is something that we have observed for reaction with many neutral molecules. Our proposed explanation is that as the cluster size increases, for clusters with 3-d structures, the number of bonds/atom increases. This as we have seen, makes the cluster ions more stable as they increase in size, but also leads to coordinative saturation, and reduced reactivity.

B_9^+ is special in that it behaves more like a small cluster than a large one. This behavior is also seen in other reactions, as shown below. B_8^+ is also unusual in that its B_8O^+ formation cross section shows both a high energy component with threshold (typical of big clusters) and also a small component peaking at low energy (typical of small clusters). This bimodal behavior could be due to isomers of B_8^+ in the beam, or less plausibly it could indicate that B_8^+ different O atom biding sites.

5. CID of B_nO^+.

Obviously if we are to really understand the oxidation reaction mechanisms for boron cluster ions, it will be essential to gain some insight into the properties of the clusters and the ways they interact with ad-atoms. Hopefully theoretical work will eventually lead to detailed understanding of these problems, but meanwhile, we are left to devise other experiments to get at the desired information. One obvious approach is to attempt to learn something about the oxidation products -- the B_nO^+.

As a first attempt we have carried out collision induced dissociation (CID) studies of selected monoxide cluster ions. These can be prepared by simply adding some CO_2 to the nitrogen buffer gas in the cooling trap. A few percent of the boron cluster ions

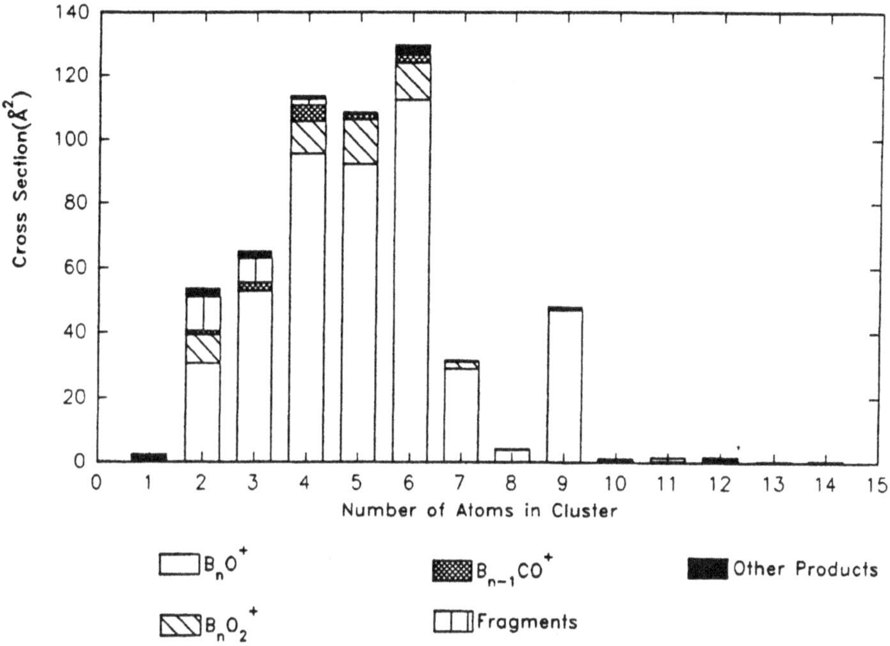

Fig. 4. Size Effects on $B_n^+ + CO_2$ reaction cross sections at 0.1 eV collision energy. Each segment represents a product channel, and the total height of each bar equals the total cross section.

are converted to B_nO^+, which can then be selected in the Wien filter and collided in the scattering cell. Because of the cluster size dependence of the CO_2 reaction, only a limited size range (n = 5-9) of oxide cluster ions can be prepared in this way, though presumably we can make others by using different oxidizing gases. Figure 5 shows the CID cross sections for two representative oxides -- B_6O^+ and B_7O^+.

For B_6O^+ the dominant fragmentation channel is loss of BO to form B_5^+, which has a threshold of only about 1.3 eV, and a relatively large cross section. Other fragmentation channels such as O loss or B^+ loss (which is the dominant CID channel

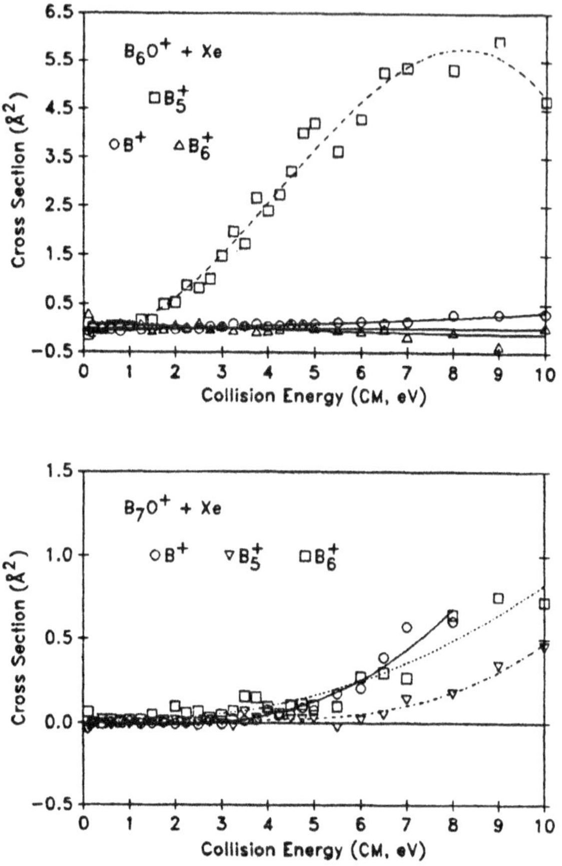

Fig. 5. CID cross sections for B_nO^+ (n = 6 and 7).

for B_6^+) are negligible even at high collision energies. This suggests that B_6O^+ might have a structure in which a BO subunit is loosely bonded to the rest of the cluster ion. Similar behavior is observed for B_8O^+ and B_9O^+

For B_7O^+, the story is quite different. The CID cross sections are much smaller, the thresholds are much higher, and three fragmentation channels are significant: B^+ loss, BO loss, and B_2O loss. The thresholds for BO and B^+ loss are comparable, and are similar to the CID thresholds observed for the B_7^+ cluster ion. Similar behavior is observed for B_5O^+, suggesting that these two oxides have structures where the O atom is strongly bound to the rest of the cluster.

B_8O^+ is unusual in that in addition to a facile BO loss channel, it shows a tendency to fission at low energies into 3 and 5 boron atom fragments, where the O atom and the charge can end up on either the B_3 or B_5 half. This behavior is suggestive of there being two isomers of B_8O^+, which in turn is consistent with our observation of bimodal reactivity of B_8^+ with CO_2.

6. Reaction of B_n^+ with N_2O.

In principle, oxidation of boron cluster ions by N_2O should be quite similar to what we have observed with CO_2. The two oxidizers are similar in many ways. Removal of one oxygen atom is relatively easy, while decomposition of the remaining CO or N_2 molecule requires much more energy. We might also expect that CO and N_2 (which are isoelectronic) should interact with boron cluster ions in a similar fashion, ie. that they should adsorb molecularly. Note that N_2 does not react significantly even with hot boron cluster ions, which is why we can use it as the buffer gas in the cooling trap. The major difference between N_2O and CO_2 would appear to be the energy for removal of the first O atom. This is only 1.7 eV for N_2O, and thus the reaction to form B_nO^+ should be more exoergic, and we might expect to see significant fragmentation of the oxide products, especially for those cluster sizes where we have observed facile elimination of BO from the oxide.

Figure 6 shows reaction cross sections for typical small, medium, and large size boron cluster ions with N_2O. The first surprise is that the dominant reaction for small clusters is:

$$B_n^+ + N_2O \quad \text{-->} \ B_nN^+ + NO,$$

rather than the expected O atom transfer reaction. Nitride formation dominates for B_n^+ (n = 2 - 6, 9). For the larger clusters (n = 7, 8, 10-15) we see predominantly B_nO^+ + N_2 formation as we expected, and the branching to B_nN^+ is negligible. Finally, we have preliminary data for reaction with even larger clusters (n = 16 - 24) where we observe nearly equal branching to oxide and nitride products.

As expected, the reaction with N_2O is more exoergic than for CO_2. This is seen in figure 7 which plots the cross sections for all product channels at 0.1 eV collision energy. Note that the total cross sections are essentially at the collision limit for small clusters, then drop off at B_{12}^+ (compared to B_8^+ for CO_2). Another indication of substantial exoergicity is that in sequential reactions at higher scattering cell pressure, fragmentation of the cluster framework (which does not occur for CO_2) is an important process. For the larger clusters, we observe thresholds for both oxide and nitride formation which suggests that there are probably activation barriers, since we estimate that the B_n^+-O bond energies should still be large enough to make the oxidation reaction exoergic.

Figure 7 also shows how the product branching changes with cluster size. Note the anomalous behavior of B_9^+, where nitride formation is the major channel. The origin of the propensity to form nitrides vs. oxides is unknown, but presumably is related to the cluster structures. One possible explanation is that in order to break the strong N-N bond

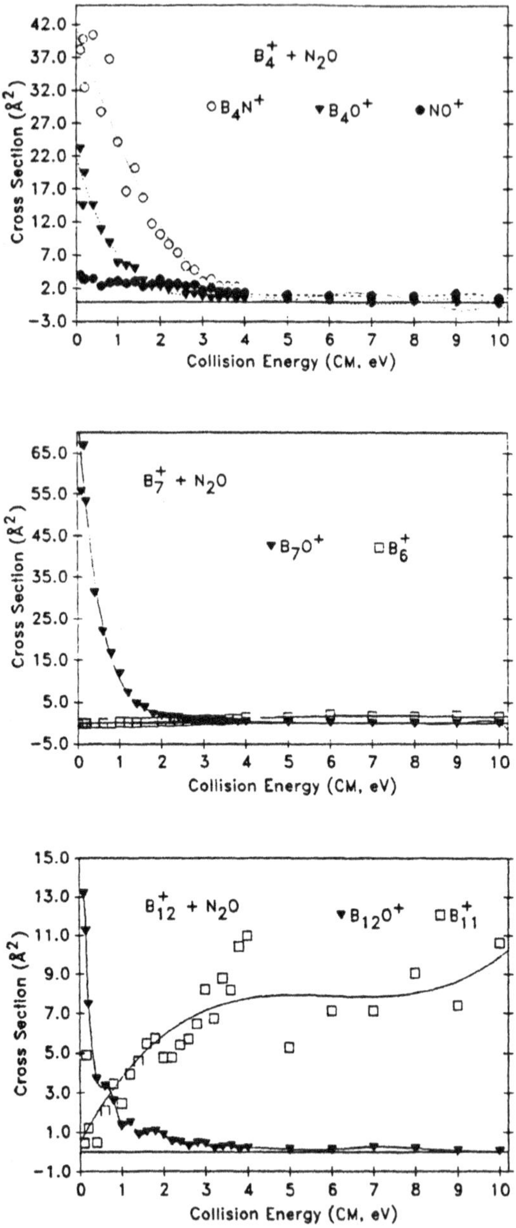

Fig. 6. Cross sections for reaction of typical B_n^+ with N_2O.

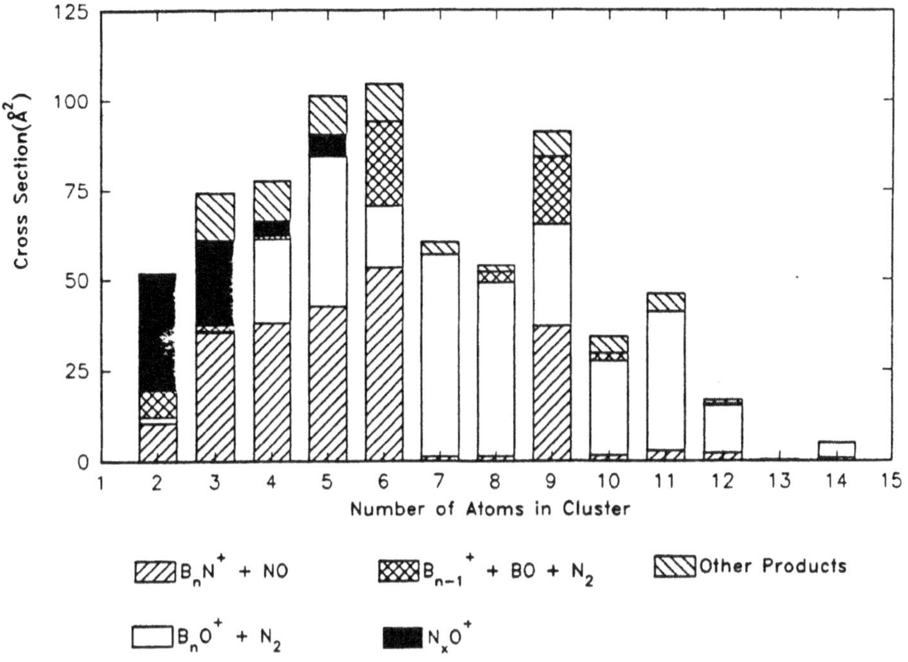

Fig. 7. Cross section for reaction of B_n^+ with N_2O at 0.1 eV collision energy as a function of cluster size.

to form nitrides it is necessary to form B_nN^+ where the N atom is strongly bound to three or more B atoms. For the small clusters which are not coordinatively saturated, this may be relatively easy to achieve. For cluster ions in the range from 7 - 13 atoms, where little nitride formation is observed, we have proposed rigid, high coordination structures building towards a complete, filled icosahedron at B_{13}^{+}. Finally for clusters in the 14 - 24 atom range, there may be regions of the structures where coordination is low allowing some nitride production.

7. Conclusions.

I have attempted to give an overview, with a few specific examples, of our efforts to experimentally work out the physical and chemical behavior of boron cluster ions. We are able to obtain some semi-quantitative information (stabilities, ad-atom binding energies, and ionization energies), and can work out the gross reaction mechanisms with some degree of confidence. What is lacking at this point is the detailed structural information that we need to get a more complete and secure handle on this problem. Currently we are working on some experimental approaches to structure (vibrational spectroscopy), and we hope that our results will stimulate the necessary theory to unravel at least some of this complex problem. Together, I believe that a judicious combination of theory and experiment could turn the boron cluster ion problem into a benchmark system for calculations on clusters.

References

1. see for example: Adams, R. M., ed. (1964) Boron, Metallo-boron Compounds, and Boranes, Interscience, New York; Emin, D., Aselage, T. L., and Wood, C. (1987) Novel Refractory Semiconductors, Materials Research Society, Pittsburgh.

2. Hanley, L., Whitten, J. L., and Anderson, S. L. (1988) "Collision Induced Dissociation and Ab Initio Studies of Boron Cluster Ions: Determination of Structures and Stabilities", J. Phys. Chem. 92 5803-5812.

3. Hanley, L., Ruatta, S. A., and Anderson, S. L. (1987) "Collision Induced Dissociation of Aluminum Cluster Ions: Fragmentation Patterns, Bond Energies, and Structures for Al_2^+ - Al_7^+", J. Chem. Phys. 87 260-268.

4. Hanley, L. and Anderson, S.L. (1988) "Oxidation of Small Boron Cluster Ions B_{1-13}^+ by Oxygen", J. Chem. Phys. 89 2848-2860; Ruatta, S. A., Hanley, L., and Anderson, S. L. (1989) "Dynamics of Boron Cluster Ion Reactions with Deuterium: Adduct Formation and Decay", J. Chem. Phys. 91 226-39; Hintz, P. A., Ruatta, S. A., and Anderson, S. L. (1990) "Interaction of Boron Cluster Ions with Water: Single Collision Dynamics and Sequential Etching", J. Chem. Phys. 92 292-303.

GAS-PHASE METAL ION CHEMISTRY: SUMMARY OF THE PANEL DISCUSSION

Robert R. Squires
Department of Chemistry
Purdue University
West Lafayette, IN 47907

Since the last NATO meeting in Les Arcs, gas phase metal ion chemistry has continued to attract new investigators and new experimental techniques. Especially notable are the use of specialized ion sources for investigating the reactions of metalloporphyrins and other low-volatility organometallic compounds, further development of the use of kinetic energy release measurements for studying the energetics and dynamics of metal ion reactions, new methods for investigating state-selected metal ion reactions such as the "metal ion chromatography" technique just announced by Kemper and Bowers, new mechanistic information about metal ion reactions obtained through multiple isotope-labelling experiments, and the determination of thermochemical data for atomic metal ion clusters and ligated metal ion complexes by CID threshold energy measurements. By all accounts, as summarized in the following, gas phase metal ion chemistry is alive and well!

Metal-Carbonyl Bond Strengths in Anionic and Neutral Transition Metal Carbonyl Complexes (Robert R. Squires and Dingneng Wang, Department of Chemistry, Purdue University, West Lafayette, IN 47907, USA)

Reliable measurements of the sequential M—CO bond energies in mononuclear binary metal carbonyl complexes are highly desirable for both fundamental and practical reasons.[1] We have undertaken a program of measurement of these quantities for simple anionic transition metal carbonyl complexes using a flowing afterglow-triple quadrupole instrument.[2] The method relies on the precise determination of the translational energy onset for reaction (1) with a

$$M(CO)_n^- \longrightarrow M(CO)^-_{n-1} + CO \quad (1)$$

thermalized metal carbonylate ion undergoing collisional activation in a triple quadrupole mass analyzer. In such cases where it can be shown that the dissociation has no reverse activation energy nor a kinetic shift, the derived threshold will be equivalent to the M—CO bond energy of the ion.

We illustrate our method here with results for the $Fe(CO)_4^-$ and $Fe(CO)_3^-$ ions that can be made in abundance in the flowing afterglow by electron ionization of $Fe(CO)_5$. Figure 1 shows an appearance plot for

K. R. Jennings (ed.), Fundamentals of Gas Phase Ion Chemistry, 131–146.

the $Fe(CO)_3^-$ fragment from $Fe(CO)_4^-$ undergoing energy-resolved CID in the triple quadrupole collision cell with argon target gas at $5x10^{-5}$ torr. The solid line is a fit of the steeply rising portion of the curve with a linear function which has been

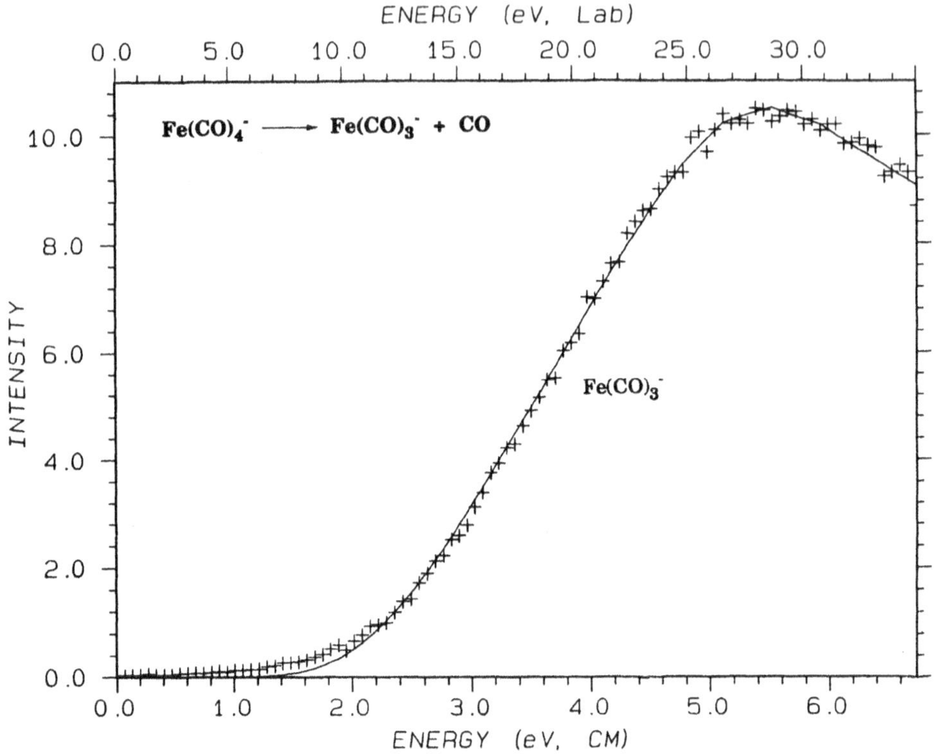

convoluted with Doppler broadening (298 K) and reactant ion beam (2 eV fwhm, lab frame) distributions according to published procedures[3]. For this data set a threshold energy of 1.5 eV is obtained; replicate measurements produce an average value of 1.65 ± 0.17 eV (38 ± 4 kcal/mol). Similar analysis of the appearance plots for $Fe(CO)_2^-$ from $Fe(CO)_3^-$ produces an average threshold energy of 42 ± 4 kcal/mol. The observation of facile ^{13}CO/CO exchange reactions with $Fe(CO)_3^-$ and other unsaturated metal carbonylates under thermal conditions supports the proposal that the M–CO dissociation thresholds are close to the thermodynamic bond energies.

The Fe–CO bond energies in $Fe(CO)_n^-$ ions have been previously estimated from negative ion appearance energy measurements with $Fe(CO)_5$.[4] A particularly disturbing feature of these earlier estimates is the exceptionally low value for D[$Fe(CO)_3^-$–CO] = 18 kcal/mol that was assigned from the energy difference between the appearance curves for $Fe(CO)_4^-$ and $Fe(CO)_3^-$.[5] However, it should be noted that from independent measurements of D[$Fe(CO)_4$–CO] = 42 ± 2 kcal/mol[6] and EA($Fe(CO)_4$) = 55 ± 7 kcal/mol[5], it is known that dissociative electron capture by $Fe(CO)_5$ is an exothermic process (eq 2).

$$Fe(CO)_5 + e^- \longrightarrow Fe(CO)_4^- + CO + 13\ kcal/mol \quad (2)$$

Therefore, the appearance curve for $Fe(CO)_4^-$ cannot be used to assign the $(CO)_3Fe^-$–CO bond energy, and the earlier estimate must be incorrect. Our new measurements reconcile the discrepancy. The two bond energies $D[Fe^- -CO] = 46$ kcal/mol and $D[Fe(CO)^- -CO] = 23$ kcal/mol from the earlier appearance energy measurements should be reliable.[4,5] Combining these quantities with the present measurements for $D[Fe(CO)_2^- -CO]$ and $D[Fe(CO)_3^- -CO]$ gives a total (149 kcal/mol that is essentially equal to the known bond disruption enthalpy ΔH_{dis} for $Fe(CO)_4^-$, $D[Fe^- -4CO] = 150$ kcal/mol.

Binding Energies and Clustering Kinetics; Co^+ with CH_4 and C_2H_6 (Petra A.M. van Koppen, Paul R. Kemper, Gert von Helden and Michael T. Bowers, Department of Chemistry, University of California, Santa Barbara, Santa Barbara, CA 93106, USA)

The apparent second order rate coefficient for clustering between

$$Co^+ + CH_4 \xrightarrow{k_{app}^{(2)}} Co^+ \cdot CH_4 \quad (1)$$

cobalt ions and CH_4 (or C_2H_6) in He was studied both as a function of temperature (from 250 K to 450 K) and as a function of He pressure (from 1 to 5 torr). The Co^+ ions were formed by electron impact on $C_5H_5Co(CO)_2$, mass selected, decelerated and injected into a high pressure drift cell.[7] Pressures in the cell were typically 1×10^{-3} torr CH_4 (or C_2H_6) and 1-5 torr He. Ions exiting the cell were quadrupole mass analyzed. Both ground and excited state Co^+ are produced in the source, however only the ground state clusters with CH_4 (and C_2H_6)[8] The fraction of ground state Co^+ was measured[9] and used in deriving the rate coefficients. The following reaction mechanism was

$$Co^+ + CH_4 \underset{k_b}{\overset{k_f}{\rightleftarrows}} (Co^+ \cdot CH_4)^* \xrightarrow{\beta k_s[He]} Co^+ \cdot CH_4, \quad k_{app}^{(2)} = \frac{k_f \beta k_s[He]}{k_b + \beta k_s[He]} \quad (2,3)$$

assumed where k_f is the collision rate (Langevin) between Co^+ and CH_4, k_b is the unimolecular back dissociation rate of the excited complex, k_s is the collision rate between $Co \cdot CH_4^+$ and He, and β is the stabilization efficiency ($0 \le \beta \le 1$). The variation of $k_{app}^{(2)}$ with He at different temperatures is shown in Figures 1 and 2. The previous measurement of Tonkyn et al.[10] is also shown as the open circle.

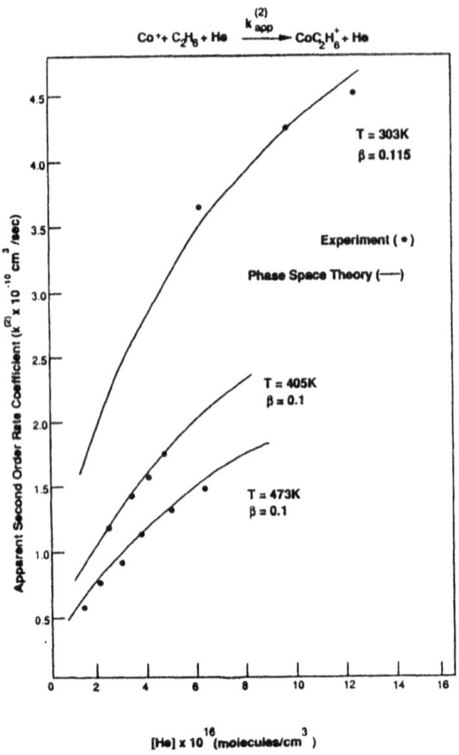

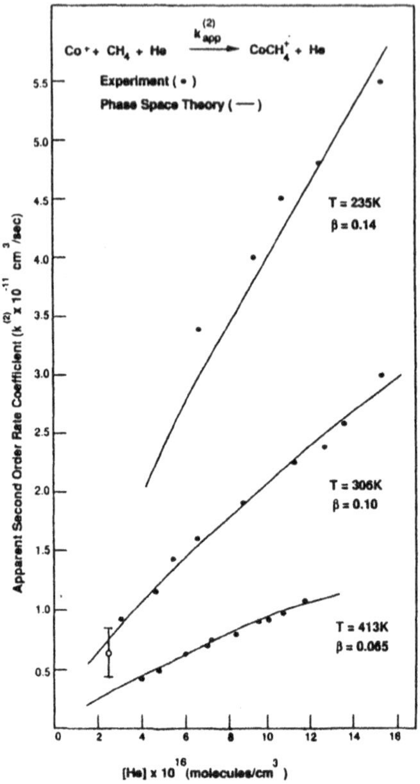

The curvature in the lines is due to the βk_s[He] term in the denominator of equation 3. At low pressures the third order rate coefficient is given by

$$k^{(3)} = \frac{k_f \beta k_s}{k_b} \tag{4}$$

Third order rates were measured to be

Co^+/CH_4		Co^+/C_2H_6	
T(K)	$k^{(3)}$ (cm^6/sec)	T(K)	$k^{(3)}$ (cm^6/sec)
234	2.4×10^{-28}	405	2.5×10^{-27}
306	1.8×10^{-28}	473	2.3×10^{-27}
413	0.79×10^{-28}		

A lower limit on the Co^+–CH_4 bond energy can be obtained by calculating K_p (and ΔG^o) from the highest temperature data. (Equilibrium is not established, thus the true ΔG^o will be more negative). If ΔS^o is taken to be -15 eu (in agreement with other similar systems[11]) then the bond energy is > 15.5 kcal/mol for Co^+/CH_4 and > 22.6 kcal/mol for Co^+/C_2H_6. The Co^+–H_2O bond energy is known to be 40.1 kcal/mol[12] and, therefore, H_2O would displace CH_4 in $Co^+ \cdot CH_4$.

Figures 1 and 2 show the surprising result that $k_{app}^{(2)}$ is >0 at [He] = 0. This might be due to stabilization of the complex by CH_4 in addition to He. This was ruled out by measuring $k_{app}^{(2)}$ as a function of CH_4 pressure from ≈$3x10^{-4}$ to $3.2x10^{-3}$ torr at 306 K. The change in $k_{app}^{(2)}$ was too small to measure (<$2x10^{-13}$ cm^3/s), and far too small to account for the observed intercept (≈$2.3x10^{-12}$ cm^3/s at 306 K). Radiative stabilization was also shown not to be a viable explanation[13]. The origin of the non-zero intercept has yet to be determined.

Modeling was done using a phase space approach in which energy and angular momentum are rigorously conserved. Physically reasonable values of complex vibrational frequencies, bond distance and binding energy are used to calculate a $k_{app}^{(2)}$ which is averaged over the thermal E,J distribution of the reactants. The binding energy is adjusted to give the best fit with experiment. The other parameters are varied to establish error limits on the calculated ΔH. A small non-zero intercept for $k_{app}^{(2)}$ was assumed in order to best fit the data. The calculated bond energies were 20.5 ± 2 kcal/mol for Co^+-CH_4 and 31.5 ± 2 kcal/mol for Co^+-C_2H_6.

A stabilization efficiency (β) of 0.1 at 300 K was used (in agreement with previous studies). A significant negative temperature dependence was found for β in the CH_4 reaction (β = 0.065 at 413K and 0.14 at 235 K). The effect in the C_2H_6 reaction was smaller (β = 0.095 at 473 K and 0.115 at 303K).

Finally, insertion of Co^+ into the C–H bond of CH_4 is known to be endothermic and therefore does not occur at thermal energies. Consideration of the bond energies derived in this study makes C–H bond activation very unlikely for C_2H_6. This is consistent with the Co^+/H_2O and Co^+/NH_3 systems where no insertion was found.[12]

Unimolecular Decomposition of $Fe^+ \cdot C_3H_8$: Probing Potential Energy Surfaces. (Richard H., Schultz and P. B. Armentrout, Department of Chemistry, University of Utah, Salt Lake City, UT 84112, USA)

The bimolecular reaction of Fe^+ with propane, reaction 1, has been the subject of considerable research during the past decade, spanning

$$Fe^+ + C_3H_8 \longrightarrow \text{products} \qquad (1)$$

a wide variety of gas-phase experimental techniques.[14,15] Our work on this system[15] has revealed surprisingly rich behavior, especially in the exothermic channels, loss of H_2 and CH_4. We have recently developed a new way of probing the potential energy surface (PES) of this reaction: observation of the energy dependence of the unimolecular decomposition of the reaction's presumed intermediate, an ion-induced dipole bound complex of the ion and neutral. We make this observation by energetically exciting the complex by collision with an "unreactive" collision partner such as a rare gas, reaction 2.

$$Fe^{+}\cdot C_3H_8 + Xe \longrightarrow \text{products} \quad (2)$$

The adduct is formed in a meter-long flow tube by termolecular collisions between Fe^{+} (created in a DC discharge), propane, and the He carrier gas. The complexes then undergo on the order of 10^5 further thermalizing collisions. We tested whether the ions were thermalized by observing their reaction with C_3D_8. Since exothermic ligand exchange was not seen, we infer that the $Fe^{+}\cdot C_3H_8$ ions are thermalized.

The results of the collisional activation experiment are shown in Figure 1. The major product is collision-induced dissociation (CID) to form Fe^{+} and C_3H_8. The two products that are exothermic in reaction 1, $FeC_2H_4^{+}$ and $FeC_3H_6^{+}$, are observed at low energy in reaction 2 as well. These two channels have thresholds below and rise faster than the onset for CID, consistent with their being exothermic in reaction 1 and arising without a barrier in that reaction. All of the endothermic products from reaction 1 (except for $C_3H_7^{+}$) are seen as endothermic products from reaction 2 as well, with thresholds consistent with the known thermochemistry. In addition, ligand exchange to form $FeXe^{+}$ is seen as well. This product begins decomposing at the threshold for CID.

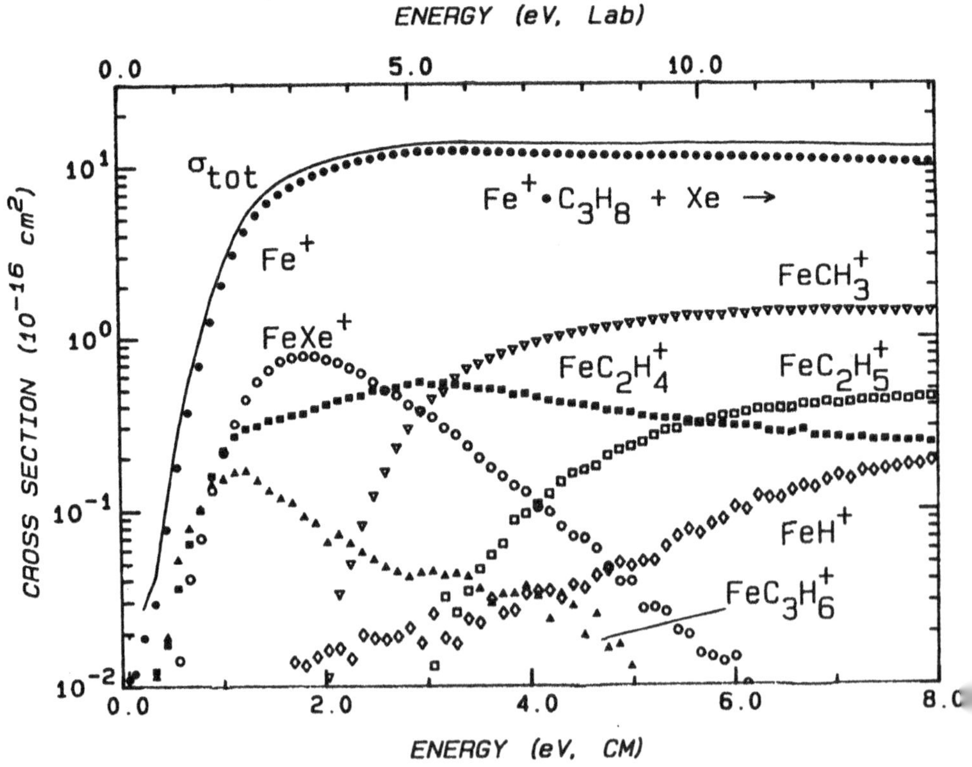

These data can now be used to give thermodynamic information about the PES directly. The depth of the initial ion-induced dipole well is determined directly from the CID threshold of 0.85 eV (20 kcal/mol). The threshold for formation of $FeC_2H_4^+$, 0.65 eV (15 kcal/mol), gives the height of the barrier between the ion-induced dipole complex and the product. This intermediate barrier thus lies 5 kcal/mol lower in energy than the separated reactants. Similar results were obtained by completely different techniques for the related Co^+ + propane system.[16] The energy of the final product ($FeC_2H_4^+$ + CH_4), 0.9 eV (22 kcal/mol) below the separated reactants, was determined by CID on the $FeC_2H_4^+$ ion.[17] We are currently attempting[17] to produce the C–C and C–H bond insertion intermediates directly and measure their thermochemistry, as well as determine whether or not barriers from these intermediates to the final products exist.

This method shows great promise as a means of elucidating the thermodynamics of the PES. In addition to this experiment, further studies are underway[17] of the reverse reactions to reaction 1, namely $FeC_2H_4^+$ + CH_4 and $FeC_3H_6^+$ + H_2, to gain further insight into this reaction system.

The Chemistry of 17 Electron Anionic Complexes of Ni (H. L. Chen, Y. H. Pan and D. P. Ridge, Department of Chemistry and Biochemistry and Center of Catalytic Technology, University of Delaware, Newark, DE 19716 USA)

Anionic metal complexes with 17 electrons are important in a variety of contexts including the electrochemistry of metal carbonyls and biochemical charge transfer catalysts. We have recently examined the chemistry of $Cr(CO)_5^-$, $Fe(CO)_4^-$ and $Ni(CO)_3^-$ and several Ni complexes of tetradentate ligands.[18,19] We comment here on several features of the Ni chemistry that reflect interesting aspects of charge transfer between ligand and metal.

The reaction of $Ni(CO)_3^-$ with electrophilic ligands gives either electron transfer or electron transfer mediated ligand substitution:

$$Ni(CO)_3^- + L \longrightarrow \underset{I}{[Ni(CO)_3^- \cdots L} \longrightarrow \underset{II}{Ni(CO)_3 \cdots L^-]} \qquad (1)$$

$$\longrightarrow Ni(CO)_2L^- + CO$$

$$\longrightarrow L^- + Ni(CO)_3$$

The electrostatically bound complex I has a 17 electron metal center, but II has a 16 electron metal center. Ligand substitution through II obviates an unfavorable 19 electron intermediate. Evidence for this mechanism comes from the correlation between the rate of ligand substitution and the electron affinities (EA) of the ligand illustrated in Fig. 1. Below a particular electron affinity the rate constant drops dramatically suggesting that the ligand must be sufficiently electronegative to convert I to II in order for ligand substitution to occur.

The reaction of benzophenone proceeds further with loss of a second CO:

$$Ni(CO)_3^- + (C_6H_6)_2{}^*CO \xrightarrow[-CO]{} \underset{\text{III}}{Ni(CO)_2(C_6H_5)({}^*COC_6H_5)^-}$$

$$\xrightarrow[-CO]{M} Ni(CO)({}^*CO)(C_6H_5)_2^- \qquad (2)$$

The loss of the second CO requires a thermal collision with $Ni(CO)_4$ suggesting the oxidative addition mechanism outlined in (2). The initial product III has a 17 electron Ni and moving the second phenyl group to the metal would give a unfavorable 19 electron species. The electrostatic energy of attraction to a collision partner, however, can provide energy to loosen a CO so that the phenyl group can shift to the metal displacing the CO completely. The retention of the labelled carbon as indicated in (2) is consistent with this mechanism.

Finally we consider the electron affinities of the species IV, V and Ni(tetraphenylporphyrin) as determined by measuring equilibrium constants for charge transfer reactions. The EA of IV is 1.06 eV, nearly the same as that of $Ni(CO)_3$ (1.08 eV). $Ni(CO)_3$, $Ni(CO)_3^-$, IV and the anion of IV are all planar and both anions have 17 metal valence electrons. These factors seem to be more important in determining the electron affinity than is the oxidation state. The system of conjugated double bonds in V has a marked effect on its electron affinity which we found to be 1.37 eV, significantly greater than that of IV. The importance of delocalization from the metal to the ligand is evident. The electron affinity of Ni(tetraphenylporphyrin) is 1.51 eV, suggesting delocalization of charge in the anion just as in the anion of IV. We believe this to be the first determination of the electron affinity of a metal porphyrin.

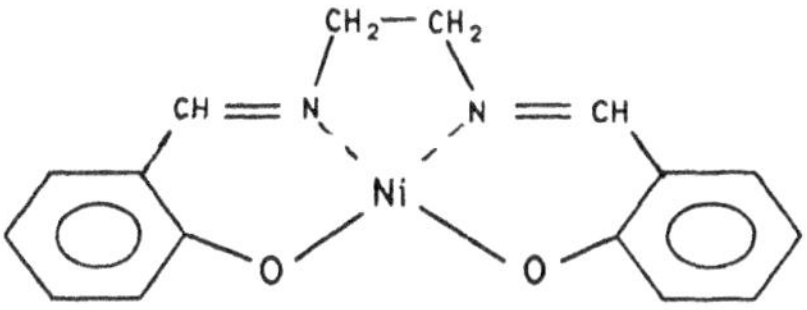

IV

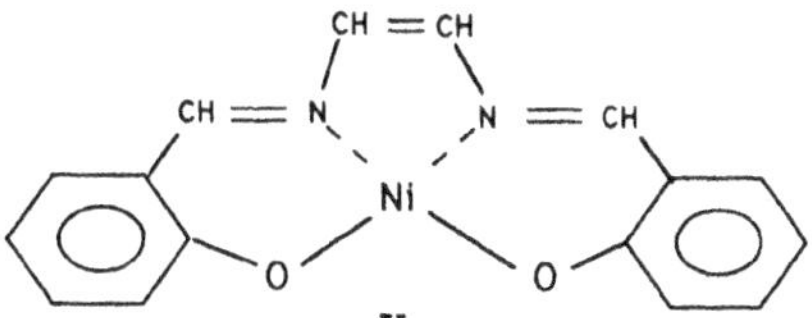

V

Fig. 1

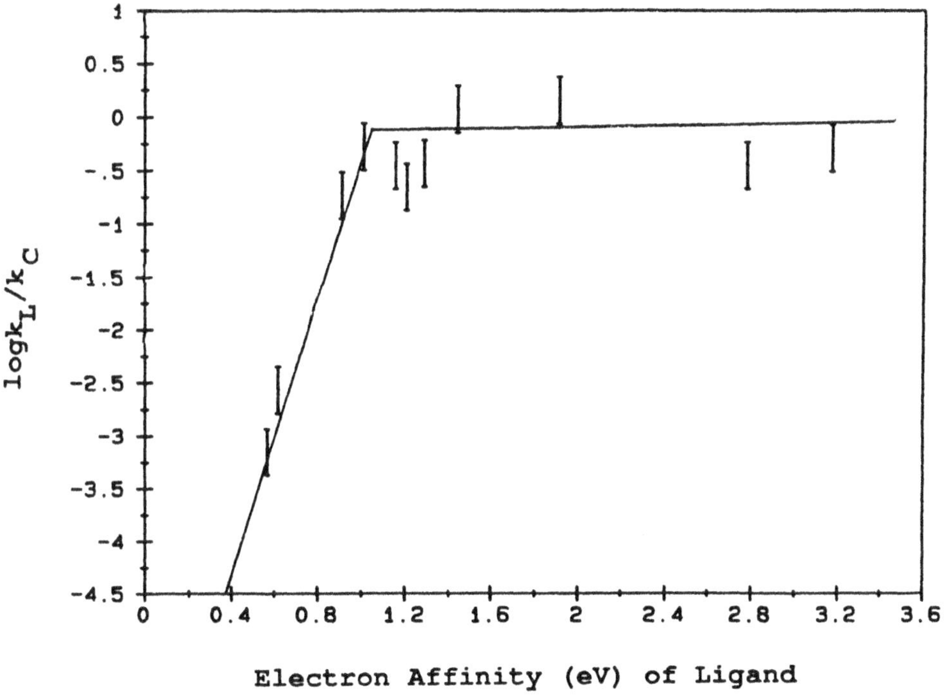

Are Reactions Between Fp^+ and R–H Reversible? Results from Solution and Gas Phase Experiments. (Dag Ekeberg, Georg Hvistendahl, Yngve Stenstrom and Einar Uggerud, Department of Chemistry, University of Oslo, P. O. Box 1033 Blindern, N-0315 Oslo 3, Norway)

Reactions between Fp^+ ($Fp = \eta^5\text{-}C_5H_5(OC)_2Fe$) and R–H (R = H, CH_3, C_2H_5) have been studied both in the gas phase and in solution. The solution experiments were carried out by adding $FpCH_3$ to concentrated sulfuric acid:

$$H_2SO_4 + FpCH_3 \longrightarrow HSO_4^- + Fp^+ + CH_4 \quad (1)$$

Deuterium labelling experiments showed that the reaction is irreversible. The reason for this is probably that Fp^+ is deactivated by adduct formation with HSO_4^-. This prevents addition of CH_4 to Fp^+.

In the gas phase (reactions were performed in a high pressure ion source) the situation is different:

$$AH^+ + FpCH_3 \longrightarrow Fp(CH_3)(H)^+ + A \quad (2a)$$

$$Fp(CH_3)(H)^+ \rightleftharpoons Fp^+ + CH_4 \quad (2b)$$

Reaction (2b) is reversible as indicated. This has been shown by deuterium labelling. In the gas phase no deactivating adduct of the 16-

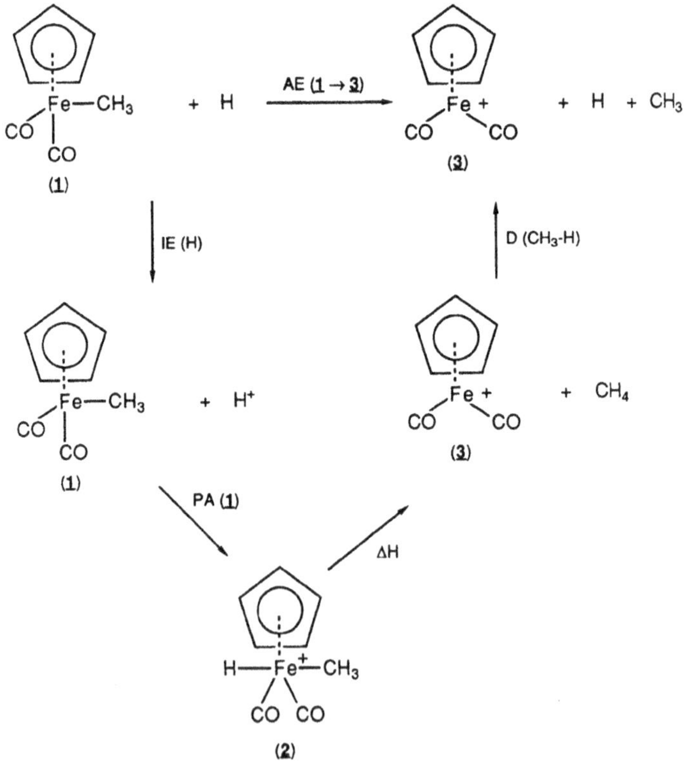

electron species Fp^+ is formed and the reverse reaction (2b) readily takes place.

In order to obtain ΔH^o for reaction (2b) the thermochemical cycle depicted in Scheme 1 was constructed. Except for the appearance energy for Fp^+ from $FpCH_3$, which we measured to be AE = 10.1 ± 0.4 eV, the data were taken from the literature (the proton affinity of $FpCH_3$ is 814 kJ/mol,[20] the ionization energy of H is 13.6 eV,[21] and the bond dissociation of methane is 436.6 kJ/mol[22]). This leads to ΔH^o = 40 kJ/mol and ΔG^o = 0 kJ/mol which is in line with the observed reversibility.

In addition we have also observed reactions between Fp^+ and H_2 or ethane:

$$Fp^+ + H_2 \longrightarrow Fp(H_2)^+ \quad (3)$$
$$Fp^+ + C_2H_6 \longrightarrow Fp(C_2H_6)^+ \quad (4a)$$
$$\longrightarrow Fp(C_2H_4)^+ + H_2 \quad (4b)$$

The latter reaction is most likely the result of an initial C–H activation followed by β-elimination. It is illuminating that $Fp(C_2H_4)^+$ also is observed to be formed in the following reactions:

$$AH^+ + FpC_2H_5 \longrightarrow Fp(C_2H_5)(H)^+ + A \quad (5a)$$
$$Fp(C_2H_5)(H)^+ \longrightarrow Fp(C_2H_4)^+ + H_2 \quad (5b)$$
$$C_2H_5^+ + FpCD_3 \longrightarrow Fp(C_2H_5)(CD_3)^+ \quad (6a)$$
$$\longrightarrow Fp(C_2H_4)^+ + CD_3H \quad (6b)$$

These observations indicate that a common mechanism leads to $Fp(C_2H_4)^+$ in the three reactions (4)-(6). The authors wish to thank VISTA (The Norwegian Academy for Science and Letters, and Statoil) for generous support of this work.

Complexes of Metal Ions with the Macrocyclic Ligand TBC

(R. C. Dunbar, D. Solooki, C. Tessier, W. Youngs, Chemistry Department, Case Western Reserve University, Cleveland, Ohio 44106 and B. Asamoto, BP Research and Development, 4440 Warrensville Road, Cleveland, Ohio 44128 USA)

The planar macrocyclic ligand tribenzocyclyne (TBC) vaporizes from a 120°C probe to give a pressure of 10^{-8} torr in the Nicolet FTMS-2000 Fourier transform mass spectrometer.

Laser desorption/ionization produces metal ions M^+ from various metal

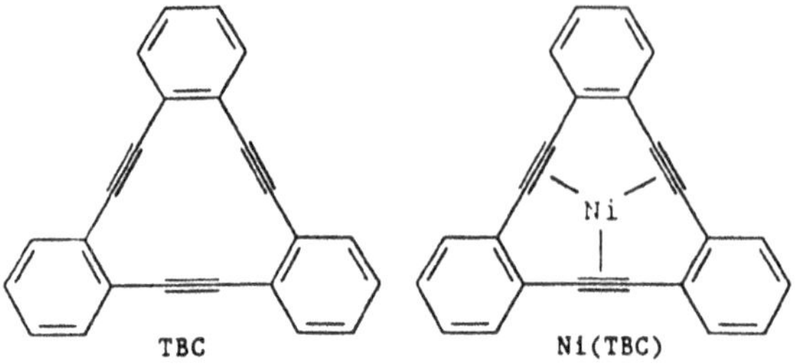

targets, which initiates formation of radiatively stabilized monomer and dimer complexes

$$M^+ \xrightarrow[k_1]{TBC} M^+(TBC) \xrightarrow[k_2]{TBC} M(TBC)_2{}^+$$

The rate k_1 for monomer complex formation is comparable and rather large for all the metals in the first transition series, with a complex being formed on every two or three collisions of M^+ with TBC. The rate k_2 for formation of dimer complexes is 25-40% of k_1 for the early transition metals, but drops off rapidly to rates below 1% collisional efficiency at the right-hand side of the first-row transition metal series.

The large k_1 values indicate efficient radiative stabilization of M^+(TBC), which suggests binding energies >1.3 eV. Similarly, the substantial k_2 values for metals on the left side of the first row indicate strong binding, but the inefficient formation of $Ni^+(TBC)_2$ and $Zn^+(TBC)_2$ suggests that these are bound by less than 1 eV.

The falloff in formation of the dimer (sandwich) complexes around iron and cobalt may be due to electron-count considerations (since the valence electron count reaches 18 at $Mn^+(TBC)_2$), or it may be due to the excellent fit of the Ni^+, Cu^+ and Zn^+ ions into the ligand cavity in the M^+(TBC) monomer complexes so that the metal is unavailable for binding by a second ligand. Study of additional metal ions and ligand displacement chemistry in the dimer complexes is expected to clarify these possibilities.

Some Speculation on the Structure of the $AlS_y{}^+$ Ions (Denise C. Parent, Chemistry Division, Naval Research Laboratory, Washington, D.C. 20375-5000 USA)

Our group has recently completed an extensive study of the Group 13 metal-oxide cluster cations.[23] Now we have turned our attention to the aluminum sulfide ions. Direct laser vaporization (DLV) of a pressed pellet of Al_2S_3 powder in the FTMS produces the ions $Al_xS_y{}^+$ (x=1-7). Of particular interest is the series of ions $AlS_y{}^+$ (y=1-6). The relative intensities of these ions for y-1-6 are .16:.03:.20:.51:1:.17. Analogous ions were not observed in the $Al_xO_y^+$ spectrum produced by DLV of Al_2O_3. Observation of the AlS_y^+ ions is perhaps to be expected since S_y clusters are known stable species and the cations were observed in the FTMS by electron ionization of the background.

A question arises as to the structure of these ions. Are they S_y clusters attached to the Al^+, or are they an Al^+ with one or more sulfur chain ligands attached? The reactivity with NO_2 was used to probe the structure. The following reaction channels were observed: Al^+ (dissociation); AlS_yO^+ and NO^+ (oxidation); AlS_zO^+ and $AlS_zNO_2^+$ (z<y, ligand exchange). The results for y=3-6 are given in the table below.

y	Branching Ratio (diss:ox:lig ex)	#S's lost
3	43: 47: 10	1 or 2
4	47: 19: 34	2
5	35: 26: 39	2
6	35: 65: 0	0

This suggests ion structures such as

```
                                                 S
                                                /
S=Al–S=S          S=S–Al–S=S          S=S–Al
                                                \
                                                 S=S
```

i.e. Al^+ with S and S_2 ligands.

AlS_6^+ is unique in its reactivity. The lack of ligand exchange reactions may indicate a cyclic S_6 attached to the aluminum cation. However AlS_7^+ and AlS_8^+, which should be just as stable, were not observed. This is especially curious as S_7^+ and S_8^+ were seen in the EI spectrum of the background. An alternative explanation is that $Al^+(S_2)_3$

```
          S=S
         /
S=S–Al
         \
          S=S
```

may be particularly stable and resistent to exchange. In studies of iron sulfides, MacMahon et al.[24] found that FeS_y^+ (y=0-5) ions reacted with ethylene sulfide to add sulfur, but that FeS_6^+ was not reactive. They concluded that either coordinated S_2 units or small sulfur rings describe the larger iron sulfide ions, though reactivity arguments point towards S_2 units as the coordinating ligands.

Experiments are underway to further characterize the oxidized aluminum sulfide ions and provide additional information on the structures.

Reactions of Ionized AsI_3 and SbI_3 with Benzene and Alkyl Benzenes
(Hansjorg Grutzmacher, Institut fur Anorganische Chemie, Universitat Heidelberg, Detlef Tholmann and Hans-Fr. Grutzmacher, Fakultet fur Chemie, Universitat Bielefeld).

Ionization of mixtures of benzene with AsI_3 and SbI_3, respectively, in a CI ion source give rise to $[C_6H_6XI_2]^+$ (X = As, Sb). The formation of these complexes has been studied in detail by FT-ICR spectrometry and corresponds to:

$$XI_3^+ \quad + \quad C_6H_6 \quad \longrightarrow \quad [C_6H_6XI_2]^+ + I \qquad X = As, Sb$$

The following rates were obtained:

AsI_3^+: k_{exp} = 5.45×10^{-10} [cm^3 $molecule^{-1}s^{-1}$]; efficiency 58%

SbI_3^+: k_{exp} = 3.8×10^{-10} [cm^3 $molecule^{-1}s^{-1}$]; efficiency 66%

In particular, the reaction of SbI_3^+ is very sensitive to excess kinetic energy, resulting in a decrease of k_{exp} and the appearance of the ions $C_6H_6^+$ and XI_2^+.
The ion $[C_6H_6AsI_2]^+$ formed in the external ion source of the FT-ICR spectrometer reacts cleanly with alkylbenzenes within the ICR cell by arene exchange.

$$[C_6H_6AsI_2]^+ \quad + \quad arene \quad \longrightarrow \quad [(arene)AsI_2]^+ \quad + \quad C_6H_6$$

The following exchange rates were measured:

arene	k_{exp} (cm^3 $molecules^{-1}s^{-1}$)	Efficiency
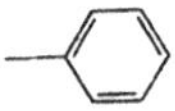	5.95×10^{-10}	63%
	5.16×10^{-10}	54%
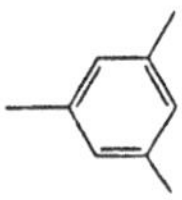	5.43×10^{-10}	58%
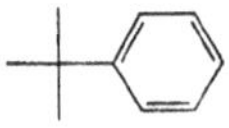	5.85×10^{-10}	58%
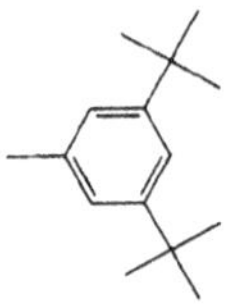	$<7.6 \times 10^{-10}$	<72%

The exchange rate is independent of the number and bulkiness of the alkyl substituent, but in the case of 3,5-di-tert-butyltoluene the formation of $C_4H_9^+$ is observed.

These results would indicate an attachment of the AsI_2^+ and SbI_2^+ to the π-cloud of the benzenes by electrostatic forces and only in special cases the insertion of XI_2^+ (isoelectronic with carbenes) into a C–C bond. However, the exchange reaction of $[C_6H_6AsI_2^+]$ with C_6H_5F and C_6H_5Cl, respectively, is very slow or absent despite of the dipole moment of C_6H_5F and C_6H_5Cl which should favor an electrostatic interaction with XI_2^+.

References

1. J. A. Connor, Top. Cur. Chem., **1977**, 71, 71.
2. S. T. Graul; R. R. Squires Mass Spectrom. Rev., **1988**, 7, 263.
3. S. T. Graul; R. R. Squires J. Am. Chem. Soc., **1990**, 112, 2517.
4. R. N. Compton; J. A. D. Stockdale Int. J. Mass Spectrom. Ion Phys., **1976**, 22, 47.
5. P. C. Engelking; W. C. Lineberger J. Am. Chem. Soc., **1979**, 101, 5569.
6. K. E. Lewis; D. M. Golden; G. P. Smith J. Am. Chem. Soc., **1984**, 106, 3905.
7. P. R. Kemper and M. T. Bowers, J. Am. Soc. Mass Spectrom., **1990**, 1, 197.
8. P. A. M. van Koppen; P. R. Kemper; G. von Helden and M. T. Bowers, to be submitted.
9. P. R. Kemper and M. T. Bowers, J. Am. Chem. Soc., **1990**, 112, 3231.
10. R. Tonkyn; M. Ronan and J. C. Weisshaar, J. Phys. Chem., **1988**, 92, 92.
11. R. G. Keesee and A. W. Castleman, Jr, J. Phys. Chem. Ref. Data, **1986**, 15, 1011.

12. P. J. Marinelli and R. R. Squires, J. Am. Chem. Soc., **1989**, 111, 4101.

13. K. I. Irikura and J. L. Beauchamp, private communication.

14. (a) P. B. Armentrout; L. F. Halle; J. L. Beauchamp J. Am. Chem. Soc., **1981**, 103, 6501. (b) L. F. Halle; P. B. Armentrout; J. L. Beauchamp Organometallics, **1982**, 1, 963. (c) G. D. Byrd; R. C. Burnier; B. S. Freiser. J. Am. Chem. Soc., **1982**, 103, 3365. (d) D. B. Jacobson; B. S. Freiser, Ibid. **1983**, 105, 5197. (e) R. Houriet; L. F. Halle; J. L. Beauchamp, Organometallics **1983**, 2, 1818. (f) B. S. Larsen; D. P. Ridge J. Am. Chem. Soc., **1984**, 106, 1912.

15. R. H. Schultz; P. B. Armentrout J. Phys. Chem., **1987**, 91, 4433. R. H. Schultz; J. L. Elkind; P. B. Armentrout J. Am. Chem. Soc., **1988**, 110, 411.

16. P. A. M. van Koppen; J. Brodbelt-Lustig; M. T. Bowers; D. V. Dearden; J. L. Beauchamp; E. R. Fisher; P. B. Armentrout J. Am. Chem. Soc., **1990**, 112, 5663; Ibid., submitted for publication.

17. R. H. Schultz; P. B. Armentrout work in progress.

18. Y. H. Pan; D. P. Ridge J. AM. Chem. Soc., **1989**, 111, 1150.

19. Y. H. Pan, Ph.D. Dissertation, University of Delaware, 1989.

20. A. E. Stevens; J. L. Beauchamp J. Am. Chem. Soc., **1981**, 103, 190.

21. S. G. Lias; J. F. Liebman; and R. P. Levin J. Phys. Chem. Ref. Data, **1984**, 13, 695.

22. A. L. Castelanho; D. Griller J. Am. Chem. Soc., **1982**, 104, 3655.

23. F. L. King; M. M. Ross; B. I. Dunlap and D. C. Parent, J. Chem. Phys., submitted.

24. T. J. MacMahon; T. C. Jackson and B. S. Freiser, J. Am. Chem. Soc., **1989**, 111, 421.

CLUSTER ION CHEMISTRY: SUMMARY OF THE WORKSHOP

P. B. ARMENTROUT
Department of Chemistry
University of Utah
Salt Lake City, UT 84108
USA

ABSTRACT. One of the exciting areas in ion chemistry that has flourished in the the past decade is studies of cluster ions. This encompasses a broad range of species from proton bound molecules to aggregates of atoms. This diversity is reflected in the contributions to the workshop on cluster ion chemistry which are summarized below. In addition to these seven reports, the workshop included oral presentations from A. J. Stace on rare gas cluster ions, T. Baer on ion complexes, and M. T. Bowers on reactions of niobium cluster ions.

LOW-ENERGY COLLISION-INDUCED DISSOCIATION OF PROTON BOUND CLUSTERS

A. G. Harrison, Department of Chemistry, University of Toronto, Toronto, Ontario M5S 1A1, Canada

A. G. Harrison described a high pressure ion source constructed for a BEQQ hybrid mass spectrometer [1]. Ion source pressures of ~5 torr can be achieved, thus allowing preparation of a variety of proton-bound cluster ions. These cluster ions can be mass-selected by the BE double-focussing mass spectrometer and introduced into the r.f.-only quadrupole collision cell where the collision-induced dissociation mass spectra can be obtained as a function of collision energy. This allows the construction of breakdown graphs expressing qualitatively the dependence of the fragmentation pathways on internal energy. Results for the methanol clusters $H^+(CH_3OH)_n$ (n = 4 to 6) showed sequential fragmentation by elimination of CH_3OH and, further, showed that the trimer ion, $H^+(CH_3OH)_3$, was the most prominent ion over much of the breakdown graph in each case. This suggests a special stability for this ion which was not noted in earlier cluster ion intensity measurements or in thermodynamic measurements [2]. The mixed cluster ion $H^+(CH_3OH)_3(H_2O)$ fragments at low collision energy primarily by elimination of H_2O; this is in contrast to the mixed cluster ion $H^+((C_2H_5)_2O)_3(H_2O)$ which does not show elimination of H_2O on collisional activation. The latter ion would be expected to have a core H_3O^+ ion and the different behavior of $H^+(CH_3OH)_3(H_2O)$ indicates that this ion does not have a core H_3O^+ ion. This conclusion is in agreement with that reached by Graul and Squires [3] from similar low energy CID experiments.

K. R. Jennings (ed.), Fundamentals of Gas Phase Ion Chemistry, 147–154.

ISOTROPY OF ION INTERACTIONS: NH_4^+ vs K^+

Joel F. Liebman and Mitchell J. Romm (University of Maryland), Michael Meot-Ner (National Institute of Standards and Technology, current address: University of Cantebury, Christchurch, New Zealand), S. M. Cybulski and Steve Scheiner (Southern Illinois University), USA

High pressure mass spectrometry measurements show that the clustering energies $\Delta H_{n-1,n}$ (generally for $1 \leq n \leq 4$) of NH_4^+ and K^+ are very similar for the solvents H_2O, NH_3, CH_3CN, and C_6H_6. This seems surprising since "strong hydrogen bonds" are expected to be influential in the NH_4^+ system but do not exist for K^+. Beyond their identical net charge and similar size, how similar are NH_4^+ and K^+?

From literature sources, we find that ΔH_{soln} $(K^+) \approx \Delta H_{soln}$ (NH_4^+) for both water and liquid ammonia. Remembering that $\Delta H_f(MX,s) = \Delta H_f(M^+,g) + \Delta H_f(X^-,g) - U_{latt}(MX)$ (normalizing charges to 1:1), we can consider the relationship: $\Delta H_f(NH_4X,s) = m\ \Delta H_f(KX,s) + b$. Statistical analysis of literature thermochemical data results in $m = 1$ and $b = \Delta H_f(NH_4^+,g) - \Delta H_f(K^+,g)$ within experimental error and with the correlation coefficient $r^2 = 0.9997$. Thus, NH_4^+ seems nearly isotropic. As long as T_d symmetry is maintained and the distance between the N and the solvent is > 2.5 Å or so, point charge-dipole or -induced dipole calculations also show isotropy regardless of the charge distribution of NH_4^+. This may be understood by classical electrostatics: The interaction energy of an ion with external fields goes as $1/R^{1+n}$ where $n = 0$ if there is another charge present, $n = 1$ if there is a dipole present, etc.. Since $r(NH_4^+) \approx r(K^+)$, $R = r_{ion} + r_{solvent}$, the electrostatic potential is nearly the same for NH_4^+ and K^+. By symmetry, the dipole (μ) and quadrupole (Q) moments of NH_4^+ vanish. As such, the first new, nonzero term for the NH_4^+ ion arises from its octupole (Ω) moment which interacts only weakly with a $1/R^{4+n}$ potential.

Ab-initio calculations also show that NH_4^+ is quite isotropic in its interactions with solvents. If the classical 1-hydrogen bond structure for $NH_4^+ \cdot H_2O$, 1, is assigned an energy of 0 kcal/mol, it is found that the 2 and 3-hydrogen bond structures, 2 and 3, respectively, are higher by only 4 and 5 kcal/mol, respectively.

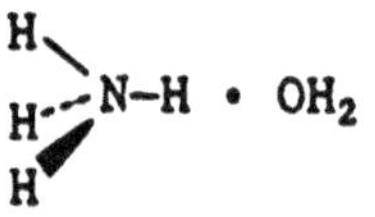

1

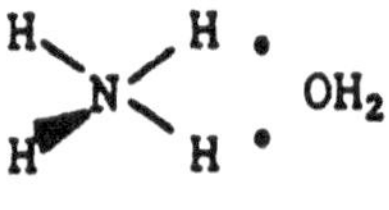

2

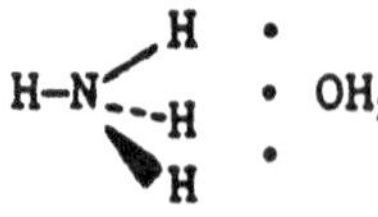

3

METASTABLE AND PHOTODISSOCIATION DYNAMICS OF THE $(OCS)(C_2H_2)^+$ CLUSTER ION, Susan T. Graul and Michael T. Bowers, Department of Chemistry, University of California, Santa Barbara, CA 93106, USA

In the present study, we have examined the dynamics of dissociation of the $(OCS)(C_2H_2)^+$ cluster ion. This system is of interest to us for several reasons: 1) this ion represents a model for the intermediate in the ion-molecule reaction of OCS^+ with C_2H_2 (or of $C_2H_2^+$ with OCS); 2) the ground state reaction exhibits mode-specific vibrational energy effects [4]; 3) photodissociation in the wavelength range 735-375 nm (1.69-3.31 eV) has been suggested to involve repulsive upper states. A photoinitiated sulfur atom reaction has been reported [5].

We have measured the kinetic energy release distributions (KERDs) for products of the metastable decomposition and photodissociation of the $(OCS)(C_2H_2)^+$ cluster ion. From analysis of the kinetic energy released in the dissociation of this cluster, we derive an upper limit for the bond energy, suggest probable structures, and obtain information about features of the ground state and relevant excited state potential energy surfaces.

Experimental. These experiments were carried out using a reverse-geometry sector mass spectrometer (VG ZAB-2F) with an argon ion laser and dye laser. The $(OCS)(C_2H_2)^+$ cluster ions were generated in a cooled high-pressure ion source (pressure = 0.04-0.08 torr; temperature = 280 K). The mass-selected ion beam is crossed at its focal point in the second field-free region (2FFR) by the output beam of the argon ion laser/dye laser. Ionic photoproducts and 2FFR metastable dissociation products are energy-analyzed at the electrostatic analyzer and detected using pulse-counting. Up/down counting is used to distinguish photodissociation products from products of CID or metastable dissociation.

Results and Discussion. Metastable decomposition of the $(OCS)(C_2H_2)^+$ ion in the 2FFR (ion lifetime ≈ 10 μsec) yields OCS^+ and $C_2H_2S^+$. The $C_2H_2S^+$ ion is the major product and is formed with a large kinetic release ($\langle E_t \rangle$ = 0.525 eV), which is consistent with the strongly exothermic sulfur atom transfer reaction to form thioketene ion or ethynethiol ion. (Our data does not permit definite identification of the structure of the $C_2H_2S^+$ product). The kinetic energy release distribution (KERD) is nonstatistical, indicating that a barrier exists on the reaction coordinate and that energy released in the exothermic sulfur atom transfer is not completely randomized. The kinetic energy release for the OCS^+ product is small and the KERD peaks near zero, consistent with a simple back-dissociation of the $(OCS)(C_2H_2)^+$ cluster ion.

Photodissociation cross sections were measured in the range of 529-357 nm (2.3-3.5 eV) from integrated photoproduct yields relative to yields from the $(N_2)_2^+$ reference. The cross section increases smoothly from about 0.3 x 10^{-18} cm^2 at 529 nm to 4 x 10^{-18} cm^2 at 357 nm, in rough agreement with the cross sections measured by Orlando et al.[2] Two ionic products, $(OCS)^+$ and $(C_2H_2)^+$, are observed throughout this energy region in an approximately constant ratio of 5:1 $(OCS)^+$:$(C_2H_2)^+$. Due to noise resulting from up/down counting of the relatively intense $C_2H_2S^+$ product from metastable dissociation, a $C_2H_2S^+$ photoproduct could not be identified with confidence, and we assign an upper limit of about 15% to the fraction of this product formed by photodissociation. No S^+ was detected as a photoproduct.

The angular distributions of both the OCS^+ and the $C_2H_2^+$ photoproducts are strongly anisotropic, which is evidence that the photoexcited $[(OCS)(C_2H_2)^+]^*$ species has a short lifetime with respect to molecular rotation. This indicates that the excited state surface(s) accessed by photoexcitation is (are) repulsive.

Kinetic energy release distributions for photoproducts are obtained using a laser polarization angle of 54.7°. The KERDs for both photoproducts are roughly triangular and peak at energies that are ~25-35% of the photon energy. The average kinetic energy released for OCS^+ and $C_2H_2^+$ scales linearly with photon energy. The dissociation can therefore be modeled using impulsive theory, which relates the average kinetic energy released to the total available energy according to the equation, $\langle E_t \rangle = (E_{av} - E_{vib})\mu_{bc}/\mu_f$ where $\langle E_t \rangle$ is the average kinetic energy released, μ_{bc} is the reduced mass of the subunits involved in the bond, μ_f is the reduced mass of the dissociation products, E_{av} is the available energy ($E_{av} = h\nu + E_{internal} - DE$, where DE is the cluster ion bond energy), and E_{vib} is the vibrational energy deposited in the products. A linear least-squares fit of the average energy released versus ($h\nu + E_{internal}$) gives the value of (μ_{bc}/μ_f) as the slope of the line and the value of $(\mu_{bc}/\mu_f)(DE + E_{vib})$ from the y-intercept. Thus, we can derive an upper limit for DE of 1.1 eV (if $E_{vib} = 0$) and a value ~9.4 for μ_{bc}. The latter value for μ_{bc} would be consistent with breaking a bond either between the sulfur atom of OCS and a carbon atom of C_2H_2, or between the oxygen atom of OCS and the pi-bond of acetylene. The former structure also seems a logical intermediate for the ground-state sulfur atom transfer reaction that yields $C_2H_2S^+$.

STRUCTURAL EFFECTS ON CHEMICAL PROPERTIES OF CARBON CLUSTER IONS

P. A. Hintze and S. L. Anderson, Department of Chemistry, State University of New York, Stony Brook, NY 11794, USA

Detailed product cross sections for reactions of C_n^+ (n = 2-12) with D_2 as a function of collision energy were measured using a guided ion beam tandem mass spectrometer. The results for C_{6-8}^+ are shown below.

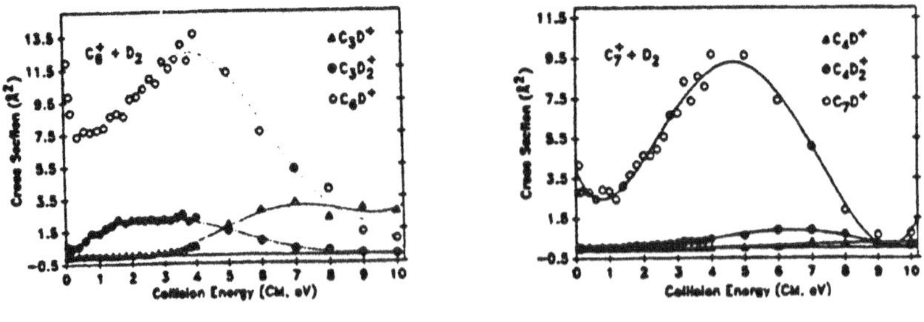

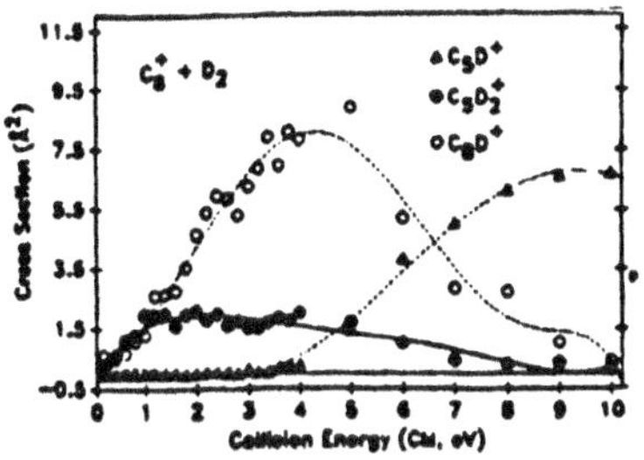

For most of the cluster sizes studied (n = 2-5,7 and 9), the only product at low collision energies (<1 eV) is C_nD^+ + D. For clusters larger than 10 atoms, energy thresholds are observed for product formation. For C_6^+ and C_8^+, formation of $C_{n-3}D^+$ + C_3 is observed at low collision energies. Above 3 eV collision energy, formation of $C_{n-3}D^+$ is observed, resulting most likely from decay of the $C_{n-3}D_2^+$ product as a result of increased collision energy. Our observation of neutral C_3 fragments is consistent with previous studies of carbon cluster ions including FTMS reaction studies [6] and photodissociation studies [7].

The C_nD^+ cross sections for n = 3-7 shows a pronounced second feature as the collision energy is increased, which may be attributed to a second structural isomer which is unreactive at low collision energy. The results for C_8^+ and C_9^+ show no evidence of a second isomer. The lack of reactivity for clusters 10 atoms and larger at low collision energy is in agreement with the proposed linear to cyclic structural change between C_9^+ and C_{10}^+.

DETERMINATION OF THE IONIZATION POTENTIALS OF ALUMINUM OXIDES VIA CHARGE TRANSFER, Stephan B. H. Bach, Stephen W. McElvany and Denise C. Parent
Chemistry Division, Naval Research Laboratory, Washington, D.C. 20375-5000, USA

The use of charge transfer to determine ionization potentials (IP) has been demonstrated for carbon cluster ions formed by direct laser vaporization (DLV) [8]. Here we show the first application of this technique to secondary ions formed by reaction of primary ions with a neutral gas. The ions of interest are the aluminum oxides, $Al_xO_y^+$ (x = 2-4, y = 1-4). Al_2O^+, $Al_3O_2^+$ and $Al_4O_4^+$ are produced by DLV of a pressed pellet of Al_2O_3 powder. The other ions studied are formed by reaction of Al_2O^+ and $Al_3O_2^+$ with N_2O or NO_2 [9].

The experiments were carried out in a FTMS. The pulse sequence utilized in these experiments is depicted in the figure. The charge transfer reagent (we used nickelocene, ferrocene and various substituted benzenes with IPs in the range of 6.28-9.25 eV) was leaked into the cell at a constant low pressure. The oxidizer and buffer (Ar) gasses were pulsed in. DL1 is the formation/relaxation time after which the ion of interest is isolated. It then undergoes reaction with the charge transfer compound during DL5. Both the DL1 and DL5 times are variable. The occurrence/non-occurrence of charge transfer with a variety of compounds is used to bracket the IP.

IPs (in eV) determined in this way are: Al_2O - 8.35; Al_2O_2 - 8.9; Al_2O_3 - 8.9; Al_2O_4 - 6.85; Al_3O_2 - 6.85; Al_3O_3 - 6.85; Al_3O_4 - 8.35; Al_4O_4 - 6.85. All values are good to 0.2 eV. The large decrease in IP between Al_2O_3 and Al_2O_4 may be related to the expected change in structure[2], from linear to cyclic, shown below. Experiments to extend the size range of oxides for which IPs are determined are in progress.

```
                                        O
                                       / \
        O=Al-O-Al=O               O-Al     Al-O
                                       \ /
                                        O
```

REACTIVITY OF SMALL IONIC CLUSTERS OF TRANSITION METALS WITH CYCLOHEXANE: FRONTIER ORBITAL THEORY, Y. H. Pan, K. Sohlberg, and D. P. Ridge, Department of Chemistry and Biochemistry and Center for Catalytic Science and Technology, University of Delaware, Newark, DE 19716, USA

$M_4(CO)_n^+$ (M = Re [10], Ir [11] and Co) have been found to react with cyclohexane to produce dehydrogenation products such as $M_4(CO)_nC_6H_6^+$. Clusters with more than some critical number of CO ligands are unreactive. For Re this number is 11 while for Ir and Co the critical numbers are 6 and 7, respectively. Since all the tetramers are expected to be tetrahedral in structure and of a similar size, this effect is clearly not steric. On the other hand the critical numbers correspond to similar numbers of cluster valence electrons. Indeed, it seems that clusters with more than 48 valence electrons will be unreactive [11]. This is just the number of electrons required to fill the d and s shells of a transition metal tetramer. We have suggested on this basis that a lowest unoccupied orbital (LUMO) in the d or s shells of the cluster is required for reactivity. Since no reactivity was found for anionic tetramers, it would appear that the LUMO must be low in energy. This is consistent with a generally accepted picture of oxidative addition of a C-H bond to a transition metal. This picture involves the donation of the bonding pair of electrons from the C-H bond to an empty s orbital on the metal. The absence of a suitable acceptor orbital can be expected to reduce activity.

There are several puzzling observations about the reactivity of Co clusters that can be rationalized in terms of our basic postulate. These observations are: (1) $CO_4(CO)_{2-4}^+$ are unreactive, while $Co_4(CO)_n^+$ with n = 0, 1, 5-7 are reactive; (2) While CO_n^+ n = 1 [12,13] 3, 4 are reactive Co_2^+ is unreactive [14]; (3) While Co_2^+ is unreactive, $Co_2(CO)^+$ is reactive [15]. Simple MO theory predicts that each of the unreactive clusters has either filled s and d band orbitals or an s band antibonding LUMO of unusually high energy. This is illustrated for Co_n^+ in Figs. 1 and 2. In Fig. 1 is shown the rate constants of Co_n^+ reacting with cyclohexane. The remarkable inertness of Co_2^+ is evident. In Fig. 2 is shown an energy level diagram for the s band orbitals of Co_n^+ obtained from a smiple extended Huckel calculation. The M-M bond lengths for the clusters were obtained from elementary bond-length bond-

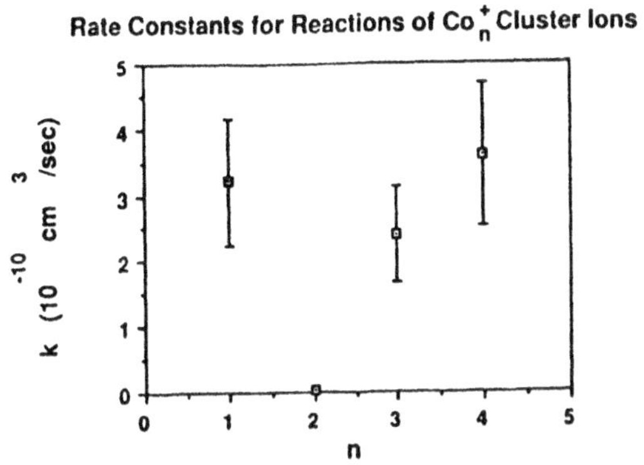

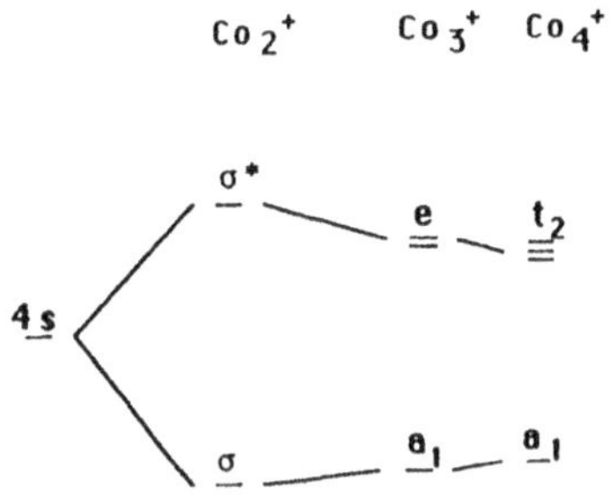

order relations. The anti-bonding orbital in Co_2^+ is clearly unusually high in energy. A probable electronic configuration for the diatomic ion would put 14 electrons in the d band of orbitals and 2 electrons in the bonding 4s sigma orbital. Hence the high energy antibonding orbital would be the LUMO and hence the cluster is unreactive. The ligand in $Co_2(CO)^+$ weakens the metal-metal interaction lowering the energy of the LUMO which leads to reactivity. By a similar treatment the $Co_4(CO)_{2-4}^+$ ions have unusually high energy s band LUMOs and are hence unreactive. To account for the reactivity of $Co_4(CO)_{5-7}^+$ it is necessary to assume that the additional CO ligands stabilize low spin electronic ground state configurations in these complexes. The lower spin states that vacancies in the s and d bands of orbitals making it possible for the clusters to react.

Similar arguments can be applied to other species accounting for seemingly idiosyncratic reactivity patterns. It is also possible to make tenative predictions using the theory. Two of these are (1) Dimer ions of early first row transition metals with their relatively small s orbitals and relatively few s and d electrons may react with alkanes; (2) $M_n(CO)_m^+$, m = Fe, Co or Ni will show discontinuities in the variation of their reactivity with m as a result of switching from high spin to low spin.

BOND ENERGIES OF TRANSITION METAL CLUSTER IONS

P. B. Armentrout, D. A. Hales, and L. Lian, Department of Chemistry, University of Utah, Salt Lake City, UT 84112, USA

Recent work in our laboratory has extended our published studies [16,17] of the thermochemical stabilities of transition metal clusters. By using collision-induced dissociation, we have now systematically obtained bond energies for Ti_n^+ for n = 2 - 22, Fe_n^+ for n = 2 - 19, Co_n^+ for n = 2 - 18, Nb_n^+ for n = 2 - 11, and for smaller clusters of V, Cr, and Ta. In all cases, the bond energies, $D(M_{n-1}^+-M)$, vary nonmonotonically with cluster size. Strong maxima are observed in these bond energies and are speculatively interpreted in terms of the structures of the clusters. For example, titanium clusters show that

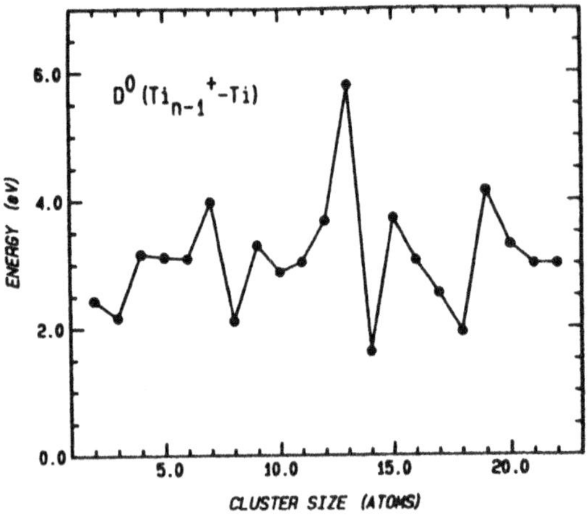

the most strongly bound cluster is Ti_{13}^+, with Ti_7^+ and Ti_{19}^+ being minor maxima, see Figure. This pattern suggests an icosahedral arrangement of Ti is the most favored packing scheme. Other metals show different patterns. Also of interest is the observation that the cohesive binding energies of all metal clusters, $\sum D(M_n^+/n)$, is represented fairly well by a simple spherical drop model with no adjustable parameters [17].

REFERENCES

1. Harrison, A. G.; Mercer, R. S.; Reiner, E. J.; Young, A. B.; Boyd, R. K.; March, R. E.; and Porter, C. J. (1987) *Int. J. Mass Spectrom. Ion Process.* **74**, 13.
2. Keesee, R. G. and Castleman, A. W. (1989) in J. P. Maier (ed.), Ion and Cluster Ion Spectroscopy and Structure, Elsevier, New York.
3. Graul, S. T. and Squires, R. R. (1989) *Int. J. Mass Spectrom. Ion Processes* **94**, 41.
4. Orlando, T. M.; Yang, B.; and Anderson, S. L. (1989) *J. Chem. Phys.* **90**, 1577.
5. Orlando, T. M.; Friedmann, A. and Maier, J. P. *J. Chem. Phys.* in press.
6. McElvany, S. W.; et al. (1987) *J. Chem. Phys.* **86**, 715.
7. Geusic, M. E.; et al. (1987) *J. Chem. Phys.* **86**, 3862.
8. Bach, S. B. H. and Eyler, J. R. (1990) *J. Chem. Phys.* **92**, 358.
9. King, F. L.; Ross, M. M.; Dunlap, B. I.; and Parent, D. C. *J. Chem. Phys.* submitted.
10. Wronka, J.; Forbes, R. A.; Laukien, F. H. and Ridge, D. P. (1987) *J. Phys. Chem.*, **91**, 6420.
11. Pan, Y. H. and Ridge, D. P. (1989) *J. Phys. Chem.*, **93**, 3375.
12. Armentrout, P. B. and Beauchamp, J. L. (1981) *J. Am. Chem. Soc.* **103**, 6628.
13. Jacobsen, D. B. and Freiser, B. S. (1983) *J. Am. Chem. Soc.* **105**, 7492.
14. Freas, R. B. and Ridge, D. P. (1980) *J. Am. Chem. Soc.*, **102**, 7129.
15. Freas, R. B. and Ridge, D. P. (1984) *J. Am. Chem. Soc.* **106**, 825.
16. Loh, S. K.; Lian, L. and Armentrout, P. B. (1989) *J. Am. Chem. Soc.* **111**, 3167. Loh, S. K.; Hales, D. A.; Lian, L. and Armentrout, P. B. (1989) *J. Chem. Phys.* **90**, 5466.
17. Hales, D. A.; Lian, L. and Armentrout, P. B. *Int. J. Mass Spectrom. Ion Processes*, accepted for publication.

Characterization of Ions via their Electronic Transitions

John P. Maier
Institut für Physikalische Chemie, Universität Basel
Klingelbergstrasse 80
CH-4056 Basel, Switzerland

ABSTRACT. Spectroscopic characterization of open-shell cations from the study of their electronic transitions, entails the development of new approaches and their application. Some of our recent advances in this area are outlined. The emission and excitation spectra of rotationally cooled cations yield their vibrational frequencies. The analysis of rotationally resolved laser excitation spectra enable the complete geometries to be obtained and this has been accomplished for a few selected polyatomic cations. The approaches used to investigate fragment ions are illustrated by means of C_2^+. Following the location of its electronic absorption spectrum in a neon matrix, gas phase characterization is carried out by the laser excitation and stimulated emission pumping techniques. To study mass-selected ions two approaches are being developed. In one, the absorption spectrum of mass-chosen ions which have been codeposited with neon to form a 5 K matrix can be measured. In the other, mass-selected ions are probed in the gas phase by a two-photon absorption-dissociation approach. The first photon induces a transition between the two bound states to be characterized whereas the second photon is there merely to produce fragment ions, which are monitored, in a further excitation process. Examples illustrating the potential of these techniques are given.

K. R. Jennings (ed.), Fundamentals of Gas Phase Ion Chemistry, 155–181.

1. Introduction

The primary aim of our spectroscopic studies on open-shell cations is the determination of the vibrational frequencies, as well as the rotational and related constants characterizing the electronic states involved in the transitions. From such data, the force constants, potentials and geometries can be established, which are often prerequisites for the analysis of the dynamical processes of ions, such as ion-molecule reactions.

There are three mainstream techniques at obtaining such spectroscopic data; via electronic, vibrational-rotational (ir) or microwave spectra. Our contributions in this area are in the first category, and in this article our more recent activity and approaches pursued to characterize open-shell cations via their electronic transitions are outlined. Other articles elsewhere cover the current advances in the ir[1] and microwave[2] investigations of ions. It should also be mentioned that the only other means of determining the geometric structure of ions (apart from calculations) is by the Coulomb-explosion technique; however the method has yet been applied only to a few ions.[3]

2. Vibrational Characterization

The vibrational frequencies of open-shell ions can be inferred from their electronic spectra, which are recorded both in emission and by laser excitation.[4] In both cases it is advantageous to cool the ions to low rotational temperatures by means of supersonic free jet expansions (T_{rot} ~ 10 K) or merely by external cooling with liquid nitrogen (T_{rot} ~ 100 K). The emission and excitation spectra are best analysed concomitantly; often the information forthcoming from one is required to interpret the details of the other. These aspects have been discussed and illustrated often in the past;[5] it suffices here to give one example with the final data on the chlorocyanide cation, $ClCN^+$.

Chlorocyanide cation is one of the 150-200 open-shell cations which are known to fluoresce from their excited electronic states[6], and the $B\ ^2\Pi \rightarrow X\ ^2\Pi$ and $A\ ^2\Sigma^+ \rightarrow X\ ^2\Pi$ band systems have been identified[7]. The spectra are complex both in emission[8] and laser excitation[9], even though the rotational temperatures are ~10 K and ~100 K, respectively, because considerable geometry changes accompany the transitions. Nevertheless, a vibrational analysis of the $B\ ^2\Pi$ - $X\ ^2\Pi$ system proved possible and leads to the vibrational frequencies summarized in figure 1.

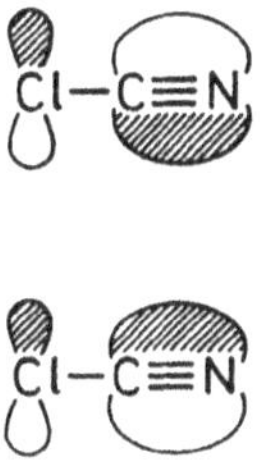

		ν_1	ν_2	ν_3
$\cdots\ \pi^4\ \sigma^2\ \pi^4$	$X\ ^1\Sigma^+$	2201	397	729
$\cdots\ \pi^4\ \sigma^2\ \pi^3$	$X\ ^2\Pi$	1915	(376)	827
$\cdots\ \pi^3\ \sigma^2\ \pi^4$	$B\ ^2\Pi$	2128	303	532

Figure 1. Experimentally determined vibrational frequencies of chlorocyanide cation (±1 cm^{-1} from emission and laser excitation spectra of the $B\ ^2\Pi$ - $X\ ^2\Pi$ transition), and molecule (ir spectrum). On the left are shown the main characteristics of the molecular orbital description of the electronic structure.

The vibrational frequencies give a clear insight into the description of the electronic structure of the ion. On the left side of the figure are depicted the topological features of the molecular orbitals involved in the ionization process[10], as well as the single electronic configuration describing the ionic states. As can be seen, the ν_1 (C≡N str.) frequency decreases with respect to the molecular value[11] ($X\ ^1\Sigma^+$ state) in both the

$X\ ^2\Pi$ and $B\ ^2\Pi$ ionic states. In contrast, the ν_3 (C-Cl str.) vibrational frequency increases in the $X\ ^2\Pi$ state but decreases in the $B\ ^2\Pi$ one, in accord with the anti-bonding, or bonding characteristics of the respective molecular orbitals in the C-Cl region (fig. 1). The analysis can be further extended in that the force constants be inferred from the corresponding values of the isotopic derivatives. Such results on vibrational frequencies, force constants and fields on the polyatomic cation $ClCN^+$, containing a heavy-atom, can now be used as bench-mark values for theoretical methods or molecular modelling approaches to simulate interactions involving ionic species.

3. Geometry Determination

Using the method of laser excitation spectroscopy, the rotational structure in the electronic transitions of linear polyatomic cations has been resolved and analysed in the past years. These include the OCS^+, XCN^+, $XCCH^+$ X-Cl, Br and $H(CC)_2H^+$ ions[5]. Among these, we chose two types as prototypes and decided to determine their complete geometric structure. The halocyanide cations $ClCN^+$ and $BrCN^+$ were chosen as examples of open-shell cations containing a heavy atom, and diacetylene cation as a large ion among the hydrocarbons.

The spectroscopic method of choice to establish the geometry of molecules in the gas-phases is the so-called r_s substitution method. In this one substitutes all the atoms in the molecule by their isotopes in turn, and determines the rotational constants. From the latter the mean distance of the substituted atom from the centre of gravity is deduced which enables the intermolecular distances to be evaluated.[12] These are known as r_s values and such a r_s structure approximates the equilibrium geometry more closely than that derived by other methods.

The geometry of diacetylene cation in its ground $X\ ^2\Pi_g$ state is considered here as an example. The $A\ ^2\Pi_u \rightarrow X\ ^2\Pi_g$ emission spectrum has been identified and rotationally analysed in 1956[13]; however only a few

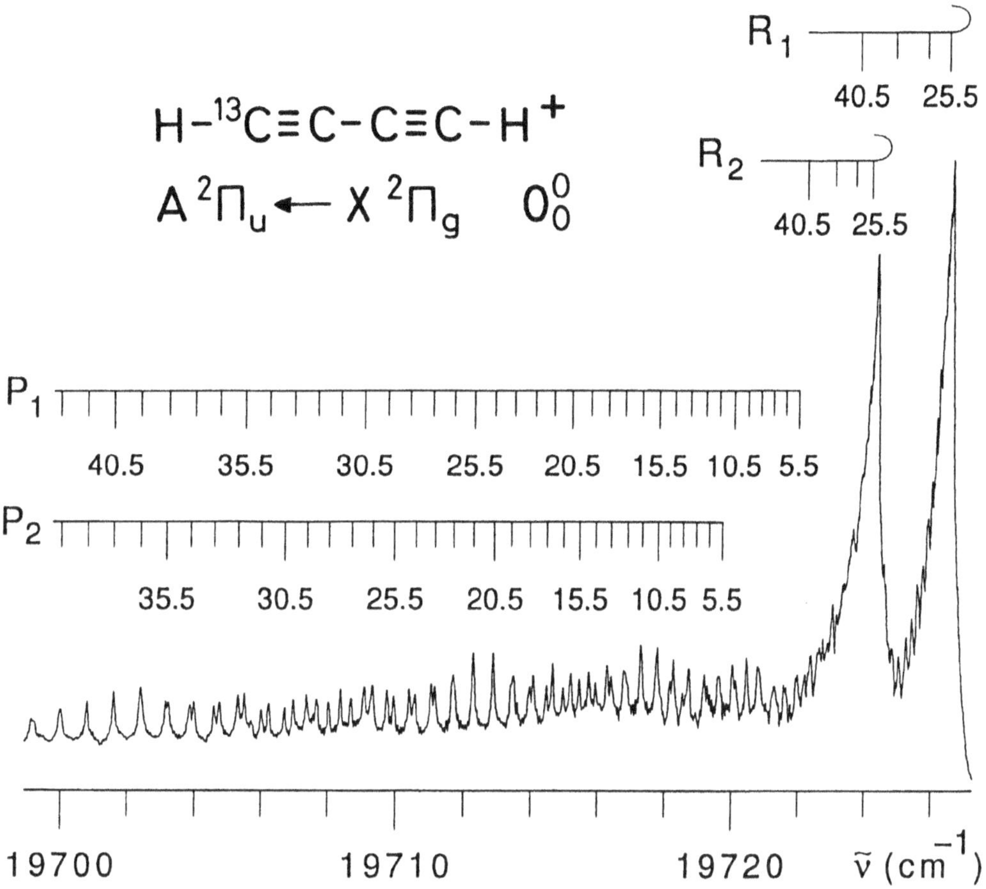

Figure 2. The rotational structure of the origin band in the A $^2\Pi_u$ - X $^2\Pi_g$ transition of an isotopically enriched derivative of diacetylene cation recorded with 0.04 cm^{-1} resolution by laser excitation spectroscopy.

years ago could the numbering (and hence the correct rotational constants) be unambiguously established[14]. This was accomplished by studying the emission spectrum at low rotational temperatures (10-20 K) by means of a supersonic free jet. Following this, the laser excitation spectrum of this transition, A $^2\Pi_u \leftarrow$ X $^2\Pi_g$, was recorded for $D(CC)_2D^+$, $H^{13}CCCCH^+$ and $HC^{13}C^{13}CCH^+$.[15] In figure 2 is shown such a spectrum for one of the isotopic ions and the rotational lines are assigned as indicated.

Finally from the inferred rotational constants the r_s structure is obtained - due to the symmetry of the species, it is not necessary to substitute all the carbon atoms by ^{13}C in turn.

The observations are summarized in fig. 3; again to the left is given the molecular orbital description of the electronic structure, configurations and state symmetry, on the right are the experimentally determined internuclear distances. The C-H bond length is practically the same in the X $^2\Pi_g$ and A $^2\Pi_u$ states of the ion as in the ground state (X $^1\Sigma_g^+$) of the molecule[16]. However, the carbon-carbon distance increases when the electron is removed from the bonding region (e.g. C≡C) but decreases when from antibonding ones (e.g. C-C in the X $^2\Pi_g$ ionic state). These trends can be correlated directly with the molecular orbital description[17] in an analogous way as discussed at the end of section 2.

With such data available, diacetylene cation can be considered as a benchmark example for the testing and development of quantum chemical calculations to predict the structure and properties of large open-shell hydrocarbon cations. In a similar vein, the complete geometries of ClCN+ [18]

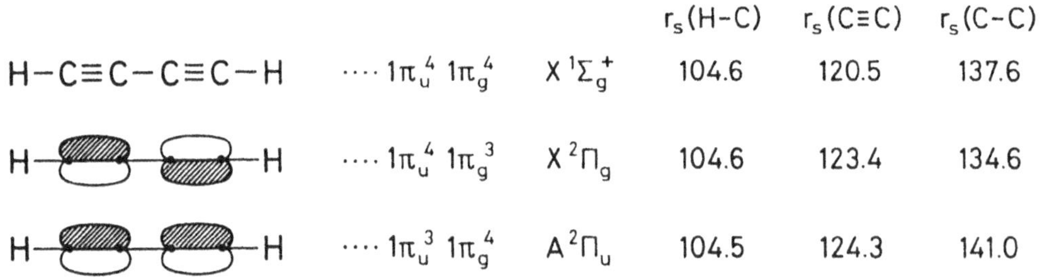

Figure 3. The experimentally determined bond lengths (r_s in pm) of diacetylene cation (from the rotational constants of its isotopic variants) and molecule (combination of electron diffraction data and rotational constants from the Raman spectrum) in relation to the molecular orbital description of the electronic structure.

and $BrCN^+$ [19] have been determined for their ground states, X $^2\Pi$, and these serve as prototypes of ions containing higher row elements.

4. Spectral Characterization of a Fragment Ion: C_2^+

Fragment ions are well known in mass-spectrometers, however, spectroscopically they are by and large uncharacterized. The two main obstacles which arise in the study of such species are the knowledge of the approximate location of the transitions (electronic or ir) to be probed with a spectroscopic technique, and secondly the ability to identify the species whose band system is being observed. The latter requirement immediately suggests the use of mass-selected ion-beams; the first hurdle remains. If one studies parent cations, the location of the transitions was found in the past by means of emission spectroscopy. Unfortunately, for fragment ions this is not a promising approach in general as the probability of forming a fragment ion (especially polyatomic) from a parent precursor in an excited electronic state is low. Thus the alternative approach which appears attractive is to study initially the absorption spectra of fragment ions in rare gas matrices. Provided that one can produce the ion in the matrix in a concentration of $> 10^{16}$ cm^{-3}, then the electronic absorption spectrum can be readily measured in the 250-1200 nm region[20]. This is the path we chose for the initial studies of C_2^+.

Two sets of spectroscopic data were available for C_2^+ prior to our investigations. A low resolution gas phase measurement on an ion beam of C_2^+ using translational energy loss spectroscopy[21], revealed that there are several transitions between doublet and quartet states of C_2^+ in the visible-uv region; tentative assignment of the broad (often eV) bands was proposed in comparison with calculated transition energies. The theoretical works themselves [22, 23] provided the other set of information on transitions of C_2^+.

In figure 4 are redrawn the potential energy curves of C_2^+ from the recent calculations[23]; as can be seen the B $^4\Sigma_u^-$ ← X $^4\Sigma_g^-$ electronic transition is expected in the visible part of the spectrum, actually in the same region as the d $^3\Pi_g$ - a $^3\Pi_u$ band system of C_2. The latter can simply be eliminated by trapping both C_2 and C_2^+ in a low temperature matrix and the a $^3\Pi_u$ state of C_2 will rapidly be depopulated, leaving the C_2^+ system unobscured.

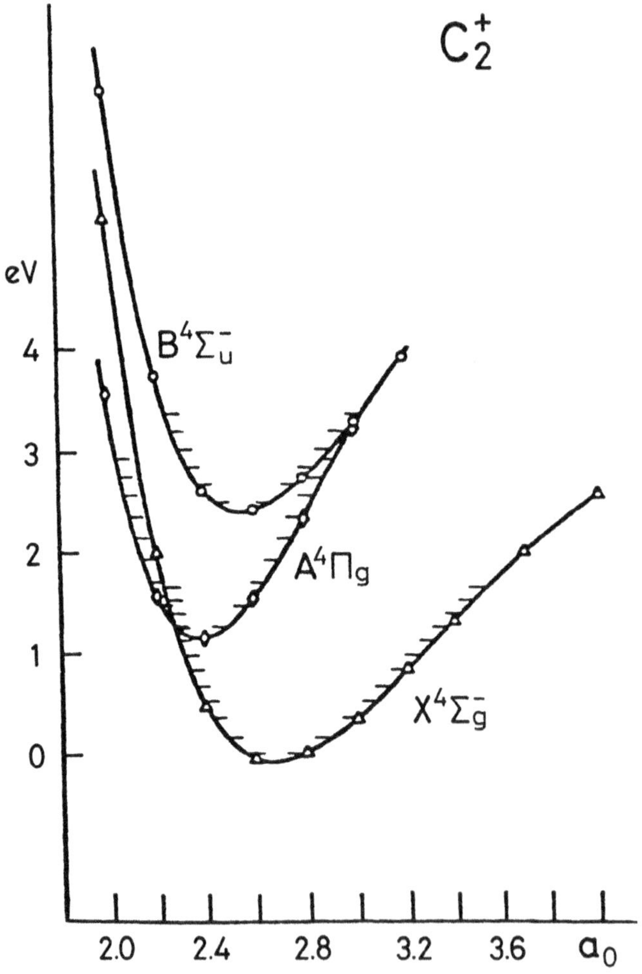

Figure 4. The calculated potential energy curves of the lowest lying quartet states of C_2^+ redrawn from ref. 23.

C_2^+ is prepared in a 5 K neon matrix in two steps; firstly C_2 is prepared by photolysis (with potentially non-ionizing radiation such as XeI-147 nm) of various precursors (XCCH, XCCX, X= H, Cl, Br) embedded there in ratios ~1:4000 (relative to Ne). The formation of C_2 is evidenced by the appearance of its two electronic absorption band systems:
$D\,^1\Sigma_u^+ \leftarrow X\,^1\Sigma_g^+$ and $A\,^1\Pi_u \leftarrow X\,^1\Sigma_g^+$. Subsequent irradiation with NeI (73.6 nm) or H Ly α (121.6 nm) photons leads to the appearance of a new band system, comprising three sharp peaks, in the expected region for the C_2^+ $B\,^4\Sigma_u^- \leftarrow X\,^4\Sigma_g^-$ absorption (fig. 5).[24] The absorption spectra of the species in the 5 K matrix are actually measured with help of the waveguide technique[25].

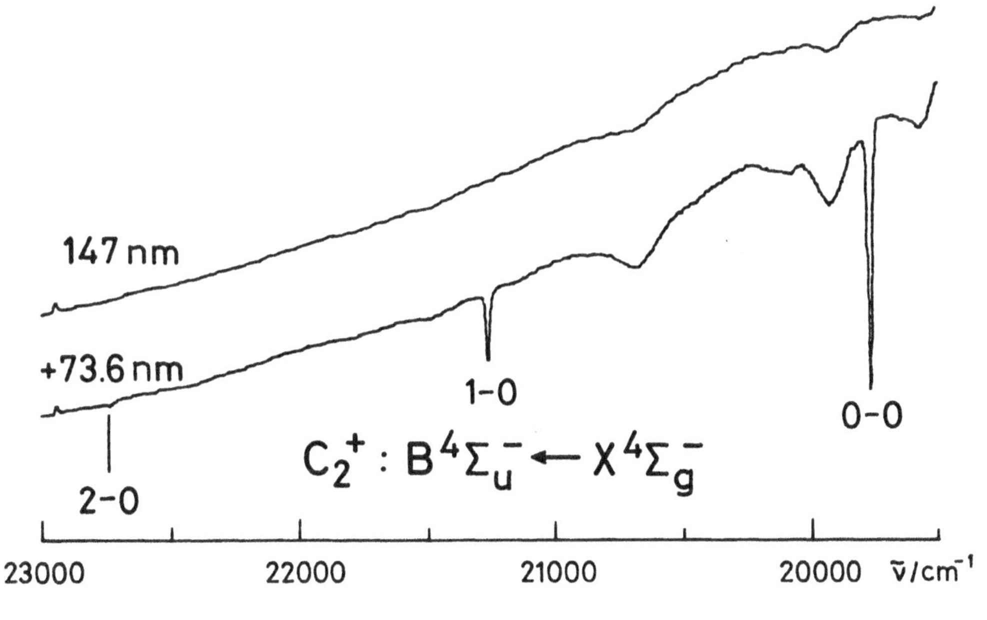

Figure 5. The $B\,^4\Sigma_u^- \leftarrow X\,^4\Sigma_g^-$ absorption bands of C_2^+ in a 5 K neon matrix recorded with 0.1 nm resolution. The top trace is observed after irradiation of a neon matrix containing chloroacetylene (ratio ~1:4000) with XeI photons and the bottom one (with bands of C_2^+) after subsequent NeI photoionization.

The three distinct peaks which are apparent in the bottom trace of fig. 5 are readily attributed to the B $^4\Sigma_u^-$ ← X $^4\Sigma_g^-$ electronic transition of C_2^+ based on 1) chemical evidence - use of different precursors, XCCH, XCCX, X- D, Cl, Br; 2) isotope effects - using $H^{13}C^{13}CH$ the shifts are as calculated for a diatomic C_2 species, 3) excellent accord with the calculated T_e and ω_e values; 4) bleaching behaviour - on irradiation of the matrix with visible light, the photodetached electrons, from counterions, recombine with cations leading to a decrease in absorption peak intensity.

Armed with the knowledge of the value of the origin of the B $^4\Sigma_u^-$ – X $^4\Sigma_g^-$ transition in the matrix (T_0- 19765(4) cm^{-1} [24]; calculated value[23] - 19920(400) cm^{-1}), characterization of C_2^+ in the gas phase was undertaken. The production of C_2^+ was achieved by the reaction of acetylene with helium metastables (2^1S, 2^3S) and He^+ (which are generated in a dc discharge). The conditions are so chosen that collisions of C_2^+ with H, H_2, and C_2H_2 are minimized. The detection of the B $^4\Sigma_u^-$ ← X $^4\Sigma_g^-$ transition is carried out by means of the laser excitation technique; wavelength cut-off and broad-band optical filters are used to reduce the stray light and the concomitantly excited C_2 $d^3\Pi_g \rightarrow a\ ^3\Pi_u$ fluorescence. A typical spectrum, recorded with 0.3 cm^{-1} resolution, is reproduced in figure 6 and shows characteristic rotational structure.[26]

The spectrum is readily rotationally assigned - it is a text-book example of a Σ_u^- - Σ_g^- transition of a homonuclear species with V zero nuclear spin. Thus all the even N rotational lines are missing; on producing $^{13}C_2^+$ all the lines become apparent but the even lines are now three times as intense as the odd ones reflecting 1/2 nuclear spin of ^{13}C. Higher resolution recordings (~0.04 cm^{-1} fwhm) result in splitting of the rotational lines into several components (up to four) as result of spin-spin and spin-rotation interactions (quartet state). In addition, the lifetimes of selected vibronic levels in the B $^4\Sigma_u^-$ state of C_2^+ can be measured, from which the oscillator strength of the respective rovibronic transition is

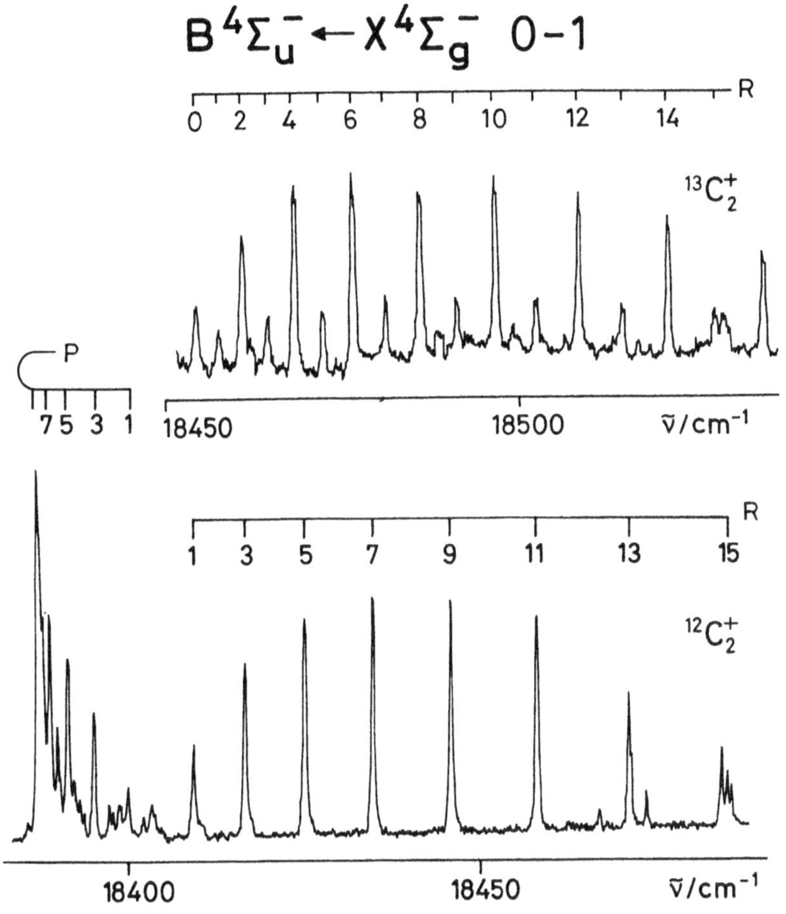

Figure 6. The rotationally resolved laser excitation spectrum ($0.3\ cm^{-1}$ fwhm) of the B $^4\Sigma_u^-$ – X $^4\Sigma_g^-$ 0-1 band of $^{12}C_2^+$ and $^{13}C_2^+$.

inferred (using the calculated Frank-Condon factors). Thus for example, $f_{0\text{-}0} = 0.023(3)$ is obtained[27].

With the knowledge of the spectroscopic constants and the oscillator strengths, C_2^+ can be identified by means of its B $^4\Sigma_u^-$ – X $^4\Sigma_g^-$ electronic transition ($T_0 = 19720.6(2)\ cm^{-1}$) and its abundance evaluated, in flames or plasmas by laser excitation spectroscopy, in comets by resonance fluorescence, and perhaps in diffuse interstellar clouds via starlight absorptio

In order to properly characterize the potential energy surface of C_2^+ in the ground state, which is necessary in considering the details of reaction dynamics of C_2^+, one needs to probe excited vibrational levels. In the production of C_2^+ from acetylene, only the v = 0,1,2 levels of the X $^4\Sigma_g^-$ state were sufficiently populated to be investigated by the laser excitation method. However, higher levels (v > 2) can be accessed by transferring at first the population to the excited state by laser excitation of a B $^4\Sigma_u^-$ ← X $^4\Sigma_g^-$ vibrational transition. Radiative transitions from the upper level to levels v > 2 in the ground state may take place. To probe such weak transitions and to enable high resolution measurements to be undertaken, instead of recording wavelength dispersed fluorescence spectra (after the laser excitation) the stimulated emission pumping technique is the more attractive[28] and has already been applied to study ions.[29] The principle is illustrated by means of the potential energy curves of C_2^+ in fig. 7.

The population is transferred from the v=0 level of the X $^4\Sigma_g^-$ state to say the v' = 2 level of the B $^4\Sigma_u^-$ state by the first laser (pump). Transitions are then stimulated back to the X $^4\Sigma_g^-$ state, say to the v = 5 level by the second laser (dump). Thus the wavelength of the pump laser is chosen to coincide with a rotational line and is kept fixed, whereas the dump laser is scanned while the fluorescence from the v' = 2 level, say to levels v > 3 in the X $^4\Sigma_u^-$ state, is monitored. When the dump laser is resonant, there is a decrease in the fluorescence intensity. Due to the double resonance nature of the experiment, the initially selected rotational line, say R(8) within the B $^4\Sigma_u^-$ ←X $^4\Sigma_g^-$ 2-0 band, results in the observation of an R(8) and P(6) lines within the B $^4\Sigma_u^-$ → X $^4\Sigma_g^-$ 2-5 band. Such examples can be seen in the bottom part of fig. 7.

The rotational structure of the whole band is gradually accumulated by changing the R(N) or P(N) line picked out by the pump laser. By this means the v = 3-6 vibrational levels of the X $^4\Sigma_g^-$ state of C_2^+ were characterized[30] and the usual vibrational-rotational constants inferred. These in turn can be used to generate a potential function for C_2^+ in the

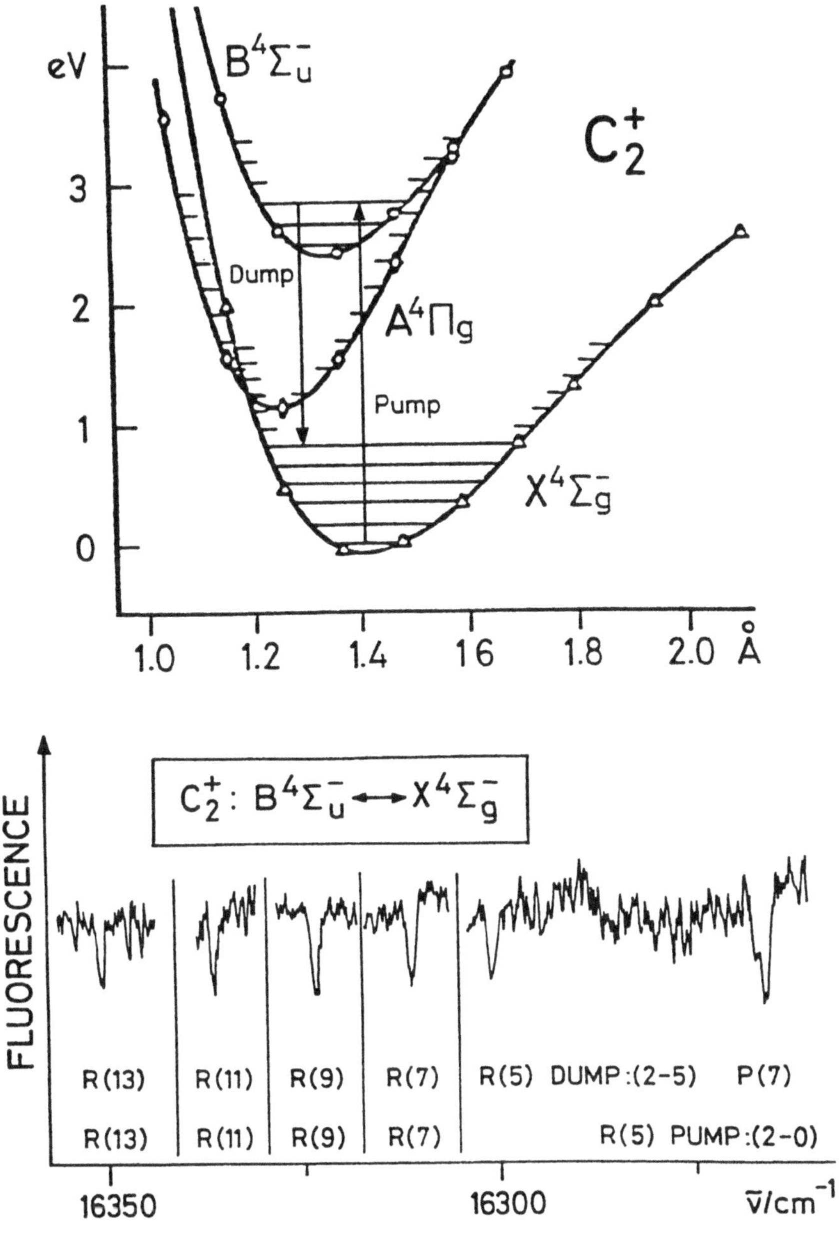

Figure 7. The spectroscopic principle of the stimulated emission pumping approach used to probe higher lying vibronic levels of C_2^+ in its ground state (top), and some of the experimentally observed transitions (bottom, 0.3 cm^{-1} resolution).

ground state. Thus one can now consider probing ion-molecule reactions of C_2^+ as both the potential energy surface and a spectroscopic method is available to follow them.

5. Absorption Spectra of Mass-Selected Ions in a Neon Matrix

Having illustrated the advantage of obtaining the initial spectral information on ions, especially those of fragments, first at low resolution in absorption prior to attempting gas phase studies with narrow band lasers (in the case of C_2^+ the gas phase origin was found to lie ~ 34 cm^{-1} to the red of the matrix value), a general approach to tackle this area materizes. Clearly one wants to combine the matrix waveguide absorption method with mass-selection. In the past years we have been developing such an approach.

In a trial set-up, an electron impact ion source was coupled via a small quadrupole mass-analyser to the matrix substrate in the fashion depicted in figure 8.[31] The ion of interest is selected by the mass-analyser and is

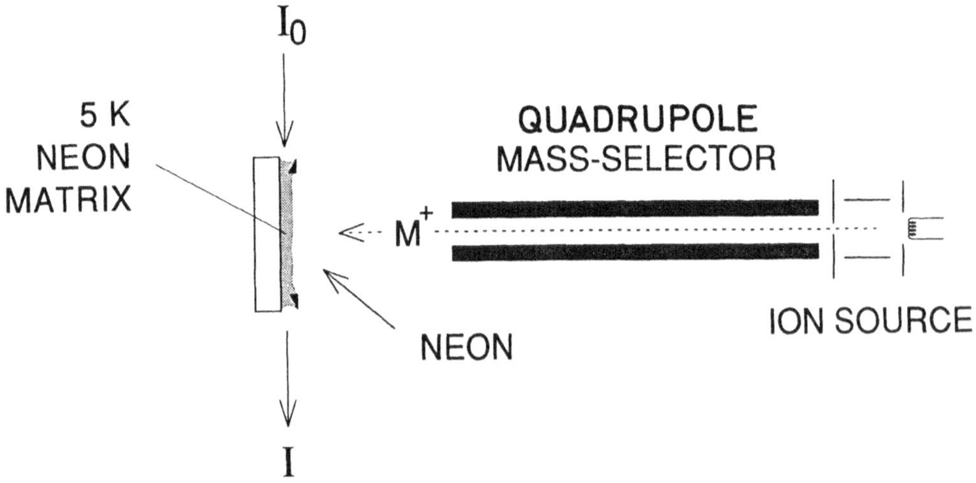

Figure 8. The set-up of the first apparatus used to measure the absorption spectra of mass-selected ions in a 5 K neon matrix.

codeposited with neon (in the usual dilute matrix isolation ratio). Typically ion currents of 10^{-8}-10^{-9} amp were used and the nominal energy of the ions exiting from the quadrupole was in the 50-150 eV range. A deposition time of 2-3 hr was required to grow a neon matrix (5K) of 100-150 μm thickness.

In figure 9 is shown the observed electronic absorption spectrum observed with such an apparatus when diacetylene was introduced into the ion source and $C_4H_2^+$ ions were mass-selected and codeposited with

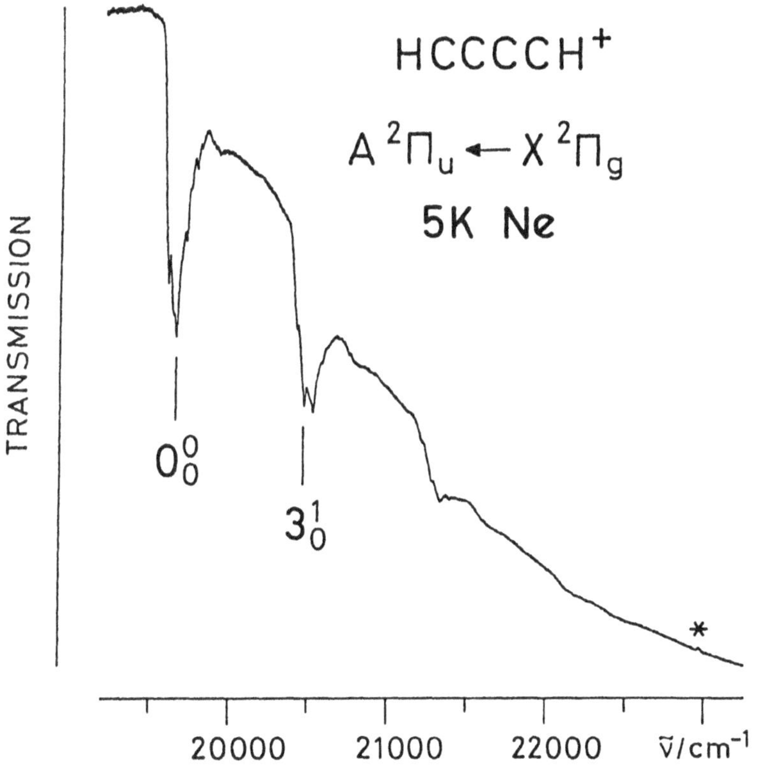

Figure 9. The A $^2\Pi_u \leftarrow X\ ^2\Pi_g$ electronic absorption spectrum of diacetylene cation (0.1 nm resolution) recorded after codepositing $C_4H_2^+$ (~ 150 eV, ~ 10^{-8} A) with neon.

neon. The band system seen is the A $^2\Pi_u \leftarrow X\ ^2\Pi_g$ electronic transition of diacetylene cation, which is well known from earlier studies both in the gas-phase and in the matrix. In a second experiment, the fragment ion C_2^+, which is produced in the ion source from various precursors such as ClCCCl, NCCN, is codeposited and the resulting spectrum (fig. 10) shows the absorption bands of C_2^+ identified previously following photoionization of C_2 (cf. fig. 5). The marked difference between the spectra is evident; in

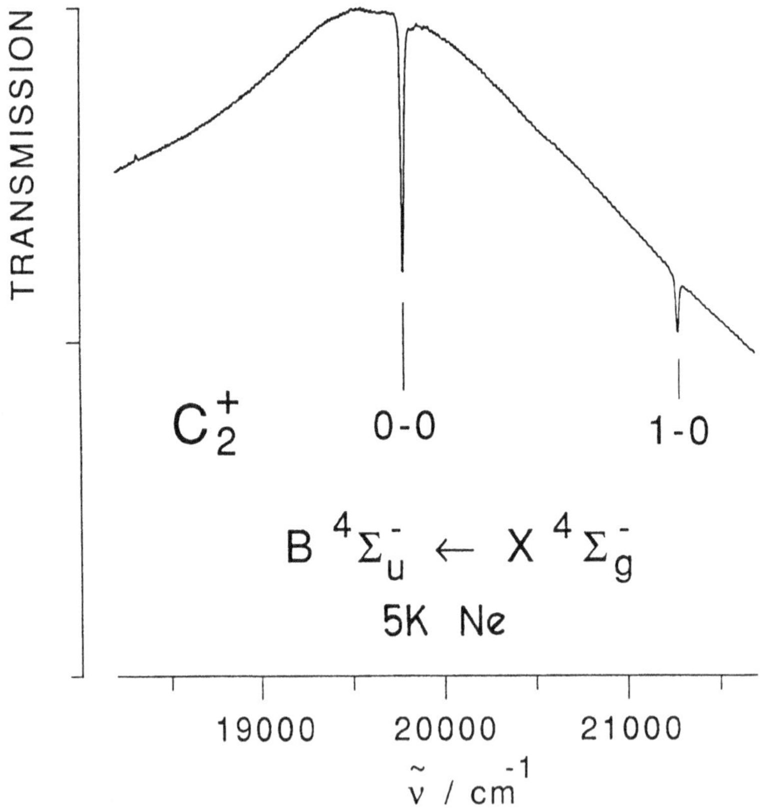

Figure 10. The B $^4\Sigma_u^- \leftarrow X\ ^4\Sigma_g^-$ absorption bands of C_2^+ in a 5 K neon matrix observed after C_2^+ (~ 150 eV, ~10^{-9} A) was codeposited with neon.

the mass-selected approach not only are the numerous unidentified bands due to side products absent, but so is the absorption of the parent precursor ion.

In order to overcome several technical problems and to make improvements - adequate differential pumping of the ion source, mass- selector and matrix chamber, increase of ion current and variability of the ion deposition translational energy - a second version of the apparatus was constructed. This is depicted schematically in fig. 11. The ion source is now coupled to the quadrupole mass-selector by an octopole ion-beam guide. The application of this apparative set-up has resulted in an order of magnitude increase in the available ion current.

In summary, these results have shown that a potentially powerful new spectroscopic approach is available for the characterization of mass-selected ions. Once fully developed it should be applicable to a variety of

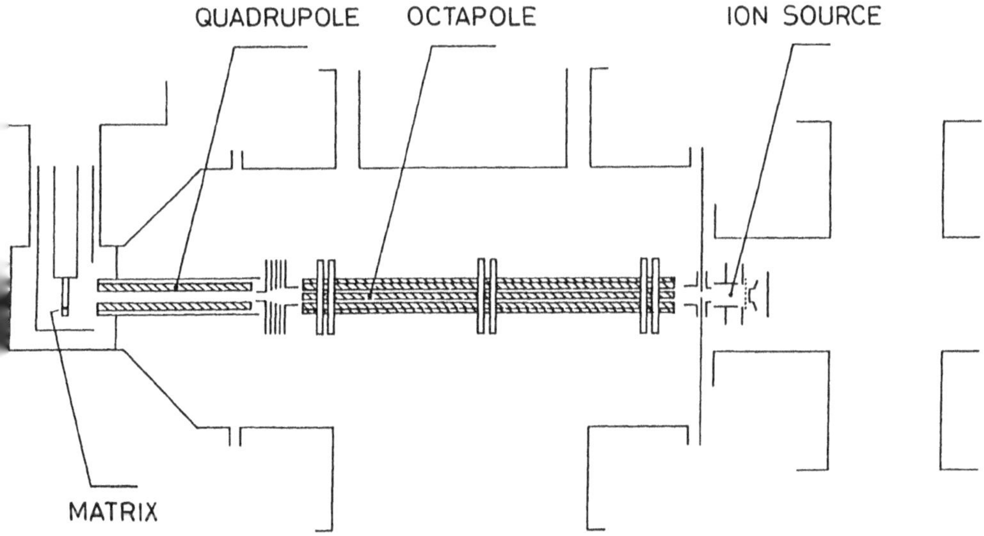

Figure 11. The basic features of the second version of the apparatus built for the codeposition of mass-selected ions with neon.

ions known hitherto only in mass-spectrometers. The ions, both positive and negative, once embedded in the 5 K neon matrix can be characterized at leisure by established techniques used in conjunction with matrix isolation - uv/vis, ir and esr for the paramagnetic ions. The technique may also prove amenable for the initial characterization of cluster ions, as well as various mass-selected neutral entities if ways can be found to efficiently neutralize the ions constrained in the matrix..

6. Absorption-Dissociation Spectroscopy of Ion-Beams in the Gas Phase

Assuming that one can locate the electronic transitions of the mass-selected ion in neon matrices, one then needs to have high resolution gas-phase methods available for the detailed characterization. When dealing with diatomic cations, the laser excitation approach is attractive (c.f. C_2^+). On the other hand, many of the interesting polyatomic cations are not likely to fluoresce and therefore alternative techniques are sought. One such is the one we have been developing based on the one photon absorption plus one photon dissociation of a mass-selected ion beam.[32] In this, two bound electronic states of the ion are to be probed and the transition between them is induced by the first photon $h\nu_1$. The electronically excited ion is then made to absorb a second photon $h\nu_2$ to a repulsive or predissociative state which produces a fragment ion. The bound-bound transition is observed by monitoring the intensity of the fragment ion as function of the laser wavelength $h\nu_1$ while that of $h\nu_2$ is unaltered. The scheme of this approach and the demonstration of this will be discussed for the case of CS_2^+, whose relevant potential energy curves are shown in figure 12.

The outlay of the apparatus constructed for the studies of mass-selected ion beams in the gas-phase is drawn schematically in fig. 13. Ions are produced at low rotational temperatures by means of electron impact

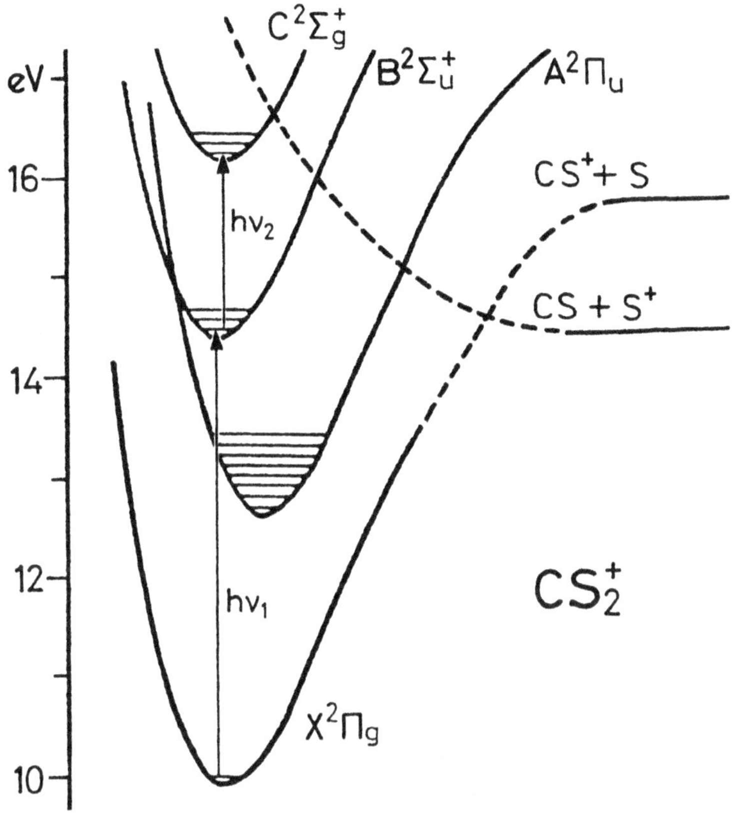

Figure 12. A schematic representation of the potential energy curves of CS_2^+ and the thermodynamic limits. In the actual experiments, the $B\ ^2\Sigma_u^+ \leftarrow X\ ^2\Sigma_g^+$ transition is probed by the $h\nu_1$ laser photons, while the $C\ ^2\Sigma_g^+ \leftarrow B\ ^2\Sigma_u^+$ one by $h\nu_2$ photons.

ionisation of a pulsed supersonic free jet and are extracted by a lens system into the first quadrupole mass-analyser (Q_1). There the ion of interest is selected (eg. CS_2^+) and is passed on into the middle quadrupole (Q_2) where only an rf field is applied so that ions within a broad range of mass are constrained at the same time. The selected ion (CS_2^+) absorbs in this region the first photon ($h\nu_1$ from a pulsed dye laser passed coaxially into the instrument), and the bound-bound transition is induced (CS_2^+: $B\ ^2\Sigma_u^+ \leftarrow X\ ^2\Pi_g$). Some ns later, the second photon ($h\nu_2$ from a second dye

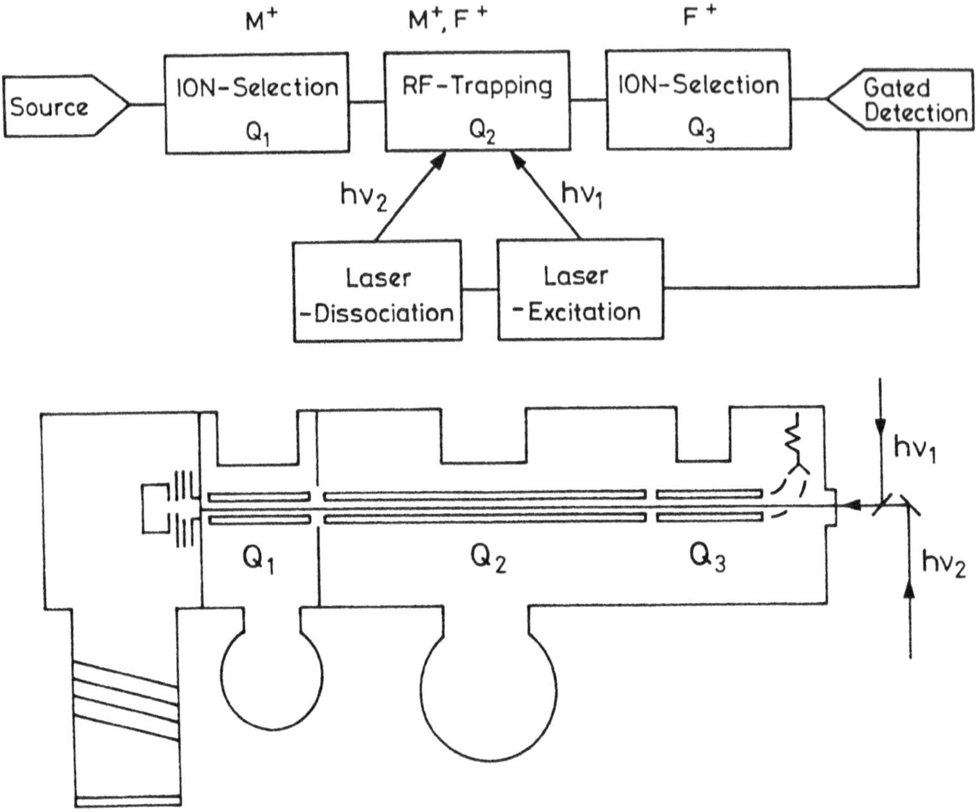

Figure13. The essential features of the tandem triple quadrupole apparatus constructed for the two-photon absorption-dissociation spectrosocopy measurements on mass-elected ions in the gas-phase.

laser but pumped by the same excimer laser as for $h\nu_1$) causes a transition from the excited state of the ion to a predissociative or repulsive final state (CS_2^+ $C\,^2\Sigma_g^+ \leftarrow B\,^2\Sigma_u^+$) which leads to fragment ions (CS^+ or S^+). Both the parent (CS_2^+) and fragment ions (CS^+, S^+) are confined by the rf field in Q_2 and are transmitted to the third tandem quadrupole mass-analyser

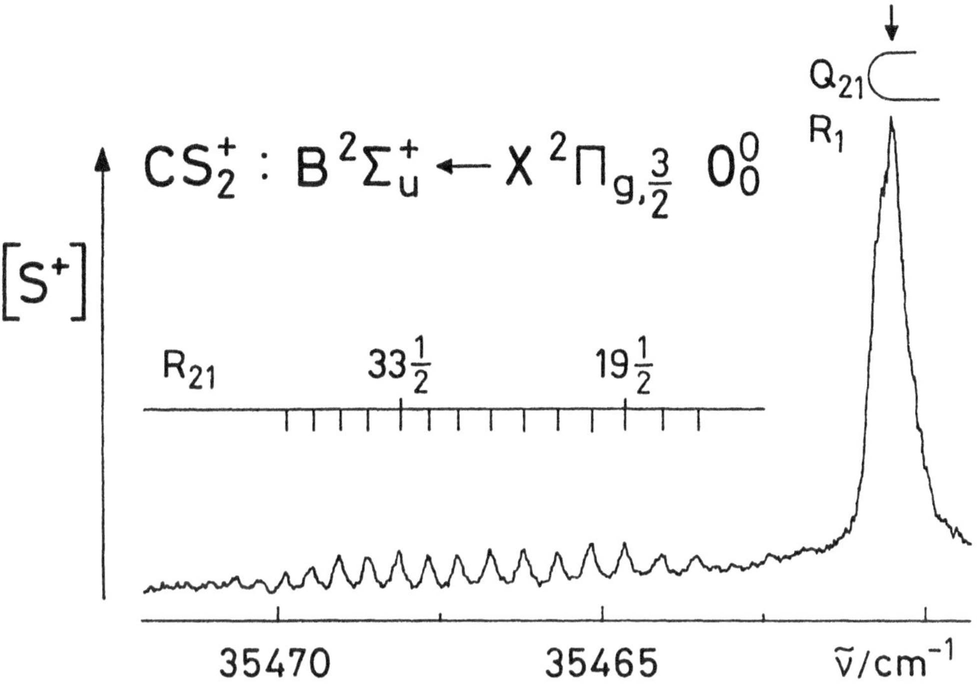

Figure 14. The rotationally resolved B $^2\Sigma_u^+ \leftarrow$ X $^2\Pi_g$ origin band of CS_2^+ observed by two photon absorption ($h\nu_1$ scanned with 0.04 cm^{-1} resolution) - dissociation ($h\nu_2$ fixed) while S^+ is monitored.

(Q_3) which is tuned to select only a particular fragment ion (e.g. S^+). The latter are counted while $h\nu_1$ is scanned.

The result of such a measurement on CS_2^+ is reproduced in fig. 14. The rotational structure of the B $^2\Sigma_u^+ \leftarrow$ X $^2\Pi_g$ transition is resolved and the indicated assignments are taken from an earlier analysis of this band system in emission[33]. This illustrates the potential fo the technique to characterize the ground and an excited bound state of a mass-selected ion.

In case of an unknown system, the envisaged procedure would be 1) locate the transition in the matrix (cf. section 5), 2) find a wavelength at which the mass-selected ion (in the ground state) in the gas-phase absorbs a photon to produce a fragment ion (either in a one-photon $h\nu_D$, or two photon process), 3) combine the information from steps 1) and 2) to perform analogous measurements as outlined for CS_2^+ (i.e. $h\nu_1+h\nu_2 = h\nu_D$). An alternative, but less sensitive approach would be to monitor the intensity change in the fragment ions produced in step 2 while $h\nu_1$ is scanned; a decrease occurs when it is resonant with the bound-bound transition. Such an experiment has been in fact demonstrated on the B $^2\Sigma^+ \leftarrow$ X $^2\Pi$ transition of N_2O^+ while observing the NO^+ product.[34]

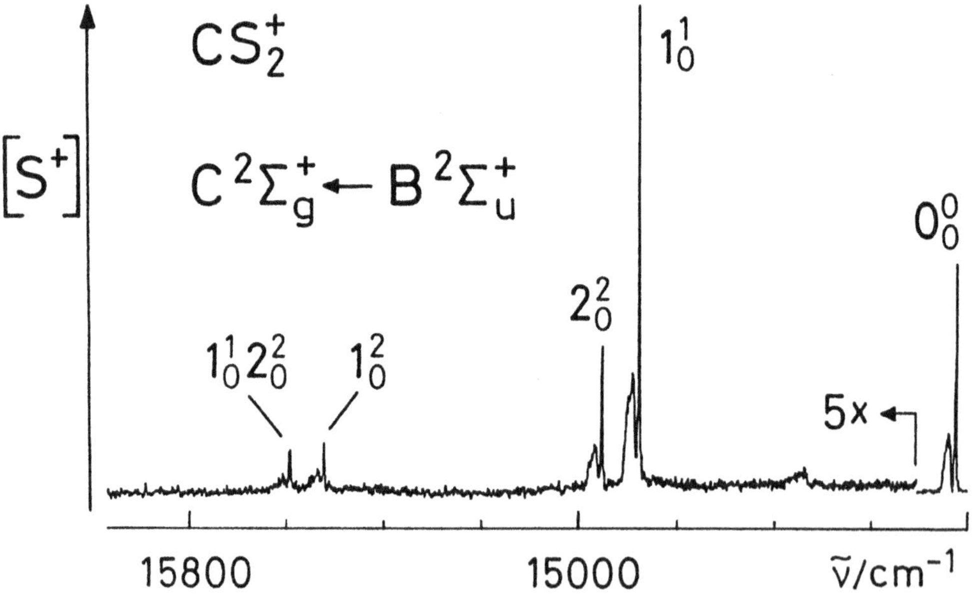

Figure 15. The vibronic structure of the C $^2\Sigma_g^+ \leftarrow$ B $^2\Pi_u^+$ v=0 transition of CS_2^+ detected by the two photon absorption ($h\nu_1$ fixed at the maximum of the B $^2\Sigma_u^+ \leftarrow$ X $^2\Pi_g$ origin band - see arrow in fig. 14) - dissociation ($h\nu_2$ scanned with 0.6 cm^{-1} bandwidth) while S^+ is monitored.

When the final excited state ($C\,^2\Sigma_g^+$ in case of CS_2^+) predissociates at a slow enough rate, the spectroscopic structure of the transition induced by the second photon ($h\nu_2$) is not smeared out. This can be exploited to deduce the spectroscopic constants of the ion in the predissociative state.

In fig. 15 is shown the result of such a measurement for the $C\,^2\Sigma_g^+ \leftarrow B\,^2\Sigma_u^+$ transition of CS_2^+. The CS_2^+ ions are prepared at first in several rotational states in the $B\,^2\Sigma_u^+$ state by exciting (with $h\nu_1$) the

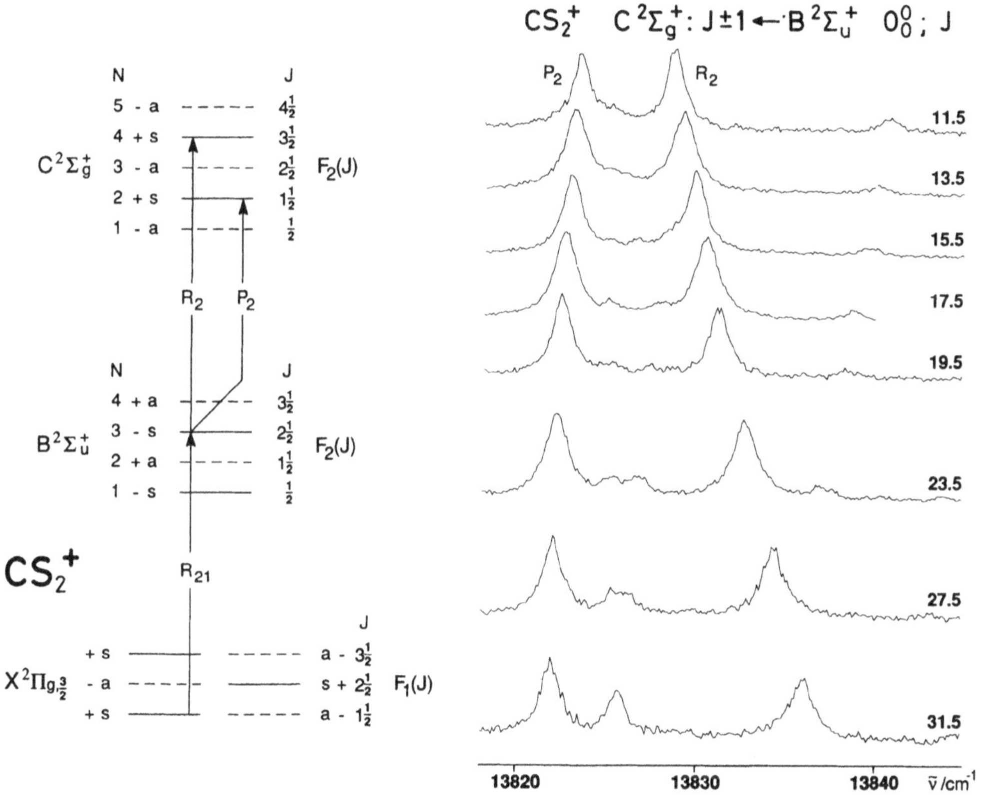

Figure 16. The rotational structure in the $C\,^2\Sigma_g^+ \leftarrow B\,^2\Pi_u$ origin band of CS_2^+. The first laser photon ($h\nu_1$) picks out a single rotational line within the $B\,^2\Sigma_u^+ \leftarrow X\,^2\Pi_g$ transition (with 0.04 cm^{-1} resolution), while the second laser photon ($h\nu_2$) is scanned and two rotational transitions are then observed.

$B\,^2\Sigma_u^+ \leftarrow X\,^2\Pi_g^+$ transition at the band head of the origin band (see arrow in fig. 14). The wavelength of the second photon $h\nu_2$ is then tuned and the fragment ion S^+ is monitored. The peaks observed in the spectrum correspond to the vibronic transitions within the $C\,^2\Sigma_g^+$ state of CS_2^+.[32]

In order to determine the geometry of the ion in the excited state, the rotational constant has to be evaluated (for the $X\,^2\Pi_g$, $A\,^2\Pi_u$ and $B\,^2\Sigma_u$ states these are known from the rotational analyses of the emission spectra). The rotational lines of the $C\,^2\Sigma_g^+ \leftarrow B\,^2\Sigma_u^+$ band system can be, observed[35] by initially populating a single rotational level in the $B\,^2\Sigma_u^+$ state; i.e. with $h\nu_1$ in the $B\,^2\Sigma_u^+ \leftarrow X\,^2\Pi_g$ transition. This is further explained by figure 16 (left); on the right some of the measured spectra are plotted. The complete rotational structure is gradually accumulated by chosing different rotational levels with $h\nu_1$. In the case of CS_2^+, and in analogous.measurements on CO_2^+ [36], the rotational analyses prove that the ions remain linear in their $C\,^2\Sigma_g^+$ states.

To conclude, the recent developments aimed at spectral characterization of ions, mass-selected in neon matrices as well as in the gas-phase have been outlined. The potential of these techniques has now been illustrated by test examples and their further development and application promises to open up new pathways to determine the spectroscopic parameters, potential energy surfaces and geometries of a variety of ions known hitherto only in mass-spectrometers.

7. Acknowledgements

The research projects carried out at Basel have only been possible because of the efforts of the various PhD students (most recently - D. Forney, M. Rösslein, T. Wyttenbach) and Postdoctoral fellows (most recently - Dr. P.O. Danis, Dr. F.G. Celii, Dr. D.D. Evard, Dr. J. Fulara, Dr. M.A. Hanratty and Dr. T.M. Orlando) The project as a whole could be realized because of the generous financial support by the "Swiss Science Foundation" throughout the years.

References

1. T. Oka, Philos. Trans. R. Soc. London Ser. A., **324**, 1298 (1987); R.J. Saykally, Science **239**, 157, (1988).
2. R.C. Woods in "Ion and Cluster Ion Spectroscopy and Structure", J.P. Maier (ed.), Elsevier, Amsterdam, 1989, p. 27.
3. Z. Vager, R. Naaman and E.P. Kanter in "Ion and Cluster Ion Spectoscopy and Structure", J.P. Maier (ed.),Elsevier, Amsterdam, 1987, p. 1.
4. D. Klapstein, J.P. Maier and L. Misev, in"Molecular Ions: Spectroscopy, Structure and Chemistry", T.A. Miller and V.E. Bondybey (eds.), North-Holland, New York, 1983, p. 175.
5. J.P. Maier, Chem. Soc. Rev. **17**, 45 (1988) and references therein.
6. J.P. Maier, O. Marthaler, L. Misev and F. Thommen, in "Molecular Ions", J. Berkowitz and K.O. Groeneveld (eds.), Plenum Press, New York 1983, p. 125.
7. M. Allan and J.P. Maier, Chem. Phys. Lett. **41**, 231 (1976).
8. J. Fulara, D. Klapstein, R. Kuhn and J.P. Maier, J. Phys. Chem. **89**, 4213 (1985).
9. F.G. Celii, J. Fulara, J.P. Maier and M. Rösslein, Chem. Phys. Lett. **131**, 325 (1986).
10. E. Heilbronner, V. Hornung and K.A. Muszkat, Helv. Chim. Acta **53**, 347 (1970).
11. G. Herzberg, "Molecular Spectra and Molecular Structure", Vol. 2, van Nostrand, Princeton 1945.
12. J. Kraitchman, Am. J. Phys. **21**, 17 (1953).
13 J.H. Callomon, Can. J. Phys. **32**, 1046 (1956).
14. R. Kuhn, J.P. Maier and M. Ochsner, Mol. Phys. **59**, 441 (1986)
15. J. Lecoultre, J.P. Maier and M. Rösslein, J. Chem. Phys. **89**, 6081 (1988).

16. J.H. Callomon and B.P. Stoicheff, Can. J. Phys. **35**, 373 (1957).
17. F. Brogli, E. Heilbronner, V. Hornung and E. Kloster-Jensen, Helv. Chim. Acta **56**, 2171 (1973).
18. M. Rösslein and J.P. Maier, J. Phys. Chem. **93**, 7342 (1989).
19. M. Rösslein, M.A. Hanratty and J.P. Maier, Mol. Phys. **68**, 823 (1989).
20. V.E. Bondybey and L.E. Brus, Adv. Chem. Phys. **41**, 269 (1980); S. Leutwyler, J.P. Maier and U. Spittel, Mol. Phys. **71**, 437 (1984).
21. A. O'Keefe, R. Derai and M.T. Bowers, Chem. Phys. **91**, 161 (1984).
22. C. Petrongolo, P.J. Bruna, S.D. Peyerimhoff and R.J. Buenker, J. Chem. Phys. **74**, 4594 (1981).
23. P. Rosmus, H.-J. Werner, E.-A. Reinsch and M. Larsson, J. El. Spectr. Rel. Phenom. **41**, 289 (1986).
24. D. Forney, H. Althaus and J.P. Maier, J. Phys. Chem. **91**, 6458 (1987).
25. V.E. Bondybey and T.A. Miller, in "Molecular Ions: Spectroscopy, Structure and Chemistry", T.A.Miller and V.E. Bondybey (eds.), North-Holland, New York, 1983, p. 125;
S. Leutwyler, J.P. Maier and U. Spittel, Chem. Phys. Lett. **96**, 645 (1983).
26. M. Rösslein, M. Wyttenbach and J.P. Maier, J. Chem. Phys. **87**, 6770 (1987).
27. J.P. Maier and M. Rösslein, J. Chem. Phys. **88**, 4614 (1988).
28. C.E. Hamilton, J.L. Kinsey and R.W. Field, Ann. Rev. Phys. Chem. **37**, 493 (1986).
29. F.G. Celii, J.P. Maier and M. Ochsner, J. Chem. Phys. **85**, 6230 (1986).
30. F.G. Celii and J.P. Maier, Chem. Phys. Lett. **166**, 517 (1990).
31. D. Forney, M. Jakobi and J.P. Maier, J. Chem. Phys. **90**, 600 (1989).

32. P.O. Danis, T. Wyttenbach and J.P. Maier, J. Chem. Phys. **88**, 3451 (1988).
33. J.H. Callomon and F. Creutzberg, Philos. Trans. R. Soc. London Sect. A, **277**, 157 (1974).
34. P.O. Danis, D.D. Evard, D. Forney, M. Rösslein, T. Wyttenbach and J.P. Maier, Faraday Discuss., Chem. Soc. **86**, 85 (1988).
35. D.D. Evard, T. Wyttenbach and J.P. Maier, J. Phys. Chem. **93**, 3522 (1989).
36. T. Wyttenbach, D.D. Evard and J.P. Maier, J. Chem. Phys. **90**, 4645 (1989).

VIBRATIONAL MODE EFFECTS IN POLYATOMIC ION REACTIONS

SCOTT L. ANDERSON
Chemistry Department
State University of New York at Stony Brook
Stony Brook, NY 11794-3400 USA

ABSTRACT. This paper discusses use of multiphoton ionization (MPI) to prepare ions with excitation in selected vibrational modes, and guided ion beam studies of the effects of different modes of nuclear motion on reactions of small polyatomic ions.

1. Introduction

Over the years a technique which has been useful in probing the dynamics of chemical reactions is examination of the effects of different types of reactant excitation on reactivity, product branching ratios, and recently on product state distributions. This state-selection technique has been particularly successful in helping understand diatom-plus-atom reactions, where it has been possible complement experiment with theoretical modeling, and in some cases with accurate scattering calculations.

For reactions of polyatomic molecules things are complicated by the increased number of vibrational degrees of freedom. It is not at all obvious how different types and symmetries of vibrations will couple to the reaction coordinate (ie. how effectively and by what mechanism they influence the course of a reactive collision). The increased number of vibrational modes clearly makes theoretical work much more difficult, but also raises some experimental problems.

First it is necessary to prepare reactants with pure excitation in the desired vibrational states. From a fundamental perspective, this is more difficult for polyatomic molecules due to the increased density of states. From a practical perspective there is the additional (BIG!) problem that the spectroscopic ground work which is needed to devise state selection schemes is frequently not available or complete, even for small (3 to 4 atom) molecules.

If the effects of exciting different vibrational modes are to be probed, it is important that the excitation stay in the desired vibrational mode during the time up to the initial phases of the collision event. Loss of mode selectivity could occur by non-radiative (IVR) processes, however this is only a possibility for high levels of excitation and/or large molecules. Even for low level excitation of small polyatomic molecules it is possible that interactions during the initial phase of a collision could mix different vibrational modes, so that the reactants lose memory of their initial excitation mode before coming into intimate contact. This might be a particularly severe problem for low energy ion-molecule collisions where the ion-induced dipole potential has a long range.

K. R. Jennings (ed.), Fundamentals of Gas Phase Ion Chemistry, 183–196.

This paper will cover several aspects of this problem. Methods for preparing mode selectively excited ions using multiphoton ionization will be briefly mentioned. Several reactions showing mode specific effects and one example where only the vibrational energy appears to be important will then be presented. The paper closes with some discussion of where this field may be going.

2. Preparation of Mode Selectively Excited Ions

There are basically two photoionization methods which can be used to prepare small polyatomic ions in particular and variable vibrational states. Photoion-photoelectron coincidence (discussed in the contribution by O. Dutuit in this volume) simply relies on selecting the internal energy of the ions, which of course allows selection of the vibrational mode if the energy resolution is high enough. The coincidence method is in principle very general, however for any given molecule the range of vibrations which can be produced is limited to those which are Franck-Condon active in ionization. In addition, resonance autoionization allows generation of ions with very high levels of vibrational excitation (although in this scheme no mode selection is possible).

Multiphoton ionization (MPI) in its simplest form simply relies on Franck-Condon factors to determine the ion vibrational states produced, however as shown below, multiphoton excitation makes it possible to tune the Franck-Condon factors to favor the desired ion vibrational state. A complete discussion of the ins and outs of MPI state-selection is given in a recent review article[1]. The basic scheme for MPI is shown in Fig. 1. We start with ground electronic and vibrational state molecules, and excite with one or more photons to a particular vibrational (and in some cases rotational and fine structure) level of a Rydberg intermediate electronic

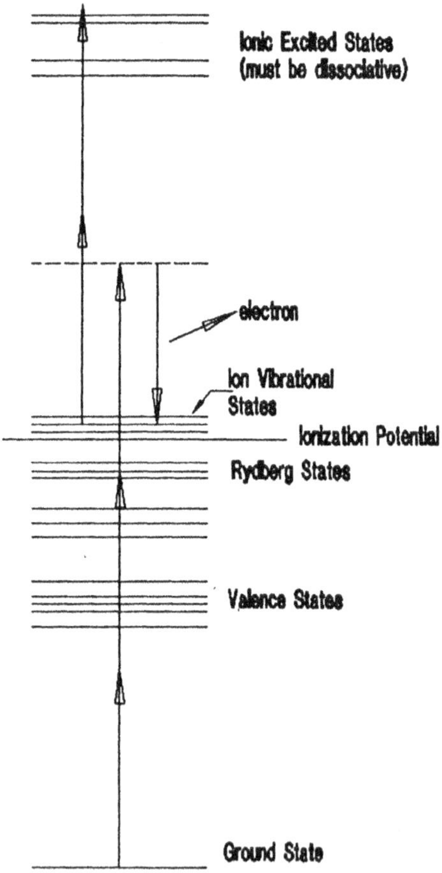

Figure 1 Typical MPI scheme.

state. This state must be chosen so that absorption of a single additional photon provides sufficient energy to ionize the molecule (otherwise accidental double resonances with higher energy levels are likely to perturb the ion internal energy state distribution). The intermediate Rydberg state basically consists of an excited electron weakly coupled to an ion core, and thus ionization of the Rydberg electron tends to leave the ion core state unchanged. In other words, the Franck-Condon factors for ionization of a good Rydberg state are diagonal -- $\Delta v = 0$ processes dominate. This means that by selecting a particular vibrational level of the Rydberg state we favor production of ions in the same vibrational state. There is also a propensity to conserve the fine structure and rotational states of the ion core, giving selection of these degrees of freedom as well. In particularly favorable cases, it is possible to achieve pure selection of fine structure, vibrational, rotational, and Λ-doublet level of the ion[2].

Table **1.** Accessible vibrational modes for $C_2H_2^+$ and OCS^+

Ion	**State**	**Energy** (cm^{-1})
$C_2H_2^+$	Ground State	0
	$\nu_2=1$ (C-C str.)	1814
	$\nu_2=2$	3624
	$\nu_{cis\text{-}bend}=2$	1253
OCS^+	$^2\Pi_{3/2}$	0
	$^2\Pi_{1/2}$	360
	(001) (C-S str.)	700
	(002)	1400
	(020) (bend)	1000
	(100) (C-O str.)	2041

Figure 1 also shows a potential problem with MPI state-selection. The ions are produced in an intense laser field, and if there are low-lying ionic excited states there is some probability for absorption of additional photons. Regardless of whether these states are radiatively or non-radiatively coupled to the ground state, a distribution of highly excited ions are produced. Since these energetic species may be much more reactive than ions in the selected vibrational state, even a small fraction can cause serious problems for ion chemistry studies. Working with 1+1 photon ionization schemes at low power, one can minimize (but not eliminate) this effect, however it is best if the excited states are dissociative. In this case, any unwanted ionic photoexcitation simply results in fragmentation, and the fragments can be easily removed by mass spectrometry.

MPI state-selection methods have been worked out for a number of small molecules[1], however only for the cases of $C_2H_2^+$[3][4] and OCS^+[5] have excitation routes been found which allow production of mode selectively excited ions. This situation partly is due to the difficulties involved in mode-selective MPI, but also reflects the fact that most work on MPI photoelectron spectroscopy has focussed on the spectroscopy and ionization dynamics, and very few concerted efforts to find state-selection routes have been mounted. For the cases of $C_2H_2^+$ and OCS^+, table 1 summarizes the vibrational modes which can be excited by MPI.

3. Experimental Setup

For our ion-molecule reaction experiments we use a tandem guided beam mass spectrometer as shown in figure 2. The neutral precursor gas is injected as a pulsed, skimmed, supersonic beam, which passes through an octapole ion guide in the center of the main chamber. The beam is canted at a 15° angle to impart some momentum along the axis of the guide. The ionizing laser is focused to intersect the molecular beam in the center of the ion guide, thus the ions are formed in the field free region. After reflecting from the rf "wall" of the ion guide, the ions are extracted by a 0.1 V/cm field and formed into a beam. To filter out fragment ions produced in the MPI process, the beam is passed through a home-made Wien filter, and injected into a second set of octapole ion guides where the desired collision energy is set. The ions are then guided through a scattering cell containing ca. 2.5 x 10^{-5} Torr of the neutral reactant, and products and unreacted primary ions are collected by the guides, mass analyzed, and counted.

4. Selected Experimental Results

To date we have studied reactions of $C_2H_2^+$ with CH_4[6], D_2[7], and OCS[8], and reaction of OCS^+ with C_2H_2[9] and rare gases[10]. For this paper I will present just a few examples of the types of effects which we observe. I begin with a reaction (the only one we have seen so far) where the vibrational mode effects are consistent with chemical intuition, then give several examples of systems where the origins of the vibrational effects are unclear.

The product channels observed for reaction of $C_2H_2^+$ with methane are:

$$
\begin{aligned}
C_2H_2^+ + CH_4 \dashrightarrow\ & C_3H_5^+ + H & \Delta H &= -0.92\ \text{eV}\\
& C_3H_4^+ + H_2 & \Delta H &= -1.33\ \text{eV}\\
& C_3H_3^+ + H + H_2 & \Delta H &= +0.41\ \text{eV}\\
& C_2H_3^+ + CH_3 & \Delta H &= +0.02\ \text{eV.}
\end{aligned}
$$

Figure 3 shows the cross sections we measure for the $C_3H_5^+$ and $C_2H_3^+$ channels, plotted as a function of collision energy for four different $C_2H_2^+$ reactant vibrational states. ν_2 is the C-C stretch with energy of 1814 cm^{-1}, and two quanta of ν_{bend} (probably the cis-bend) is 1253 cm^{-1}. The cross sections for the $C_3H_4^+$ channel are not plotted because they are simply a factor of 3.1 lower than those for $C_3H_5^+$, but otherwise have identical collision and vibrational energy dependence. $C_3H_3^+$ is a minor channel that primarily results from decomposition of $C_3H_5^+$ and $C_3H_4^+$ at high collision energies.

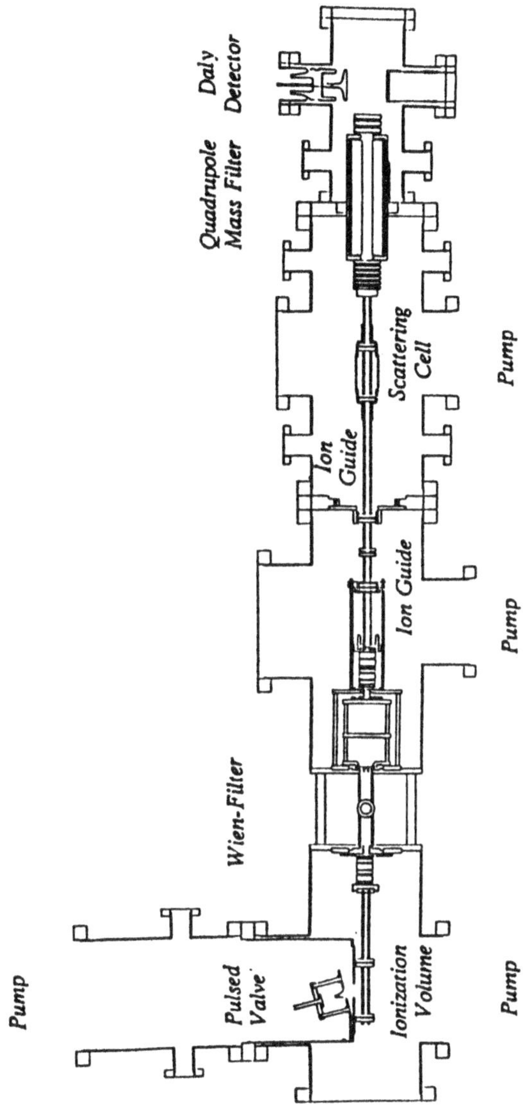

Figure 2 Side view of the Stony Brook MPI - guided beam mass spectrometer system.

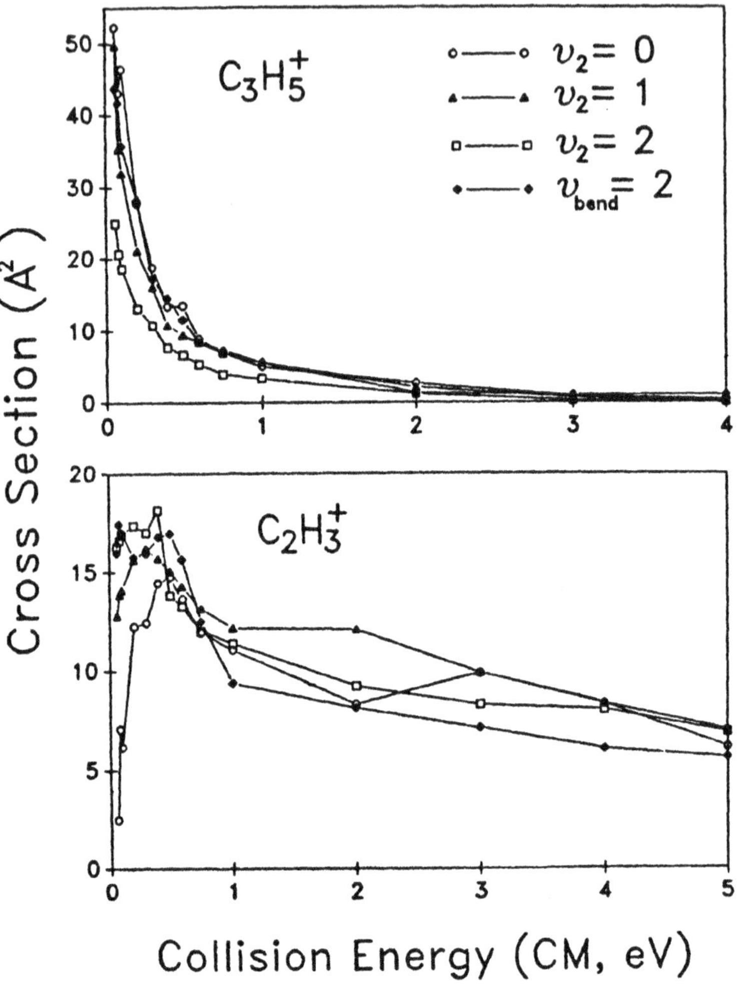

Figure 3 Cross sections for $C_2H_3^+$ and $C_3H_5^+$ products.

Isotopic labeling experiments show that the $C_3H_n^+$ channels all involve a common intermediate complex. This suggests that any mode specific nuclear motion effects (collision or vibrational) must come from effects on the complex formation cross section, since presumably the complex rapidly loses memory of the initial mode of excitation.

What we observe is that both collision and vibrational energy suppress the reaction, however different modes are not equally effective. Close examination of the figure shows that the inhibition from adding 1 or 2 quanta of ν_2 (C-C stretch) is similar to the reduction due to an equivalent increase in collision energy. On the other hand, 1253 cm^{-1} of bending excitation has essentially no effect -- in other words, bending excited ions are more reactive than would be expected based on their energy content.

The $C_2H_3^+$ channel does not show much isotopic scrambling, and thus appears to be mostly due to direct H-atom transfer from methane to $C_2H_2^+$. This interpretation is supported by measurements of the product velocity distributions which are strongly forward peaked[11]. $C_2H_3^+$ formation is slightly endoergic and the cross section for ground state $C_2H_2^+$ has an almost resolved threshold. At low energies, H-atom transfer is strongly enhanced by both collision energy and vibration. The effects from adding one or two quanta of ν_2 excitation are about 20% and 30% larger that the enhancements that result from adding equivalent amounts of collision energy for ground state $C_2H_2^+$. Bending excitation is much more effective. Note that 2 quanta of the bend (1253 cm^{-1}) has nearly the same effect as two quanta of the C-C stretch (3624 cm^{-1}) even though the energy is three times lower. At high energies, bending actually decreases reactivity slightly.

At least at low energies, bending excited acetylene cations appear to be generally more reactive than ions with equivalent amounts of collision energy, while C-C stretching has an effect similar to collision energy. Our rationalization for this mode selectivity is as follows. For any of the reaction channels it is necessary to make additional bonds to $C_2H_2^+$, and this requires rehybridization of one or both acetylenic carbon atoms. C-C stretch and collision energy are not particularly helpful in this electronic promotion, however bending energy goes directly into lowering the energy gap between the sp and sp^2 states. Bending is thus exactly the type of energy which can promote reactivity. Another way of stating this mechanism is to note that the transition state for making bonds to $C_2H_2^+$ must have one or both of the CCH bonds bent, and therefore bending vibration tends to drive the system toward the transition state.

This appears to be a nice example where the vibrational mode effects are consistent with a 'Freshman Chemistry' picture of the reaction mechanism, and we were quite happy to have such a straightforward explanation for our results. Unfortunately, the behavior of the next system we looked at is not consistent with this picture. Figure 4 shows the cross sections for the three product channels observed in reaction of $C_2H_2^+$ with D_2. These are m/e 27 (C_2HD^+ - isotope exchange), m/e 29 ($C_2HD_2^+$ - $C_2H_3^+$ formation), and m/e 28 (combination of $C_2D_2^+$ and $C_2DH_2^+$ - isotope exchange and $C_2H_3^+$ formation). Both isotope exchange and $C_2H_3^+$ formation are roughly thermoneutral (depending on isotope) and as the figure shows, both have small and weakly energy dependent cross sections. The fact that isotope exchange occurs suggests that a $C_2H_4^+$ complex must form in at least some collisions, and decay of this complex may also be the origin of the $C_2H_3^+$ product.

Thus we have another system not unlike $C_2H_2^+$ + CH_4 in principle, where complex formation and H atom transfer to $C_2H_2^+$ are prominent in the mechanism.

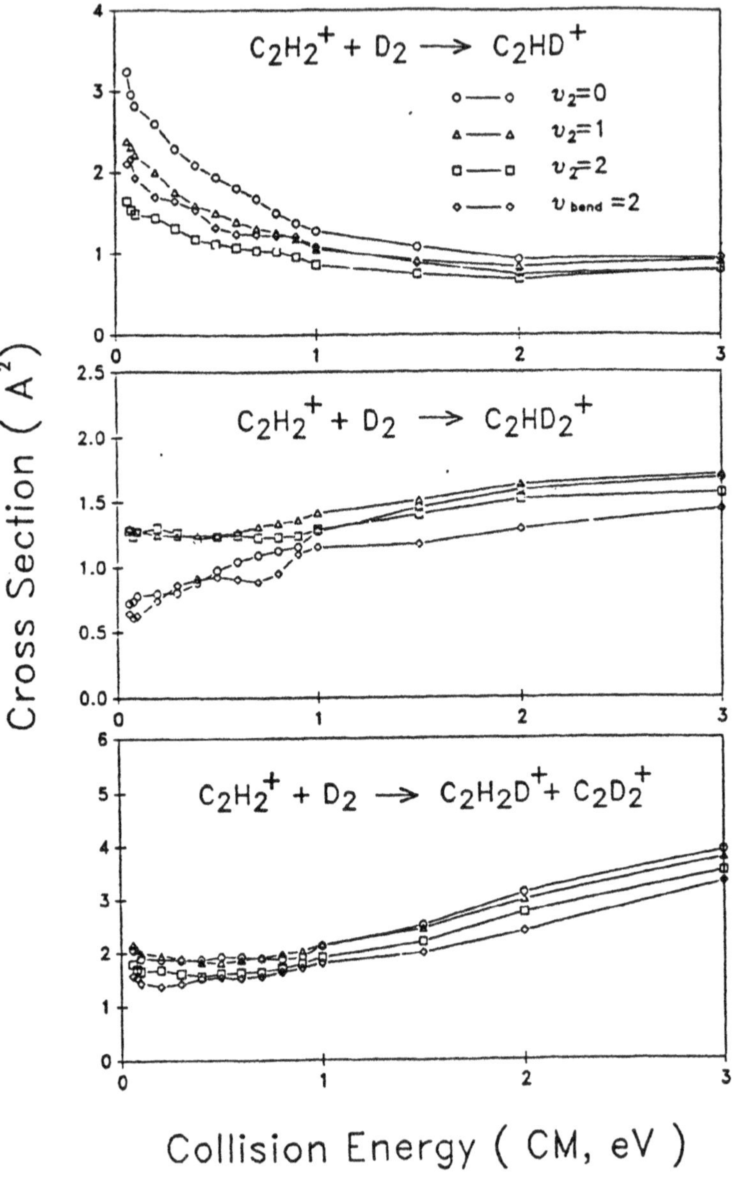

Figure 4 Product cross sections for reaction of $C_2H_2^+$ with D_2.

Using the same rehybridization or bent transition state argument, we would expect this reaction to be enhanced by bending vibration. Instead, we find that bending excitation inhibits reaction more than would be expected from the energy (ν_2=1814 cm^{-1}; $2\nu_{bend}$=1253 cm^{-1}). This apparent inconsistency may indicate that the basic ideas behind the mechanism are hopelessly oversimplified, or simply that the $C_2H_2^+$ + D_2 system reacts by a different mechanism where different dynamical factors control reactivity.

Some suggestion that there might be something special about reaction with hydrogen is given by comparison with Zare and co-workers' studies of reactions of NH_3^+ with NH_3[12], CH_4[13], and D_2[14]. For H-atom transfer from both NH_3 and CH_4, they observe that exciting the umbrella bend enhances reactivity. (For NH_3^+, only the bend can be populated by MPI due to Franck-Condon factors). This was ascribed to the idea that bending drove the system toward the transition state, which must be bent. In contrast, the H-atom transfer reaction from D_2 was not strongly enhanced by the bending vibration. The similarities between the $C_2H_2^+$ and NH_3^+ system in this regard may of course just be coincidental. Clearly it would be interesting and very helpful to have some theoretical insight into the mechanisms for these simple polyatomic reactions.

As a final example of the sort of mode effects we have observed, consider charge transfer (CT) in the $[C_2H_2{:}OCS]^+$ system. Since the ionization potential of C_2H_2 is 0.23 eV higher than that of OCS, CT from $C_2H_2^+$ to OCS is exoergic, and the reverse process is of course endoergic. Figure 5 shows the cross sections for both directions. For reference the ion-dipole capture cross section for $C_2H_2^+$ + OCS is plotted along with the data. It is important to note that for reaction of both initial charge states, production of $C_2H_2S^+$ is a major channel at low collision energies, and this discussion of the CT channels is really only telling part of the story.

For the exoergic CT from $C_2H_2^+$ to OCS, the cross sections are large and relatively collision energy independent. At energies above around 1.5 eV, the CT cross section is larger than the collision cross section. This is the classic signature of CT via a long range electron hop mechanism. At lower collision energies, the CT cross section increases as the ion-dipole forces allow larger impact parameter collisions to reach the curve-crossing radius where CT becomes likely. Finally, at very low energies, the CT cross section begins to decrease. This is simply due to competition with the $C_2H_2S^+$ production channel which becomes dominant below about 0.1 eV collision energy.

For the high energy part of the problem, the vibrational effects are nearly independent of collision energy, as might be expected for a mechanism involving electronic transitions in weakly interacting molecules. In the limit of long electron hopping range, it is inefficient to transfer momentum in the CT process, thus the cross section is maximized if the internal energy of the initial state is resonant with some final state. In addition, it is reasonable to expect that CT will be favored between initial and final states which have good vibrational overlaps. This suggests that we should be able to understand the observed vibrational effects in terms of Franck-Condon factors and energy resonances between the different initial states and the available final states, at least for high collision energies where the CT cross section is large compared to the collision

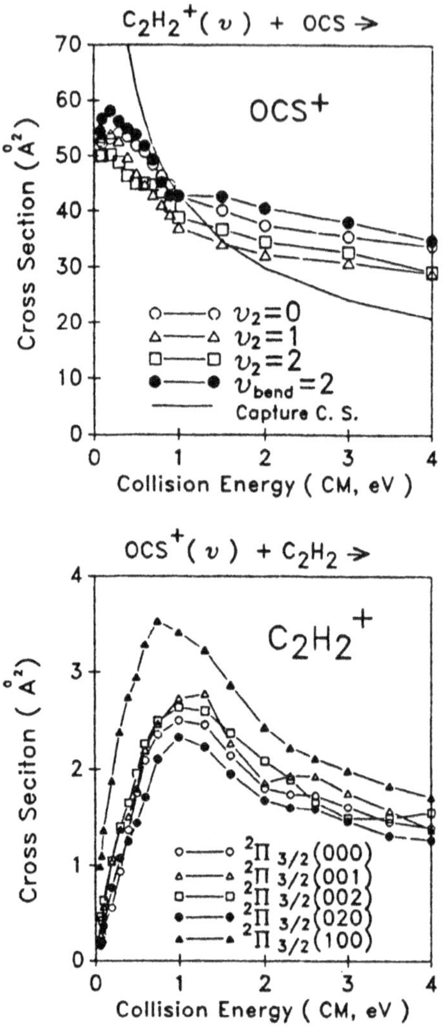

Figure 5 Charge transfer cross sections for both directions in the $[C_2H_2{:}OCS]^+$ system.

cross section.

The results for $C_2H_2^+$ + OCS are consistent with this model[8]. The modest enhancement for bending excited $C_2H_2^+$ is consistent having a final state in very close

(3 cm^{-1}) resonance. Likewise, the inhibition observed for $C_2H_2^+$ in the $\nu_2=1$ state is evidently due to lack of near resonant final states. Unfortunately, while we can estimate the Franck-Condon factors for ionization of OCS, the neutralization Franck-Condon factors for various vibrational states of $C_2H_2^+$ are not known, thus the analysis is strictly qualitative.

Charge transfer from OCS^+ to C_2H_2 is endoergic, and thus requires conversion of vibrational or collision energy to drive the electron transfer. This conversion of nuclear kinetic energy requires an intimate collision, and not unexpectedly, the CT cross section is small - with a maximum value around an order of magnitude smaller than for the reverse reaction. The reaction also shows a threshold which is not completely resolved due to ion beam energy spread and target gas thermal motion. Note that for collision energies below about 0.3 eV, all vibrational modes appear to enhance the CT cross section, as does collision energy. Enhancement is not surprising for an endoergic reaction, but note that all modes of nuclear motion appear to be equally efficient - the observed enhancement is simply proportional to the energy content in the vibrational modes. This non-specific behavior suggests that the principle CT mechanism at low collision energies may involve an intermediate complex where the initial nuclear kinetic energy mode is randomized.

Very quickly as the collision energy is raised a few hundred meV, the vibrational effects become non-monotonic. The most dramatic effect is that excitation of the bend begins to inhibit reaction. In addition, the CO stretch appears to cause a larger enhancement than might be expected from its energy content (2041 cm^{-1}). Above 1 eV collision energy, a curious thing happens. The CT probability begins to be somewhat inhibited by collision energy, while the effects of vibrational excitation remain unchanged (ie. stretching vibrations enhance, bending inhibits).

One curious thing is the effect of the CO stretch. As pointed out already, this charge transfer reaction is 0.23 eV endoergic for ground state reactants. For excitation of one or two quanta of the CS stretch (ν_3) or of two quanta of the bend (ν_2), the reactant energy is still below the ground state of the products, thus precluding long range electron transfer. For the ν_1 excited ions however, the CT reaction becomes nearly resonant - just 20 meV <u>exoergic</u>. Under these circumstances we might expect a very large enhancement since there should be a reasonable CT probability at long range. The fact that we see only a modest enhancement indicates that this vibrational mode does not couple well to drive the CT process.

Figure 6 suggest why this might be the case. This is a slice through the entrance channel for $[C_2H_2:OCS]^+$ collisions along the ν_1 (CO stretch) coordinate. For CT from OCS^+ to C_2H_2 we start on the lower potential surface. For ground state ions, there is insufficient energy to cross to the upper surface via a long-range curve crossing. For $\nu_1=1$ the system presumably does indeed undergo CT via the crossing seam at long intermolecular separations, however because the seam is located near the classical turning point for the vibration, it presumably has a high probability to simply transfer back to the lower energy OCS^+ + C_2H_2 surface as the molecules separate and the <u>net</u> CT probability is small.

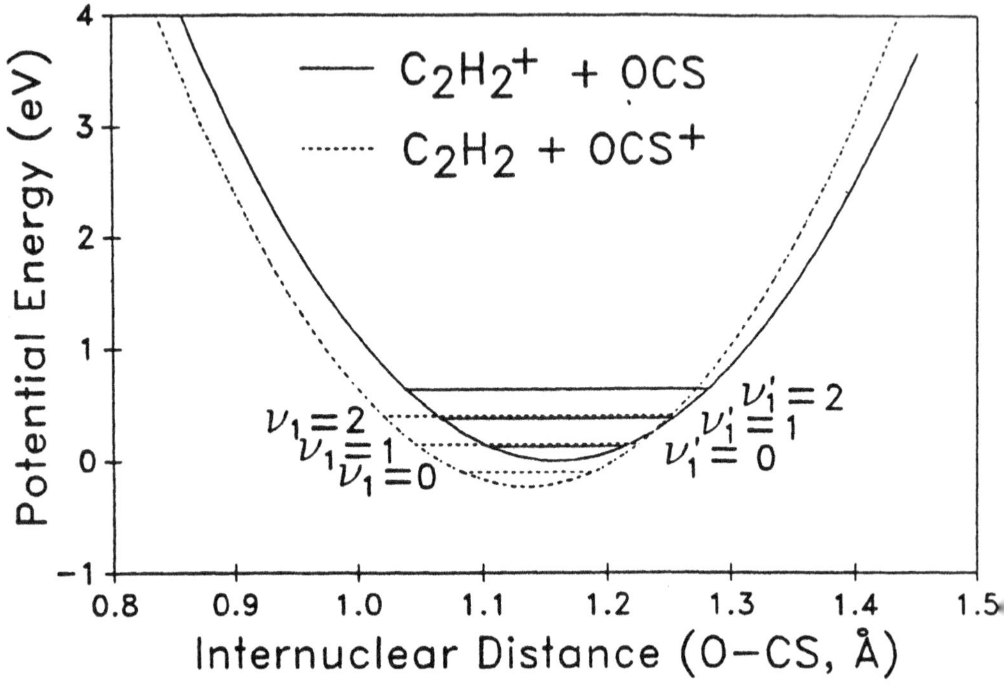

Figure 6. Slice along ν_1 through the entrance channel for $[C_2H_2{:}OCS]^+$collisions

5. Conclusions and future directions.

Hopefully I have demonstrated some of the variety of mode specific effects that we have observed in our work to date. This effort is really just getting started and as should be clear, we have not accumulated a large enough database to allow us to draw general conclusions about the dynamical origins of the mode effects. It is clear however that in nearly every reaction we have studied to date, there are significant vibrational mode effects meaning that the initial energy is not randomized before the system reaches a transition state.

There are several things we are doing to advance this work. First, we need more integral cross section data to be able to draw any firm conclusions about the origins of mode effects, and therefore we will continue to crank out results on a variety of reactions. The instrument is currently being modified to improve the collision energy resolution. We believe that we should be able to routinely work with beams of about 50 meV laboratory energy spread. This will enable us to get down to lower collision energies where the energy of the vibrations is large compared to the total energy in the collision. At the same time, the new ion optics will enable us to measure the product energy and angular distributions. This will provide information on the reaction mechanisms and also allow us to determine energy disposal and how it is affected by reactant vibrational mode.

References

1. Anderson, S. L. (submitted) "Multiphoton ionization state-selection: Vibrational mode and rotational state control" Adv. Chem. Phys. (to appear in 1991).

2. Xie, J. and Zare, R. N. (1989) 'Rotationally state-selected HBr^+: Preparation and characterization', Chem. Phys. Lett. 159, 399-405.
Xie, J., Leahy, D. J., and Zare, R. N. (1990) private communication.

3. Orlando, T. M., Anderson, S. L., Appling, J. R., and White, M. G. (1987) "Photoelectron Spectroscopy of Ungerade Excited States of Acetylene: Intermediate State Mixing and Ion State Selection" J. Chem. Phys. 87 852-860.

4. Ashfold, M. N. R., Tutcher, B., Yang, B., Jin, Z., and Anderson, S. L. (1987) "Gerade Rydberg States of Acetylene Studied by Multiphoton Ionization and Photoelectron Spectroscopy", J. Chem. Phys. 87 5105-5115.

5. Yang, B., Eslami, M. E., and Anderson S. L. (1988) "Multiphoton Ionization Photoelectron Spectroscopy Study of OCS: Rydberg Vibronic Structure and Ion State Selection", J. Chem. Phys. 89 5527-5534.

6. Orlando, T. M., Yang, B., and Anderson, S. L. (1989) "The Effects of Stretching and Bending Vibration on the Reaction of Acetylene Cations with Methane", J. Chem. Phys. 90 1577-1587.

7. Yang, B., Chiu, Y., Fu, H., and Anderson, S. L. (1990) "Comparison of Bending and Stretching Vibrational Effects on the Reaction of $C_2H_2^+$ with D_2", (in preparation).

8. Orlando, T. M., Yang, B., Chiu, Y., and Anderson, S. L. (1990) "The Effects of Different Vibrational Modes and Translational Energy on the Reaction of Acetylene Cations with Carbonyl Sulfide", J. Chem. Phys. 92 7356-7364.

9. Yang, B., Chui, Y., and Anderson S. L. (1990) "Effects of C-O Stretch, C-S Stretch, Bending Vibration, and Fine Structure Excitation on the Reaction of OCS^+ with Acetylene", (in preparation).

10. Yang, B., Chiu, Y., Fu H., and Anderson, S. L. (1990) "Effect of different vibrational modes on collision-induced-dissociation of OSC^{+}" (in preparation).

11. These measurements performed in the laboratory of Dieter Gerlich at Freiburg using thermalized $C_2H_2^+$.

12. Conaway, W. E., Ebata, T., and Zare, R. N. (1987) "Vibrationally state-selected reactions of ammonia ions III. $NH_3^+(v) + ND_3$ and $ND_3^+ + NH_3$" J. Chem. Phys. 87 3453-3460.

13. Conaway, W. E., Ebata, T., and Zare, R. N. (1987) "Vibrationally state-selected reactions of ammonia ions II. $NH_3^+(v) + CH_4$" J. Chem. Phys. 87 3447-3452.

14. Morrison, R. J. S., Conaway, W. E., and Zare, R. N. (1985) "Effect of internal and translational energy on the $NH_3^+(v) + D_2$ ion-molecule reaction" Chem. Phys. Lett. 113 435-440.

WORKSHOP: THE ROLE OF ENERGY IN ION-MOLECULE PROCESSES

DENISE C. PARENT
Chemistry Division
Naval Research Laboratory
Washington, D.C. 20375-5000

ABSTRACT. A summary of the workshop on the role of energy in ion-molecule processes is given. Contributions cover the dependence of simple bimolecular reactions on kinetic and internal energy. Examples of kinetic energy release in unimolecular dissociation and internal excitation in charge transfer reactions are also discussed.

Energy is one of the critical parameters affecting the outcome of a chemical process. It is also a powerful probe of reaction dynamics. Thus the "Holy Grail" of chemical dynamics is the complete state-to-state specification of a chemical reaction, where the kinetic energy and internal states of the reactants are known, and the products are completely characterized as to kinetic energy and internal state. Such an experiment is not yet feasible for most reactions. However progress has been made in implementing various aspects of the completely specified experiment. Experiments which study the dependence of the reaction rate and product branching ratios on the kinetic energy are common. Fewer experiments which probe the effect of internal excitation have been carried out, due to the greater difficulty of producing reactants in selected quantum states. Experiments which measure the kinetic energy release and determine the internal state distributions of products have also been reported. Again, it has proven easier to measure the kinetic energy of the products than to determine their internal energy or state distribution.

In this workshop on the role of energy in ion-molecule processes, seven participants gave reports of their work on various aspects of this topic. Many of the contributions looked at the effect of kinetic or internal (both vibrational and rotational) energy on the process under study. Kinetic energy release in unimolecular dissociations and internal excitation in ion-molecule reactions or collisions were the subject of other reports. The remainder of this article consists of summaries of the presentations.

1. Ion-Molecule Reactions at Very Low Translational Energies

P. Tosi (with F. Eccher and D. Bassi, Universita degli Studi di Trento, Italy) described their efforts to study ion-molecule reactions at very low translational energies.

A powerful approach for dealing with low energy ion beams is based on the use of radio-frequency octopole ion guides [1]. Guided ion beam tandem mass spectrometers have allowed measurements of integral reactive cross sections down to a nominal collision energy of 5 meV [2].

K. R. Jennings (ed.), Fundamentals of Gas Phase Ion Chemistry, 197–207.

Unfortunately, since the observed cross section is actually the convolution of the true cross section with the relative energy distribution of the reactants, at very low energies the thermal motion of the target gas obscures the behavior of the true cross section. As noted by Armentrout [2] the observed cross section tends toward $\sigma_{eff} \propto E^{-0.5}$ as $E \rightarrow 0$ regardless of the low energy behavior of the true cross section. In other words for nominal collision energies $E << k_B T$, where T is the scattering cell temperature, because of the Doppler broadening, beam-cell experiments will always yield a cross section which behaves as $E^{-0.5}$. For energies comparable with $k_B T$ (*i.e.* energies below ~ 100 meV) the Doppler broadening hides any sharp features of the cross section. As an example, at a nominal collision energy of 50 meV a beam-cell measurement has an energy spread characterized by a FWHM of about 130 meV [2].

This problem may be overcome by replacing the static reaction cell with a kinematically well defined beam of reactant molecules so as to reduce Doppler broadening effects. The main problem of this approach is the much lower density of molecules in a beam with respect to a static cell. A partial solution is to use a pulsed neutral beam that is much more intense than a continuous beam [3]. Since the kinetic energy of the neutral beam is of the order of a few $k_B T_0$, the temperature of the source T_0 will determine the kinetic energy of the neutral beam. For this reason very low energies can be reached only by cooled beams. Unfortunately the temperature of a pulsed source cannot be precisely controlled. Therefore in our apparatus we use a continuous supersonic beam which, although less intense than a pulsed beam, can be easily cooled down to liquid nitrogen temperature. The price paid in terms of intensity is compensated by the possibility of reaching a lower collision energy. The cooled beam is crossed at the center of an octopole guide with an ion beam. This crossed beam configuration enables us to obtain much better energy resolution than in experiments using a room temperature scattering cell and to investigate the detailed structure of the integral reactive cross section as a function of the collision energy [4].

High resolution measurements of the integral cross section for the reaction of Ar^+ with H_2 to produce ArH^+ show structure as a function of the collision energy, as seen in Figure 1. *Preliminary* measurements for the reaction of Ar^+ with D_2 indicate an energy shift of this structure, so that it appears at the same velocity. This seems to exclude internal energy effects (vibration and rotation) and to suggest a purely kinematic effect. We speculate about the relationship with non-adiabatic couplings between the two 3/2 surfaces of the Ar^+/H_2 system and the H_2^+/Ar surface. This subject is currently under investigation.

2. Rotational and Vibrational Energy Effects in Simple Ion-Molecule Reactions

A. Viggiano (with R. Morris, J. Van Doren and J. Paulson, Geophysics Laboratory, USA) presented their results on the internal energy dependence of several ion-molecule reactions.

By measuring rate constants and/or branching ratios as a function of average kinetic energy ($\langle KE_{cm} \rangle$) at several temperatures in a variable temperature-selected ion flow drift tube we are able to derive information on the internal energy dependences of ion-molecule reactions. $\langle KE_{cm} \rangle$ in a VT-SIFDT is given by [5]

$$\langle KE_{cm} \rangle = (m_i+m_b)v_d^2/2(m_i+m_n) + 3k_BT/2 \tag{1}$$

where m_i, m_b and m_n are the masses of the reactant ion, buffer gas and reactant neutral respectively, v_d is the ion drift velocity and T is the temperature. Comparison of data at the same average kinetic energy but with different neutral gas temperatures yields information on the internal temperature and/or internal energy dependences of the rate constants and branching ratios.

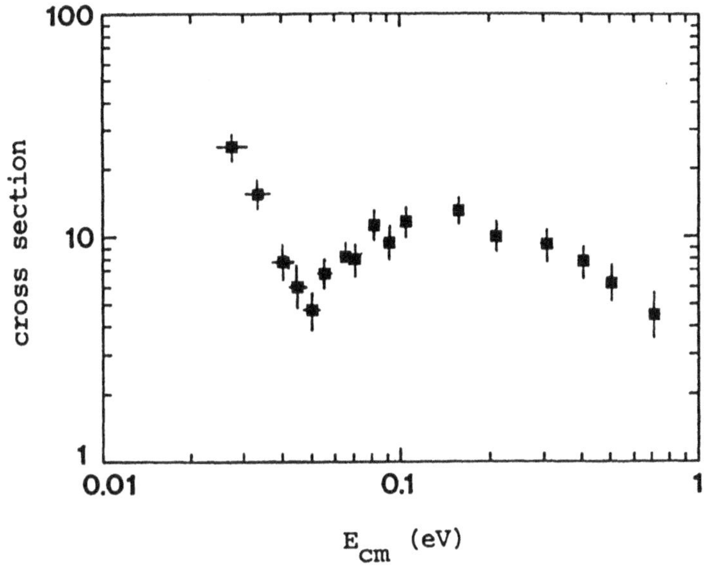

Figure 1. Measured reaction cross section (in arbitrary units) as a function of the center-of-mass collision energy for the reaction $Ar^+ + H_2 \rightarrow ArH^+ + H$.

The simplest situation to interpret is that of monatomic ions reacting with neutrals having high vibrational frequencies. In this case the dependence of the rate constants and/or branching ratio on the rotational temperature of the neutral is derived. We have studied several reactions for which rotational temperature has no effect on the rate constant, *i.e.* the reactions of O^- with SO_2 [6], CH_4 and N_2O [6], as well as others. For the charge transfer reaction of Kr^+ with HCl, a negative rotational temperature dependence of $T^{-0.37}$ that is approximately independent of kinetic energy was found [8]. A similar dependence is seen for DCl, even though the rate constant is smaller. The cause of the rotational temperature dependence is most likely due to an orientation effect that is more pronounced at lower rotational velocities due to more efficient dipole locking.

We have recently investigated the influence of N_2 rotational temperature on the rate constant for the charge transfer reaction of Ar^+ with N_2 and found a strong positive rotational temperature dependence [9]. This rotational temperature dependence is explained as follows. At the temperatures of this study, the reaction is known to proceed to form $N_2^+(v=1)$. This channel is slightly endothermic. Comparing the data at a constant total (kinetic and rotational) average energy shows rotational and translational energy are approximately equally efficient in driving the reaction since no dependence is found.

We have extended the technique to more complicated systems. In summary, our results show that vibrational energy of CH_4 drives the reaction of O_2^+ with CH_4, vibrational excitation of N_2O and CO_2 enhances the collisional detachment rate of NO^- [10], and vibrational excitation of CO_2^+ increases the rate of the reaction of CO_2^+ with O_2.

3. Orientation Effects in Ion-Molecule Reactions

J.B. Marquette (with B. Rowe and C. Rebrion, CNRS Meudon, France) related their results on the effect of rotational excitation and dipole moment on various ion-molecule reactions.

The CRESU apparatus [11,12] was designed to study kinetics of ion-molecule processes in the range of interstellar temperatures (10-100 K). Among the most significant results obtained, the reaction of ground state atomic nitrogen ions with *n*-H_2, *p*-H_2 and D_2 was shown to be very slightly endothermic by 18 meV [13]. Moreover, the comparison of data on both normal and para hydrogen shows that the rate coefficient critically depends on the amount of available rotational energy in the neutral reactant. The spin-orbit energy of the monatomic primary ions also promotes the processes.

Another interesting example is the reaction of atomic nitrogen ions with the three structural isomers of dichloroethylene [14] which differ from each other by the value of their dipole moment. Thus, these processes are particularly suited to examine the intrinsic importance of that parameter in ion-dipole reactions. The reaction involving the non-polar isomer is in nearly perfect agreement with the ion-induced dipole model, establishing that these reactions are capture processes. But more surprisingly, the increase of the rate coefficient with decreasing temperature is well below theoretical predictions over the complete temperature range (27-300 K) while the reaction probability remains temperature independent. It is tentatively suggested that the collinear approach is not the most favorable.

4. Effects of Internal Energy on Endothermic Ion-Molecule Reactions

P. Armentrout (with L. Sunderlin, D. Clemmer and D. Hales, University of Utah, USA) described experiments in which rotational and vibrational energy effects are observed.

Guided ion beam instruments are a powerful means of examining the kinetic energy dependence of ion-molecule reactions. Recently, we have made several additions to this apparatus which now permit us to perform different types of experiments designed to address the role of internal energy in ion-molecule chemistry.

The first type of experiment is enabled by the recent addition of a cooled gas cell to our instrument [15]. This allows the neutral reagent to be cooled from room temperature to about 100 K. This cooling yields a dramatic sharpening of the threshold for the reaction $C^+ + H_2 \rightarrow CH^+ + H$, an effect which is due primarily to the decreased translational energy distribution of the H_2 gas [16]. A similar effect is noted for the near-thermoneutral reactions of N^+ with H_2, D_2 and HD, although here the reduction in rotational temperature is also influential [16].

A second instrumental addition is the development of a flowing afterglow ion source designed to produce thermalized polyatomic ions [17]. One of the early applications has been the examination of the charge transfer reaction between NO_2^+ and $C_6H_5CF_3$, reported by Ausloos and Lias to be exothermic [18]. We obtain a similar result when the NO_2^+ is formed in a He discharge [19]. However, when CO_2 is added to the flow, the exothermic reactivity is fully quenched and the reaction is clearly endothermic, although only slightly. It seems clear that small amounts of NO_2^+ vibrational energy enhances the charge transfer reaction probability by over an order of magnitude at thermal kinetic energies.

Another recent study compares the reactivity of Co_2^+ ions formed by electron impact on $Co_2(CO)_8$ and in a laser vaporization/supersonic expansion source. The latter are shown to be much colder internally via collision-induced dissociation and ligand exchange experiments [20].

5. Kinetic Energy Release Distributions in Unimolecular Dissociation

E. Uggerud (with T. Helgaker, G. Hvistendahl and C. Schulze, University of Oslo, Norway and H. Jensen, Århus University, Denmark) discussed the kinetic energy release observed in the unimolecular dissociation of organic ions.

Unimolecular reactions are accompanied by kinetic energy release, T_R, in the form of relative motion of the two products formed. In general, the kinetic energy release has two sources, the kinetic shift (kinetic energy of the nuclei at the transition state) and the reverse activation energy of the reaction [21].

When there is a large barrier for the reverse reaction, a substantial part of the kinetic energy release in metastable ion fragmentation stems from the potential energy difference, ΔV_b^* between the transition state and the separated products. The fraction of ΔV_b^* converted into relative translation is strongly dependent on the detailed reaction dynamics (the motion of the individual atoms) *en route* from transition state to products.

A new method for calculating classical trajectories on molecular *ab initio* potential energy surfaces is presented [22]. The equations of motion are integrated on a sequence of model surfaces constructed from analytically calculated molecular gradients and Hessians from *ab initio* wave functions. The method avoids the explicit calculation of the total surface and is therefore applicable whenever the gradient and Hessian can be calculated.

Peak profiles corresponding to fragmentation in the first field free region were obtained by scanning the ion source high voltage using an MS 902 mass spectrometer [23]. In order to remove instrumental effects each peak was deconvoluted according to the method of Rumpf and Derrick [24]. This procedure gives rise to a distribution function $n(T_R)$ of the kinetic energy release.

The first reaction studied was $CH_2OH^+ \rightarrow CHO^+ + H_2$ [25]. Analysis of the result of the trajectory calculation showed that 83% of ΔV_b^* is released. This is in good agreement with our experiment which shows that the $n(T_R)$ distribution has a maximum corresponding to T_R=1.80 eV which is 77% of the experimental ΔV_b^*. It is interesting to notice that at the transition state only 44% of the potential energy difference is predicted to end up as relative translational energy. The system therefore "picks up" relative translation along the trajectory from transition state to products.

The second reaction studied was $CH_3CH_3^+ \rightarrow CH_2CH_2^+ + H_2$ [26]. In this case there is also good agreement between experiment and the trajectory calculation. The calculated value is 85% kinetic energy release, while the maximum of the distribution corresponds to T_R=0.30 eV, which means that 86% of the experimental ΔV_b^* is released.

6. Vibrational Excitation in Collisions of Symmetric Diatomics

D. Ridge (with K. Sohlberg, University of Delaware, USA) discussed the conversion of translational energy into vibrational energy in collisions of nitrogen ions with nitrogen.

Both McAfee *et al.* [27] and Howard and Futrell [28] have examined collisions between N_2^+ and N_2 and found a striking pattern of vibrational excitation. The vibrational frequencies of N_2^+ and N_2 are close enough that it is not possible to distinguish between vibrational excitation of these two species in experiments where the measured quantity is kinetic energy loss. The results show a marked preference for kinetic energy loss corresponding to an even number of quanta of vibrational energy. In the case of the ion and neutral initially in the ground state,

$$N_2^+(v_i{=}0) + N_2(v_i{=}0) \rightarrow N_2^+(v_f) + N_2(v_f') \qquad (2)$$

with initial total translational energy T_i and final total translational energy T_f, the dominant features in the scattering correspond to

$$(T_i - T_f)/h\nu = v_f + v_f' = 2n \tag{3}$$

where ν is the average of the N_2^+ and N_2 frequencies and n is an integer.

We propose that because of the symmetry of the system both collision partners must be equally vibrationally excited. That necessarily gives an even number of vibrational quanta of excitation. To test this notion we calculated classical trajectories for hard sphere collisions of two harmonic oscillators in a collinear geometry at 5 eV collision energy. We assumed the harmonic oscillators M1-M2 and M3-M4 had the force constants and masses of N_2. The interaction between M2 and M3 was the hard sphere interaction. The initial vibrational energy of the oscillators corresponded to the zero point energy of N_2. Under these conditions if the initial phases of the oscillators are equal the final energies of the oscillators must be equal by symmetry. The question, of course, is whether the oscillators will be equally excited if the initial phases are unequal. To answer this question we calculated 10^5 trajectories with different initial phases, sampling all initial phases.

Some results are illustrated in Figure 2 which shows the final energy of each oscillator as a function of the initial phase of oscillator M1-M2 for two initial phases of oscillator M3-M4. It is obvious that the final energies of excitation tend to be the same for the two oscillators. Analysis of the individual trajectories indicates that for 64% of the collisions the difference between the final vibrational energies was less than one vibrational quantum, for 22% of the collisions the difference was between one and two quanta and for 14% of the collisions the difference was between two and three quanta. The probability that the difference exceeded three quanta was negligible even though total excitation of as much as 16 quanta was found. The final energies of the two oscillators are clearly closely correlated.

The sum of the probabilities of a difference of 0-1 quanta and a difference of 2-3 quanta is 78%. Even if we reduce this by the fraction of collisions in which the total excitation was less than one quantum (32%) we get 46% which is 2.1 times the 22% of collisions where the final difference is between 1 and 2 quanta. This 2.1-fold excess may be taken as an indication of the relative probability that $v_f - v_f'$ is even and that hence $v_f + v_f'$ is even.

More extensive classical trajectory calculations on more realistic three dimensional surfaces indicate that this correlation between the final energies of the two oscillators is a feature of more realistic systems [29]. Semi-classical calculations are underway to see if the effect survives the introduction of quantum mechanics [29].

7. Internal Excitation of the Products in Charge Transfer Reactions

J. Futrell (with S. Howard and A. Rockwood, University of Delaware, USA) presented results on the product internal states in the charge transfer of argon to nitric oxide.

The crossed molecular beam method utilized in our laboratory for the investigation of ion-molecule reaction dynamics provides valuable information on the energetics of reactions in addition to providing clues regarding fundamental mechanisms by which the underlying ion-atom exchange processes occur. We characterize the reactant ion and neutral velocity vectors experimentally and measure the velocity vectors of the product ion; the velocity vectors of the product neutral can be calculated from energy and momentum conservation. The transformation of the laboratory velocities into the center-of-mass frame using standard methods provides the information required for constructing contour diagrams of relative

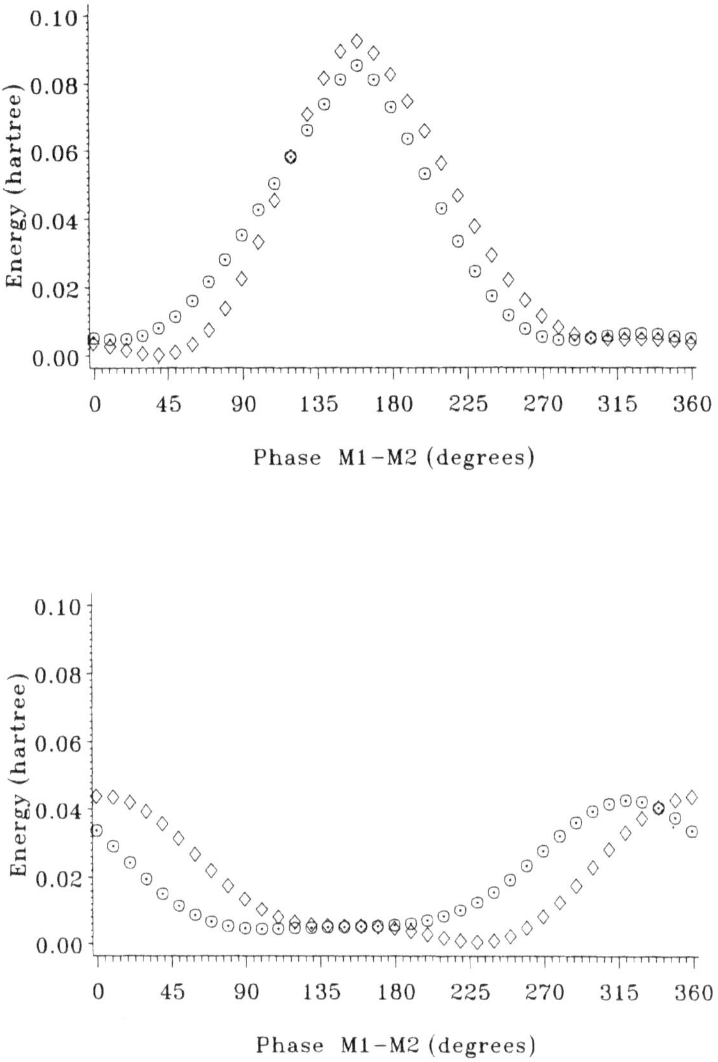

Figure 2. The final energy of each oscillator (M1-M2: circles and M3-M4: diamonds) is plotted as a function of the initial phase of the M1-M2 oscillator. In the top panel the initial phase of the M3-M4 oscillator is 120° while in the bottom panel it is 340°.

intensity of product ions as a function of velocity, often referred to as Newton diagrams. The graphs display the probability that a product ion is scattered at any scattering angle with a particular velocity with respect to the center-of-mass.

One area which has occupied our attention for several years is the dynamics of charge transfer processes. Our experimental data for the charge transfer of argon ions with NO at 4.75 eV and at the very low collision energy of 0.43 eV are shown in the top and bottom diagrams respectively of Figure 3. The charge transfer process is translationally exothermic, and the product ions leave the collision center with slightly higher velocity than the initial NO neutral velocity. The velocity of the argon ion is indicated by the vertical bar to the right of the center-of-mass, while the velocity of the NO neutral reactant is indicated by the vertical bar to the left of the center-of-mass. The circumscribed circle in the bottom diagram shows the velocity the particles would have if the charge transfer reaction is

$$Ar^+(^2P_{3/2}) + NO(^2\Pi, v=0) \rightarrow Ar(^1S_0) + NO^+(a^3\Sigma^+, v=0). \quad (4)$$

We infer from the fact that the positions of maximum intensity in the reactive scattering contour diagram match the energetics of reaction 4 rather well that this is the reaction which actually occurs. (Our reactant ions are generated in a plasma source which relaxes the initial distribution to the $Ar^+(^2P_{3/2})$ level.) It appears that NO^+ is formed in the first electronic excited state which is nearly resonant with the $Ar^+_{3/2}$ recombination energy, rather than in the ground electronic state, which is a much more exothermic process,

$$Ar^+(^2P_{3/2}) + NO(^2\Pi, v=0) \rightarrow Ar(^1S_0) + NO^+(X^1\Sigma^+, v=0) + 6.399 \text{ eV}. \quad (5)$$

This general conclusion was reached both in an earlier beam study [30] and in a drift tube study [31] of the same reaction. The interesting additional feature in Figure 3 is the development of nearly complete backward-forward symmetry about the center-of-mass at very low collision energy. This is the characteristic dynamic signature of an orbiting complex.

Figure 3 (top diagram) also shows our data for this reaction at higher translational energy. As before the energetics correspond to reaction 4 rather than to reaction 5. The most dramatic change in the scattering patterns is the complete disappearance of backward-forward symmetry when the collision energy is increased tenfold. Clearly the reaction mechanism is direct at this collision energy. We deduce from these experiments that the charge transfer reaction occurs at relatively large internuclear distances with little exchange of angular momentum, an "electron jump" mechanism. We infer that a curve-crossing mechanism of the Demkov type occurs as the reactants approach each other, since the asymptotic energy levels of the reactants and products of reaction 4 are quasi-resonant. This occurs in the incoming reactant channel. The strongly attractive potential well of $ArNO^+$ (isoelectronic with NOCl) causes orbiting trajectories at low collision energies. However, orbital and angular momentum exchange is modest in the collision complex (which is presumably relatively short-lived, *i.e.* a few rotational periods) and the products separate without energy randomization.

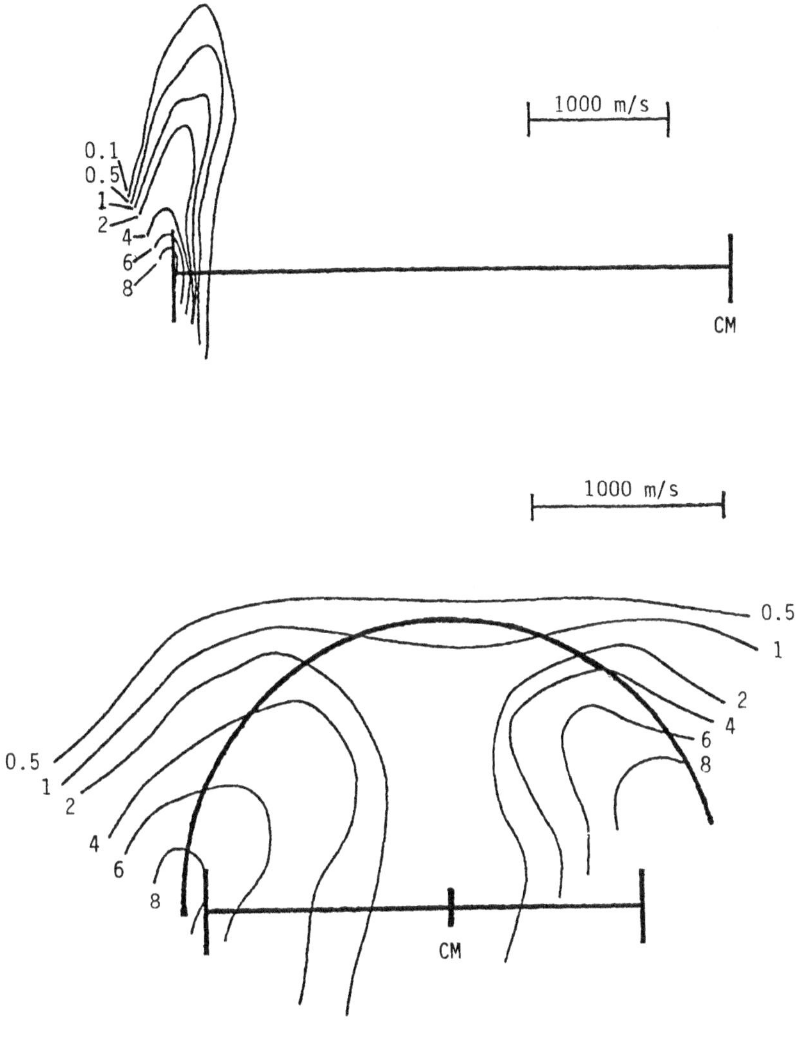

Figure 3. Center-of-mass scattering contour diagrams for the charge transfer reaction of $Ar^+(^2P_{3/2})$ with NO at 4.75 eV (top) and 0.43 eV (bottom) collision energy. Velocity contours indicate relative intensity. Vertical bars to the right of the CM denote the tip of the Ar^+ velocity vector while the vertical bar to the left denotes the neutral velocity vector in both diagrams. The circle indicates the loci of product ion velocity vectors calculated for the state-specific reaction $Ar^+(^2P_{3/2}) + NO(^2\Pi, v=0) \rightarrow Ar(^1S_0) + NO^+(a^3\Sigma^+, v=0)$.

References

[1] Tosi, P., Fontana, G., Longano, S. and Bassi, D. (1989) 'Transport of an ion beam through an octopole guide operating in the r.f.-only mode', *Int. J. Mass Spectrom. Ion Proc.* **93**, 95-105.

[2] Ervin, K. M. and Armentrout, P. B. (1985) 'Translational energy dependence of $Ar^+ + XY \rightarrow ArX^+ + Y$ ($XY=H_2,D_2,HD$) from thermal to 30 eV c.m.', *J. Chem. Phys.* **83**, 166-189.

[3] Gerlich, D., Disch, R. and Scherbart, S. (1987) '$C^+ + H_2(j) \rightarrow CH^+ + H$: The effect of reagent orientation on the integral cross section in the threshold region', *J. Chem. Phys.* **87**, 350-359.

[4] Tosi, P., Boldo, F., Eccher, F., Filippi, M. and Bassi, D. (1989) 'High-resolution measurements of the integral cross-section for the reaction $Ar^+ + H_2 \rightarrow ArH^+ + H$ as a function of collision energy', *Chem. Phys. Lett.* **164**, 471-474.

[5] McFarland, M., Albritton, D. L., Fehsenfeld, F. C., Ferguson, E. E. and Schmeltekopf, A. L. (1973) 'Flow-drift technique for ion mobility and ion-molecule reaction rate constant measurements. I. Apparatus and mobility measurements', *J. Chem. Phys.* **59**, 6610-6619.

[6] Morris, R. A., Viggiano, A. A. and Paulson, J. F. (1990) 'Temperature and kinetic energy dependence of the rate constant for the reaction of O^- with SO_2', *J. Phys. Chem.* **94**, 1884-1886.

[7] Morris, R. A., Viggiano, A. A. and Paulson, J. F. (1990) 'Branching ratios and rate constants for the reaction of $^{16}O^-$ and $^{18}O^-$ with N_2O and $^{14}N^{15}N^{16}O$', *J. Chem. Phys.* **92**, 3448-3452.

[8] Viggiano, A. A., Morris, R. A., Dale, F., Paulson, J. F., Giles, K., Smith, D. and Su, T. (1990) 'Kinetic energy, temperature, and derived rotational temperature dependences for the reactions of $Kr^+(^2P_{3/2})$ and Ar^+ with HCl', *J. Chem. Phys.* **93**, 1149-1157.

[9] Viggiano, A. A., Van Doren, J. M., Morris, R. A. and Paulson, J. F. (1990) *J. Chem. Phys.* **93**, in press.

[10] Viggiano, A. A., Morris, R. A. and Paulson, J. F. (1990) 'Kinetic energy and temperature dependences of the rate constants for electron detachment of NO^- by N_2O, CO_2, N_2, CH_4, C_2H_6, and C_3H_8', *J. Phys. Chem.* **94**, 3286-3290.

[11] Dupeyrat, G., Marquette, J. B. and Rowe, B. R. (1985) 'Design and testing of axisymmetric nozzles for ion-molecule reaction studies between 20 °K and 160 °K', *Phys. Fluids* **28**, 1273-1279.

[12] Rowe, B. R. and Marquette, J. B. (1987) 'CRESU studies of ion/molecule reactions', *Int. J. Mass Spectrom. Ion Proc.* **80**, 239-254.

[13] Marquette, J. B., Rebrion, C. and Rowe, B. R. (1988) 'Reactions of $N^+(^3P)$ ions with normal, para, and deuterated hydrogens at low temperatures', *J. Chem. Phys.* **89**, 2041-2047.

[14] Rebrion, C., Marquette, J. B. Rowe, B. R., Chakravarty, C., Clary, D. C., Adams, N. G. and Smith, D. (1988) 'Reactions of N^+ and H_3^+ with the structural isomers of $C_2H_2Cl_2$', *J. Phys. Chem.* **92**, 6572-6574.

[15] Sunderlin, L. S. and Armentrout, P. B. (1990) 'Temperature dependence of the reaction of O^+ with HD', *Chem. Phys. Lett.* **167**, 188-192.

[16] Sunderlin, L. S. and Armentrout, P. B. (1990) work in progress.

[17] Schulz, R. H. and Armentrout, P. B. (1990) in preparation.

[18] Ausloos, P. and Lias, S. G. (1978) 'Reactions of NO_2^+ and solvated NO_2^+ ions with aromatic compounds and alkanes', *Int. J. Chem. Kin.* **10**, 657-667.

[19] Clemmer, D. E. and Armentrout, P. B. (1990) in preparation.

[20] Hales, D. A. and Armentrout, P. B. (1990) 'Effect of internal excitation on the collision-induced dissociation and reactivity of $Co_2^{+\cdot}$', *J. Cluster Science* **1**, 127-142.

[21] Cooks, R. G., Beynon, J. H., Caprioli, R. M. and Lester, G. R. (1973) *Metastable Ions*, Elsevier, Amsterdam.

[22] Helgaker, T., Uggerud, E. and Jensen, H. J. Aa. (1990) *Chem. Phys. Lett.*, accepted.

[23] Barber, M. and Elliot, R. M. (1964) in *Proceedings of the 12th Annual Conference on Mass Spectrometry and Allied Topics*, Montreal.

[24] Rumpf, B. and Derrick, P. J. (1988) 'Determination of translational energy release distributions through analysis of metastable peaks', *Int. J. Mass Spectrom. Ion Proc.* **82**, 239-257.

[25] a. Hvistendahl, G. and Uggerud, E. (1985) 'Secondary isotope effect on kinetic energy release and reaction symmetry', *Org. Mass Spectrom.* **20**, 541-2.

b. Rickard, G. J., Cole, N. W., Christie, J. R. and Derrick, P. J. (1978) 'Energy partitioning in unimolecular decomposition. Isotope effect on the kinetic energy release in the loss of H_2 from $(CH_2{=}OH)^{+\cdot}$', *J. Am. Chem. Soc.* **100**, 2904-2905.

c. Wijenberg, J. H. O. J., Lenthe, J. H. van, Ruttink, P. J. A., Holmes, J. L. and Burgers, P. C. (1987) 'On the dissociation characteristics of H_2C^+OH ions and their participation in the H_2-catalysed isomerisation of C^+OH to HC^+O', *Int. J. Mass Spectrom. Ion Proc.* **77**, 141-154.

[26] a. Hvistendahl, G. and Williams, D. H. (1975) 'Partitioning of reverse activation energy between kinetic and internal energy in reactions of some simple organic ions', *J. Chem. Soc. Perkin Trans. II* **2**, 881-885.

b. Bombach, R., Dannacher, J. and Stadelmann, J.-P. (1984) 'The rate/energy functions for the competitive fragmentation processes of ethylene and ethane cations', *Int. J. Mass Spectrom. Ion Proc.* **58**, 217-231.

[27] a. McAfee, K. B. Jr., Szmanda, C. R., Hozack, R. S. and Johnson, R. E. (1982) 'Dynamics of energy transfer during charge-exchange in symmetric diatomic systems', *J. Chem. Phys.* **77**, 2399-2407.

b. McAfee, K. B., Szmanda, C. R. and Hozack, R. S. (1975) 'Vibrational energy transfer during charge exchange in N_2^+ on N_2 collisions', *J. Phys. B.* **14**, L243-L247.

[28] Howard, S. and Futrell, J. H. (1990) private communication.

[29] Sohlberg, K. and Szalewicz, K. (1990) private communication.

[30] Herman, Z., Pacák, V., Yencha, A. J. and Futrell, J. (1976) 'Crossed beam study of charge transfer of Ar^+ with NO', *Chem. Phys. Lett.* **37**, 329-334.

[31] Dotan, I., Fehsenfeld, F. C. and Albritton, D. L. (1979) 'On the product ratios of the reactions of Ar^+ with NO and $NO^+(a^3\Sigma^+)$ with Ar at 300 K', *J. Chem. Phys.* **71**, 3289-3294.

RADIATIVE ASSOCIATION: SUMMARY OF WORKSHOP PRESENTATIONS

ROBERT C. DUNBAR
Chemistry Department
Case Western Reserve University
Cleveland, OH 44106

It became evident early in the meeting that a number of people had been thinking about radiative association of ions with neutral molecules, and the workshop gave a timely opportunity to bring some of these thoughts together. The groups of Bartmess, Dunbar, McMahon and Ridge have all gathered data on radiative association reactions in the ICR ion trap. The theoretical scheme presented by the author suggests a semi-quantitative framework for interpretation, correlation and prediction of radiative association reactions of this type, involving fairly large molecules near room temperature. Gerlich's work represents some of the work being done on smaller systems of particular astrophysical interest, and Bierbaum's results consider the possible importance of radiative processes under flowing-afterglow conditions.

THEORY OF RADIATIVE ASSOCIATION FOR POLYATOMIC IONS. COMPARISON OF PREDICTION AND EXPERIMENT FOR NO^+ + C_6H_5CN.

(Robert C. Dunbar, Case Western Reserve University, Cleveland, OH 44106.)

Radiative association (for cations) follows the scheme

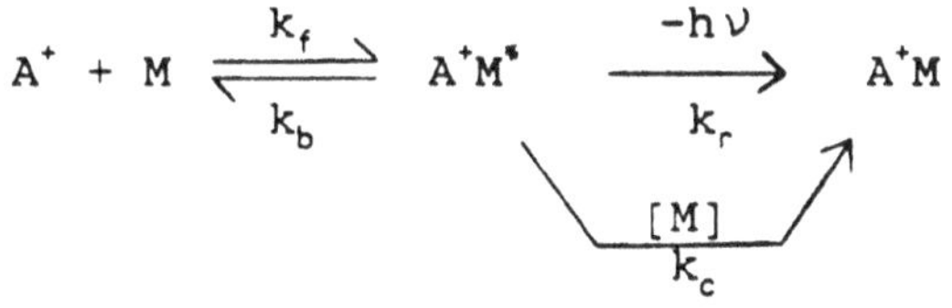

The vibrationally/rotationally excited A^+M^* ion will redissociate unless it emits a photon or undergoes a collision. We have developed a semi-quantitative estimation scheme for the class of processes involving polyatomic systems stabilizing by infrared radiation at low pressures. The modelling used

K. R. Jennings (ed.), Fundamentals of Gas Phase Ion Chemistry, 209–217.

- k_f from simple Langevin orbiting collision rates
- k_b from RRKM unimolecular dissociation theory
- k_r from the well established radiative emission calculations using the thermodynamic formulation.

The modelling resulted in a map of the probability of radiative association (expressed as the ratio of stabilization to redissociation, $R=k_r/k_b$) as a function of N (the number of internal degrees of freedom of A^+M), T (the temperature of the reactants) and E_b, (the binding energy of the A^+M complex in eV). This is shown in the figure, and can also be approximated by the equation

$$\log_{10}R = E_b\left[\frac{0.123}{\log_{10}N - 1.076} + 0.000182(T - 300)(1 + \frac{30}{N})\right]^{-1} - 10.2$$

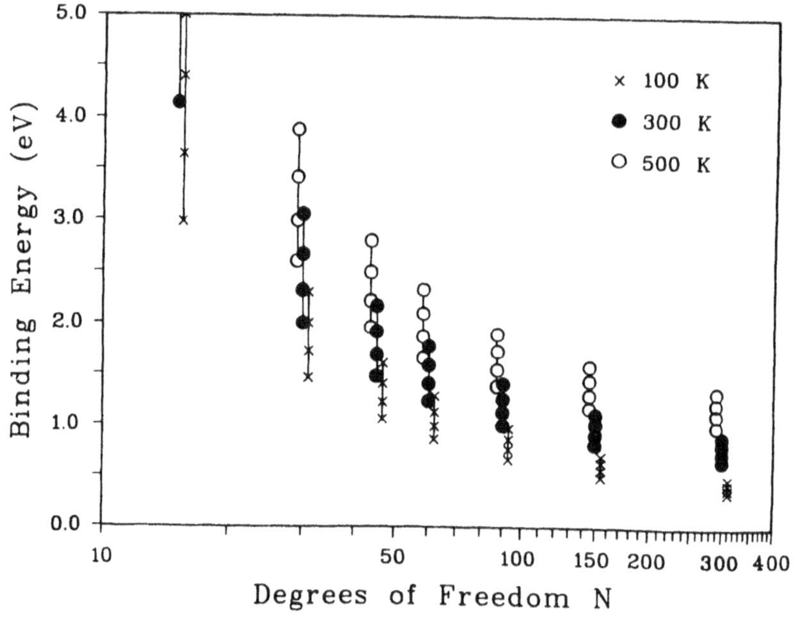

Figure 1. Semiquantitative estimation scheme for radiative association in polyatomic systems. Along the vertical axis is plotted the binding energy of the ion-neutral complex needed to give a specified ratio R of association to redissociation of the collision complex. Each set of four connected points corresponds to R values of 1, 0.1, 0.01 and 0.001 (reading from top to bottom).

Comparing a number of literature results with predictions of this scheme suggests that the association efficiency can generally be predicted within much better than a factor of 10, or conversely the binding energy can be estimated to much better than 10% from a measured radiative association efficiency.

The association reaction

$$NO^{+} + C_6H_5CN \longrightarrow NO^{+}\cdot C_6H_5CN$$

($N = 39$, $E_b = 2.01$ eV, $T = 375$ K) was studied in our ICR spectrometer. A radiative association probability of 0.014 per collision was determined, in good agreement with the predicted radiative association efficiency of ~0.01.

A more detailed description of this theory is to appear in Int. J. Mass Spectrom. Ion Proc. (in press).

DETERMINATION OF RATE CONSTANTS FOR LOW PRESSURE ASSOCIATION REACTIONS BY FOURIER TRANSFORM ION CYCLOTRON RESONANCE SPECTROMETRY (J. J. Fisher and T. B. McMahon, Chemistry Department, Guelph-Waterloo Centre, University of Waterloo, Waterloo, Canada N2L 3G1)

The best conditions for the determination of radiative association rate constants are those where collisional stabilization of the chemically activated intermediate initially formed cannot completely dominate over emission. The major limitation of techniques used to date to study these processes is that at the operating pressures of 10^{-6} torr or greater the time between ion-molecule collisions is usually much less than the radiative lifetime of the excited intermediate. However, Fourier Transform Ion Cyclotron Resonance (FTICR) has the advantages of excellent signal to noise at pressures as low as 10^{-9} torr and trapping times as long as thousands of seconds to greatly facilitate the observation of very slow reactions at low pressures.

Using the general mechanism outlined in Scheme 1 for formation of proton bound dimers, application of the steady state approximation to the excited intermediate leads to the expression given by equation (1) for the apparent bimolecular rate constant for the reaction of a protonated species with the corresponding neutral base. In the limit of low pressure of the third body, M, this expression further simplifies to equation (2). Thus it is apparent that from a plot of the apparent bimolecular rate constant *vs.* pressure a slope and intercept may be obtained which relate k_f, k_b, k_C and k_{RA}. The value of k_f can be obtained from high pressure mass spectrometric measurements of the saturated clustering rate at high pressures and the value of k_C can be obtained from trajectory calculations, as outlined by Su and Chesnavich, by invoking the strong collision assumption; that is every collision of the excited intermediate with a third body leads to a collisionally stabilized adduct. In the present work the third body is the parent base itself and it is therefore expected that the strong collision assumption will be valid. In this way the known values of k_f and k_C can be combined with the slope and intercept to obtain both the rate constant for unimolecular decomposition of the activated adduct, k_b, as well as that for emission, k_{RA}.

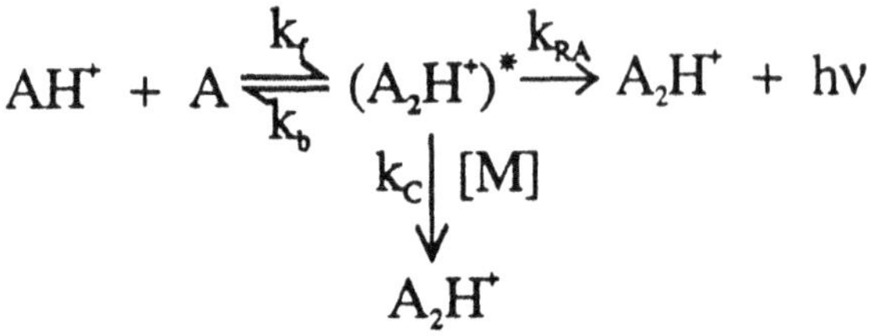

Scheme 1

$$k_{app} = \frac{k_f \cdot (k_{RA} + k_C[M])}{k_b + k_{RA} + k_C[M]} \tag{1}$$

$$k_{app} = \frac{k_f \cdot k_{RA}}{k_b + k_{RA}} + \frac{k_b \cdot k_f \cdot k_C[M]}{(k_b + k_{RA})^2} \tag{2}$$

The data obtained for a number of proton bound dimers is summarized in Table 1. In each case the extent of chemical activation is essentially the same (30 ± 1 kcal/mol) but the number of normal modes differs significantly. As expected, with increasing molecular size the unimolecular decomposition rate constant decreases markedly. No significant trend in radiative rate constant with number of modes is noted however and it appears that the nature of the mode plays a far more dominant role. For example, acetone may owe its somewhat larger value of k_{RA} to the C=O functionality.

Table I Summary of Rate Data for Low Pressure Association Reactions

	Acetone	Acetonitrile	Dimethyl Ether	Diethyl Ether
k_f ($\times 10^{-9}$ cm^{3}molecule^{-1}s^{-1})	2.2	4.3	0.9	0.8
k_C ($\times 10^{-9}$ cm^{3}molecule^{-1}s^{-1})	1.9	3.5	1.5	1.2
k_b ($\times 10^4$ s^{-1})	1.55	21.4	2.2	0.055
k_{RA} ($\times$ s^{-1})	65.2	40.7	48.3	18.0
τ_b ($\times 10^{-6}$ s)	64.3	4.7	45.5	1810
τ_{RA} ($\times 10^{-3}$ s)	15.3	24.6	20.7	55.5

Radiative association in negative ion systems has also been investigated. Of particular interest was the reaction of fluoride ion with BF_3 which has previously been reported by Babcock to have a significant radiative component at low pressures. Our investigation showed however that this was likely due to the presence of a small amount of impurity, most probably BF_2OH, which leads to the formation of BF_4^- by a sequence of exothermic bimolecular reactions. Our studies indicated that any radiative association reaction which might occur was too slow to be detected by our technique ($k_{app} < 10^{-14}$ cm^3 molecule^{-1} sec^{-1}). This points to the extreme care which must be taken in investigations of potential radiative association reactions to ensure that the apparent reaction is not occurring via a sequence of reactions involving a low level of impurity.

* For a detailed account see J.J. Fisher and T.B. McMahon Int. J. Mass Spectrom. Ion Proc. 100, in press, 1990.

T. Su and W.J. Chesnavich *J. Chem. Phys.* **76**, 5183 (1982)
C.R. Herd and L.M. Babcock *J. Phys. Chem.* **91**, 2372 (1987)

RADIATIVE ASSOCIATION REACTIONS OF $SI(CH_3)_3^+$ (Y. Lin, M. S. B. Munson and D. P. Ridge, Department of Chemistry and Biochemistry, University of Delaware, Newark, DE 19716).

The trimethyl silyl cation undergoes radiative association with acetone.[1] The association is very efficient proceeding at nearly the collision rate. We have found that the trimethylsilyl cation reacts in a similar fashion with a number of oxygen containing nucleophilic organic molecules. At the low pressures and long trapping times accessible in an FTICR spectrometer it can be established unequivocally that the mechanism for the reaction is radiative association,

$$Si(CH_3)_3^+ + A \longrightarrow [Si(CH_3)_3A^+]^+ \longrightarrow Si(CH_3)_3A^+ + h\nu, \quad (1)$$

and reaction rate constants and efficiencies can be determined. Efficiencies for (1) with various nucleophiles A as determined using FTICR are listed in Table I.

These reactions offer an opportunity to examine the connection between the ion-neutral binding energy and the efficiency of radiative association. Trimethyl silyl cation transfer reactions,

$$Si(CH_3)_3A^+ + B \longleftrightarrow Si(CH_3)_3B^+ + A, \quad (2)$$

are sufficiently facile that equilibrium constants for them can be measured using high pressure mass spectrometry or FT-ICR. A ladder of relative $Si(CH_3)_3^+$ affinities was constructed and anchored to the value of $D(Si(CH_3)_3^+$ - acetone) obtained by Stone, et al.[2] The resulting values of $D(Si(CH3)_3^+$ - A) are given in Table I.

The theory discussed above by R. Dunbar[3] provides a functional relationship between the ion-neutral binding energy, the number of vibrational modes in the ion netural complex and the temperature. It is thus possible to deduce ion-neutral bond energies from our measured radiative association efficiencies. These are listed in Table I as values of Eb and compare quite well with the more directly measured values of D also listed in Table I. The values of Eb are systematically less than the values of D suggesting that the reactant ions in the association reactions may have a slightly higher effective temperature than the cell wall temperatures. Nevertheless the comparison is encouraging and suggests that measurement of radiative association rate constants may provide binding energies in systems where other techniques are difficult or unavailable. We have for example, obtained preliminary data on the association of imidazole with an iron tetraphenyl porphyrin that suggests that the imidazole binds to the porphyrin with a binding energy greater than 0.8 eV. This suggest that radiative association may provide useful results pertinent to problems in biochemistry.

1. R. Orlando, D. P. and B. Munson, J. Am. Soc. Mass Spectrom. 2, 144 (1990).

2. J. A. Stone et al., Int. J. of Mass Spectrom. Ion Proc. 74, 59 (1986).

3. R. Dunbar, Int. J. Mass Spectrom. Ion Proc., To be published. See also the contribution to this workshop.

TABLE I

A	Eff.(exp.)[a]	$D((CH_3)_3^+-A)$/ev	Eb/ev
c-hexanone	1.1+.1	2.06	>1.77
Acetone	0.68+0.07	1.95[b]	1.80
2,5-dimethylfuran	0.40+.007	1.85	1.52
Dimethylether	0.25+.006	1.81	1.71
Anisole	1.2+0.1	1.80	>1.75
2-methylfuran	1.18+0.03	1.78	1.55
Furan	0.029+.004	1.64	1.52

(a) Ratio of experimental rate constant to collision rate constant for reaction 1.

(b) Obtained from ref. 2.

ASSOCIATION REACTIONS IN THE TANDEM FLOWING AFTERGLOW/SIFT
(Veronica Bierbaum, Department of Chemistry and Biochemistry, Boulder, CO 80309

Kinetic data were obtained in the tandem flowing-afterglow selected ion flow tube for ten association reactions in helium buffer gas: 1) HO^-+H_2O; 2) DO^-+D_2O; 3) $CH_3O^-+H_2O$; 4) $CH_3O^-+D_2O$; 5) $CD_3O^-+H_2O$; 6) $CD_3O^-+D_2O$; 7) $CH_3O^-+CH_3OH$; 8) $CH_3O^-+CH_3OD$; 9) $CD_3O^-+CD_3OH$; and 10) CD_3OD. Plots of the effective bimolecular rate constant versus helium pressure are linear for all reactions. Replacement of OH by OD has no measurable effect on reactivity, whereas replacement of CH_3 by CD_3 increases the reactivity significantly. The data for reactions (1) and (2) have zero intercepts as expected but there is an increasing indication of a non-zero intercept along the list of reactions. For reactions (9) and (10), an intercept of 8 x 10^{-11} cm^3/s would be inferred from a linear least squares fit of the data. Non-zero intercepts from similar flowing afterglow data have previously been interpreted by other workers as evidence for radiative stabilization. That explanation was seriously considered here but discarded; both the bond energy (28.8 kcal/mol) and the number of vibrational degrees of freedom are too small to allow for efficient radiative association. Other explanations, including formation of $CH_3O^-{\cdot}He$ clusters which undergo switching with CH_3OH, or stabilization by CH_3OH as the third body, are similarly untenable. The kinetic data, although showing no measureable curvature, must represent a portion of the fall-off region.

RADIATIVE ASSOCIATION IN BORATE AND BORONATE ESTERS
(J. Bartmess and J. P. Kiplinger, Department of Chemistry, University of Tennessee, Knoxville, TN 37996)

Based on the general theory of radiative association [R. C. Dunbar, this volume], it was predicted that molecules with intense ir stretches in the center of the ir spectrum [1000-1400 cm^{-1}] should be especially good at radiative emission [Caldwell & Bartmess, J. Phys. Chem. 85, 3571 (1981)]. The B-O bond satisfies these conditions.

In a trapped ICR spectrometer, methoxide adds to $B(OMe)_3$ in a bimolecular, pressure independent process for which $k=1.010^{-10}$ cc/molecule-s. Similarly, methoxide adds in a bimolecular process to the cyclic boronate ester $MeB(OCH_2)_2CMe_2$ even though deprotonation should be exothermic here by ca. 11 kcal/mole. The radical cation of $(MeO)_3B$ reacts with neutral $(MeO)_3B$ to give no addition product, but rather $[2M-15]^+$. In contrast, the radical cation of the cyclic boronate ester reacts with its neutral to give primarily the $2M^+$ addition ion. This difference in the cyclic boronate vs open borate systems is interpreted as due to ring opening in the addition product. This creates a floppy side chain, free to rotate, that should increase the intermediate lifetime with respect to dissociation back to reactants. The longer lifetime favors radiative emission.

RADIATIVE ASSOCIATION STUDIES (Dieter Gerlich, Albert-Ludwigs-Universitat Freiburg, Hermann-Herder-Strasse 3, D-7800 Freiburg)

Goals of our radiative association experiments:
1. Determination of rate coefficients for small molecular systems which are of astrophysical interest: $C^+{\cdot}H_2$, $CH_3^+{\cdot}H_2$; work on $C_2H_2^+{\cdot}H_2$ and $C_3H^+{\cdot}H_2$ is in progress. 2. Study of the model system $H^+{\cdot}H_2$. 3. Spectroscopic studies on the intermediate complex using lasers for stabilizing it via induced emission. So far no success with a Nd:YAG laser (1 mu); work on H_3^+ with a CO_2 laser is in progress. 4. Direct determination of the radiative lifetime of highly excited molecules by using CO_2 laser induced fragmentation.

Experimental: Our apparatus uses a radio frequency ring electrode ion trap, which differs from the usually-used ICR cells or Paul traps. The dominant advantage is that the ions are stored in a wide field free region which allows us to work with very cold ions and to simulate conditions of interstellar space[1].

Results: We have recently[2] reported ternary (k_3) and radiative (k_r) association rate coefficients for the aggregation of cold hydrogen molecules on hydrogen ions at very low temperatures:

	k_3 [$10^{-29}cm^6/s$]			k_r [$10^{-16}cm^3/s$]
	25k	80K	350K	80K
$H^+ + 2H_2$		5.4±1		1.3
$D^+ + 2D_2$		7.6±1	3.5	1
$H^+ + H + He$		10.7±5		–
$H_3^+ + 2H_2$	7	2.5		< 0.5
$D_3^+ + 2D_2$		3.7		< 0.5

Under certain assumptions for complex formation and complex stabilization we obtain the following $H^{\cdot}H_2$ complex radiative lifetime:

$$1/k_d > 0.65.10^{-10} \text{ s} \quad \text{and} \quad 1/A_r > 1.2 \cdot 10^{-3} \text{ s}.$$

The first value deviates significantly from $6.8 \cdot 10^{-10}$ s derived from classical trajectories. Also the radiative lifetime is larger than the calculated value of $0.14 \cdot 10^{-3}$ s and larger than our measured LID result of $0.37 \cdot 10^{-3}$ s. The difference between our two experimental values for the radiative lifetime are most probably due to significant differences in the angular momentum distributions of the two H_3^+ molecules studied, the collision complex $H^+ \cdot H_2$ and the highly excited molecules $(H_3^+)^{**}$.

Going from H_3^+ complexes to H_5^+ leads to an increase of the degrees of freedom, and one would expect an increase of the complex lifetime. However this is more than counterweighted by the large change in the binding energy (from 4.35 eV to about 0.3 eV) as can be seen from a comparison of the association rate coefficients (see Table).

Results for the stabilization of $CH_3^+ \cdot H_2$ and $CD_3^+ \cdot D_2$ at 80 K have been published recently[1]. The derived upper bound on the radiative rate is in good agreement with older predictions [Bates 1983] and with the assumption that the CH_5^+ complex is stabilized by spontaneous emission of infrared photons, but in serious conflict with recent theoretical conclusions [Bates 1987]. From the point of view of our experiment, inclusion of the value $A_r = 3.5 \cdot 10^4$ s^{-1} in calculations modeling dense interstellar clouds is not recommended.

The radiative association of $C^+ \cdot H_2$ is the important first step in the interstellar gas phase formation of hydrocarbons. In a liquid nitrogen cooled ion trap we have obtained $k_r = (7+3) \times 10^{-16}$ cm^3/s. From the ternary rate coefficient we have deduced a radiative rate of 5×10^4 s^{-1}, being indicative for an optical transition. Very recent measurements in our laboratory at a temperature of about 25 K resulted in a value of $(5.5+2) \times 10^{-16}$ cm^3/s. The main problems in the determination of these values arise from reactions with gas impurities.

1. D. Gerlich and G. Kaefer, Ap. J. 347, 849 (1989).

2. D. Gerlich, G. Kaefer and W. Paul, Symposium on Atomic and Surface Physics, Obertraun, Austria, 18-21 March 1990.

Workshop "Spectroscopy of Ions"

Ch. Ottinger
Max–Planck–Institut für Strömungsforschung
3400 Göttingen, Germany

Throughout the programme of the NATO ASI on Fundamentals of Gas Phase Ion Chemistry, frequent reference was made, both in the lectures and the discussions, to spectroscopic properties of molecular ions. The lectures by R. Donovan, P. Sarre, O. Dutuit and J. Maier were devoted entirely to spectroscopic techniques and problems. In many other lectures, the dynamical properties of ions in electronically and vibrationally excited states were also addressed, e.g. by T. Baer, A.J. Stace and M. Bowers. In addition, the workshop on Radiative Association focussed on the emission of molecular ions, both large and small. All this testifies to the great importance which spectroscopy has assumed in an area of research which formerly used to be a branch of mass spectrometry. It seemed thus timely to organize a workshop within this conference expressly on the subject of "Spectroscopy of Ions", in order to provide a forum for an even broader representation of this field.

The contributions which follow are arranged in the order in which they were presented at the meeting. They are devoted to three main topics:

- The study of electronic transition band systems and the information which can be derived from them concerning the nature of the chemical bonding in an important class of molecular ions.
- Measurements of IR radiative relaxation rates of vibrationally excited molecules, diatomic and polyatomic; the knowledge of this quantity is a prerequisite for the interpretation of bimolecular and unimolecular reaction dynamics of such species.
- Ab initio calculations of the structure and IR spectra of small polyatomic (in the present cases triatomic) molecular ions. The few examples given demonstrate that modern quantum chemistry can provide reliable information in situations where experiments are either not precise enough or cannot be done at all.

K. R. Jennings (ed.), Fundamentals of Gas Phase Ion Chemistry, 219–235.

A new class of UV molecular ion spectra

B. Müller and Ch. Ottinger
Max–Planck–Institut für Strömungsforschung
3400 Göttingen, Germany

and

P. Rosmus, University of Frankfurt, Germany

Previous work in Göttingen has shown that the products of ion–molecule reactions are often formed in electronically excited states, some of which emit hitherto unobserved band systems. This ion chemiluminescence method was used in the present experiment to study the spectra of the ionic monohalides of group IIIa atoms. These species (e.g. $AlCl^+$) are isoelectronic with the well–known alkaline earth halides (e.g. MgCl), yet very little information exists on their spectra. We have observed emission from 10 out of the 12 species AX^+ with A=B, Al, Ga, In and X=F, Cl, Br; in most cases two or even three band systems could be identified [1]. Fig.1 shows the results from reactions of In^+ with F_2, Cl_2, Br_2 . Apart from some atomic lines, the spectra are ascribed to InF^+, $InCl^+$, $InBr^+$. The identification is based on comparison with related species (especially the isoelectronic neutral halides MX) , on a few published photoelectron and optical spectra and on some accurate ab initio calculations.

The broad quasicontinuum (coarse hatching in fig.1) is due to emission from a $^2\Sigma$ state, arising from $A^+(^3P)+X(^2P)$, into the $^2\Sigma$ ground state of AX^+ , having the dissociation asmptote $A^+(^1S)+X(^2P)$. Similar to the situation with systems such as BH^+ and AlH^+ [2], the excited state potential minimum is shifted outward, with respect to that of the ground state, due to an avoided crossing. This explains the broad, unstructured emission spectrum with a sharp cut–off to the red. The neutral MX molecules do <u>not</u> have an excited $^2\Sigma$ state of this type.

The narrow feature marked by fine hatching in fig.1 is the emission from a $^2\Pi$ state into the ground state. This band system does have a counterpart in MX , namely the well–known A–X systems. For MX , the $^2\Sigma$ ground state correlates diabatically with $M^+(^2S)+X^-(^1S)$, on account of the ionic nature of the MX

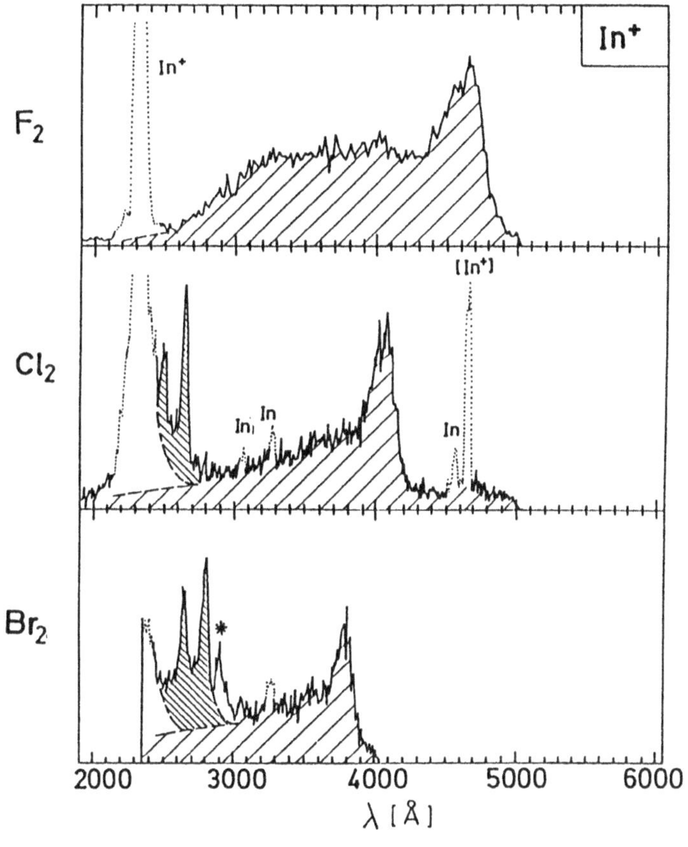

Fig. 1: Emission spectra observed from chemiluminescent reactions of In^+ ions with F_2, Cl_2, Br_2, at 2.5 , 3.8 and 5.8 eV_{CM} , respectively. The resolution was 100 Å FWHM with F_2 and 50 Å with Cl_2 and Br_2 (from ref.[1]). The two molecular band systems discussed in the text are indicated by hatching. A third, marked by the asterisk, may be the analog of the MX(B $^2\Sigma$–X $^2\Sigma$) transition.

chemical bond. Similarly the MX $^2\Pi$ state correlates diabatically with $M^+(^2P)+X^-(^2S)$. Thus the $MX(^2\Pi-^2\Sigma)$ transition can be considered as the molecular "mirror image" of the simple (alkali–like) atomic transition of M^+, $^2P-^2S$. In fact, reviewing the literature data, a close quantitative parallelism is found between the MX and the M^+ transition energies as well as the corresponding fine structure splittings ($^2\Pi_{1/2,3/2}$ and $^2P_{1/2,3/2}$, respectively). The nature of the halogen, X , has very little effect. This systematic parallelism should also obtain for the AX^+ species. Here the ground state dissociates diabatically into $A^{++}(^2S)+X^-(^1S)$, and the $^2\Pi$ state into $A^{++}(^2P)+X^-(^1S)$ Based on the known A^{++} $^2P-^2S$ transition energies and their fine structure splittings and comparing with the trends in the MX species, the "narrow" features of the newly observed AX^+ spectra could then be unambiguously identified as the postulated $^2\Pi-^2\Sigma$ transition. Its spin–orbit splitting is clearly resolved in fig.1 and could also be observed for the Ga halide ions.

Quantitative confirmation of much of the above arguments was obtained from ab initio calculations done for AlF^+ and $AlCl^+$. Using a very efficient CASSCF program developed by Rosmus, Werner and Knowles, the AlF^+ potential curves shown in fig.2 were calculated on a CRAY–XMP machine. Full CI is included. GTO basis sets with the following Huzinaga type wave functions were used:

Al:	15(12)s ,	13(10)p ,	3d ,	1f
F :	12(9)s ,	8(7)p ,	3d ,	1f

(in parentheses: contracted basis set). 20 reference configurations were used for the Σ states and 17 for the Π states. The curves labeled B $^2\Sigma$ and C $^2\Pi$ represent the states discussed above. Clearly the B–X emission will occur as a broad band system with a maximum close to its "red" limit. An AlF^+ transition with these characteristics was, in fact, observed [1]. Its contour resembled that of the "broad system" of InF^+ (and also $InCl^+$ and $InBr^+$) shown in fig.1. A spectrum similar to the "narrow systems" of $InCl^+$ and $InBr^+$, on the other hand, would be expected to result from the C–X transitions of AlF^+ . The position of the C–state potential

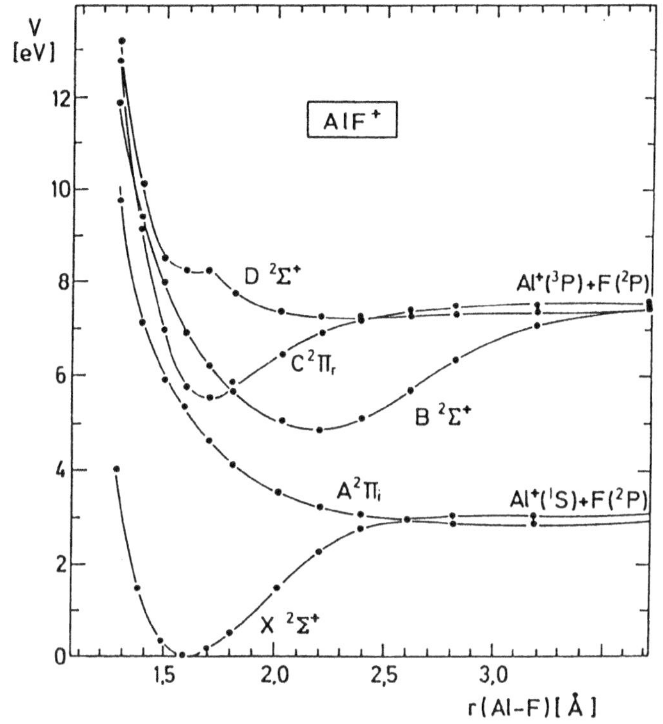

Fig. 2: Calculated potential energy curves of AlF^+.

curve in fig.2 suggests emission confined to a small wavelength range on the "blue" side of the B–X system. However, no such band system was observed for AlF^+ [1]. The reason for this is very likely the A $^2\Pi$ state shown in fig.2. It interacts strongly with the C $^2\Pi$ state near its minimum and probably causes it to predissociate.

Similar conditions may obtain for BF^+, GaF^+ and InF^+, for which also no narrow band system was observed ([1] and fig.1). The ab initio calculations for $AlCl^+$, on the other hand, show a much wider gap between the two $^2\Pi$ states, and accordingly a "narrow system" was, in fact, observed in this case [1].

[1] Th. Glenewinkel–Meyer, A. Kowalski, B. Müller, Ch. Ottinger and W.H. Breckenridge, J.Chem.Phys. 89, 7112 (1988).

[2] Ch. Ottinger and J. Reichmuth, J.Chem.Phys. 74, 928 (1981); B. Müller and Ch. Ottinger, J.Chem.Phys. 85, 232 (1986); R. Klein, P. Rosmus, and H.–J. Werner, J.Chem.Phys. 77, 3559 (1982).

CONTRIBUTION TO THE PANEL DISCUSSION ON SPECTROSCOPY

RADIATIVE LIFETIMES OF VIBRATIONALY EXCITED AND METASTABLE IONS

S. Fenistein, M. Heninger, G. Mauclaire, Y.M. Yang and R. Marx

Laboratoire de Physico-Chimie des Rayonnements, Université Paris-Sud, 91405 Orsay, France

To follow radiative decays in the millisecond to second range, too slow for spectroscopic investigation, one can use the so called "monitor ion" technique where the energy content of an ion is probed via its chemical reactivity. In principle any reaction may be used provided its rate changes by at least an order of magnitude in a narrow internal energy range. Charge or proton transfers are good monitors since at thermal energy endothermic reactions are much slower than exothermic ones.

Reaction $A^{+}(v)+M \longrightarrow M^{+} + A$ will probe the A^{+} with internal energy IE ≥ IP(M)-IP(A).

Reaction $AH^{+}(v) + M \longrightarrow MH^{+} + A$ will probe the AH^{+} with internal energy IE ≥ PA(A)-PA(M).

IP and PA are respectively the ionization potentials and proton affinities.

The triple cell ICR spectrometer used in our experiments consists of three differentially pumped ICR cells communicating through two ion funnels. In the first cell the ions are produced by electron impact or ion-molecule reactions. They are drifted into the second cell where the pressure is kept as low as possible and the ions are stored for variable relaxation times. In the third cell the ions react with the monitor gas at constant pressure and reaction time. Finally the reactant and product ions are drifted back to the low pressure cell to undergo FT detection.

The amount of monitor ions M^{+} (MH^{+}), normalized to the total ion intensity, is an exponential function of the relaxation time and its decay gives the apparent lifetime of the reactant ion A^{+}(AH^{+}). Correction for radiative cascade when several level are populated, and (or) for quenching by the gases leaking from the source and reaction cells, are needed to get the radiative lifetimes.

Our results for NO^{+}(v=1-4) [1-2] are in very good agreement with the theoretical values [3]. For $H(D)Cl^{+}$ [4] and HCO^{+}, the theoretical values [5-7] differ by 20 to 50% but there is quite a good agreement for the isotope effect (see table).

Preliminary results obtained for $NO^{+}(a^{3}\Sigma^{+})$: (80 ± 20) ms for v≥1 and (120 ± 25) ms for v≥0, are much shorter than the two experimental values published for v≥0: 530^{+300}_{-100} ms [8] and 1450^{+1150}_{-450} ms [9] for v≥0. There is no, to our knowledge, any reliable theoretical calculation.

Since this is so far the only available experimental method to determine lifetimes in the millisecond to second time range one has to be aware of its limitations and possible problems.

Two types of problems have to be considered:

1) Problems related to the FT-ICR technique itself, mainly mass discrimination. In normal experimental conditions this does not affect very much the lifetime except if discrimination depends on trapping time (oscillations on the decay curve). We will not discuss this point here.

2) Reliability of the thermodynamic data (IP and PA). In order to identify the radiating states, the energetics of the monitor reaction must be known with an accuracy of at least half of a vibrational spacing (~0.05 (eV) for the bending mode of HCO^{+} for example) of course if the ions can be produced in selected states one has simply to check that the reaction does not occur with the state just below. However as there are very few such cases one really needs accurate thermodynamic data.

Concerning the IP's very accurate data are reported in the tables. However those are vertical IP's obtained by photo electron spectroscopy which may be somewhat higher than the adiabatic IP's involved in charge transfer reactions occuring through a complex which leaves long enough to allow nuclear motion. In this case the monitor reaction may probe level(s) lower than expected from the vertical IP's.

The proton affinity scale is certainly less accurate (≥ 1-2 kcal.mole^{-1}). Besides, it has been subject recently to quite a few controversial discussions and experimental re-determinations (see pannel discussion on PA).

In conclusion one must be very carefull and use, if possible, several monitors to cross check the lifetimes measurements.

IONS			EXPERIMENTAL VALUES	THEORETICAL VALUES
NO^+		v=1	80 ± 15 ms [2]	88 ms [3]
		v=2	42 ± 8 ms [2]	45 ms [3]
		v=3	26 ± 5 ms [2]	31 ms [3]
		v=4	21.5 ± 4 ms [2]	24 ms [3]
HCl^+		v=1	3.4 ± 1 ms [4]	4.6 ms [5]
DCl^+		v=1	13 ± 3 ms [4]	19.6 ms [6]
HCO^+	ν_1 (H-C)		(4 ± 1 ms) ?	3.8 ms [7]
	ν_2 (Bending)		> 500 ms	626 ms [7]
	ν_3 (C-O)	v=1	20 ± 4 ms	17 ms [7]
		v=2	12 ± 2 ms	
		v=3	(4 ± 1 ms) ?	

REFERENCES

1. M. Heninger, S. Fenistein, M. Durup-Ferguson, E.E. Ferguson, R. Marx and G. Mauclaire, Chem. Phys. Lett. **131** (1986) 439
2. S. Fenistein, M. Heninger, R. Marx, G. Mauclaire and Y.M. Yang Chem. Phys. Lett., **172** (1990) 89
3. G. Chambaud and P. Rosmus, Chem. Phys. Letters **165** (1990) 429
4. M. Heninger, S. Fenistein, G. Mauclaire, R. Marx and Y.M. Yang, J. Chem. Soc. Faraday Trans 2, **85** (1989) 1705
5. H.-J. Wener, P. Rosmus, W. Schatzl and W. Meyer, J. Chem. Phys., **80** (1984) 831
6. P. Rosmus, private communication
7. P. Botchwina, private communication
8. C.H. Kuo, T. Wyttenbach, C.G. Beggs, P.R. Kemper and M.T. Bowers J. Chem. Phys., **92** (1990) 4849
9. A. O'Keefe and J. McDonald, Chem. Phys., **103** (1986) 425

Acknowledgement: This work has been supported by CNRS and AFORS under grant n°87-0390

Radiative Lifetime Measurements of Vibrationally Excited States of $NO^+(X^1\Sigma^+)$

Thomas Wyttenbach, Cindy G. Beggs and Michael T. Bowers

Department of Chemistry
University of California
Santa Barbara, CA 93106

The nitric oxide cation is one of the most important and abundant atmospheric ions.[1] In the upper ionosphere the ion-molecule reaction

$$O^+(^4S) + N_2(X^1\Sigma_g^+,v) \rightarrow NO^+(X^1\Sigma^+,v) + N(^4S)$$

produces vibrationally excited NO^+ ions which are responsible, via dissociative attachment, for the removal of free electrons and for the formation of $N(^2D)$ atoms.[2] In order to estimate the vibrational state distribution in the NO^+ ions, the transition probability coefficients of spontaneous emission are needed. Further, Einstein coefficients (i.e., radiative lifetimes) for the vibrational states of ions in their ground electronic states are needed to determine the importance of radiative association reactions in interstellar molecule synthesis, and also to calculate the ion concentration responsible for spectra observed in planetary atmospheres.

Our experiments are done in a single cell ICR machine with Fourier transform detection. It is equipped with two skimmed, differentially pumped valves, capable of delivering 100 μs wide bursts of gas. The first valve delivers a 100 μs pulse of NO. The output of an excimer pumped dye laser system intercepts the gas burst in the center of the ICR cell and creates state selected $NO^+(X^1\Sigma^+, v)$ via 2 + 1 resonant-enhanced multiphoton ionization.[3] The vibrational radiative lifetime of these state selected NO^+ ions is then determined using the monitor ion technique.[4] The second pulsed valve is loaded with a gas that will react with $NO^+(v)$ but not $NO^+(v-1)$. By observing the intensity of the monitor gas signal as a function of the delay time of the second pulsed valve, the radiative lifetime is directly determined.

The results are summarized in Table I.

Table I: Radiative lifetimes (ms) of $NO^+(X^1\Sigma^+, v)$

	Experiment		Theory	
	This work	Fenistein et al.(a)	Chambaud and Rosmus(b)	Partridge et al.(c)
v=1	90 ± 10	83 ± 16	88	88
v=2	49 ± 5	43 ± 8	45	45
v=3	30 ± 3	26 ± 5	31	31
v=4	25 ± 3	23 ± 5	24	
v=5	16 ± 4(d)		20	

(a) Reference 5. The uncertainties are those quoted by the authors.
(b) Reference 7. No uncertainties given.
(c) Reference 8. No uncertainties given.
(d) This is a preliminary number. More experiments are being done to verify it.

It is clear that our results are in excellent agreement with the most recent theory. They are also in good agreement with the experimental results of Fenistein et.

al.[5] who used a triple cell ICR technique and rather complex kinetic analysis to deduce the various radiative lifetimes. It is important that our results agree with theory because these represent direct and independent measurements and thus provide a direct test of the approximations used in the theoretical calculations. Presumably, similar approximations could be used to predict radiative lifetimes in other systems where measurements cannot easily be made.

Finally, having measured the radiative lifetimes, we can now proceed to measure state specific bimolecular reactivity of $NO^{+}(v)$. Such measurements are underway[3,6] and show some surprisingly large effects on simple systems. These results will be presented in due course.

References

1. A.T. Stair and H.P. Gauvin, in Aurora and Afterglow, B.M. McCormac (ed.) Reinhold, N.Y., (1967) p. 365.

2. D.G. Hopper, *J. Chem. Phys.* 72, 3679 (1980).

3. C.G. Beggs, C.-H. Kuo, T. Wyttenbach, P.R. Kemper and M.T. Bowers, *Int. J. Mass Spectrom. Ion Proc.* 100, xxxx (1990).

4. M. Heninger, S. Fenistein, M. Durup-Ferguson, E.E. Ferguson, R. Marx and G. Mauclaire, *Chem. Phys. Lett.* 131, 439 (1986).

5. S. Fenistein, M. Heninger, R. Marx, G. Mauclaire and Y.M. Yang, *Chem. Phys. Lett.* (in press).

6. C.G. Beggs, *Ph.D. Dissertation*, University of California at Santa Barbara (1990).

7. G. Chambaud and P. Rosmus, *Chem. Phys. Lett.* 165, 429 (1990).

8. H. Partridge, S.R. Langhoff and C.W. Bauschlicher Jr. (Private Communication).

IR-Radiative Relaxation of n-Butylbenzene Ions

Guy Uechi and Robert C. Dunbar
Chemistry Department
Case Western Reserve University
Cleveland, OH. 44106

The aim was to find the rate of infrared radiative relaxation of n-butylbenzene ions containing a small amount of internal energy in excess of thermal energy:

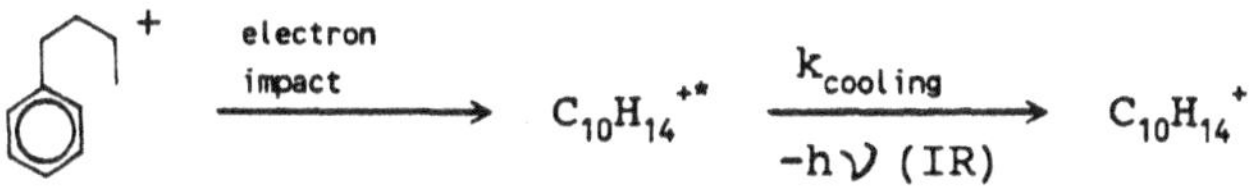

The excess internal energy of the excited ion is of the order of 0.3 eV, contained in the form of vibration.

The probe of internal energy (the internal temperature thermometer) was the competitive photodissociation at 458 nm into two products,

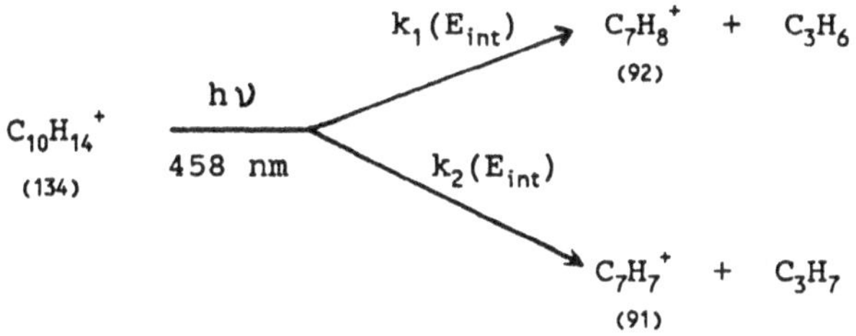

The branching ratio for this photodissociation is strongly dependent on E_{int}, and is known with confidence, so that an observation of k_1/k_2 provides a well calibrated measure of E_{int} at the moment of the laser pulse.

The figure shows the 91/92 photoproduct ratio as a function of the delay time between the electron beam pulse and the laser pulse. Using the known calibration, this is converted to a plot of E_{int} versus time, as shown by the energy scale on the right side of the figure.

It is seen that the internal energy declines on a time scale of seconds. The contribution of collisional relaxation was removed by making such curves at 3 pressures and extrapolating to zero pressure. The result was a value of 0.3 eV for the average energy (in excess of the 0.2 eV of equilibrium thermal energy) deposited by electron impact ionization (~13 eV electrons), and an IR radiative cooling rate constant

$$k_{cooling} = 0.5 \ s^{-1}.$$

This cooling rate has the same order of magnitude as those for other ions (dioxane ion[1], styrene ion[2]) studied in our laboratory in the energy regime a few tenths of an eV above room temperature.

1. F. S. Huang and R. C. Dunbar, J. Am. Chem. Soc. **111**, 6497 (1989).

2. R. C. Dunbar, J. Chem. Phys. **91**, 6080 (1989).

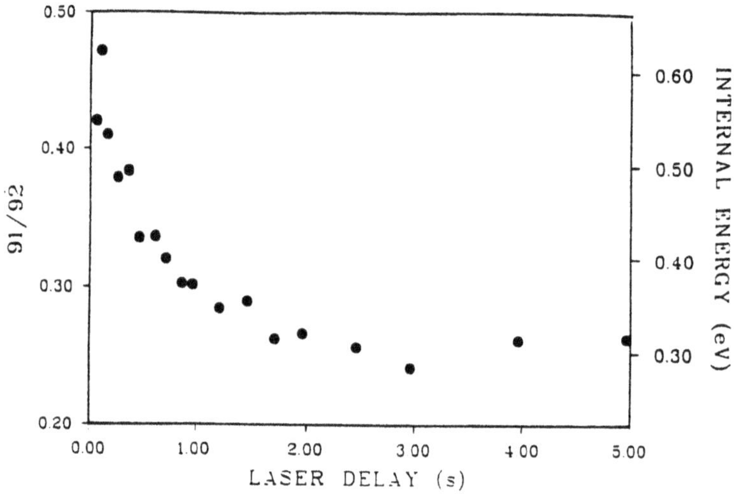

Ab initio calculations for HCC^- and Cl^- - - - HF

P. Botschwina (Inst. für Physikal. Chemie, Universität Göttingen, D–3400 Göttingen) and
P. Sebald (FB Chemie, Universität Kaiserslautern, D–6750 Kaiserslautern)

The acetylide anion, HCC^-, is the simplest polyatomic carbanion and thus of general interest. Nevertheless, precise spectroscopic information on gas–phase HCC^- is still not available. In 1987 Gruebele et al. [1] claimed to have observed 12 lines within the ν_3 band (-CC stretch) of this species. The derived spectroscopic parameters, in particular the band origin and – much more pronounced – the vibration–rotation coupling constant α_3 turned out to be in poor agreement with our ab initio results and the correctness of the assignments was seriously questioned [2,3]. In the mean–time we have considerably extended our previous work and some new results are presented here (see also [4]).

Making use of the Coupled Electron Pair Approximation, CEPA [5], and a basis set of 98 contracted Gaussian–type orbitals (11s, 6p, 3d, 1f / 8s, 3p contracted to [8,4,3,1 / 5,3]), calculations have been carried out for 198 different nuclear configurations. These energy points were fitted by a polynomial function of type

$$V(S_1,S_2,S_3) = \sum_{ijk} C_{ijk} S_1^i S_2^j S_3^k \qquad \text{(j: even)}$$

where the symmetry coordinates S_1, S_2, and S_3 are defined as follows:

$S_1 = 2^{-\frac{1}{2}} (R_1 + R_2 - R_{1e} - R_{2e})$; $S_2 = 2^{-\frac{1}{2}} (R_1 - R_2)$; $S_3 = R_3 - R_{3e}$.

R_1 and R_2 are internuclear separations $C_{(1)}H$ and $C_{(2)}H$, R_3 is the $C_{(1)}C_{(2)}$ distance, and $R_{1e} = R_{2e}$ and R_{3e} are the optimum distances in C_{2v} symmetry (T–shaped form). 60 non–redundant coefficients C_{ijk} were included in the fit and a standard deviation of 1.7 cm^{-1} (0.005 kcal mol^{-1}) was obtained. The equilibrium geometry of HCC^- was calculated to be linear with r_e (HC) = 1.0705 Å and R_e (CC) = 1.2493 Å. A contour plot for the movement of the H atom around the CC framework is shown in Fig. 1. We find a shallow local energy minimum in the C_{2v} configuration which has not yet been described in the literature although it appears already in the SCF approximation (provided that p–functions are available at the hydrogen nucleus). It lies 7587 cm^{-1} above the absolute energy minimum at linear geometry. The two equivalent saddle points to hydrogen–shift are calculated to be 271 cm^{-1} higher in energy.

Calculated band origins of a number of rovibrational transitions of HC_2^- and DC_2^- are listed in Table 1. The bending vibrational frequencies are remarkably small (512.5 and 408.4 cm^{-1} for HC_2^- and DC_2^-, respectively). The ν_3 band origin is calculated at 1814.9 cm^{-1} and after consideration of the remaining errors in the calculations we recommend ν_3 = 1807 $\pm$ 5 cm^{-1}. This differs markedly from the value of 1758.6 cm^{-1} reported by Gruebele et al.(1). The calculated α_3 value of 0.0100 cm^{-1} is at complete variance with the value of Gruebele et al. (0.0193 cm^{-1}) and we are thus led to the conclusion that the observed lines are not due to HCC^-. So there is still the challenge to the experimentalist to obtain a high–resolution spectrum of HCC^-.

Table 1. Band origins of HC_2^- and DC_2^- (in cm^{-1}).

Band	$H^{12}C^{12}C^-$	$D^{12}C^{12}C^-$
ν_2^1	512.5	408.4
$2\nu_2^0$	1007.1	805.0
$3\nu_2^1$	1509.9	1206.0
ν_3	1814.9	1719.4
$4\nu_2^0$	1995.9	1595.2
$\nu_2^1+\nu_3$	2320.5	2126.7
$2\nu_2^0+\nu_3$	2807.9	2522.8
ν_1	3216.7	2482.2
$2\nu_3$	3611.3	3423.8
$\nu_1+\nu_2^1$	3709.6	2874.8
$\nu_2^1+2\nu_3$	4109.6	3829.6
$\nu_1+2\nu_2^0$	4181.9	3254.3
$2\nu_2^0+2\nu_3$	4588.7	4224.5
$\nu_1+\nu_3$	5018.7	4177.9
$3\nu_3$	5389.6	5113.5
$2\nu_1$	6319.3	4913.7
ZPE [a]	3085.3	2543.1

a) Zero–point energy.

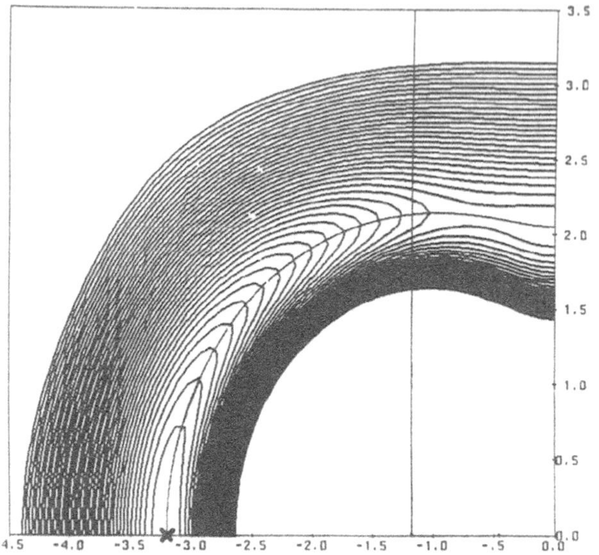

Fig. 1. Contour plot for hydrogen migration in HCC^- (at optimum values for CC separation). Contour lines in inter vals of 500 cm^{-1}.
The absolute energy minimum (linear HCC^-) is marked by a cross. Scaling of axes is in bohr.

During the past few years the bihalide anions have attracted much attention. They are prototypes of strongly hydrogen–bonded complexes with large anharmonicity effects and may be also the starting point for performing "spectroscopy of the transition state" [6–9]. Following our previous work on $ClHCl^-$ [10], we have recently investigated the asymmetric species Cl^- – – – HF in considerable detail. Using CEPA–1 and a basis set of 126 cGTOs, 148 points of the potential and electric dipole moment surface have been calculated. The equilibrium geometry is linear with R_{1e} (ClH) = 1.9240 Å and R_{2e} (HF) = 0.9627 Å, in close agreement with the "full CI estimates" of Sannigrahi and Peyerimhoff [11]. Cuts through our analytical 3 D potential surface [4] at 3 different orientation angles are shown in Fig. 2.

The fundamentals of Cl^- – – – HF are predicted to occur at ν_1 = 2814, ν_2^1 = 863, and ν_3 = 250 cm^{-1}. The latter two values should be accurate to ca. 10 cm^{-1} whereas the ν_1 band origin (-FH stretch) may be in error by ca. 30 cm^{-1}. Compared to analogous calculations for free HF a substantial red–shift of 1177 cm^{-1} is observed. The transition moment of the ν_1 band is calculated to be 0.52 D, five times as high as in free HF. Such a high value will facilitate detection of this band by means of high–resolution IR spectroscopy.

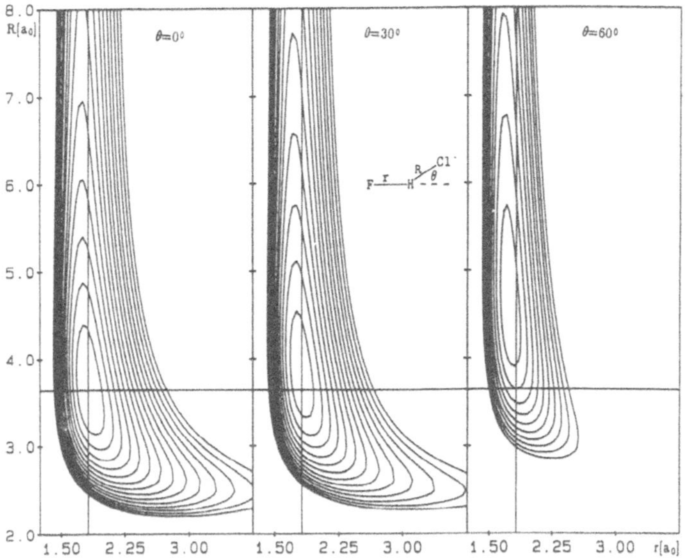

Fig. 2. Contour plots of the potential surface of Cl^- – – – HF at different orientation angles. Contour lines in intervals of 1000 cm^{-1}. The contour line with highest energy always corresponds to 14000 cm^{-1}.

References

[1] M. Gruebele, M. Polak, and R.J. Saykally, J. Chem. Phys. 87, 1448 (1987).

[2] P. Botschwina, J. Chem. Soc. Faraday Trans. 2, 84, 1559 (1988).

[3] P. Botschwina, in "Ion and Cluster Ion Spectroscopy and Structure, edited by J.P. Maier (Elsevier, Amsterdam, 1989).

[4] P. Sebald, Dissertation (Kaiserslautern, 1990).

[5] W. Meyer, J. Chem. Phys. 58, 1017 (1973); 64, 2901 (1976).

[6] R.B. Metz, T. Kitsopoulos, A. Weaver, and D.M. Neumark, J. Chem. Phys. 88, 1463 (1988).

[7] A. Weaver, R.B. Metz, S.E. Bradforth, and D.M. Neumark, J. Phys. Chem. 92, 5558 (1988).

[8] R.B. Metz, A. Weaver, S.E. Bradforth, T.N. Kitsopoulos, and D.M. Neumark, J. Phys. Chem. 94, 1377 (1990).

[9] S.E. Bradforth, A. Weaver, D.W. Arnold, R.B. Metz, and D.M. Neumark, J. Chem. Phys. 92, 7205 (1990).

[10] P. Botschwina, P. Sebald, and R. Burmeister, J. Chem. Phys. 88, 5246 (1988).

[11] A.B. Sannigrahi and S.D. Peyerimhoff, Chem. Phys. Lett. 112, 267 (1984).

ON THE STRUCTURE OF THE O_3^+ RADICAL CATION

T. Schmelz, G. Chambaud and P. Rosmus
Fachbereich Chemie der Universität,
D - 6000 Frankfurt, Germany

The order of the electronic states of the ozone radical cation has been controversial[1,2], due to the lack of high resolution spectra and accurate electronic structure calculations for this species. Recently, an agreement[2,3] has been achieved that the electronic ground state of O_3^+ is of $^2A_1$ symmetry. The photoelectron spectrum[1] shows that this state is almost degenerate with another doublet state. Depending on the symmetry of the first electronically excited state, both states could form a conical intersection and couple *via* the asymmetric stretching mode.

We have mapped[4] the three-dimensional potential energy functions of several O_3^+ states by the CASSCF method and calculated the vertical ionisation energies of ozone by a large scale MRCI approach. The MRCI ionisation energies were calculated to be: X^2A_1 12.44 eV, A^2B_2 12.49 eV, C^2A_2 13.17 eV and D^2B_1 14.11 eV (the total MRCI energy for X^1A_1 of O_3 was calculated to be: -225.089740 a.u.). Hence, the $^2A_1$ ground state is found to be almost degenerate with the $^2B_2$ state. Definitely, both states are to be assigned to the first band of the photoelectron spectrum[1] of O_3 between 12.5 and 13.2 eV. The second band around 13.5 eV corresponds to the C^2A_2 state. The C_{2v} two-dimensional contour plots of the CASSCF total energies for the X^2A_1 and A^2B_2 states are displayed in Figure 1.

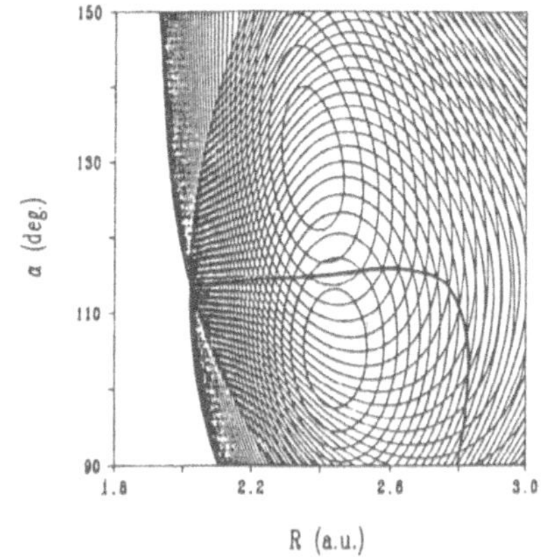

Figure 1:

Contour plots of the CASSCF potential energy functions for the X^2A_1 and A^2B_2 electronic states of O_3^+. On these C_{2v} cuts the X^2A_1 state has a minimum around 130°, the A^2B_2 around 105°, the contours are plotted in steps of 1000 cm^{-1} relative to the minimum energy of the X^2A_1 state, the crossing seam is indicated by a full line.

At the equilibrium structure of the neutral ozone (R_e = 2.40 bohr; α_e = 116.8°) the CASSCF approach yielded a separation of both states of 0.13 eV, the MRCI approach an even smaller difference of 0.05 eV. The crossing seam lies in the Franck - Condon region of the photoelectron spectrum and, there, it is almost independent of the bond angle. In C_s symmetries the two states interact. As a result, the lower component could possess an unsymmetrical equilibrium structure. Our adiabatic CASSCF potential energy function for the lower A' state was found[4] to be extremely shallow along the O_2^+ + O dissociation path.

For a similar reason, the C^1B_1 state of SO_2 has an unsymmetrical equilibrium structure and the adiabatic potential energy function a double minimum along the asymmetric stretching coordinate[5]. However, the barrier[5] to the C_{2v} molecular geometry of 141 cm^{-1} is rather small.

It has been known[6,7] that is is extremely difficult to calculate accurately the dissociation energy of ozone and the asymmetric part of its ground state potential energy function by *ab initio* theoretical methods. Due to the low lying conical intersection the situation is even more intriguing for O_3^+.

In our calculations we have not yet reached the precision needed for an unequivocal distinction between the unsymmetrical or symmetrical minima on the O_3^+ adiabatic ground state potential energy function. At the present stage of our study[4], the theoretical CASSCF results suggest that the asymmetric stretching mode in O_3^+ is a large amplitude mode which couples the X^2A_1 and the A^2B_2 states. The rotationally resolved spectra of O_3^+ could answer the question whether or not this ion has an unsymmetrical equilibrium structure.

1. S. Katsumata, H.Shiramu and T. Kimura, Bull. Chem. Soc. Jpn. 57, 1784 (1984) and references therein.
2. P. Decleva, G. De Alti and A. Lisini, J.Chem. Phys. 89, 367 (1988) and references therein.
3. P. Malquist, H. Argren and B. O. Roos, Chem. Phys. Lett. 98, 444 (1983).
4. T. Schmelz, G.Chambaud, P.Rosmus, H. Köppel, L.Cederbaum and H.-J. Werner, to be published.
5. A.R. Hoy and J.C.D. Brand, Mol. Phys. 36, 1409 (1978).
6. J.F. Stanton, W.N. Lipscomb, P.H. Meyers and R.J. Bratlett, J. Chem. Phys. 90, 1077 (1989).
7. M.V. Rama Krishna and K.D. Jordan, Chem. Phys. 115, 423 (1987).

THE SEARCH FOR ELUSIVE POTENTIALLY INTERSTELLAR NEUTRALS BY NEUTRALIZATION-REIONIZATION MASS SPECTROMETRY: POLYCARBON CHALCOGENIDES $X(C_n)Y$ (X, Y: O, S)

Detlev SÜLZLE and Helmut SCHWARZ*
Institut für Organische Chemie der
Technischen Universität Berlin
Straße des 17. Juni 135
D-1000 Berlin 12
Federal Republic of Germany

ABSTRACT. Many elusive polycarbon chalcogenides $X(C_n)Y$ (X, Y: O, S; $n \geq 2$) having both **odd** and **even**-numbered carbon atoms can easily be generated and characterized in the gas phase by electron transfer processes using the technique of neutralization-reionization mass spectrometry (NRMS). Collision-induced dissociations serve as a powerful means to unambiguously assign the connectivities of these cumulene molecules, many of which do also exist as stable radical cations and anions. The reported "instability" of most polycarbon chalcogenides in solution or in a matrix does not reflect an intrinsic property of these molecules; rather it points to facile intermolecular reactions.

1. INTRODUCTION

The extraordinary recent interest [1-29] in linear and/or quasi-linear molecules of the general structure $\mathbf{X(C_n)Y}$ (X, Y: lone electron pair, H_2, O, S; $n \geq 2$) containing polycumulated double bonds is due to several factors. These species possess and/or are predicted to exhibit unique spectroscopic and chemical properties; in addition, their reactivity/stability properties as well as their electronic ground states (singlet versus triplet) follow an "odd/even" pattern. In fact, for the **odd**-numbered analogues of $\mathbf{X(C_n)Y}$ (n = 3, 5) many combinations are known for long to exist. For example, among the polycarbon dithiones S(Cx)S, propadienedithione (SC_3S) has been known for nearly a century to be a relatively stable molecule [30]. Similarly, several dioxides of **odd**-numbered carbon cumulenes are known to exist, and typical examples are OC_3O and OC_5O [19,20]. In distinct contrast, the **even**-numbered systems, which according to Hund's rule, are predicted to correspond to triplets in their ground electronic states, exhibit a reactivity such that - except for two cases [31] - they cannot even be produced as a transient under matrix conditions. However, this "instability" is not likely to re-

K. R. Jennings (ed.), Fundamentals of Gas Phase Ion Chemistry, 237–248.

flect an intrinsic property. **ab initio** MO calculations [2] leave no doubt that these molecules should be "makeable" molecules as the spin-allowed dissociations of the triplet ground states are predicted to be quite endothermic. The failure to detect many of these molecules in the condensed phase is perhaps due to two factors:
(i) The species have not been generated in the bond triplet state but rather in a singlet state, which is predicted for several cumulenes to dissociate spontaneously. (ii) Extremely facile **inter**molecular reactions with the environment prevent an experimental verification. In fact, in the following it will be demonstrated that many of these elusive molecules are easily accessible in the diluted gas phase, where they are readily generated in beam experiments from their ionic counterparts by using the technique of **neutralization-reionization mass spectrometry** (NRMS) [32-38].

It is also worth mentioning that some of these linear polycarbon molecules are believed to play a crucial role in the genesis of interstellar molecules. A typical example corresponds to the triplet of :CCS which was recently detected [15-17] in interstellar clouds (Taurus Mol. Cloud-1, Sagittori B2, IRC + 10216) and later characterized by NRMS [39] as well as **ab initio** MO calculations [28,39] as a perfectly stable gas-phase molecule. Obviously, NRMS experiments supported by appropriate MO calculations are quite likely to aid in the radioastronomical search for these elusive molecules. In this Review we will confine ourselves to recent results obtained in our laboratory, and, after a very brief description of the NRMS-methodology, we shall discuss the following species: (i) polycarbon disulfides, (ii) polycarbon dioxides, and (iii) polycarbon oxide sulfides.

2. EXPERIMENTAL SECTION

Most of the mass spectrometric experiments were carried out using a modified VG Instruments ZAB machine; this is a four-sector mass spectrometer of **BEBE** configuration (**B** stands for magnetic and **E** for electric sector). Positive ions are generated by electron impact ionization (70 - 100 eV) of a suitable precursor; radical anions were generated in the negative ion mode with NH_3 as an electron moderation gas. Collision-induced dissociations (CIDs) [40-42], used to characterize the primary ions $M_1^{+/-}$, were brought about by mass-selecting a 8-keV beam of $M_1^{+/-}$ by means of B(1)/E(1) and colliding it with a target gas (usually helium) in a collision chamber located in the third field-free region (transmittance, T, 60 - 90 %). Ionic products were recorded by scanning B(2). Charge reversal (CR) experiments [40-44] were conducted by colliding the beam of mass-selected anionic precursors M_1^- with oxygen (c. 80 % T) and recording the positively charged fragment ions M_i^+ by scanning B(2). In the **NRMS** experiment, a beam of mass-selected M_1^+ ions (having 8-keV translational energy) was neutralized in the first cell of a differentially pumped tandem collision cell (the latter is located in the field-free region between E(1)/B(2)) by colliding the beam with xenon (c. 80 % T). Unreacted ions were deflected away from the beam of neutral species M_i by putting a voltage on a deflector electrode; subsequent reionization $M_i \longrightarrow M_i^+$ occurred in the second collision cell by collision with oxygen (c. 80 % T). The resulting mass spectra were obtained by scanning B(2). If anionic precursors were subjected to a CR/NRMS experiment the combination oxygen/oxygen as collision gases turned out to be the most efficient one. If not prevented on sensitivity grounds, a further revealing experiment was performed by taking advantage of the unique properties of a multi-sector mass spectrometer. To this end, mass-selected M_1^+ were first neutralized ($\longrightarrow M_1$), reionized ($\longrightarrow M_1^+$), and the resulting "recovery ions" M_1^+ were subjected to a kinetic energy analysis by means of E(1), and the interference-free beam of M_1^+ species was collisionally activated. The CA mass spectra obtained, which basically contain no contribution from reionized neutral fragments which may be formed in the neutralization step, permit a structural assignment ("connectivity") of reionized M_1. The minimal lifetimes of the neutral molecules M_1 are identical with the transit time from the first to the second collision cell; t is of the order of 10^{-6} s.

3. RESULTS AND DISCUSSION

3.1 Polycarbondithiones SC_nS (n = 2 - 6)

Theory predicts [2] that ethylenedithione SC_2S, the dimer of carbon monosulfide, should be "makeable", since it corresponds to the global minimum of the C_2S_2 hypersurface and its spin-allowed decomposition into ^1CS and ^3CS is endothermic by 39 kcal/mol. While it became only very recently possible to generate SC_2S in an argon matrix [31], its gas-phase synthesis and characterization was already achieved in 1988 [45] by using NRMS. In fact, the high yields of ions permitted to perform **all** experiments described in Chapter 2, thus leaving no doubt as to the connectivity of C_2S_2 and confirming Schaefer's prediction.

More recently [46] the study of **1** [47] provided an easy, efficient entry for the generation of the whole series $C_nS_2^{+\cdot}$ (n = 2 - 6), (Scheme 1). The connectivity of these cumulenes is clearly reflected in their CA mass spectra. Typical fragments, for example for $C_6S_2^{+\cdot}$, are due to losses of a sulfur atom, the elimination of C_nS (n = 1, 2, 3), the formation of $C_n^{+\cdot}$ (n = 1 - 6), and charge-stripping to generate a doubly-charged ion (Scheme 2). An identical fragmentation pattern, though with varying intensities, is observed for the whole series $C_nS_2^{+\cdot}$ (n = 2 - 6), and the base peak in the CA mass spectra of $C_nS_2^{+\cdot}$ invariably corresponds to the loss of CS.

If a beam of mass-selected $C_nS_2^{+\cdot}$ species is subjected to an NRMS experiment (xenon//oxygen) one obtains in all cases only abundant recovery signals (70 - 100 %); moreover, the fragmentation pattern is often close if not identical with the one observed in the CA mass spectra. **Thus, polycarbondithiones SC_nS as well as their radical cations are perfectly stable molecules.** It is obvious, that the difficulty to preparing these species under more conventional conditions, i. e. in solution or a matrix, does not reflect any intrinsic instability.

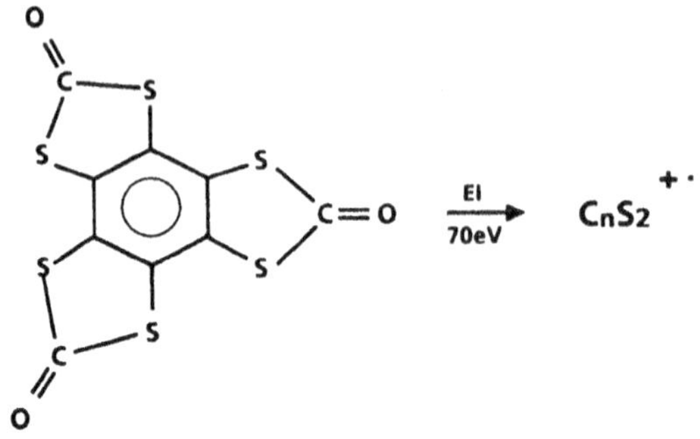

n	m / z	% rel. Intensität
2	88	60
3	100	65
4	112	55
5	124	75
6	136	50

1

Scheme 1

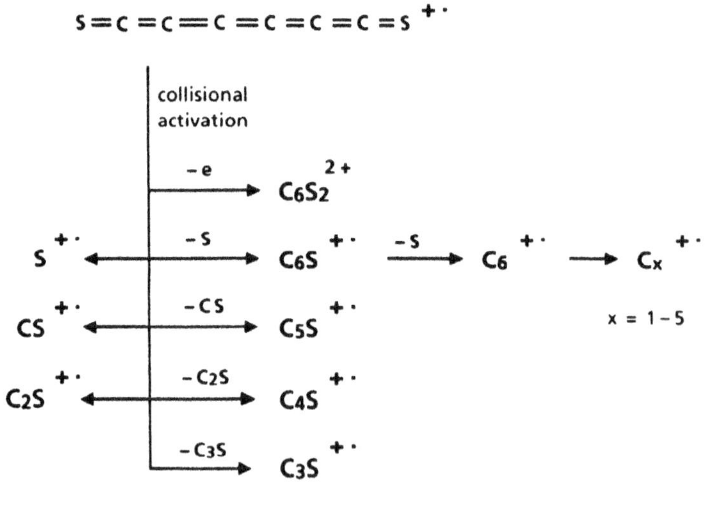

Scheme 2

3.2 **Even**-Numbered Polycarbondioxides OC_nO (n = 2, 4)

As already mentioned in the Introduction, **odd**-numbered polycarbondioxides OC_nO (n = 3, 5) are known to exist in distinct contrast to their **even**-numbered analogues OC_nO (n = 2, 4) [48]. In contrast to SC_2S, all efforts, beginning as early as 1913, to prepare a chemically bound O=C=C=O species have proven unsuccessful [49-57], although theory [2] clearly indicates that the spin-allowed dissociation of the $^3\Sigma_g^-$ state of C_2O_2 is quite endothermic (74 kcal/mol). This failure is perhaps due to two factors: (i) The C_2O_2 species have not been generated in the bound triplett state but rather in the singlet state $^1\Sigma_g^+$, which is predicted [2,58] to dissociate spontaneously to two ground-state CO molecules. (ii) The lifetime of the otherwise stable $^3\Sigma_g^-$ C_2O_2 species may be drastically shortened by an efficient spin-orbit coupling if a potential-curve crossing is available not too far from the minimum (e. g. by a $^1\Sigma_g^+$ state that is repulsive with respect to C-C bond stretch [2]).

The situation is different with OC_4O (1.2.3-butatriene-1.4-dione, **2**). Maier's group recently succeeded [31] in generating **2** in an argon matrix via photochemically induced decarbonylation. That the existence of **2** is not the result of a matrix effect but rather reflects an inherent stability of OC_4O, is clearly demonstrated by its gas-phase synthesis and characterization using NRMS [59]. A brief summary of this exciting finding is given below.

The radical anion $C_4O_2^{-\cdot}$ ($2^{-\cdot}$) is accessible by dissociative electron attachment to the heterocycle shown in Scheme 3.

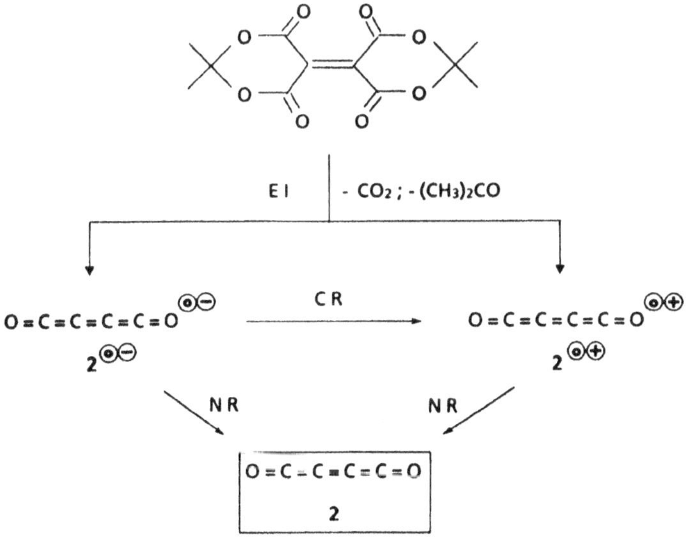

Scheme 3

If the mass-selected beam of $C_4O_2^{-\cdot}$ is subjected to a CA experiment, one obtains the spectrum shown in Fig. 1a. All signals of this spectrum can be accounted for in terms of direct bond cleavage processes, and the spectrum is in keeping with a connectivity $OCCCCO^{-\cdot}$ ($\mathbf{2}^{-\cdot}$). We also conclude from this spectrum that, except for CO, the neutral species corresponding to C_4O_2, C_4O, C_3O, C_2O, C_2 and C_1 must possess positive electron affinities.

Charge reversal of $C_4O_2^{-\cdot}$ with oxygen gives rise to the spectrom shown in Fig. 1b. We not only obtain an intensive "recovery signal" for $C_4O_2^{+\cdot}$, which indicates that $C_4O_2^{+\cdot}$ must also reside in a potential minimum, the fragmentation pattern is, again, typical for what one would expect for a cumulene structure.

As indicated by the NR spectra given in Fig. 1c and Fig. 1d, neutralization of both $C_4O_2^{-\cdot}$ (oxidation) and $C_4O_2^{+\cdot}$ (reduction) can be brought about. We note two features: (i) The quite intensive "recovery signals" point to a favourable Franck-Condon factor, and (ii) the practically identical fragmentation patterns which are, again, compatible with a linear structure **OCCCCO** for the neutral (2), the radical anion ($\mathbf{2}^{-\cdot}$), and the radical cation ($\mathbf{2}^{+\cdot}$), respectively. Further proof that 1.2.3-butatriene-1.4-dione **2** is a perfectly stable molecule is provided by tandem experiment of the following type: If reionized $C_4O_2^{+\cdot}$, generated in the sequence $\mathbf{2}^{+\cdot} \longrightarrow \mathbf{2} \longrightarrow \mathbf{2}^{+\cdot}$ is mass-selected and collisionally activated, the interference-free CA spectrum obtained (not shown) is identical with the CA mass spectrum of directly generated $\mathbf{2}^{+\cdot}$.

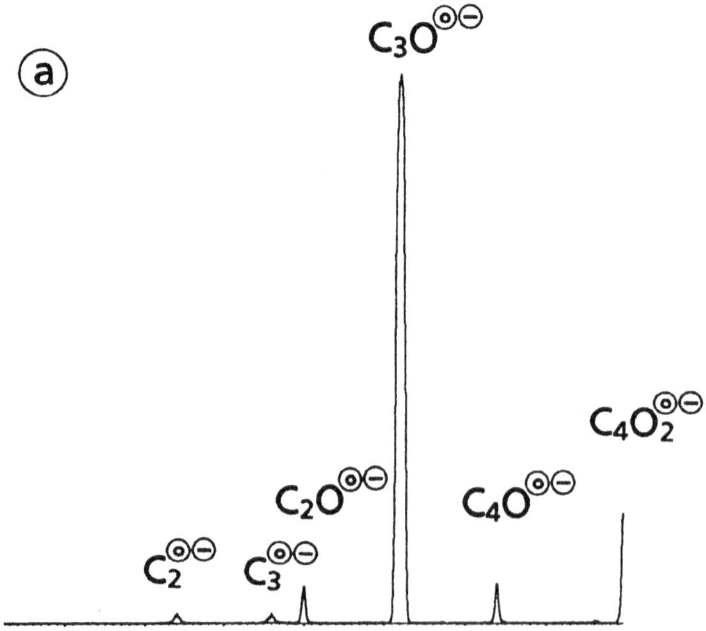

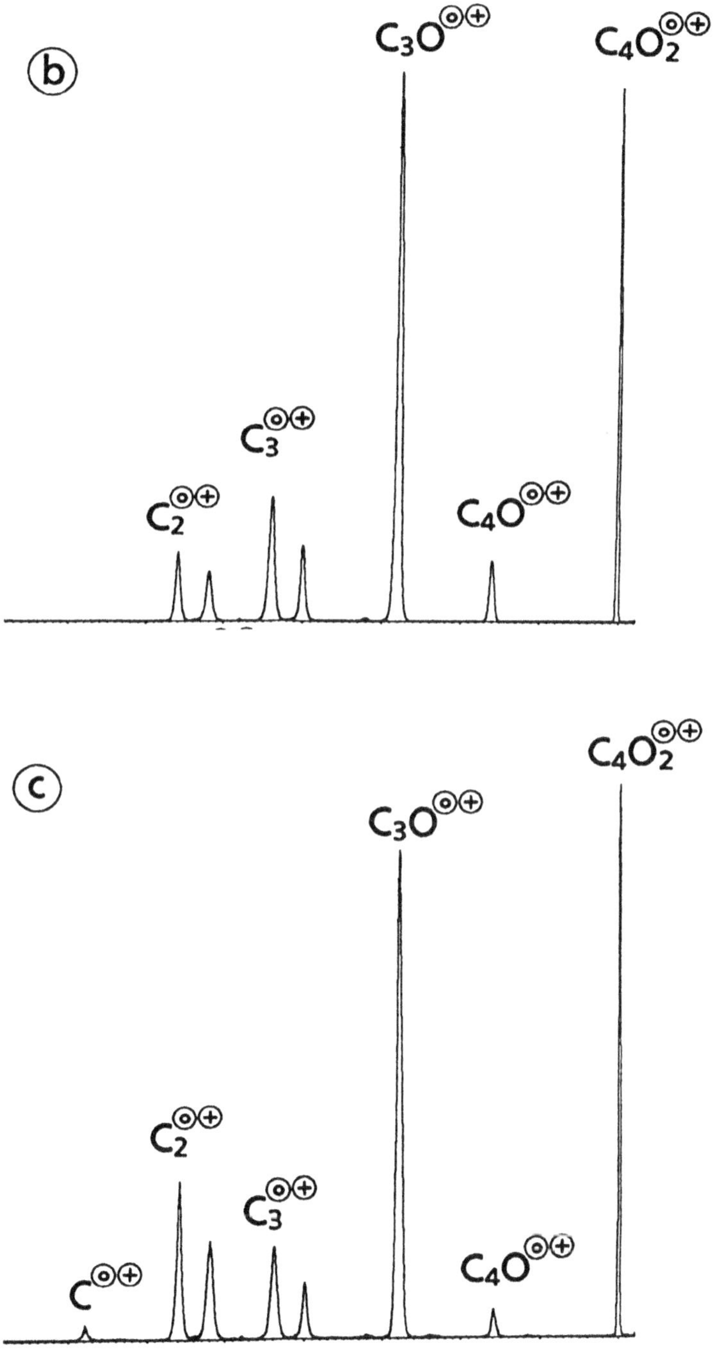
(b)
$C_3O^{\odot\oplus}$
$C_4O_2^{\odot\oplus}$
$C_3^{\odot\oplus}$
$C_2^{\odot\oplus}$
$C_4O^{\odot\oplus}$
(c)
$C_4O_2^{\odot\oplus}$
$C_3O^{\odot\oplus}$
$C_2^{\odot\oplus}$
$C_3^{\odot\oplus}$
$C^{\odot\oplus}$
$C_4O^{\odot\oplus}$

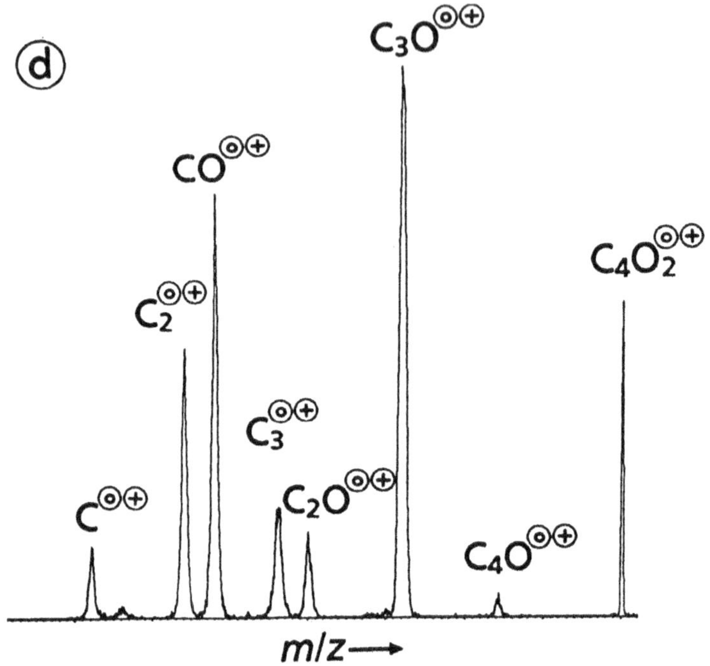

Fig. 1. (a) CA mass spectrum of $C_4O_2^{-\cdot}$ (oxygen, 80 %T); (b) CR mass spectrum of $C_4O_2^{-\cdot}$ (oxygen, 80 %T); (c) NR mass spectrum of $C_4O_2^{-\cdot}$ (oxygen, 80 % T//oxygen, 80 %T); (d) NR mass spectrum of $C_4O_2^{+\cdot}$ (xenon, 70 %//oxygen, 80 %T)

3.3 Polycarbon Oxide Sulfides $O(C_n)S$ (n = 2 - 5)

In contrast to the "symmetric" cumulenes $S(C_n)S$ and $O(C_n)O$, the mixed analogues $O(C_n)S$ possess a permanent dipole moment; thus these molecules are ideal candidates for a radioastrochemical search. In fact, the tentative identification of 3-thioxo-1.2-propadiene-1-one, **OCCCS** (**3**) in interstellar medium by Turner [1] triggered intense research activity using "earth-bound facilities": The first synthesis and microwave characterization of **3** by the Winnewisser group [8,11,60-62] was followed by its gas-phase synthesis via pyrolysis combined with a PE-spectroscopic identification by Bock et al. [12]. In fact, **OCCCS** (**3**) and its radical cation $\mathbf{3}^{+\cdot}$ and radical anion $\mathbf{3}^{-\cdot}$ were also recently characterized using NRMS [63]. The mass-selected species $C_3OS^{+\cdot/-\cdot}$ which together with other $C_nOS^{+\cdot/-\cdot}$ cumulenes are available from appropriate heterocyclic precursors have fragmentation pattern in keeping with the linear connectivity. A typical example is given for the C_3OS system in Scheme 4.

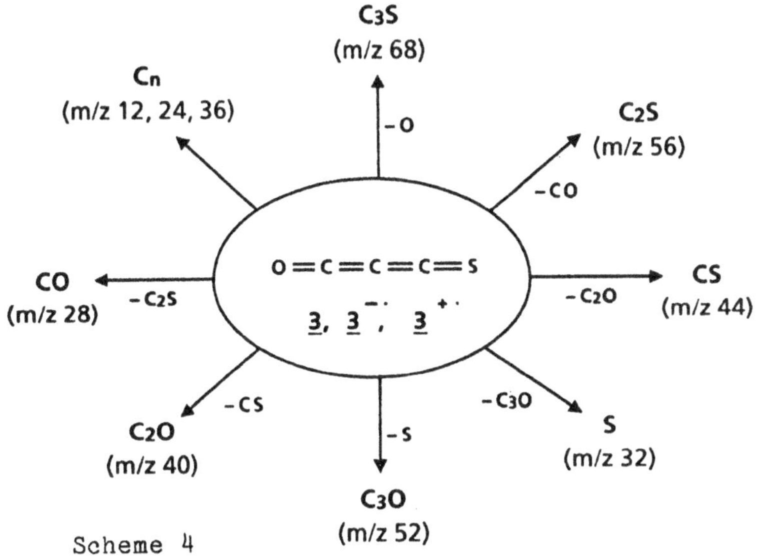

Scheme 4

Reduction of $OC_3S^{+\cdot}$ or oxidation of $OC_3S^{-\cdot}$ in NR experiments gives extremely abundant recovery signals together with fragment ions from which the connectivity can be derived [63]. In addition, a CA spectrum of reionized OC_3S was practically identical with the CA spectrum of directly generated $OC_3S^{+\cdot}$. Thus, **OCCCS** (**3**), and its radical cation ($\mathbf{3}^{+\cdot}$) and radical anion ($\mathbf{3}^{-\cdot}$) are indeed stable molecules.

Similarily, experimental evidence was also recently reported [63] for the existence of the higher homologues **OCCCCS** and **OCCCCCS** (tetra- and pentacarbon oxide sulfides) and their corresponding radical ions.

This holds true for the smallest member of this series, the thioxoethylenone **OCCS** [64]. For this system the experimental findings clearly point to linear structures for both the neutral and the radical cation having the connectivity **OCCS**$^{o/+\cdot}$. Isomeric forms are not compatible with the various collision-induced dissociation reactions observed, and the experimental findings were supported by **ab initio** MO studies. These indicate that the neutral OC_2S molecule generated in the NR experiment corresponds to the triplet state ($^3\Sigma$) (the singlet is predicted to spontaneously dissociate to CO and CS). In addition, significant energy barriers prevent the triplet of OC_2S as well as the cation radical $OCCS^{+\cdot}$ from spontaneous dissociation.

4. ACKNOWLEDGEMENTS

The continuing support of our work by the Deutsche Forschungsgemeinschaft, the Fonds der Chemischen Industrie, the Gesellschaft von Freunden der Technischen Universität Berlin, and the Graduiertenkolleg Chemie Berlin is appreciated. We are particularly grateful to **Prof. Egon Fanghänel** and **Prof. Johan K. Terlouw** for enlightening discussions as well as a fruitful collaboration. Technical assistance by **Waltraud Zummack** is gratefully appreciated.

5. REFERENCES AND NOTES

[1] B. E. Turner, **Astrophysics Lett. 23** (1983) 217.
[2] G. P. Raine, H. F. Schaefer III, R. C. Haddon, **J. Am. Chem. Soc. 105** (1983) 194.
[3] A. B. Raksit, D. K. Bohme, **Int. Mass Spectrom. Ion Processes 55** (1983) 69.
[4] M. Winnewisser, **Chem.-Ztg. 18** (1984) 1, 54.
[5] H. E. Matthews, W. M. Irvine, P. Freiberg, R. D. Brown, P. D. Godfrey, **Nature** (London) **310** (1984) 125.
[6] N. W. Broten, J. M. MacLeod, L. W. Avery, W. M. Irvine, B. Hogelund, P. Freiberg, A. Hjalmarsen, **Astrophys. J. 276** (1984) L 25.
[7] W. W. Duley, D. A. Williams, **Interstellar Chemistry**, Academic Press, London (1984).
[8] M. Winnewisser, E. W. Peau, **Acta Phys. Hung. 55** (1984) 33.
[9] R. D. Brown, E. H. N. Rice, **J. Am. Chem. Soc. 106** (1984) 6475.
[10] B. P. Winnewisser, **Mol. Spectrosc. Mod. Res.** (1985) 321.
[11] M. Winnewisser, B. P. Winnewisser, G. Winnewisser, in G. H. F. Diercksen (Ed.), **Molecular Astrophysics**, D. Reidel Publishing Company, Dordrecht (1985) p. 375.
[12] H. Bock, R. Dammel, D. Jacobi, **J. Am. Chem. Soc. 108** (1986) 7844.
[13] D. K. Bohme, **Nature** (London) **319** (1986) 128.
[14] R. D. Brown, **J. Rev. Phys. Chem. 5** (1986) 101.
[15] J. Cernicharv, M. Guelin, H. Hein, C. Kahane, **Astron. Astrophys. 181** (1987) L 9.
[16] S. Saito, K. Kanaguchi, S. Yamamoto, M. Ohishi, H. Suzuki, N. Kaifu, **Astrophys. J. 317** (1987) L 115.
[17] S. Yamamoto, S. Saito, K. Kawaguchi, N. Kaifu, H. Suzuki, M. Ohishi, **Astrophys. J. 317** (1987) L 119.

[18] G. Winnewisser, E. Herbst, **Top. Curr. Chem. 139** (1987) 119.
[19] F. Holland, M. Winnewisser, G. Maier, H. P. Reisenauer, A. Ulrich, **J. Mol. Spectroscop. 130** (1988) 470.
[20] R. J. Van Zee, G. R. Smith, W. Weltner, Jr., **J. Am. Chem. Soc. 110** (1988) 609.
[21] G. Maier, H. P. Reisenauer, U. Schäfer, H. Balli, **Angew. Chem. Int. Ed. Engl. 27** (1988) 566.
[22] R. D. Brown, K. G. Dyall, P. S. Elmes, P. D. Godfrey, D. McNaughton, **J. Am. Chem. Soc. 110** (1988) 789.
[23] K. D. Bohme in T. J. Millar, D. A. Williams (Eds.) **Rate Coefficients in Astrochemistry**, Kluwer Academic Publishers, Amsterdam (1988).
[24] R. F. C. Brown, K. J. Coulston, F. W. Eastwood, M. J. Irvine, A. D. E. Pullin, **Austr. J. Chem. 41** (1988) 225.
[25] R. J. Van Zee, R. F. Ferrante, K. J. Zeringue, W. Weltner, Jr., D. E. Ewing, **J. Chem. Phys. 88** (1988) 3465.
[26] D. W. Ewing, **J. Am. Chem. Soc. 111** (1989) 8809.
[27] D. Sülzle, H. Schwarz, **Chem. Phys. Lett. 156** (1989) 307.
[28] D. J. Peeso, D. W. Ewing, T. T. Curtis, **Chem. Phys. Lett. 166** (1990) 307.
[29] R. F. C. Brown, K. J. Coulston, F. W. Eastwood, A. D. E. Pullin, A. C. Staffa, **Austr. J. Chem. 43** (1990) 561.
[30] B. V. Lengyel, **Ber. Dtsch. Chem. Ges. 26** (1893) 2960.
[31] G. Maier, H. P. Reisenauer, H. Balli, W. Brandt, R. Janoschek, **Angew. Chem. Int. Ed. Engl. 29** (1990) in press.
[32] J. K. Terlouw, P. C. Burgers, B. L. M. van Baar, T. Weiske, H. Schwarz, **Chimia 40** (1986) 357.
[33] C. Wesdemiotis, F. W. McLafferty, **Chem. Rev. 87** (1987) 485.
[34] J. K. Terlouw, H. Schwarz, **Angew. Chem. Int. Ed. Engl. 26** (1987) 805.
[35] H. Schwarz, **Pure Appl. Chem. 61** (1989) 685.
[36] J. K. Terlouw, **Adv. Mass Spectrom. 11** (1989) 984.
[37] J. L. Holmes, **Mass Spectrom. Rev. 8** (1989) 513.
[38] F. W. McLafferty, **Science 247** (1990) 925.
[39] D. Sülzle, H. Schwarz, **Chem. Ber. 122** (1989) 1803.
[40] K. Levsen, H. Schwarz, **Angew. Chem. Int. Ed. Engl. 15** (1976) 509.
[41] R. G. Cooks (Ed.), **Collision Spectroscopy**, Plenum, New York (1978).
[42] K. Levsen, H. Schwarz, **Mass Spectrom. Rev. 3** (1983) 77.
[43] J. H. Beynon, **Proc. Royal Soc. Ser. A.** (London) (1981) 378.
[44] J. H. Bowie, **Mass Spectrom. Rev. 3** (1984) 161.
[45] D. Sülzle, H. Schwarz, **Angew. Chem. Int. Ed. Engl. 27** (1988) 1337.
[46] D. Sülzle, N. Beye, E. Fanghänel, H. Schwarz, **Chem. Ber. 123** (1990) in press.
[47] Incidentally, **1** also serves as an ideal precursor to generate $C_6S_6^{+\cdot}$, which upon neutralization gives rise to the theoretically most interesting benzo-tri(1.2-dithiet): D. Sülzle, N. Beye, E. Fanghänel, H. Schwarz, **Chem. Ber. 122** (1989) 2411.
[48] $(CO)_2$ does, of course, exist as a weakly bound van der Waals complex with a dissociation energy of 0.1 kcal/mol: J. O. Hirschfelder, C. F. Curtiss, R. B. Bird, **Molecular Theory of Gases and Liquids**, Wiley, New York (1964).

[49] H. Staudinger, B. Anthes, **Ber. Dtsch. Chem. Ges. 46** (1913) 1426.
[50] D. M. Hirst, J. D. Hopton, J. W. Linnett, **Tetrahedron Suppl.** 2 (1963) 15.
[51] J. J. Blomfield, J. R. S. Irelan, A. P. Marchand, **Tetrahedron Lett.** (1968) 5647.
[52] D. Bryce-Smith, A. Gilbert, **J. Chem. Soc. Chem. Commun.** (1968) 1319.
[53] J. Straling, B. Zwanenburg, A. Wagenaar, A. C. Udding, **Tetrahedron Lett.** (1969) 125.
[54] H.-D. Scharf, R. Klar, **Tetrahedron Lett.** (1971) 517.
[55] D. L. Dean, H. Hart, **J. Am. Chem. Soc. 94** (1972) 687.
[56] H.-D. Scharf, R. Klar, **Chem. Ber. 105** (1972) 575.
[57] M. B. Rubin, M. Weiner, H.-D. Scharf, **J. Am. Chem. Soc. 98** (1976) 5699.
[58] R. C. Haddon, D. Poppinger, L. Radom, **J. Am. Chem. Soc. 97** (1975) 1645.
[59] D. Süzle, H. Schwarz, **Angew. Chem. Int. Ed. Engl. 29** (1990) 000.
[60] M. Winnewisser, J. J. Christiansen, **Chem. Phys. Lett. 37** (1976) 270.
[61] M. Winnewisser, E. W. Peau, K. Yamada, J. J. Christiansen, **Z. Naturforsch. 36a** (1981) 819.
[62] M. Winnewisser, E. W. Peau, **Chem. Phys. Lett 71** (1982) 377.
[63] D. Sülzle, H. Schwarz, **J. Am. Chem. Soc. 112** (1990) 000.
[64] D. Sülzle, J. K. Terlouw, H. Schwarz, **J. Am. Chem. Soc. 112** (1990) 628.

THE MEASUREMENT AND INTERPRETATION OF ONSET ENERGIES

Tomas Baer
Department of Chemistry
University of North Carolina
Chapel Hill, NC 27599-3290
USA

ABSTRACT. The measurement and interpretation of ionization energies and fragment ion appearance energies are discussed in the light of new experimental methods, including multiphoton ionization. The ionization energy of very cold molecules can now be measured with a precision of 0.1 meV. In order to use these data to determine equally accurate heats of formation, it will be necessary to know the vibrational frequencies of the molecule and ion. Fragment ion onsets cannot yet be measured with such high accuracy. In addition, the use of appearance energies in determining fragment ion and neutral heats of formation require a number of assumptions. These ideas are reviewed here. The problems associated with the interpretation of fragment ion onsets are illustrated for the case of methyl loss from the ethylbenzene ion.

1. Introduction

In many experiments the only difference in the measurement of an ionization energy and a fragment ion appearance energy is in the setting on the mass dial of the mass spectrometer. However, accurate measurements of these energies requires very different approaches and involves very different considerations. These differences are explored in this paper.

2. The Determination of Ionization Energies

2.1. FAVORABLE FRANCK-CONDON CASES

The accuracy in measuring adiabatic ionization energies is determined by several factors including the band width of the ionizing radiation, in certain experiments the band pass of an electron monochromator, and the Franck-Condon factors connecting the ground state molecule with the ground state ion. The first two factors are determined by the nature of the experimental apparatus, while the latter is inherent in the molecular system to be studied. The accuracy of measured ionization energies has improved enormously during the past 30 years. Electron impact ionization with non-monochromatic electrons gave values with error limits on the order of 0.3 eV.[1] In the 1960's photoionization mass spectrometry was introduced by Watanabe[2] and Ingram[3] who measured the IE of

K. R. Jennings (ed.), Fundamentals of Gas Phase Ion Chemistry, 249–265.

numerous molecules with a resolution on the order of 0.02 eV. The resolving power of these instruments has improved during the intervening years so that with the use of a 3 m monochromator a resolution of the order of 2 meV is possible.[4] Finally, the use of laser multi-photon ionization in recent years has allowed IE's to be measured to within 0.1 meV[5,6] and still better resolution is certainly within our reach.

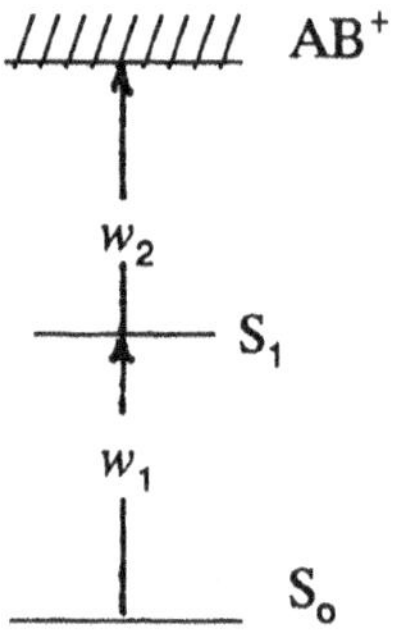

Figure 1. Diagram of two photon resonant ionization in which w_1 remains fixed, while w_2 is scanned through the ionization energy.

Among the first to measure ionization energies with the use of multiphoton ionization is Wallace *et al.*[7,8] who used a 2 photon two color experiment via a resonant intermediate state to ionize such molecules as aniline and indole. As shown in Figure 1, laser 1 was tuned to the first transition ($S_o \rightarrow S_1$) and kept fixed while laser 2 was scanned through the ionization region. The parent ion signals for aniline ions as a function of the second laser energy is shown in Figure 2. The ionization energy of aniline was measured at 7.720 ± 0.002 eV. A similar study of indole yielded an ionization energy of 7.7602 ± 0.0006 eV. The latter value represents a considerable improvement over a PES value of 7.75 ± 0.05 eV.

One reason that the ion onset in Figure 2 is so sharp is that the sample was cooled by a supersonic expansion. As a result, the rotational energy distribution is very narrow so that the transitions involved in the excitation scheme are between a very small number of rotational levels. It is important that the spectroscopy of the excited state be well understood so that the ground ro-vibrational level is reached by laser 1. If laser 1 populates an excited vibrational level (e.g. $\nu_3 = 1$) laser 2 may populate exclusively the $\nu_3 = 1$ level of the ion state. Why this might be so is easily seen in Figure 3. If the S_1 state has a geometry similar to that of the ion, the vibrational wavefunctions of the two states will be similar. Under these circumstances $\Delta v=0$ transitions between the S_1 and ion states will dominate. That is, in the limit of identical S_1 and ground state ion vibrational wavefunctions, the Franck-Condon factors for the final transition, $\langle\phi_i|\phi_j\rangle = 1$ if i=j, and zero otherwise. This results from the orthogonal property of a set of vibrational wavefunctions. It is thus imperative that the ground state of S_1 be initially excited.

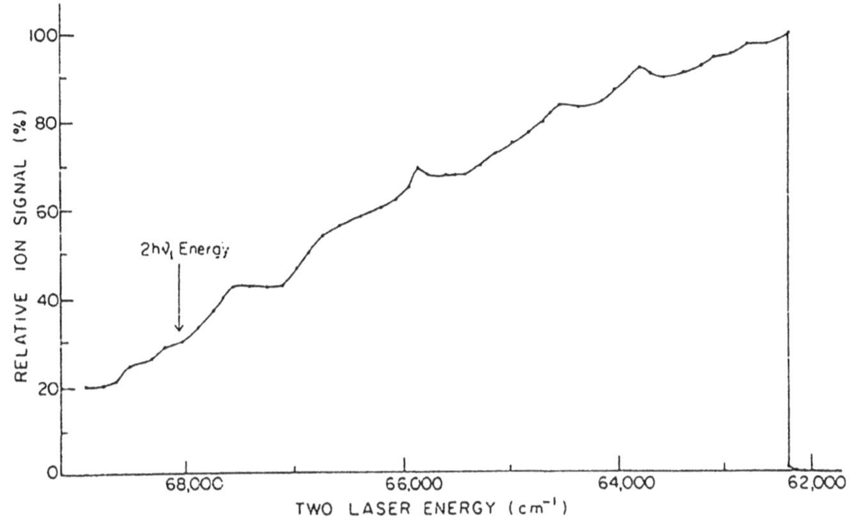

Figure 2. The laser photoionization efficiency spectrum of aniline. Laser 1 was fixed to the S_1 transition, while laser 2 was scanned. The energy scale is the sum of the two laser photons. Spectrum taken from Hager and Wallace, ref. 7.

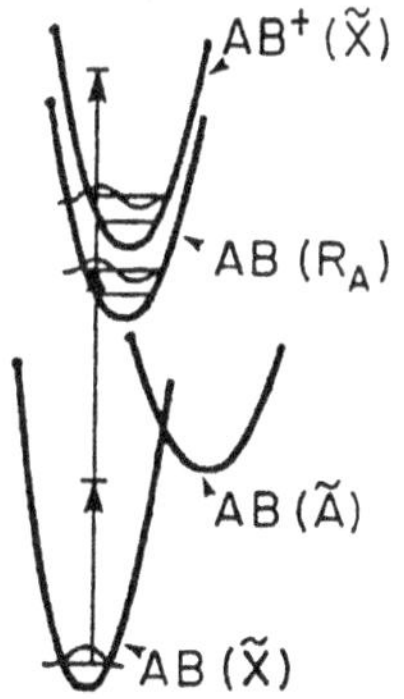

Figure 3. Schematic potential energy curves illustrating the Franck-Condon factors for a v=1 to v=1 transition from the Rydberg state to the ion state.

It is an interesting fact that even if the sample had not been cooled by supersonic expansion, the onsets might still be sharp. The reason is that the laser excitation in the first step is always between just two quantum states. Thus, laser 1 will tend to excite a single ro-vibrational level in the S_1 state, and the onset for ionization may appear sharp, but as pointed out above, it will not necessarily correspond to ionization to the ground ro-vibrational level.

Although the accuracy of the onset in Figure 2 is impressive, there is an important correction that must be applied to the onset energy. This is a result of field ionization of the high Rydberg states that are populated. As laser 2 sweeps from below to the molecule's ionization energy, high Rydberg states will be populated. Because an electric field is generally required to extract the ions, field ionization will lower the measured IE. The lowering of the IE (in atomic units) is given by $2\sqrt{\epsilon}$, where ϵ is the electric field (in atomic units). In more practical units this converts to $\Delta \mathrm{IE} = -6.121\sqrt{\epsilon}$, where the shift is expressed in cm^{-1} and the electric field in V/cm. Thus, even in the relatively low field of 1 V/cm, the shift in the ionization energy will be 6.1 cm^{-1} (0.75 meV). The experiment should therefore be carried out under very controlled extraction field conditions and the resulting IE corrected for the shift. The use of a zero field condition followed by an extraction pulse does not help the problem because the high Rydberg state molecules will simply remain in the ion source until such time when the field is turned on after which they will ionize.

The use of ionization energies to determine the heats of formation is not a straightforward task. The IE is fundamentally a spectroscopic quantity which is referenced to individual quantum states. On the other hand, the heat of formation is a thermodynamic quantity. The connection between the two is best made at 0 K where the thermal energy distribution collapses to the single ground ro-vibrational level of the molecule or ion. As an example, figure 4 shows the cycle involved in determining the heats of formation of aniline. The conversion from 298 K to 0 K requires a knowledge of the heat capacity of the molecule or the ion, which in turn requires a knowledge of the vibrational frequencies. The cycle as applied to aniline is as follows:

$$6C + 3.5\,H_2(g) + 0.5\,N_2(g) \xrightarrow{\Delta H^\circ_{f0}} C_6H_5NH_2(g)$$

$$\downarrow \Delta c_p(298\text{-}0) \qquad\qquad\qquad\qquad \uparrow \Delta c_p(0\text{-}298)$$

$$6C + 3.5\,H_2(g) + 0.5\,N_2(g) \xrightarrow{\Delta H^\circ_{f298}} C_6H_5NH_2(g)$$

The $298\Delta c_p$ term ($H^\circ - H^\circ_0$) for the elements in their standard states is 0.251 kcal/mol (C as graphite); 2.024 kcal/mol (H_2), and 2.072 kcal/mol (N_2).[9] The change in enthalpy for the elements is thus + 9.63 kcal/mol. With a knowledge of the vibrational frequencies, we can calculate $\Delta c_p(0\text{-}298) = -4.31$ kcal/mol for the enthalpy change in aniline from 298 to 0 K. The heat of formation of aniline at 298 K is listed in Pedley *et al.*[10] as 20.82 ± 0.2 kcal/mol. We thus obtain a ΔH°_{f0} (aniline) = 26.14 ± 0.2 kcal/mol.

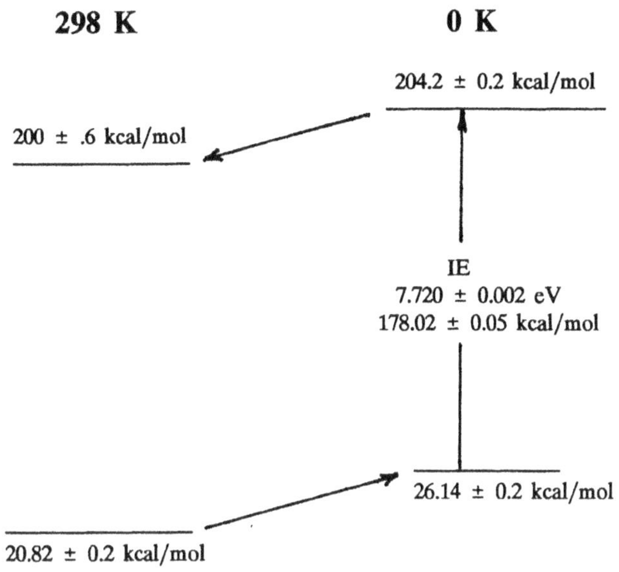

Figure 4. The heats of formation of the aniline system at 0 and 298 K.

Armed with the heat of formation at 0 K, we can now determine the 0 K heat of formation of the aniline ion simply by adding the IE to the molecule's heat of formation. This leads to ΔH^{o}_{f0} (aniline ion) = 204.0 ± 0.2 kcal/mol. We note that we have lost some of the accuracy of the ionization energy measurements which was ± 0.05 kcal/mol because the neutral molecule's heat of formation is known only to 0.2 kcal/mol. However, the major problem is encountered in converting the 0 K heat of formation of the ion into the room temperature heat of formation. Again, we need the vibrational frequencies of the ion, and these are virtually unknown. The frequencies of the aniline ion are probably better known than most ions because the geometry does not change much upon ionization and there are relatively few low frequencies modes in aniline. It is always the low frequency modes that contribute most to the Δc_p term and these are also the most poorly known. By assuming that the vibrational frequencies of the aniline ion are slightly less than those of the neutral molecule, we estimate that the ion heat of formation at room temperature is about 200 ± 0.6 kcal/mol.

From the above example it is clear that a major limitation in deriving accurate ion heats of formation is in our knowledge of the ion vibrational frequencies. These frequencies are beginning to be determined by various spectroscopic methods.[11-15] However, most of the efforts are limited to small ions. An ion such as aniline has 36 vibrational frequencies. The assignment of all modes even for stable neutral molecules is a difficult task which is accomplished only after detailed analysis of IR and Raman spectra, and after carrying out a force field calculation. This seems a prohibitively difficult undertaking for large ions. However, a knowledge of the neutral molecule frequencies can greatly aid the interpretation of the ion spectra. For instance, those vibrational states not excited in the ionization process can be presumed to have frequencies similar to those of the neutral molecule, or else to be not allowed because of symmetry restrictions.

One of the most promising new tools for measuring ion vibrational frequencies is multiphoton ionization (MPI) - threshold photoelectron spectroscopy (TPES). TPES with electron time of flight (TOF) energy analysis has a resolution that is limited only by the laser band width. It has been applied to such molecules as NO[5] and benzene,[6] but is now being extended to more complex systems such as pyrazine and di-fluorobenzene.[16,17] A particularly important feature of this method involves an initial photon absorption step to a resonant intermediate state. In this case the process is termed resonance enhanced multiphoton ionization, or REMPI. As was pointed out above, different ion frequencies can be excited by choosing various intermediate vibrational levels. Rilley and co-workers[18,19] were the first to take advantage of this effect in their study of a series of REMPI-PES of such molecules as benzene,[18] and toluene.[19] The resolution in these spectra is limited by the resolution of the electron energy analyzer rather than by the laser band width. This is because the resolution in electron TOF analysis deteriorates rapidly with electron kinetic energy. Thus, at zero energy, the resolution is laser band width limited, while at 1 eV, the resolution is on the order of 20 meV.

2.2.UNFAVORABLE FRANCK-CONDON CASES

All of the above mentioned molecules (aniline, pyrazine, di-fluorobenzene, and indole) are ones in which the 0-0 transition is a strong one. That is, these are molecules that maintain the same geometry in the ground and ionic states. However, there are many molecules that change their geometry rather drastically upon ionization. Among them is CCl_4 which was for many years assumed to be unstable as a parent ion. It is now clear that the parent ion is stable.[20] However, its geometry is so different from that of the tetrahedral neutral that it cannot be reached via a single excitation step. Another example is NO_2. This molecule is bent as a neutral. However, upon removal of an electron, it becomes isoelectronic with CO_2 and is thus linear in the ionic form. This change in geometry is immediately apparent in the shape of the PES shown in Figure 5.[21] While the vertical IE is 11.22, the onset for the adiabatic IE is lost in the noise. Killgoar *et al.*[22] determined an adiabatic IE of 9.616 eV on the basis of high resolution photoionization study. But their onset is also not very convincing.

Two approaches to the determination of an adiabatic IE for such molecules are a) measurement of appearance energies for dissociative ionization in which the

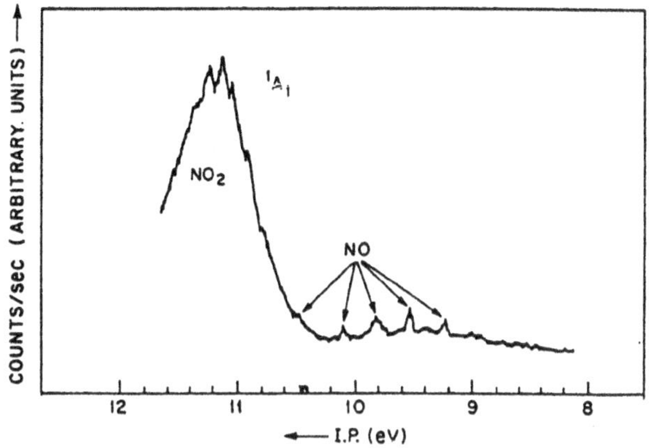

Figure 5. The PES of NO_2. Spectrum taken from Brundle *et al*. reference 21.

fragment ion is produced, and b) the use of several laser colors in a REMPI experiment in which the geometry is changed sequentially during several excitation steps. This second approach is similar to that of producing some doubly charged ions such as CH_4^{+2} which cannot be formed in a single ionization step, but can be formed by charge stripping of the CH_4^+ ion.[23]

Lifshitz et al.[24] have measured the dissociative photoionization onset for NO_2^+ from nitromethane. This ion dissociates to several fragment ions among them:

CH_3NO_2 --->	$CH_3NO_2^+$ + e^-	11.1 ± 0.05 eV
CH_3NO_2 --->	CH_2NO^+ + OH	11.75 ± 0.05
CH_3NO_2 --->	NO^+ + CH_3O	11.75 ± 0.05
CH_3NO_2 --->	NO_2^+ + CH_3	12.1 ± 0.1
CH_3NO_2 --->	$CH_2NO_2^+$ + H	11.8 ± 0.1

The onsets measured by Lifshitz *et al.*[24] are very close in energy which means that there is little effect of kinetic shift, an effect discussed fully in the next section of this paper. As a result, even though the NO_2^+ formation is energetically the fourth fragment to appear, its onset can still be related to the thermochemical dissociation limit. By using this onset and the known heats of formation of CH_3 and nitromethane, it is possible to calculate a ΔH°_{f0} (NO_2^+) = 230.1 kcal/mol. Combining this with the neutral ΔH°_{f0} (NO_2) = 8.65 kcal/mol, leads to an IE of 9.60 ± 0.1 eV. This is actually quite close to the value proposed by Killgoar *et al.*[22] However, this is still only an upper limit because of such factors as a reverse activation energy in the dissociation process of the nitromethane ion.

Haber *et al.*[25] recently determined the NO_2 IE by using a three color REMPI scheme. Figure 6 shows the sequence of laser photons used to reach the ionization energy. Laser frequencies w_1 and w_2 are used to excite the 3pσ Rydberg state. A third laser is then used to scan through the manifold of high lying Rydberg states just below the ionization limit. Because the 3pσ state has a geometry very similar to that of the ion, transitions between it and the higher lying Rydberg states will be strong for Δv=0 transitions. By reaching this 3pσ via two photons, it is possible to excite the v=0 level. Although the signal was not strong because of the weak Franck-Condon factors for the first transition, it was sufficient to obtain an ion signal from the third laser. An extrapolation of the Rydberg series to the ionization limit then allowed Haber to extract an IE (NO_2) of 9.585 ± 0.002 eV.

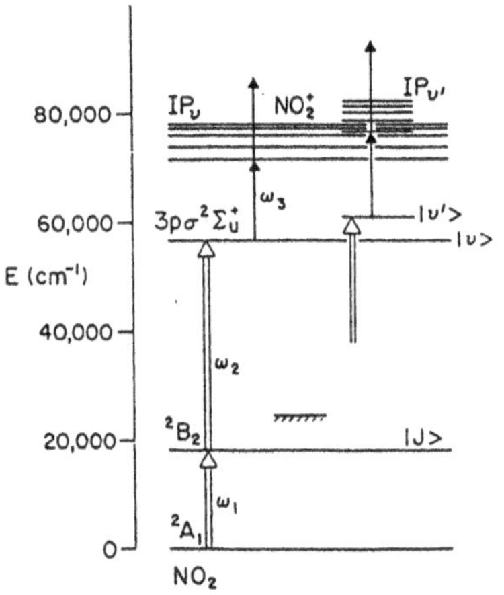

Figure 6. The laser excitation scheme used to excite the high Rydberg states of NO_2 converging to the adiabatic ionization energy of NO_2^+.

Many other molecules have weak 0-0 Franck-Condon factors. It is usually evident from the PES or the photoionization efficiency scan. Figure 7 shows this for the case of dimethyldisulfide. The slowly rising parent ion signal (m/z=94) prevents the determination of an onset, and therefore the IE. A multi-color laser REMPI process might allow the IE energy to be determined here also. However, one important requirement for the success of such a scheme is that the intermediate states are stable and do not dissociate. If they dissociate more rapidly than the laser photon can be absorbed, the ion signal will vanish. Butler et al.[26] used another trick to determine this molecule's IE. This was through the analysis of the metastable ion lifetime in which the adiabatic ionization energy is an important parameter. The principle of this will be discussed in the following section of this chapter. The results are illustrated in Figure 8 which also shows the reason for the poor 0-0 Franck-Condon factors, namely the change in the dihedral angle upon ionization.

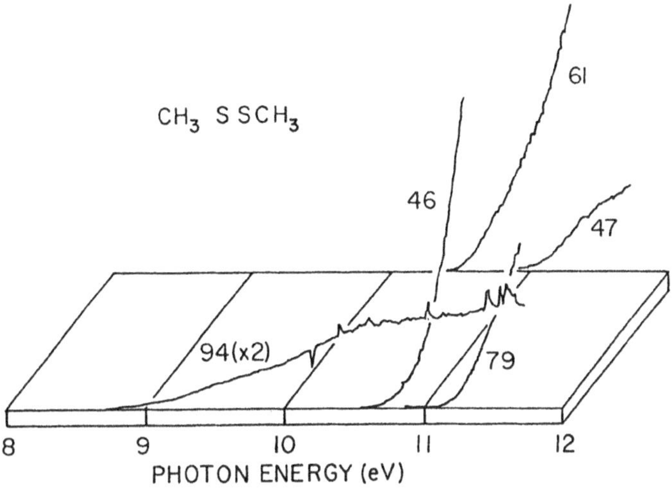

Figure 7. The photoionization efficiency curves for dimethyldisulfide. The slowly rising parent ion signal (m/z = 94) is a result of the small Franck-Condon factors for the 0-0 transition. Spectrum is taken from Butler and Baer, reference 26.

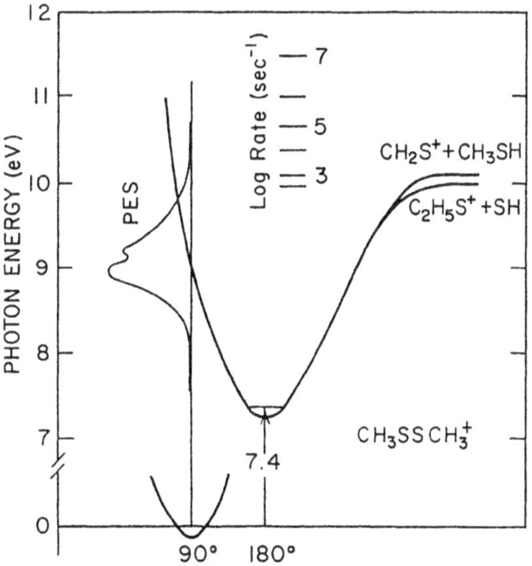

Figure 8. The potential energy diagram for the ionization of dimethyldisulfide. Taken from Butler and Baer, reference. 26.

3. Fragment Ion Dissociation Onsets

3.1. THE KINETIC SHIFT

The problems associated with determining dissociation energy onsets are considerably more sever than are ionization energies. One of the most prevalent problems is that of the kinetic shift. Any large ion with an activation energy for dissociation greater than about 1 eV will be metastable. If the dissociation rate at the onset is less than about 10^4 sec^{-1}, the onset cannot be determined by a simple photoionization spectrum in which the fragment ion signal is monitored as a function of the ionizing energy. An example for the case of CO loss from the phenol ion is shown in Figure 9. The true fragmentation onset is shown along with the calculated dissociation rates.

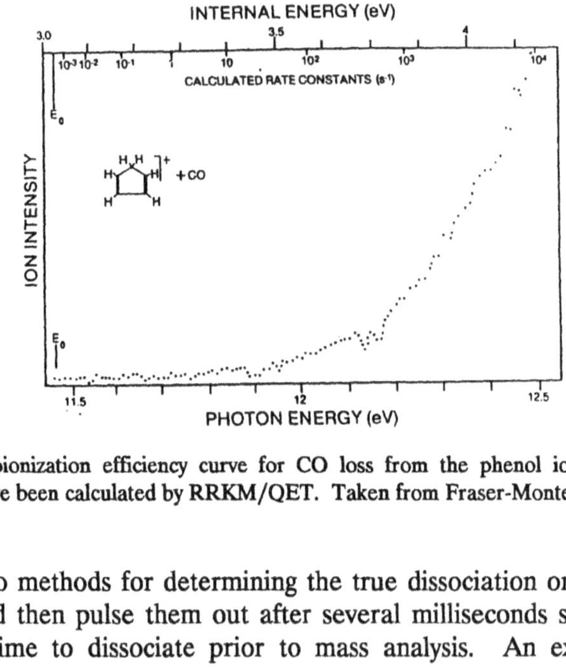

Figure 9. The photoionization efficiency curve for CO loss from the phenol ion. The indicated dissociation rates have been calculated by RRKM/QET. Taken from Fraser-Monteiro *et al.* reference 27.

There are two methods for determining the true dissociation onset. One is to trap the ions and then pulse them out after several milliseconds so that the ions have sufficient time to dissociate prior to mass analysis. An example for the dissociation of the aniline ion is shown in Figure 10. The onset decreases with ion trapping time until is reaches a limit at storage times greater than 0.5 msec.[28] However this onset is still only an upper limit to the true dissociation limit because if the dissociation rate dips below 10^2 sec^{-1}, infra-red fluorescence will become the dominant decay process.[29-31] Thus, in Figure 10 it is not certain whether the limiting appearance energy of 11.26 ± 0.2 eV is determined by the IR fluorescence time, or by the dissociation rate. In fact, RRKM calculations[32] indicate that the dissociation rates near the onset are much smaller than 10^2 sec^{-1} so that IR fluorescence most certainly dominates the limiting value in Figure 10. However, because the k(E) at low energy changes very rapidly with ion internal energy, the upper limit determined in Figure 10 is probably very close to the true dissociation limit.

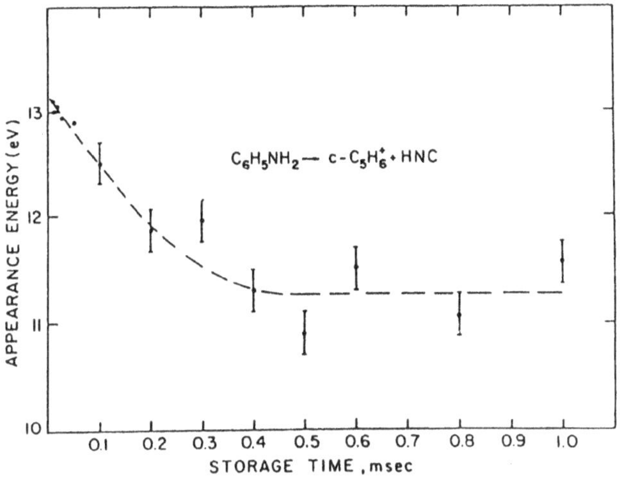

Figure 10. The onset for $C_5H_6^+$ formation from aniline as a function of the ion trapping time. Taken from Lifshitz *et al.* reference 28.

The other method of determining the true dissociation onset is through the measurement of dissociation rates over a range of ion internal energies. Extrapolation of these rates to the dissociation onset by modeling the k(E) curve with the statistical theory for unimolecular decay (RRKM/QET) then yields the onset. However, this approach also has its problems. First, we must assume that the decay mechanism is the same at high energies, where the rates are measured, as it is near the dissociation limit. Secondly, as pointed out by Bowers and co-workers,[31] the position of the transition state changes as the energy is increased. The net effect of this transition state switching is that extrapolation of the high energy rates to the region near the onset will tend to underestimate the dissociation energy. Nevertheless, these two methods are the only ones available at the present time to overcome the problem of the kinetic shift.

3.2. Measurement of Dissociation Rates

Dissociation rates in the range from $2x10^4$ to $2x10^6$ sec^{-1} of energy selected ions have been measured now for a number of years by the technique of photoelectron photoion coincidence (PEPICO). In recent years laser MPI studies have also contributed to this field by providing a higher resolution in selecting the ion energy.[34-39] Finally, photodissociation in an ICR cavity has allowed Dunbar and co-workers to measure rates as low as 3×10^3 sec^{-1}.[40,41] Experimental details can be obtained from the indicated references.

Examples of dissociation rates measured by PEPICO[32] for the aniline ion are shown in Figure 11. The loose transition state model fits the measured rates quite well. The assumed dissociation limit was 11.20 ± 0.1 eV which agrees within the margin of error with the 11.26 ± 0.2 eV measured by Lifshitz et al.[28] Since extrapolation of the k(E) curve may underestimate the onset energy because of the transition state switching mechanism while the ion trapping method overestimates it because of IR fluorescence, we can consider the agreement between these two very different techniques to indicate that the true onset energy is indeed in the vicinity of 11.23 eV. Of particular significance is that this value was extrapolated from data which were obtained between 12.7 and 13.6 eV thereby requiring the extrapolation over a 1.5 eV energy range.

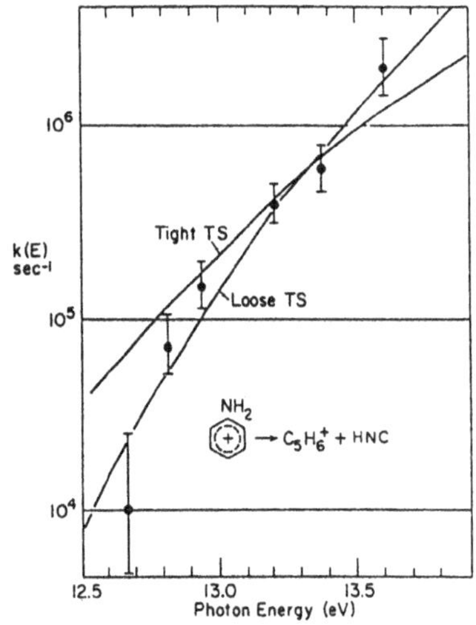

Figure 11. The measured (points) and calculated (solid lines) dissociation rates as a function of the aniline ion internal energy. Taken from Baer and Carney, reference 32.

3.3 THE DISSOCIATION OF ETHYLBENZENE IONS

The major dissociation path for the ethylbenzene ion at low ion internal energies is the loss of CH_3. A major question in this reaction concerns the structure of the resulting $C_7H_7^+$. The heats of formation of two $C_7H_7^+$ isomers are now reasonably well established.[42] These are the benzyl $[C_6H_5CH_2^+]$ and the tolyl $[C_6H_4CH_3^+]$ ions. The former has been determined by both the IE measurement of the benzyl radical[43] and by the measurement of the dissociation limit of benzyl bromide.[42] The tolyl ion heat of formation was determined[42] by the dissociative ionization of nitrotoluene which has been shown to produce pure tolyl ions.[44] The PEPICO rate studies were used to extrapolate the k(E) curve to the onset as described in the previous section. The derived heats of formation at 298 K are 214.1 ± 1.0 kcal/mol for benzyl, 251.9 ± 2 kcal/mol for meta-tolyl, and 256.7 ± 2 kcal/mol for para-tolyl.

Ethylbenzene ions have been investigated by several workers[45-48] among which were two photoionization studies. They found onsets for the $C_7H_7^+$ ion of 10.06 and 10.79 eV, respectively. On the basis of the lower onset, McLoughlin *et al.*[48] suggested that the product ion had the lower energy tropylium structure. However, CA studies have not supported this claim.[44] Rather, the $C_7H_7^+$ ion from ethylbenzene seems to be a mixture of benzyl and tropylium ions.

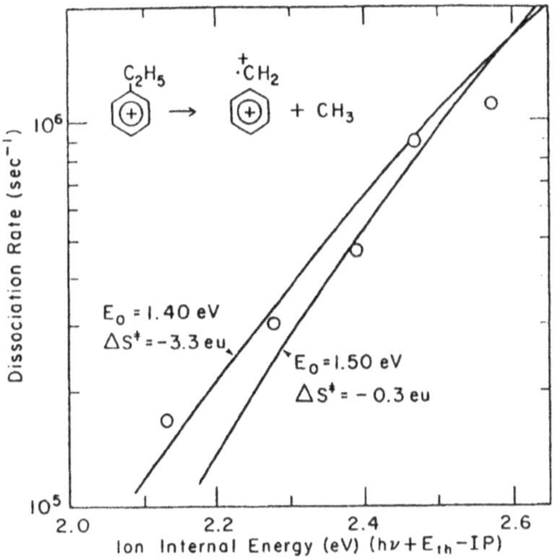

Figure 12. The dissociation rates of energy selected ethylbenzene ions. The points are measured by PEPICO. The solid lines are two RRKM/QET calculations in which different vibrational frequencies and activation energies were chosen. They serve to indicate the error involved in extracting the activation energy and entropy.

We have measured the dissociation rates of energy selected ethylbenzene ions by PEPICO. The resulting rates and RRKM/QET calculations are shown in Figure 12. They indicate that the activation energy for CH_3 loss is 1.45 ± 0.5 eV and that the activation entropy is about -1.6 cal/K. The true 0 K onset for dissociation is thus 10.23 eV. If the thermal energy at room temperature is taken into account, the 298 K onset might be in the vicinity of 10.12 eV, which is very close to the value reported by McLoughlin *et al.*[48] The extrapolated dissociation rate at the onset is below 10^2 sec^{-1} so that the fragment ion signal at threshold is very weak, as was indeed observed by McLoughlin.

The entropy of activation provides some information about the ion structure. A negative entropy of activation suggests that the transition state is "tight" and that there is a reverse activation energy associated with this reaction. Indeed, our ion time of flight peak widths were rather wide which is an indication of kinetic energy release. These considerations lead us to propose that the barrier at 10.23 eV is only an isomerization barrier which cannot be used to determine the heat of formation of the product ion. Figure 13 shows the relationship between the ethylbenzene molecule, its ion, and the transition state for further reactions which eventually lead to CH_3 loss.

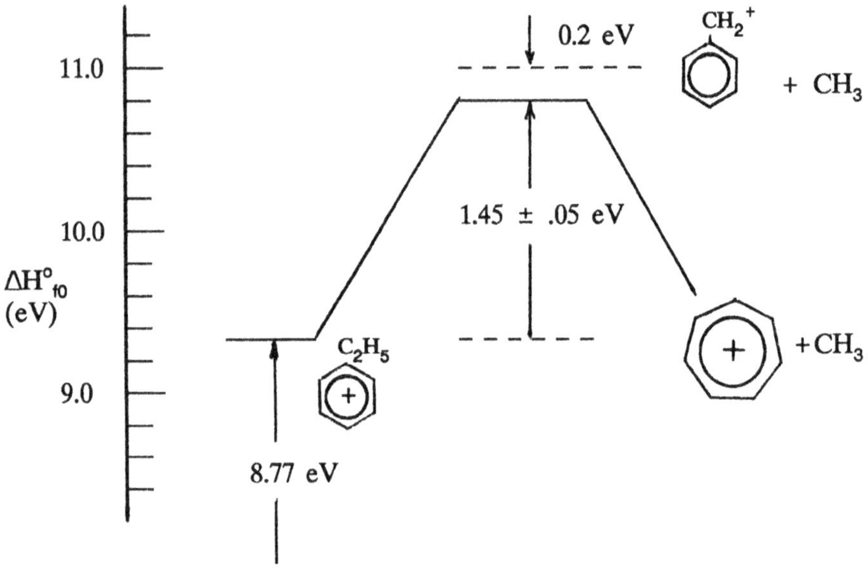

Figure 13. The energy diagram for the dissociation of ethylbenzene ions to either benzyl or tropyl product ions.

The reaction mechanism for methyl loss from ethylbenzene ions is probably considerably more complex than implied in Figure 13. Grotemeyer and Grutzmacher[45,46] have investigated this reaction with extensive labeling studies and have concluded that the ethylbenzene ion rearranges to a number of other $C_8H_{10}^+$ isomers prior to dissociation. A large fraction of the ethylbenzene ions dissociated only after complete scrambling of the C and H atoms, while the other part of the dissociation takes place with little isotopic scrambling. This conclusion is not in conflict with our findings. However, the details of the Grotemeyer/Grutzmacher mechanism probably require some modification because the mechanism suggests that tropylium ions should be formed at the dissociation onset. By raising one or more barriers for isomerization, the Grotemeyer/Grutzmacher mechanism could account for both the isotopic labeling data as well as the PEPICO and the most recent CA results.[44] The two reaction paths suggested by Grotemeyer and Grutzmacher[45] are most likely associated with production of the benzyl and tropyl product ions. The former is the result of a direct dissociation, while the latter is produced after isomerization to other $C_8H_{10}^+$ isomers. As suggested by the PEPICO results in Figure 13, the isomerization barrier is just 0.2 eV below the direct dissociation path. Thus at energies corresponding to dissociation rates typical of metastable ions, the ethylbenzene ion has energy considerably in excess of both the direct dissociation as well as the isomerization paths.

If indeed the reaction proceeds via two reaction paths, one a direct dissociation with a high activation energy but low activation entropy, and the other a rearrangement process with a lower activation energy but a negative entropy of activation, then this reaction would serve as an excellent example in which extrapolation of the k(E) curve to the dissociation onset is dangerous. Because the direct dissociation is most likely favored at higher energies where the rate measurements are carried out, the extrapolation to the onset should favor the direct dissociation. Unfortunately, the energy range over which the decay rates were measured is not sufficient to detect this double reaction path in ethylbenzene ions.

4. Conclusion

These examples have shown that onset energies are difficult to measure and more difficult to interpret. The ionization energies can be determined with very high precision using laser methods. However, in cases where the Franck-Condon factors for 0-0 transitions between the molecule and the ion are very small, multi-step excitation methods are needed to extract accurate ionization energies. The problems encountered in extracting dissociation energies are more difficult and more common. The major obstacle is the slow dissociation rate at the dissociation limit. The interpretation of the onset is even more difficult than establishing its value. That is, an onset may be associated with a direct dissociation or with an isomerization to some other more stable structure prior to dissociation. Kinetic energy release, dissociation rates measurements which provide the activation entropy, CA studies of the product ion structure, and isotopic labeling studies provide essential information which are needed before firm conclusions about the reaction mechanism can be established.

5. Acknowledgments

The U.S. Department of Energy is gratefully acknowledged for financial support of this work.

6. References

1. Field, F.H. and Franklin, J.L. (1957) *Electron Impact Phenomena and the Properties of Gasesous Ions* Academic Press, N.Y.
2. Watanabe, K., Nakayama, T. and Mottle, J. (1962) J. Quant. Spectrosc. Radiat. Transfer **2** 369
3. Murad, E. and Inghram, M.G. (1964) J. Chem. Phys. **40** 3263; and Steiner, B., Giese, C.F., and Inghram, M.G. (1961) J. Chem. Phys. **34** 189
4. Dehmer, P.M. and Chupka, W.A. (1977) J. Chem. Phys. **66** 1972
5. Muller-Dethlefs, Sander, M., and Schlag, E.W. (1984) Z. Naturforsch. Teil A **39** 1089.
6. Chewter, L.A., Sander, M., Muller-Dethlefs, K., and Schlag, E.W. (1987) J. Chem. Phys. **86** 4737
7. Hager, J.W. and Wallace, S.C., (1988) Anal. Chem. **60** 5
8. Smith, M.A., Hager, J.W., and Wallace, S.C. (1984) J. Chem. Phys. **80** 3097
9. Wagman, E.E., Evans, W.H., Parker, V.B., Schumm, R.H., Halow, I., Bailey, S.M., Churney, K.L., Nuttall, R.L., (1982) J. Phys. Chem. Ref. Data, Suppl. **11** 2
10. Pedley, J.B., Naylor, R.D., Kirby, S.P. (1986) *Thermochemical Data of Organic Compounds* 2nd Edition, Chapman and Hall, London
11. Saykally, R. (1988) Science **239** 157
12. Blake, G.A., Laughlin, K.B., Cohen, R.C., Busarow, K.L., and Saykally, R.J. (1987) Astrophys. J. **316** L45
13. Anderson, S.L., Goodman, L., Krogh-Jespersen, K., Ozkzbak, A.G., and Zare, R.N. (1985) J. Chem. Phys. **82** 5329
14. Crofton, M.W., Jagod, M.F., Rehfuss, B.D., Kreiner, W.A., and Oka, T. (1988) J. Chem. Phys. **88** 666
15. Woods, R.C. (1988) Philos. Trans. R. Soc. London **A324** 141
16. Johnson, P.M. private communication
17. Muller-Dethlefs, K. private communication
18. Long, S.R., Meek, J.T., and Rilley, J.P. (1983) J. Chem. Phys. **79** 3206
19. Meek, J.T., Long, R.S., and Rilley,J.P. (1982) J. Phys. Chem. **86** 2809
20. Hacaloglu, J. and Andrews, L. (1989) Theochem **59** 325
21. Brundle, C.R., Neumann, D., Price, W.C., Evans, D., Potts, A.W., and Streets, D.G. (1970) J. Chem. Phys. **53** 705
22. Killgoar, P.C., Leroi, G.E., Chupka, W.A., and Berkowitz, J. (1973) J. Chem. Phys. **59** 1370

23. Ast, T., Porter, C.J., Proctor, C.J., and Beynon, J.H. (1981) Chem. Phys. Lett. **78** 439
24. Lifshitz, C., Rejwan, M., Levin, I., and Peres, T. (1988) Int. J. Mass Spectrom. Ion Proc. **84** 271
25. Haber, K.S., Zwanziger, J.W., Campos, F.X., Wiedmann, R.T., and Grant, E.R. (1988) Chem. Phys. Lett. **144** 58
26. Butler, J.J. and Baer, T. (1983) J. Am. Chem. Soc. **105** 3451
27. Fraser-Monteiro, M.L, Monteiro, L., DeWit, J., and T. Baer (1984) J. Phys. Chem. **88** 3622
28. Lifshitz, C., Gotchiguian, P., and Roller, R. (1983) Chem. Phys. Lett. **95** 106
29. Dunbar, R.C., Chen, J.H., So, H.Y., and Asamoto, B. (1987) J. Chem. Phys. **86** 2081
30. Asamoto, B., and Dunbar, R.C. (1987) Chem. Phys. Lett. **139** 225
31. Asamoto, B., and Dunbar, R.C. (1987) J. Phys. Chem. **91** 2804
32. Baer, T. and Carney, T.E. (1982) J. Chem. Phys. **76** 1304
33. Chesnavich, W.J., Bass, L., Su, T., and Bowers, M.T. (1981) J. Chem. Phys. **74** 2228
34. Proch, R., Rider, D.M., and Zare, R.N (1981) Chem. Phys. Lett. **81** 430
35. Durant, J.L., Rider, D.M., Anderson, S.L., Proch. F.D., and Zare, R.N. (1984) J. Chem. Phys. **80** 1817
36. Kuhlewind, H., Kiermeier, A., Neusser, H.J., and Schlag, E.W. (1987) J. Chem. Phys. **87** 6488
37. Neusser, H., (1989) J. Phys. Chem. **93** 3891
38. Stanley, R.J., Cook, M., and Castlemann, Jr., A.W. (1990) J. Phys. Chem. **94** 3668
39. Lemaire, J. Dimicoli, I., and Botter, R. (1987) Chem. Phys. **115** 129
40. So, H.Y. and Dunbar, R.C. (1988) J. Am. Chem. Soc. **110** 3080
41. Dunbar, R.C. (1990) J. Phys. Chem. **94** 3283
42. T. Baer, J.C. Morrow, J.D. Shao, and S. Olesik, (1988) J. Am. Chem. Soc. **110** 5633
43. House, F.A., Beauchamp, J.L. (1978) J. Am. Chem. Soc. **100** 3290
44. Buschek, J.M., Ridal, J.J., and Holmes, J.L. (1988) Org. Mass Spectrom. **23** 543
45. Grotemeyer, J. and Grutzmacher, H.F. (1982) Org. Mass Spectrom. **17** 353
46. Grotemeyer, J. and Grutzmacher, H.Fr. (1982) *Current Topics in Mass Spectrometry and Chemical Kinetics* A. Maccoll, Ec. Heyden (London) pp. 29-59
47. Akopyan, M.E. and Vilesov, F.K. (1966) Russ. J. Phys. Chem. **40** 63
48. McLoughlin, R.G., Morrison, J.D., and Traeger, J.C. (1979) Org. Mass Spectrom. **14** 104

THERMODYNAMIC INFORMATION FROM DRIFT-TUBE EXPERIMENTS ? SUMMARY OF THE PANEL DISCUSSION

Michael Henchman[1]
Department of Chemistry
Brandeis University
Waltham
MA 02254-9110, U.S.A.

Throughout its 40-year history, gas-phase ion chemistry has been explored with a battery of different techniques. From the very beginning, however, a technique has been needed to bridge the energy / temperature gap between beam experiments at the lowest energies (~0.3 eV) and swarm experiments at the highest temperatures (~900 K). Today, this need has been largely met by the drift tube: the solution, however, is not entirely satisfactory. A system in a drift tube is not maintained under isothermal conditions. It cannot be characterized by a truly thermodynamic temperature. So when claims are made that thermodynamic quantities have been measured in drift tubes, several questions need to be asked. What is the thermodynamic significance of these so-called thermodynamic quantities? How can that significance be evaluated? How should the values be treated in compilations of thermodynamic data?

"Thermodynamic" properties, measured in drift tubes, can clearly be useful. Examples in this chapter demonstrate how such measurements can provide valuable information which is not otherwise available. The task is to separate the wheat from the chaff — a hard enough task for conventional thermodynamic data and very much harder indeed for *quasi*-thermodynamic data. The panel discussion attempted to initiate some discussion and analysis of these questions, without any expectation of being able to deliver, at this stage, definitive answers.

There are five contributors to this panel discussion. Dressler discusses an exciting new technique to measure directly the velocity distributions of ions drifting in an electric field. This reveals how closely the distribution function of the ions in the drift tube approximates a Maxwellian distribution. Viggiano provides a general introduction, discussing how drift-tube measurements at an effective temperature may differ from the corresponding isothermal measurements at the equivalent *true* temperature. Javahery reports measurements of proton affinities with a drift tube. Armentrout discusses drift-tube thermochemistry in the context of the extensive information available from beam experiments. In conclusion, Bartmess discusses the difficulty of determining the effective temperature for another non-isothermal technique — the quadrupole ion trap.

Ion Velocity Distributions In Drift Tubes (R. A. Dressler,[2] J. P. M. Beijers, V. M. Bierbaum, H. Meyer, S. M. Penn and S. Leone, Joint Institute for Laboratory Astrophysics, University of Colorado and National Institute of Standards and Technology, and Department of Chemistry and Biochemistry, University of Colorado, Boulder, CO 80309-0440, U. S. A)

The fact that ion velocity distributions can strongly affect ion-molecule reaction rates measured in drift tubes was first brought to light by Albritton and coworkers.[3] These authors reported rate constants for the reaction $O^+ + N_2 = NO^+ + N$ in an argon buffer that

K. R. Jennings (ed.), Fundamentals of Gas Phase Ion Chemistry, 267–280.

were up to a factor of 4 higher than those determined in a helium buffer at a given collision energy. The large dependence of the rate constants on the buffer gas was attributed to a high energy tail in the velocity distribution of the O^+ ions when drifted in argon.

The steady-state velocity distributions of ions drifting in a uniform electric field in dilute buffer gases are a function of the momentum transfer cross sections in collisions between the ions and the buffer gas atoms. The momentum transfer cross sections are directly linked to the ion/buffer-gas interaction potential. The ion/helium potential exhibits a very shallow well, ~15 meV deep, which is significantly below thermal collision energies at room temperature. Momentum transfer scattering is therefore mainly due to scattering from the repulsive part of the potential at near-thermal and higher energies, whereas the long-range attractive part of the potential only contributes marginally to momentum transfer. The ion/argon potential, on the other hand, has a significantly larger well depth of ~120 meV, which is significantly larger than thermal collision energies at room temperature. At near-thermal energies, momentum transfer will be strongly influenced by scattering from the attractive part of the potential.

The different types of scattering are reflected in the field dependences of the reduced mobilities of ions.[4] The reduced mobility μ is directly related to the average steady-state drift velocity v_D and is given by:

$$\mu = v_D \cdot N / (N_0 \cdot E) \qquad (1)$$

where N_0 is the gas density at standard pressure and temperature, N is the effective gas number density and E is the electric field strength. At room temperature it is generally observed that ion mobilities in helium exhibit only a small dependence on the applied field. This is due to the fact that the magnitude of the momentum transfer cross section does not depend on the relative velocity in the case of a hard-sphere type potential. The mobilities are observed to decrease slowly with increasing field. Such behavior can be reproduced theoretically when assuming scattering from a purely repulsive potential. In argon, the scattering is governed by the long-range ion-induced dipole interaction at low field strengths. The momentum transfer cross section from such an attractive potential is known to be proportional to $1/v$ where v is the relative collision velocity. Consequently one observes an increase in the ion mobility with increasing field at low field strengths. At field strengths at which the collision energy is comparable to the ion/argon potential well depth, the mobility reaches a maximum. At higher fields, momentum transfer is most efficient via scattering from the repulsive part of the potential, and one observes a small decline in the mobility with increasing field, as observed in helium.

A vast amount of mobility / electric-field-strength measurements has been carried out in drift tubes. Important information on the respective ion/buffer gas interaction potentials has been obtained from the experimental data. The velocity distributions of drifting ions, however, can provide more detailed information on the ion/neutral interaction. Although much theoretical effort has been conducted to calculate velocity distributions of ions drifting in a uniform electric field in dilute buffer gases,[4] almost all experimental attempts to back up the detailed theory have failed. Previous experimental efforts have consisted of electrostatic experiments, which unfortunately exhibited strong selectivity with respect to particular velocity components, and which could not avoid perturbing the uniformity of the electric field.

Recently, we have introduced a much less intrusive way of determining ion velocity distributions in drift tubes.[5,6] Doppler profiles of ions are obtained using single-frequency laser-induced fluorescence. In a first series of experiments, we chose to study the velocity distributions of an atomic ion, thus avoiding internal excitation. The ion Ba^+ exhibits a convenient excitation and detection spectrum. Figure 1 shows examples of Doppler profiles obtained in a helium flow. In this case, the tunable laser beam is propagated along the drift-tube axis in the direction of the electric field. The Doppler profiles shift to higher laser frequencies with increasing field. The frequency shift of the maximum yields directly the average drift velocity of the ions. The Doppler profiles are essentially Gaussian and therefore represent shifted Maxwellian velocity distributions. The band widths then yield the ion temperature parallel to the laser beam. The laser can also be aligned perpendicular to the electric field, thus allowing the determination of the perpendicular ion temperatures, which are found to be cooler than those parallel to the electric field. The experimental data of Ba^+ drifted in helium are in excellent agreement with the three-temperature theory of Lin, Viehland and Mason,[7] when assuming a purely repulsive potential.

Figure 2 shows examples of Doppler profiles of Ba^+ obtained in an argon flow.[8] The frequency shifts are much smaller due to the much lower mobility in argon. In this case, however, skewing of the Maxwellian velocity distribution is clearly visible at higher drift fields. The skewing is the result of the $1/v$ dependence of the momentum transfer cross sections. Fast ions experience smaller momentum transfer cross sections and will therefore tend to "run away", thus creating a high energy tail. The broadening of the Doppler profiles is also substantially more visible indicating larger amounts of random energy. The temperatures parallel to the field direction must now be determined through analysis of the three most important moments of the velocity distribution: the first moment yields the ion mobility or average drift velocity, the second central moment is related to the random energy or temperature, and the third central moment is a measure of the asymmetry or "skewness" of the distribution. The derived ion temperatures parallel and perpendicular to the field are again in good agreement with the three-temperature theory.

Figure 3 is a plot of the determined skewness parameter versus E / N, the electric field to gas number density. It is seen that at higher fields the skewness reaches a maximum. This is an indication that "ion run away" is not unlimited and that scattering from the repulsive part of the potential begins to affect the momentum transfer cross sections at higher fields. Eventually the skewness is expected to decrease and the velocity distributions become shifted Maxwellian distributions, as observed in helium.

The above results clearly confirm the theoretical work which states that three temperatures are of importance when deducing thermochemical data from drift-tube measurements: the buffer gas temperature, the temperature parallel to the field and the temperature perpendicular to the field. Knowledge of the field dependence of the ion mobility and the ion-buffer gas interaction potential allow the calculation of various moments of the velocity distribution using a parameterized three-temperature theory.[8] In the case of molecular ions, the situation becomes more complicated since molecular ions can gain rotational and vibrational energy through collisions with the buffer gas atoms.The velocity distributions may also become dependent on the rotational quantum number.Furthermore, the molecular ions may exhibit rotational alignment in drift tubes which may further affect the velocity distributions and result in discrepancies between drift-tube and beam measurements. Rotational alignment has been observed in the case of N_2^+ drifted in helium.[9] Currently, work is being completed in this laboratory on velocity distributions of the diatomic ion CO^+ and alignment of polyatomic ions.

Approaches For Estimating Thermodynamic Information From Drift-Tube Experiments (A. A. Viggiano, GL (AFSC) / LID, Ionospheric Physics Division, Hanscom AFB, MA 01731-5000, U.S.A.)

Drift-tube experiments, as a function of electric field strength, suggest three ways to measure thermodynamic properties for the reaction:

$$A^{+/-} + B = C^{+/-} + D \qquad (2)$$

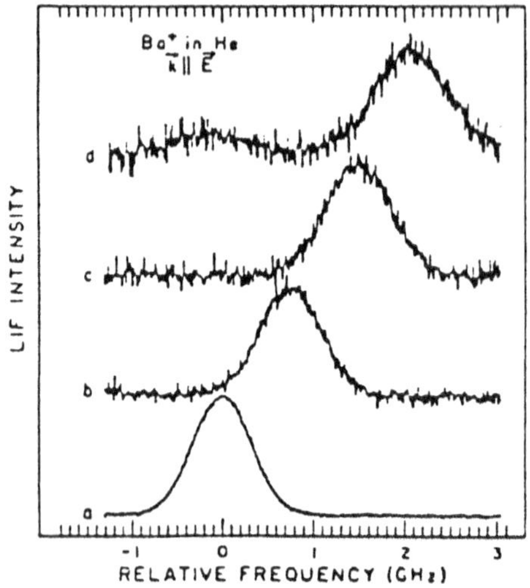

Fig. 1. Doppler-broadened LIF spectra of Ba^+ in He, as function of drift field. (a) 0 Td, (b) 7.44 Td, (c) 14.9 Td, (d) 22.3 Td. Laser (455.53 nm) in parallel direction.

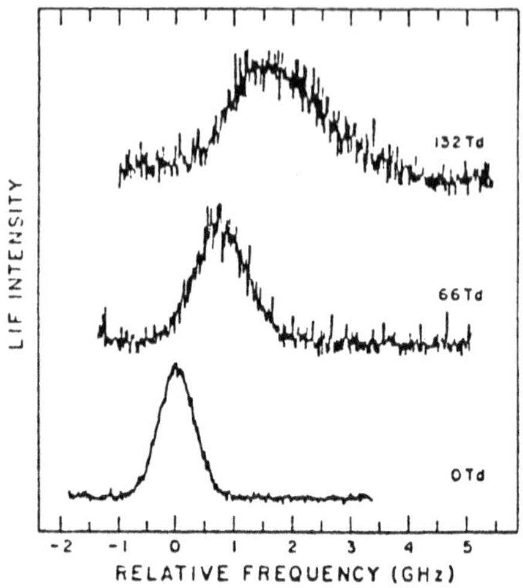

Fig. 2. Doppler-broadened LIF spectra of Ba^+ in Ar, for different drift fields. Laser in parallel direction.

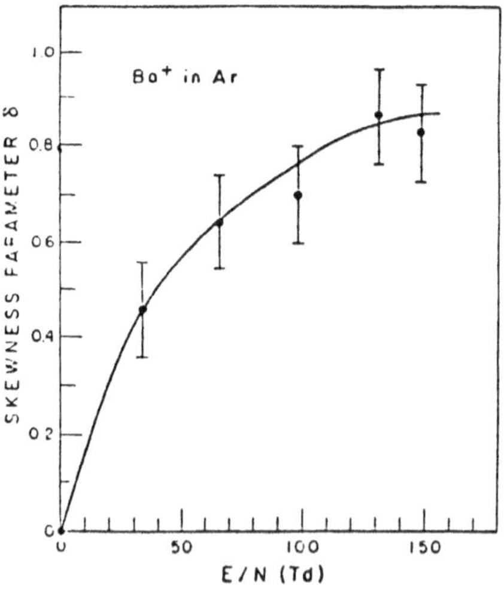

Fig. 3 . Skewness parameters for Ba^+ in Ar, measured as a function of E / N.

1). Measure the equilibrium constant $K = [C^{+/-}][D] / [A^{+/-}][B]$ directly.

2). Measure forward and backward rate constants to give the equilibrium constant $K = k_f / k_r$.

3). Measure the rate constant in the endothermic direction as a function of field strength and from a *pseudo*-Arrhenius plot, $k = A\ \exp(-E_a/RT)$, with effective temperatures estimated from the field strengths, derive an "activation energy". This estimates the endothermicity $\Delta H° = E_a$, assuming the absence of a reverse activation energy.

For several reasons, method 1) can be applied only in a few limited cases and will not be considered further here.

Drift tubes can be used to measure thermodynamic quantities only if the drift-tube distribution function closely approximates a Maxwell-Boltzmann distribution. How thermal are (a) the relative velocity distribution, (b) the rotational state distribution, and (c) the vibrational state distribution?

(a). RELATIVE VELOCITY DISTRIBUTION: This topic has been treated by the previous contributor to this chapter. Atomic ions, such as Ba^+, in a helium buffer show closely Maxwellian velocity distributions, as indicated by the Gaussian functions in Figure 1. However in an argon buffer, the distribution is not Maxwellian, the non-Gaussian plot of Figure 2 revealing the presence of a high-velocity or high-energy tail. The reason for this difference is discussed fully by Dressler. Very early in the development of the drift-tube technique, the kinetic consequence of using different buffer gases was established. For reactions for which $\sigma(v)$ does not show a v^{-1} dependence, rate constants showed a different energy dependence in the two buffer gases.[11] A further test is provided by the case of atomic ions reacting with atomic reactants: when compared appropriately, the rate constant shows the same functional dependence upon temperature as it does upon electric field in a helium buffer.[12]

(b). ROTATIONAL STATE DISTRIBUTION: The rotational state distribution may also be studied directly using laser-induced fluorescence. Rotational excitation is facile, occurring in ~10 collisions for N_2^+ drifting in helium,[13] and yields a Boltzmann distribution with an effective internal temperature T_{int}

$$3/2\ \mathbf{k}T_{int} = KE_{cm} = 3/2\ \mathbf{k}T + 1/2\ M_B v_d^2 \qquad (3)$$

where $\mathbf{k}$ is Boltzmann's constant, KE_{cm} is the mean center-of-mass collision energy of the ion with respect to the buffer gas, T is the thermal temperature of the buffer gas, M_B is the mass of the buffer gas and v_d is the drift velocity.[14] In contrast, the mean center-of-mass energy between the ion and the reactant molecule KE_{cm}^R, is given by

$$3/2\ \mathbf{k}T_{eff}^R = KE_{cm}^R = 3/2\ \mathbf{k}T + 1/2\ M_R(m + M_B)v_d^2 / (m + M_R) \qquad (4)$$

where m is the ion mass , M_R the reactant molecule mass, and T_{eff}^R is the effective reaction temperature for the drift-tube reaction.[15] For a particular set of values for E/N, T, and the masses, v_d is fixed and so is KE_{cm}^R. The rate constant measured under these conditions is k(T,E/N) and, to a first approximation, is expected to correspond to a thermal rate constant $k(T_{eff}^R)$ at an effective reaction temperature T_{eff}^R, given by (4).

It is apparent from (3) and (4) that the effective ion rotational temperature T_{int} is not in general equal to the effective reaction temperature T_{eff}^R. Even though the effective rotational temperature of the ions may differ from the effective reaction temperature, the difference may be minor. The small amount of evidence available suggests that rate constants often show only a slight dependence on rotational energy.[16] For a few reactions however, there is experimental evidence that suggests a strong dependence on rotational energy — for example, a negative $T^{-0.4}$ dependence for $Kr^+(^2P_{3/2}) + HCl = Kr + HCl^+$ [17] and a positive dependence for $Ar^+ + N_2 = Ar + N_2^+$.[18]

(c). VIBRATIONAL STATE DISTRIBUTION: Ions readily undergo vibrational excitation and de-excitation in a drift tube, but the efficiency is significantly less than that for the corresponding rotational processes. To achieve a steady-state Boltzmann vibrational distribution may require an infinitely long drift tube. In such cases, no effective vibrational temperature can be assigned. Vibrational energy, unlike rotational energy, often has a marked influence on rate constants. This is illustrated dramatically by the difference between a truly isothermal van't Hoff plot and a curved van't Hoff plot obtained with a drift tube.[19] Consequently, to obtain accurate thermodynamic data from drift-tube measurements, one must work in a regime where vibrational energy will not strongly affect the measured rate constants.

The charge-transfer reaction

$$CO_2^+ + O_2 = CO_2 + O_2^+ \quad (5)$$

illustrates how vibrational excitation of the CO_2^+ reactant can influence the rate constant.[20] Assuming efficient T $\leftrightarrow$ V coupling and the establishment of a steady state and equipartition of energy, the internal temperature, T_{int}, (both rotational and vibrational), is given by (3). The effective reaction temperature of the drift-tube experiment, T_{eff}^R, (fixed by both the thermal temperature of the buffer gas T and the drift velocity), is given by (4). As noted above, the internal temperature, T_{int}, will in general differ from the effective temperature, T_{eff}^R. The predicted difference between T_{int} and T_{eff}^R for Reaction (5) is shown in Figure 4, which can be interpreted as follows. For a fixed value of the effective reaction temperature, the internal temperature depends on the temperature of the buffer gas. In other words, for fixed T_{eff}^R, T_{int} increases with increasing T: as the drift field contributes increasingly to the fixed effective reaction temperature, so the internal temperature of the ions increases. Experimental data for Reaction (5)[20] can be discussed as follows. The rate constants — data not shown here — are independent of T for a fixed value of T_{eff}^R for values of the effective temperature $T_{eff}^R < 750$ K : rotational energy therefore does not affect the rate constant. For $T_{eff}^R > 750$ K, the rate constant *does* increase with increasing T for a fixed value of T_{eff}^R : vibrational energy of the CO_2^+ therefore does increase the rate constant.[21]

This type of experiment can now be used to reveal which reaction rate constants do not depend on internal energy. This condition must be satisfied if Method (2) above is to be used to obtain thermodynamic information from drift-tube data. A second condition is that the relative translational energy distributions be Maxwellian — and this seems to be satisfied where helium is used as a buffer. Even if both of these conditions are satisfied, considerable caution is needed in estimating thermodynamic data from drift-tube experiments.

We conclude with an example of the use of Method (3) above to obtain thermodynamic data from drift-tube measurements. A schematic Arrhenius equation may be written for drift-tube experiments as $k = A \exp(-E_a^{drift} / E)$, where E is the energy in the drift field [$E = KE_{cm}^R$ according to the nomenclature of (4)] and E_a^{drift} is the activation energy measured in the drift-tube experiment. Where the ion is atomic, the drift-field energy is exclusively translational and may be given by (4). The isothermal activation energy, E_a, has then been computed from the measured "drift-tube activation energy", E_a^{drift}, as follows, $E_a \simeq 2/3\, E_a^{drift} \simeq \Delta H°$. This analysis has been applied to the reaction $N^+ + H_2 = NH^+ + H$ and to the reactions of N^+ with HD and with D_2.[22] (See the section by Armentrout for cautionary statements about this technique).

We also have used this approach for a reaction where the ion is polyatomic and where the energy available in the drift field is translational, vibrational and rotational.

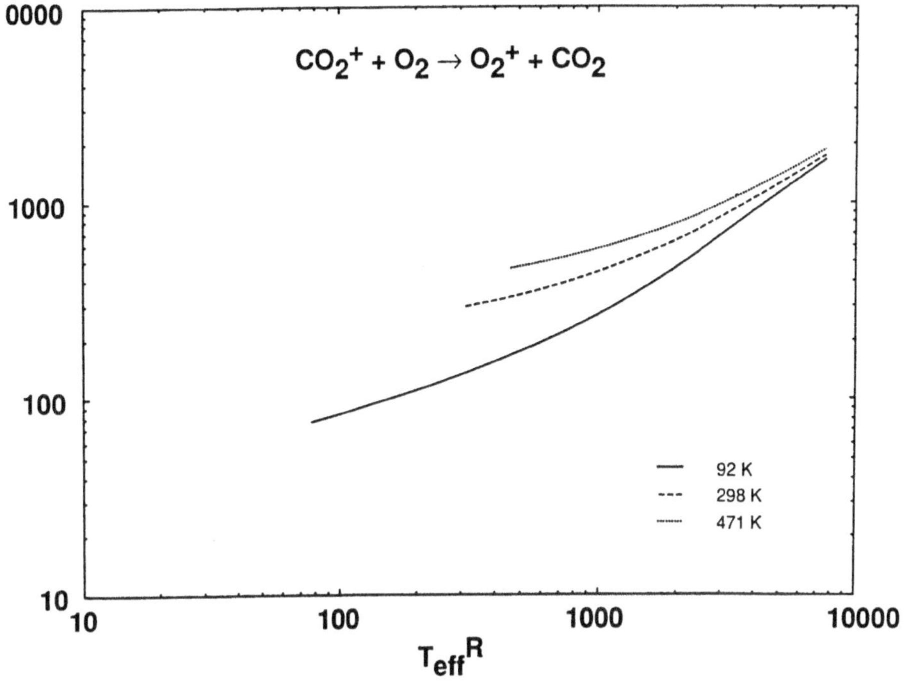

Fig. 4 . Plots of T_{int}, the internal temperature, as a function of T_{eff}^R, the effective translational temperature, for the title reaction, for various temperatures T, of the buffer gas.

Unlike the case of the atomic ions discussed in the previous paragraph, (4) cannot be used to estimate the energy available in the drift tube. Instead, we have used an empirical approach which assumes a linear relationship between the reaction endothermicity, ΔH°, and E_a^{drift}, the activation energy derived from the drift-tube Arrhenius plot.[23] Thus, in order to determine the constant of proportionality, x, in the relationship $\Delta H^\circ = x \cdot E_a^{drift}$, we measure E_a^{drift} for a calibration reaction of known ΔH°, taking care that the calibration reaction is homologous with the reaction under study. Therefore to measure the (unknown) endothermicity of reaction (6), we used reaction (7) of known endothermicity as a calibrant.

$$PO_3^- + HI = HPO_3 + I^- \tag{6}$$

$$NO_3^- + HCl = HNO_3 + Cl^-. \tag{7}$$

Our measurements yield the first experimental value for the enthalpy of deprotonation of HPO_3 of 311 kcal/mol, in agreement with theory.[23] This approach is necessitated by the failure of conventional techniques, since monomeric HPO_3 is unstable and polymerizes.

The Application Of A Selected Ion Flow Tube To Determine Proton-Affinity Differences (G. Javahery, M. Tichy and N.D. Twiddy, Department of Physics, U.C.W. Wales, Aberystwyth SY23 3BZ, Great Britain)

In 1975 the NOAA group, who had pioneered the use of the flowing afterglow/drift tube for studying ion-molecule reactions, showed convincingly that flow /drift experiments could not be used to make thermochemical measurements for chemical reactions.[19] This conclusion was based on a study of reaction (8)

$$N_2OH^+ + CO = N_2O + HCO^+. \tag{8}$$

In flowing afterglow experiments, forward and backward rate constants were measured as a function of temperature ; and the resulting van't Hoff plot was linear, as expected. In the corresponding drift-tube experiments, forward and backward rate constants were measured as a function of electric field strength ; and the resulting van't Hoff plot was non-linear.

This situation has remained this way for 14 years until recent measurements in our laboratory on the Aberystwyth SIFDT (Selected Ion Flow Drift Tube) have completely revised this view.[25] We have studied 9 proton-transfer reactions, each in the forward and backward directions, as a function of electric field strength. Of these 9, 7 produced linear Arrhenius-type and van't Hoff-type plots and 2 gave markedly non-linear plots. Both of these last two involved the same reactant ion N_2OH^+ — one of the two being the reaction (8) studied previously by the NOAA group. (The Aberystwyth and the NOAA data agree perfectly.) So clearly the NOAA choice of reaction (8) as a test case was a most unfortunate one. It has become clear that the non-linear results where N_2OH^+ is the reactant ion is due to the presence in the flow tube of two isomeric forms of N_2OH^+.

We have discovered, to our surprise, that reliable proton affinity differences can be obtained from van't Hoff-type plots of rate constant *vs.* reciprocal center-of-mass kinetic energies, in SIFDT measurements. This has important practical implications because the wide kinetic energy range (~1 eV) far exceeds the maximum temperature variation possible (<900 K ~ 0.1 eV). This allows much larger proton affinity differences to be spanned for a single reaction. Also flow tubes at elevated temperatures are notoriously difficult to operate and maintain.

The applicability of the SIFDT technique to thermochemical problems is quite surprising, as the internal degrees of freedom are not equilibrated with the translational kinetic energy. The neutral reactants maintain rotational and vibrational temperatures of the ambient temperature of the buffer gas, T. The ions attain rotational equilibrium with the relative kinetic energy in a few collisions[13] but the ion vibrations do not[26], at least for low kinetic energies and the relatively small number of collisions of the ions with the buffer in typical SIFDT experiments.

For these experiments, a van't Hoff-type plot can be represented by the expression

$$\ln K = \Delta S / R - \Delta H / RT = \Delta S / R - 3\,\Delta H / 2\,KE_{cm}{}^{R} \qquad (9)$$

where from (4), in molar units, $KE_{cm}{}^{R} = 3/2\,RT_{eff}{}^{R}$. This will only yield linear plots if ΔH and ΔS do not change significantly over the energy range of the investigation.

Presumably a significant variation of the ΔH and ΔS with $KE_{cm}{}^{R}$ would result in a non-linear van't Hoff-type plot. If significant vibrational excitation of the ions by the strong applied field occurred, this would change the value of ΔH as the energy, $KE_{cm}{}^{R}$, is varied, producing non-linearity. Therefore it can be concluded that in the case of our 7 linear van't Hoff-type plots[25] these effects are insignificant. This implies that the value of ΔS obtained from the intercept should be in good agreement with that calculated from the standard entropies of the ions and molecules in these reactions and this should be a sensitive test of the above conclusions. The good agreement between the values of ΔH obtained and the established values and the good agreement between calculated values of ΔS and the values obtained from the intercepts establish the validity of the SIFDT apparatus for such measurements.

Let us consider briefly, and speculatively, why SIFDT studies of proton-transfer reactions may yield linear Arrhenius-type plots of rate constants and linear van't Hoff-type plots of equilibrium constants...the latter yielding values for ΔH and ΔS which agree with accepted values, as we have seen.

What is clear is that rotations and vibrations of the neutral reactants are not excited in the SIFDT, i.e: they remain in a thermal distribution fixed by the ambient temperature of the buffer gas, T. It is also likely that the ions are rotationally equilibrated with the relative kinetic energy with respect to the buffer gas, KE_{cm}, consistent with theoretical prediction[27] and experiment,[13] in the case of atomic buffer gases, as for He in the present case. The major uncertainty concerns ion vibrational excitation. It is uncertain to what extent the ions are vibrationally excited in the He buffer gas. The presumptive evidence, from the validity of the kinetic-energy "van't Hoff" plots is that rotations and vibrations play no role. As a general observation, it has been commonly observed that exoergic proton transfer occurs on nearly every collision for small molecules, such as those considered here. This suggests that the sole criterion is energy and not the rotational and vibrational states involved, except as they contribute to making an endoergic reaction exoergic.

With regard to rotation, we do not expect a significant rotational effect in such reactions, in part because the rotational quanta are so small and we expect small ΔJ change propensities, a sort of angular momentum conservation. With regard to vibration, He is an extremely poor vibrational exciter, especially for the high-frequency H-stretching modes; and most of the experiments are carried out at energies below the H-stretching energy threshold. The linearity of the plots clearly indicates that there is not an onset of vibrational excitation with a consequent sharp change of reactivity. The extent of vibrational excitation in the lower-frequency bending modes is unknown, and quite possibly significant but apparently does not yield an effect.

To date the SIFDT experiments at Aberystwyth, in conjunction with temperature-variable SIFT experiments at Birmingham,[28] have been used to derive improved values for the proton affinities of simple molecules in the range 126-140 kcal/mol. The precision of the method is well illustrated by our data for the proton affinity of CF_4 which is tightly constrained by no less than four links to other molecules, NO, CO_2, CH_4 and HCl. The respective proton affinity differences of 0.6, 1.7, 4.0 and 7.0 kcal/mol fix the proton affinity of CF_4 to be 126.4 kcal/mol which may be compared with the previous best value of ~126 kcal/mol.[29]

Using Ion-Beam Data To Evaluate Drift-Tube Measurements Of Thermodynamic Quantities (P. B. Armentrout, Department of Chemistry, University of Utah, Salt Lake City, UT 84112, U.S.A.)

Two approaches can be taken to evaluate the question of whether reliable thermodynamic information can be derived from drift-tube measurements of the kinetic energy dependence of endothermic ion-molecule reactions: experimental and theoretical. Some of the issues associated with these experiments are illustrated in Figure 5, which shows a comparison of results taken from three laboratories for the reaction, $N_4^+ + D_2 = N_2D^+ + D + N_2$: drift-tube (DT) results in a He buffer,[27] flow-drift tube (FDT) results in an N_2 buffer,[30] and unpublished data from our guided ion beam instrument.[31] It is quite clear that the rates obtained in the drift tubes are higher than those in the beam data. This can be easily explained by suggesting that, as the drift field is increased, not only do the N_4^+ ions obtain increased kinetic energy, but become vibrationally excited as well (more so in the N_2 than in He). Both laboratories admit to such a possibility. In the beam data, the reactions take place under single-collision conditions such that the only energy being varied is the kinetic energy of the ion.

Figure 5 shows that, despite this unknown vibrational energy, the *pseudo*-Arrhenius plots of log k against 3 / 2E are quite linear for both drift-tube data sets. (The presumed conversion between kinetic energy and temperature is $E = 3\,k_BT / 2$). The slopes of these lines give "activation energies" of 0.24 eV (FDT) and 0.17 eV (DT) compared to 0.09 eV for the beam data (taken at the lowest energies, 3 / 2E > 15). These variations illustrate that linear *pseudo*-Arrhenius plots cannot be taken as *prima facie* evidence for thermodynamic behavior. The clear non-linearity of the beam data also demonstrates that choosing a region of linearity may be extremely difficult if high quality data are not available over an extended energy range — a situation which is generally not true for flow tubes.

For comparison, Figure 5 also includes data for the H_2 analogue of this reaction taken under conditions where the reaction temperature is varied under equilibrium conditions (Eq).[30] The similarity of the DT and Eq data is striking (and leads to a similar activation energy, 0.17 and 0.15 eV, respectively) but is probably serendipitous for the following reason. There is no *a priori* reason to anticipate that true thermal equilibrium of the vibrational and rotational degrees of freedom with the translational energy is achieved in the drift tube with a He buffer any more than with an N_2 buffer. Indeed, the Eq data actually fall between the He and N_2 buffer data, suggesting that the vibrational temperature in the He buffer is cold. The difference between the "activation energy" derived from the Eq and beam results is probably due to a difference in E_a (a thermal activation energy) and E_0, the true energy minimum required for reaction. This is undergoing further evaluation but is discussed briefly below.

To understand why some limited success has been observed for thermodynamic data obtained by drift tubes, we need to understand whether it is reasonable to expect that a *pseudo*-Arrhenius plot for an ion-molecule reaction should be linear. To consider this question, we start with the well established observation that exothermic ion-molecule reactions often proceed at the capture rate. For polarizable molecules without dipole moments, the capture cross section is given fairly accurately by the Langevin-Gioumousis-Stevenson (LGS) equation,[32]

$$\sigma(E) = \pi e\,(2\alpha / E)^{1/2} \tag{10}$$

where e is the electronic charge, α the polarizability of the neutral molecule, and E is the relative kinetic energy. To convert to a rate constant, we multiply the cross section by the velocity [$v = (2E / \mu)^{1/2}$ where μ is the reduced mass of the reactants] and integrate over

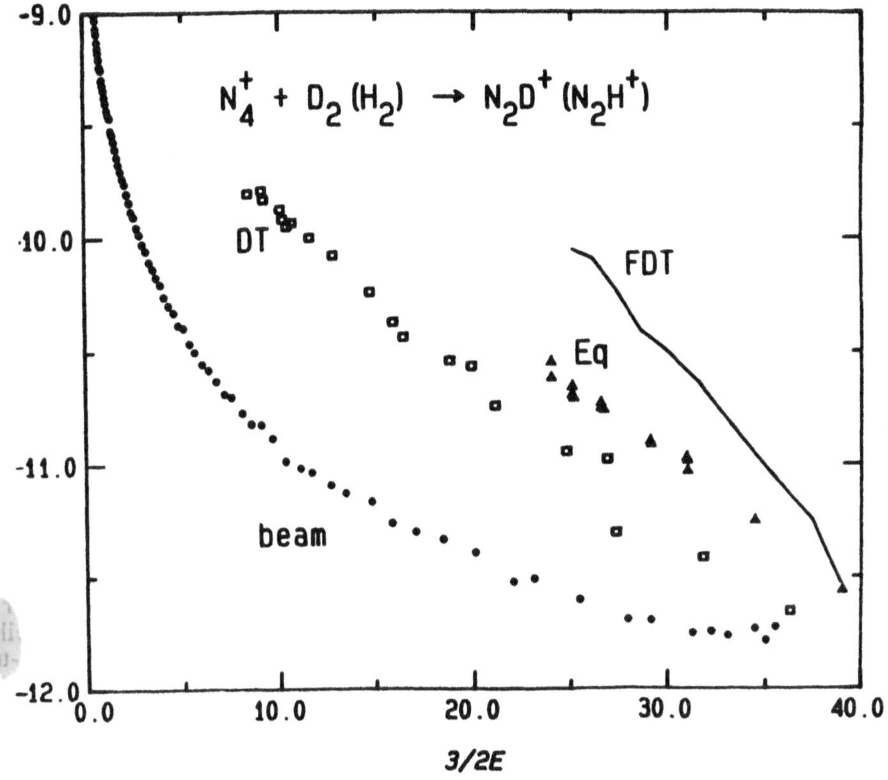

Fig. 5. *Pseudo*-Arrhenius plots of the rate constants obtained for the reaction of N_4^+ with D_2 taken from ref. 27 (DT, open squares), ref. 30 (FDT, line), and ref. 31 (beam, points). Also shown is the temperature dependent rate constant from ref. 30 for $N_4^+ + H_2$ (Eq, triangles) where the formula $E = 3 k_BT/2$ is used to convert from temperature to energy.

the Maxwell -Boltzmann distribution of energies. This yields the familiar LGS expression for an ion-molecule rate constant, $k(T) = 2\pi e (\alpha / \mu)^{1/2}$.

For endothermic ion-molecule reactions, we might guess that the expected behavior can be obtained by applying microscopic reversibility to eq (10).[33] This yields the cross section for an endothermic process as

$$\sigma(E) = \sigma_0(E - E_0)^{1/2} / E \tag{11}$$

where E_0 is the endothermicity of the reaction and

$$\sigma_0 = \pi e (2 \alpha_p)^{1/2} (\mu_p g_p / \mu_r g_r) \tag{12}$$

where g is the electronic degeneracy of the specified channel and the subscripts refer to the reactant (r) or the product (p) channel. When converted to a rate constant, one obtains the Arrhenius equation explicitly,

$$k(T) = A_0 \exp(-E_0 / \mathbf{k}T) \tag{13}$$

where the pre-exponential factor, $A_0 = 2\pi e (\alpha_p/\mu_r)^{1/2} (\mu_p g_p /\mu_r g_r)$, is independent of T! It should be noted that the energy dependent cross section form represented by eq (11) is the only form which yields the Arrhenius equation explicitly.[34] If the energy dependence differs at all from this equation, the Arrhenius activation energy, E_a, differs from E_0 and the pre-exponential factor will be temperature dependent, leading to curved Arrhenius plots (but this may only be obvious for sufficiently precise data taken over an extended energy range).

Thus, if the endothermic reaction behaves ideally, i.e. as in eq (11), then a linear *pseudo*-Arrhenius plot is fully in accord with theory. Indeed, the cross sections for the near-thermoneutral reactions of N^+ with H_2, HD, and D_2 can be described rather well by eq (11).[35] This is consistent with the success of a *pseudo*-Arrhenius analysis of drift-tube data.[22] However, in general, few of the many endothermic ion-molecule reactions studied in our laboratory actually behave as suggested by eq (11). This includes the reaction of N_4^+ with D_2, which explains the large curvature obtained in the beam data shown in Figure 5. This is also the fundamental reason why the beam data (which should yield a threshold close to E_0) gives a different threshold than the temperature dependent data (which gives E_a).

The Quadrupole Ion Trap [A Three Dimensional, Non-Flowing Drift Tube]
(John Bartmess, Department of Chemistry, University of Tennessee, Knoxville, TN 37996 U.S.A. and Gary Glish and Scott McCluckey, Oak Ridge National Laboratory, Oak Ridge, TN 37831 U.S.A.)

We are interested in determining an effective temperature for ions trapped in a quadrupole ion trap. The conditions are as follows: helium pressure ~10^{-3} torr; AC voltage 100 - 500 V; trapping times ~ ms, # of collisions during trapping time 10^3 - 10^4 collisions.

In order to determine the effective temperature of the ions within the ion trap, we are studying the dissociation of cluster ions formed in an ion spray source. For the process $AB^{+/-} = A + B^{+/-}$, we know that when $\Delta G_{diss} = 0$, $K = 1$. When this holds, the temperature T_{diss} can be evaluated as $T_{diss} = \Delta H_{diss} / \Delta S_{diss}$ using measured values of enthalpies and entropies of association.

From experiments on $H^+(H_2O)_n$, observing $0 \le n \le 5$, and on $H^+(CH_3OH)_n$, observing $0 \le n \le 3$, we estimate an effective temperature of > 600±150 K. The next stage of the investigation is to obtain a full fit to the whole range of n, to follow the fit as a function of time and to use negative ions as the thermometer.

Quo Vadis ?

Whence come we and whither are we bound? Here are statements from practitioners—one from the early days (1975): "Thus drift tube results must be used warily, particularly when they are to be applied in lieu of temperature data"[19] and another, recent one from 1989: "It thus appears that kinetic energy van"t Hoff plots are often linear and when they are they yield valid thermochemical data."[25] What caused the re-evaluation of the drift-tube thermodynamic data and is the re-evaluation valid?

The re-evaluation has many causes, which include important theoretical advances[4] and experimental measurements of distribution functions (see Dressler, above). My own view, however, is that the principal boost has been a psychological one. In several cases, the drift-tube studies seem to give results which are as good as or, if not, better than traditional methods. The operative word in the previous sentence is *seem*. As Armentrout points out above, we expect many van't Hoff-type plots to be non-linear, so the linear = acceptable and non-linear = unacceptable identifications are too simplistic a basis on which to proceed. How then should we proceed? We need now a fundamental underpinning of this subject, relating the macroscopic drift tube data to microscopic beam data. We need to show how measured excitation functions (measured to very low energies using the guided-ion beam technique) and distribution functions can generate both linear and non-linear van't Hoff-type plots which do agree with drift tube data, both for atomic ions and polyatomic ions, and for different buffers. We need, for many reactions, a basic understanding of the phenomena to create the confidence needed to apply the method to unexplored reactions. Again, in applying empirical coefficients to relate so-called Arrhenius activation energies to endothermicities, as Viggiano does effectively above, we need to establish a catalog of such values to explore periodic trends, self consistency, etc.

REFERENCES

(1). Preparation of this chapter was underwritten by a contract from the Geophysics Laboratory (Air Force Systems Command), Ionospheric Physics Division, of the U.S. Air Force.

(2). Present address: Geophysics Laboratory, GL/PHK, Hanscom AFB, MA 01731, U. S. A.

(3). Albritton, D.L.; Dotan, I.; Lindinger, W.; McFarland, M.; Tellinghuisen, J. and Fehsenfeld, F.C. **1973**, *J. Chem. Phys. 66*, 410.

(4). Mason, E.A. and McDaniel, E.W.: **1988**, *Transport Properties of Ions in Gases*, Wiley : New York.

(5). Dressler, R.A.; Meyer, H.; Langford, A.O.; Bierbaum, V.M. and Leone, S.R. **1987**, *J. Chem. Phys. 87*, 5578.

(6). Dressler, R.A.; Beijers, J.P.M.; Meyer, H.; Penn, S.M.; Bierbaum, V.M. and Leone, S.R. **1988**, *J. Chem. Phys. 89*, 4707.

(7). Lin, S.L.; Viehland, L.A. and Mason, E.A. **1979** *Chem. Phys. 37*, 411.

(8). Penn, S.M.; Beijers, J.P.M.; Dressler, R.A.; Bierbaum, V.M. and Leone, S.R. **1990**, *J. Chem. Phys.* in press.

(9). Waldman, M. and Mason, E.A. **1981**, *Chem. Phys. 58*, 121.

(10). Dressler, R.A.; Meyer, H. and Leone, S.R. **1987**, *J. Chem. Phys. 87*, 6029.

(11). Albritton, D.L.; Dotan, I.; Lindinger, W.; McFarland, M.; Tellinghuisen, J. and Fehsenfeld, F.C. **1977**, *J. Chem. Phys. 66*, 410.

(12). Morris, R.A.; Viggiano, A.A.; Paulson, J.F. and Su, T. **1990**, *Phys. Rev. A, 41*, 5943.

(13). Duncan, M.A.; Bierbaum, V.M.; Ellison, G.B. and Leone, S.R. **1983**, *J. Chem. Phys. 79*, 5448.
(14). Leone, S.R.: **1989**, in *Bimolecular Collisions* (Ashfold, M.N.R. and Baggott, J.E., Eds.) Royal Society of Chemistry : London, p. 377.
(15). Ref. (4), p.471.
(16). Relevant references are cited in : Viggiano, A.A.; Morris, R.A. and Paulson, J.F. **1988**, *J. Chem. Phys. 89*, 4848.
(17). Viggiano, A.A.; Morris, R.A.; Dale, F.; Paulson, J.F.; Giles, K.; Smith, D. and Su, T. **1990**, *J. Chem. Phys. 93*, 1149.
(18). Viggiano, A.A.; Morris, R.A. and Paulson, J.F. **1990**, *J. Chem. Phys.* in press.
(19). Lindinger, W.; McFarland, M.; Fehsenfeld, F.C.; Albritton, D.L.; Schmeltekopf, A.L. and Ferguson, E.E. **1975**, *J. Chem. Phys. 63*, 2175.
(20). Viggiano, A.A.; Morris, R.A. and Paulson, J.F.; to be published.
(21). In the range of T investigated, O_2 is not vibrationally excited.
(22). Adams, N.G. and Smith, D. **1985**, *Chem. Phys. Letters 117*, 67.
(23). Viggiano, A.A.; Morris, R.A.; Paulson, J.F.; Henchman, M.J.; Miller, T.M. and Miller, A.E.S. **1990**, *J. Phys.Chem.* in press.
(24). Lohr, L.L. and Boehm, R.C. **1987** *J. Phys. Chem. 91*, 3203.
(25). Tichy, M.; Javahery, G.; Twiddy, N. and Ferguson, E.E. **1989**, *Int. J. Mass Spectrom. Ion Processes 93*, 165.
(26). Ferguson, E.E. **1986** *J. Phys. Chem. 90*, 731.
(27). Tichy, M.; Twiddy, N.; Wareing, D.P.; Smith, D. and Adams, N.G. **1987**, *Int. J. Mass Spectrom. Ion Processes 81*, 235.
(28). Adams, N.G.; Smith, D.; Tichy, M.; Javahery, G.; Twiddy, N. and Ferguson, E.E. **1989**, *J. Chem. Phys. 91*, 4037.
(29). Tichy, M.; Javahery, G.; Twiddy, N. and Ferguson, E.E.: **1990** in *Symposium on Atomic and Surface Physics* (Märk, T. D. and Howorka, F., Eds.) Studia Studentenförderungs Ges. m.b.H: Innsbruck, p.267.
(30). Lindinger, W.; Dotan, I.; Albritton, D.L. and Fehsenfeld, F.C. **1978**, *J. Chem. Phys. 68*, 2607.
(31). Schultz, R.H. and Armentrout, P.B., work in progress.
(32). Gioumousis, G. and Stevenson, D.P. **1958**, *J. Chem. Phys. 29*, 294.
(33). Mahan, B.H. **1975**, *J. Chem. Ed. 52*, 299. Levine, R.D. and Bernstein, R.B. **1972**, *J. Chem. Phys. 56*, 281. Ervin, K.M. and Armentrout, P.B. **1984**, *J. Chem. Phys. 80*, 2978.
(34). LeRoy, R.L. **1969**, *J. Phys. Chem. 73*, 4338. Menzinger, M. and Wolfgang, R.L. **1969**, *Angew. Chem. Int. Ed. 8*, 438.
(35). Ervin, K.M. and Armentrout, P.B. **1987**, *J. Chem. Phys. 86*, 2659.

ION THERMOCHEMISTRY: SUMMARY OF PANEL DISCUSSION

JOHN E. BARTMESS
Dept. of Chemistry
Univ. of Tennessee
Knoxville TN, USA

John E. Bartmess, Dept. of Chemistry, Univ. of Tennessee, Knoxville TN, USA, presented information regarding the current status of the Gas-phase Ion Thermochemistry Data compilations, available through the National Institute for Standards and Technology, Gaithersburg MD, USA. There is both a hardcopy book (1988) and an computerized version (1989) available. The computerized version will be kept current with yearly or bi-yearly updates.

T.B. McMahon, Univ. of Waterloo, Waterloo, Ontario, Canada. "A Pulsed Electron Beam, Variable Temperature High Pressure Mass Spectrometric Study of Proton Transfer Between Two Bases A and B."

For the equilibrium $AH^+ + B \rightleftarrows BH^+ + A$, the data in the Table were obtained, via a van't Hoff plot.

A	B	ΔH	ΔS	ΔG^{600}	ΔG^{300}
i-C_4H_8	Me_2CO	-2.1	-0.5	-1.8	-2.0
i-C_4H_8	$MeCO_2Me$	-3.4	-1.3	-2.6	-3.0
i-C_4H_8	$MeCO_2Et$	-6.6	-1.7	-5.6	-6.1
i-C_4H_8	Et_2CO	-7.7	-2.9	-6.0	-6.8
i-C_4H_8	NH_3	-12.0	-7.3	-7.7	-9.9
Me_2CO	Et_2O	-5.1	-2.0	-3.9	-4.5
Me_2CO	Et_2CO	-5.4	-1.8	-4.3	-4.9
Et_2O	Et_2CO	-1.7	-2.1	-0.4	-1.1
Et_2CO	NH_3	-4.3	-4.9	-1.4	-2.8

Units Used: ΔH, kcal $mole^{-1}$; ΔG, kcal $mole^{-1}$; ΔS, cal $mole^{-1}$ K^{-1}

The free energies agree reasonably well with literature values for these basicities, but the experimental entropies reported here are considerably different from those derived from symmetry considerations alone. This places ammonia 12.0 kcal/mol larger in proton affinity than 2-methylpropene, in contrast to 8.1 kcal/mol for the current proton affinity scale. Dr. Sharon Lias of NIST (Gaithersburg MD USA) commented that Dr. Michael Meot-ner of NIST has observed comparable results.

K. R. Jennings (ed.), Fundamentals of Gas Phase Ion Chemistry, 281–287.

John L. Holmes, Univ. of Ottawa, Ottawa, Ontario, Canada.

Regarding additivity schemes for estimating the thermochemical properties, there is a need for an update of the Benson group additivity values; Dr. Steve Stein at NIST is working on such a scheme. For ions, there are several incorrect values in the 1988 "Gas-phase Ion and Neutral Thermochemistry" compilation. For example, the adiabatic ionization energy of oxirane in the compilation is ca. 0.4 eV too large. Values for other cycloalkanes and cyclic ethers also may be incorrect.

The use of RCH_2NH_2 compounds to determine appearance energies for $R^{\cdot}$ and R^+ appears to be useful. For R = cyclohepta-2,4,6-trienyl, metastable kinetic energy release indicates that there is no reverse activation energy for this process.

Renée A.L. Peerboom, Ivo L. Freriks, Leo J. de Koning, Nico M.M. Nibbering; Institute of Mass Spectrometry, University of Amsterdam, The Netherlands. "Carbanion Stabilization by Adjacent Cyanide and Isocyanide Substituents".

The goal of this project is determine how carbanions are stabilized by substituents in the gas phase. Possible effects include π/σ stabilization, dipole stabilization, and polarizability. The compounds used were substituted cycloalkanes, C_3 to C_7, with the substituents -C≡N and -N≡C, as well as the parent cycloalkanes. The variation in ring size results in some control regarding the hybridization state of the carbanion (sp^2 vs. sp^3). Gas phase acidities were determined, to give a measure of the relative carbanion stabilities. It is assumed that $\delta\Delta H_{acid} \approx \delta EA(A^{\cdot})$.

Based on literature results concerning these types of carbanions, it is expected that the cyano group should favor a planar carbanion, with charge extensively delocalized to nitrogen. The isonitrile should result in a pyramidal carbanion.

For the parent cycloalkanes, the ΔH_{acid} determination was by the kinetic method of DePuy et al:

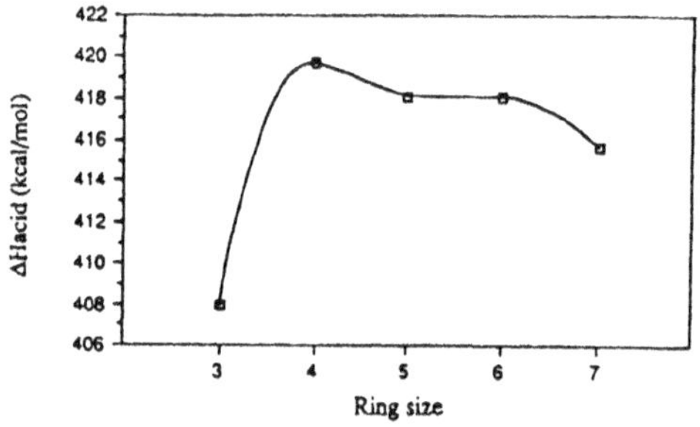

For the cycloalkylcyanides and cycloalkylisocyanides, acidity determination was by equilibrium experiments:

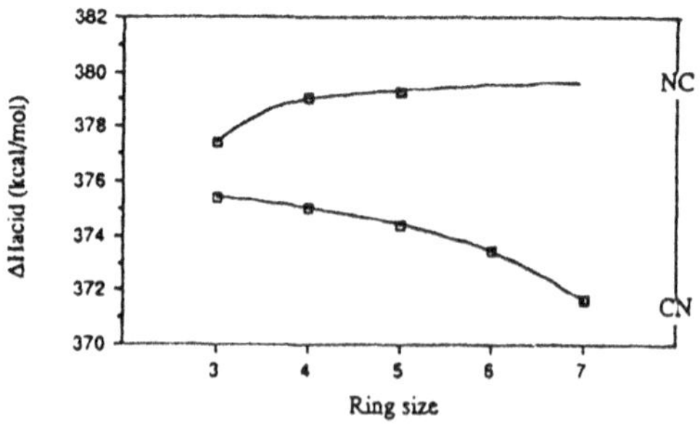

The conclusions of this study are that for the parent cycloalkanes, the three-membered ring has a large % s-character dominant, while for larger rings, there is little influence of polarizability. For the cycloalkylcyanides, acidity becomes stronger with increasing ring size. This implies that the cyano group acts predominantly as a π-acceptor. For the cycloalkylisocyanides, acidity becomes weaker as the ring size increases. This implies that the isocyano group acts predominantly as σ-acceptor.

G. Javehery, M. Tichy, J. Glosik, N.D. Twiddy. Physics Department, U.C.W. Aberystwyth, Wales, U.K.; and E.E. Ferguson, Laboratoire de Physico-Chemie des Rayonnement, Universite de Paris-Sud, Batiment 350, 91405 Orsay, France. "The application of a Non-Thermal Selected Ion Flow Drift Tube (SIFDT) to (i) The determination of proton affinity differences, and (ii) A study of the isomeric forms of N_2OH^+."

The forward and reverse rate constants for 9 proton transfer reactions have been measured as a function of average centre of mass kinetic energy, KE_{cm} using a selected ion flow drift tube (SIFDT). In all but two cases the plots of log(rate constant) against $1/KE_{cm}$ (Arrhenius plot) were linear. As a result the dependence of log(equilibrium constant K), with $1/KE_{cm}$ (van't Hoff plot) were linear, from which values of the enthalpy and entropy changes were obtained from slope and intercept respectively. Because ΔH is a measure of the difference between the proton affinities of two neutral species, the data can be used to provide a proton affinity difference ladder. This ladder agrees extremely well with the established proton affinity scale. The experimental values of entropy change agree with values calculated from the entropies of the individual ions and neutrals. The agreements of the ΔH's and ΔS's so determined establishes the validity, and utility, of a SIFDT apparatus for proton affinity studies, reversing the previously held view which was based on FDT measurements involving N_2OH^+. Measurement of dependence of the rate constant for the reaction of protonated N_2O with CH_4 as a function of KE_{cm} reveals two species of different reactivity. These two species established to be the isomers $HNNO^+$ and $NNOH^+$, the latter found to be more energetic than the more stable $NNOH^+$, compatible with theoretical calculations. Reaction with CH_4 allows a determination of the proton affinity of both forms of protonated N_2O. From equilibrium constant measurements we have determined the proton affinities for HBr(139.4), N_2O protonated on O (137.2), HCl (133.4), N_2O protonated on N (131.7), CH_4 (130.3), CO_2 (128.1), NO (126.9), and CF_4 (126.4) kcal/mol, utilizing the precise PA(CO) = 141.4 kcal/mol, obtained from a recent improved value of ionization energy and heat of formation, as a reference point.

P.B. Armentrout, E.R. Fisher, and D.E. Clemmer, Department of Chemistry, University of Utah, Salt Lake City, UT 84112, USA. "Thermodynamic Data from Endothermic Reactions Thresholds: methane, HOO·, and NO_2."

While much of the thermochemical data available for ion chemistry comes from equilibrium measurements and spectroscopic data, there is now a wide variety of molecules for which thermodynamic data can be obtained from measurements of endothermic reaction thresholds. Recently, we have used our guided ion beam apparatus[1] to use this method to determine the adiabatic ionization energies (IE) of two problematical molecules.

The first of these is methane, which undergoes a Jahn-Teller distortion upon ionization, thereby leading to unfavorable Franck-Condon factors. In a recently published study,[2] the endothermic charge transfer reaction, $O_2^{+\cdot}(^2P_g, v = 0) + CH_4 \rightarrow CH_4^{+\cdot} + O_2$, was used to measure $IE(CH_4) = 12.54 \pm 0.07$ eV. Also, the heat of formation of the peroxide radical, $\Delta H^o_{f,298}(HOO\cdot) = 3.8 \pm 1.2$ kcal/mol, was measured from the hydride transfer reaction of the same reactants.

In a second study[3], $IE(NO_2)$ has been measured in several reactions. Here, the NO_2 neutral molecule is strongly bent while the cation is linear, again leading to a very unfavorable Franck-Condon overlap near the adiabatic IE. Endothermic charge transfer between NO_2 and Zn^+ or NO^+, and between $NO_2^{+\cdot}$ and $C_6H_5CF_3$, CH_3I, or C_3H_6 leads to an average ionization energy of 9.51 ± 0.06

eV. This is in good agreement with a recent spectroscopic value of 9.586 ± 0.002,[4] the lowest value presently in the literature.

1. Ervin, K.M; Armentrout, P.B. J. Chem. Phys. 1985, 83, 166.
2. Fisher, E.R.; Armentrout, P.B.J. Phys. Chem. 1990, 94, 4396.
3. Clemmer, D.E.; Armentrout, P.B., in preparation.
4. Haber, K.S.; Zwanziger, J.W.; Campos, F.X.; Wiedman, R.T.; Grant, E.R. Chem. Phys. Lett. 1988, 144, 58.

Joel F. Liebman (with Sharon O. Yee and Carol A. Deakyne), Department of Chemistry, University of Maryland - Baltimore County; "Normalizing/Formalizing the Ion/Neutral Thermochemistry of Fluorinated Species."

It has long been known that fluorine substitution results in major changes in the energetics of organic molecules. For example, tetrafluoro- ethylene has a considerably weaker carbon-carbon double bond than ethylene; indeed, it is weaker than the carbon-carbon single bond in ethane or hexafluoroethane. This study addresses itself to this anomaly and related effects in other hydrocarbon and fluorocarbon systems. Following J.P. Simons (Nature 205, 1308 (1965)) the bond strength in C_2F_4 was "corrected" by twice the singlet-triplet energy difference of CF_2, $\Delta E_{ST}(CF_2)$, so that the ethylene and carbon correlate much as they do for C_2H_4 and CH_2. As a result, it is found that the C=C bond strengths of C_2F_4 and C_2H_4 are nearly equal. Likewise, it is found that $C_2F_4^{+\cdot}$ has a much weaker C=C bond than $C_2H_4^{+\cdot}$: $D(C_2F_4) < D(C_2H_4^{+\cdot}) << D(C_2H_4^{+\cdot})$. Before explaining this in terms of resonance structures, geometry changes, bond orders, and hyperconjugation, the bond strength in $C_2F_4^{+\cdot}$ again is corrected by $\Delta E_{ST}(CF_2)$ to correspond to $C_2H_4^{+\cdot}$, and find that the new bond energy for the two radical cations are nearly equal. Our understanding also applies to classical ethyl cations and the two dissociation modes $CX_3^+ + :CY_2$ and $CX_3^\cdot + CY_2^{+\cdot}$ (X, Y = H and/or F) and to both acetylenes and their radical cations, by careful examination of the neutral and cationic methylidyne fragments. Finally we note that bond strength order $C_2F_6 > C_2H_6 > C_2H_6^+ > C_2F_6^+$ which appears quite sensible, except for the fact that if one uses current IPs and ΔH_f values, $D(C_2F_6^+)$ is less than zero!

Charles H. DePuy, Dept. of Chemistry, University of Colorado, Boulder CO, "Bond Dissociation Energies for $CH_2{=}CH_2$, HC≡CH, $CH_2{=}CH^\cdot$ and $HC_2^\cdot$".

The title bond dissociation energies were obtained by combining the gas phase acidities of the corresponding conjugate bases (H_2CCH^-, HCC^-, $CH_2{=}C^-$ and C_2^-) with the electron affinities of the corresponding radicals obtained by Carl Lineberger's and Barney Ellison's groups at Boulder. The following results, in kcal/mol, were obtained:

	ΔH_{acid}	BDE(C-H)
HC≡CH	376.4 ± 0.6	131.3 ± 0.7
$CH_2{=}CH_2$	408.0 ± 0.6	109.7 ± 0.8
$CH_2{=}CH^\cdot$	383.3 ± 3.4	81.0 ± 3.5
HC≡C·	>346.9 ± 3.0	>108.6 ± 3.0

These values now make it possible to calculate the energy required to break all the bonds beginning with ethylene and ending with carbon and hydrogen.

In the process of redetermining the gas phase acidity of acetylene we were able to use HF as a primary standard acid to increase the accuracy by which its acidity and those of the simple alcohols are known. For example, we found, for CH_3OH and C_2H_5OH, respectively, ΔG_{acid} at 298 = 375.1 ± 0.6 and 377.0 ± 0.6 kcal/mol. Earlier measurements have been ±2.0 kcal/mol.

Michael Henchman; Department of Chemistry, Brandeis University, Waltham MA 02254-9110, "Thermodynamic Conventions: Stationary or Thermal Electron?"

Many judge thermodynamics to be the most boring part of chemistry and thermodynamic conventions to be the most boring part of thermodynamics. Yet, like death and taxes, thermodynamic conventions cannot be avoided; and they raise questions which are surprisingly interesting.

We are greatly indebted to Sharon Lias and to the late Henry Rosenstock for continuing thermodynamic compilations, which are invaluable for ongoing research, and for raising and clarifying the matter of the two thermodynamic conventions. This was discussed at the NATO Study Institute at La Baule[1] and again in the introduction to the 1988 NIST Thermodynamic Compilation.[2] The two conventions are presented as arbitrary conventions to be chosen at will - - the thermal electron convention, adopted by the thermodynamicists, or the stationary electron convention, adopted by ion physicists and chemists. According to the second convention, heats of formation of positive ions at 298 K are 6.197 kJ/mol lower, and those of negative ions correspondingly higher, than the values according to the first convention. Standard enthalpy changes for ionic reactions

$$AB^{+/-} + CD \rightleftarrows AD + BC^{+/-} \tag{1}$$

must be independent of the choice of convention and are readily derived from data, provided the data are based on a common convention. Which convention is better? The question, in terms of the above discussion, seems irrelevant.

The question is not irrelevant where the electron is a reactant or product in (1). (Examples include dissociative attachment and associative detachment.) Consider reaction (2) as one such example

$$e^- + HI \rightarrow H + I^- \qquad \Delta H = +3 \pm 1 \text{ kJ/mol} \tag{2}$$

where I have used the data for (2) and (3) in the NIST compilation[2] (stationary electron) to derive vH from

$$H \rightarrow H^+ + e^- \qquad \Delta H = +1312 \text{ kJ/mol} \tag{3}$$
$$HI \rightarrow H^+ + I^- \qquad \Delta H = +1315 \text{ kJ/mol} \tag{4}$$

In contrast, the reaction enthalpy of (2), according to the thermal electron convention, is shown to be -3 kJ/mol - to be compared with the value of +3 kJ/mol by the stationary electron convention, the difference being the thermal energy of the electron, 6 kJ/mol.

Experimental measurements of ΔH for (2) therefore suggest a means of choosing between the two conventions. Measurements, as a function of temperature, of dissociative attachment and associative attachment involving hydrogen atoms, are far from straightforward but have been achieved by Smith and Adams.[3] Elsewhere I show how the experimental data are consistent with the thermal electron convention and inconsistent with the stationary electron convention.[4]

What conclusions are to be drawn? The limited data available suggest that the choice of a convention is not arbitrary but point to the thermal electron convention. In practice, what should be done? Nothing would be achieved by changing the heats of formation of all the positive and negative ions in the NIST thermodynamic compilation because in standard applications, for example to (1), the tabulated values can be used directly. Only where the electron appears in the equation must its thermal energy be included. In practice, only one change is required in the NIST compilation, The first entry in the Table 2. Negative Ion Table should therefore read:

e^-: $\Delta H_{acid}(H) = 1318$ kJ/mol; $\Delta G_{acid}(H) = 1311$ kJ/mol replacing the respective values of 1312 and 1308 given in the present compilation. The value for $\Delta G_{acid}(H)$ = 1311 kJ/mol [reaction (3)] is derived from ΔG = -17 kJ/mol for the reaction (2) and $\Delta G_{acid}(H)$ = 1294 kJ/mol [reaction (4)]. The value for (2) comes from the experimental equilibrium constant K(300) = 1.0 ± 0.3 x 10^3 ;[3] and the value for (4) comes from the NIST compilation.[2]

The self consistency of these values is readily checked. An alternative value for $\Delta G_{acid}(H)$ is 1313 kJ/mol, derived from $\Delta H_{acid}(H)$ = 1318 kJ/mol and values of S^o_{298} for H(g), H^+(g) and e^-(g) from the JANAF Tables.[5] The difference of 2 kJ/mol lies within the accuracy of the ΔG and ΔH values of ±1 kJ/mol.

This discussion was made possible by the pioneering experiments of David Smith and Nigel Adams. I thank them for ongoing collaboration and John Bartmess for helpful discussion.

References:

1. S. G. Lias *NATO ASI Ser., Ser. B*, 1979, 40, 223. H. M. Rosenstock ibid., pp. 246-249.
2. S. G. Lias, J. E. Bartmess, J. F. Liebman, J. L. Holmes, R. D. Levin and W. G. Mallard 1988 *J. Phys. Chem. Ref. Data*, 1988, 17, Suppl. 1, pp. 8-10.
3. D. Smith and N. G. Adams 1987 J. Phys. B: At. Mol. Phys., 1987, 20, 4903.
4. M. Henchman, "Symposium on Atomic and Surface Physics '90," T. D. Märk, F. Howorka, Eds., Studia Studentenförderungs G.m.b.H, Innsbruck, 1990, p. 127.
5. M. W. Chase, Jr., C. A. Davies, J. R. Downey, Jr., D. J. Frurip, R. A. McDonald, and A. N. Syverud, *J. Phys. Chem. Ref. Data*, 1985, 14 Suppl. 1, pp. 1010, 1211-2.

Fu-Shiuan Huang and Robert C. Dunbar, Department of Chemistry, Case- Western Reserve University, Cleveland OH 44106. "Rates of Dissociation and Kinetic Shift in Methylnaphthalene Ions."

Various threshold measurements have placed the appearance energy of the $C_{11}H_9^+$ ion from the radical cation of 1-methylnaphthalene, via benzylic hydrogen atom loss, at around 4 eV. The apparently analogous dissociation of the toluene radical cation to the benzyl cation and a hydrogen atom has a threshold below 2 eV. This difference seems unreasonably large.

In fact, because of its large size and tight transition state, methylnaphthalene ion has a major kinetic shift which invalidates even very careful appearance threshold measurements. To determine the true activation energy reliably, it is necessary to measure the dissociation rate as a function of ion internal energy, and extrapolate the resulting rate-energy curve to the energy at which the dissociation rate goes to zero.

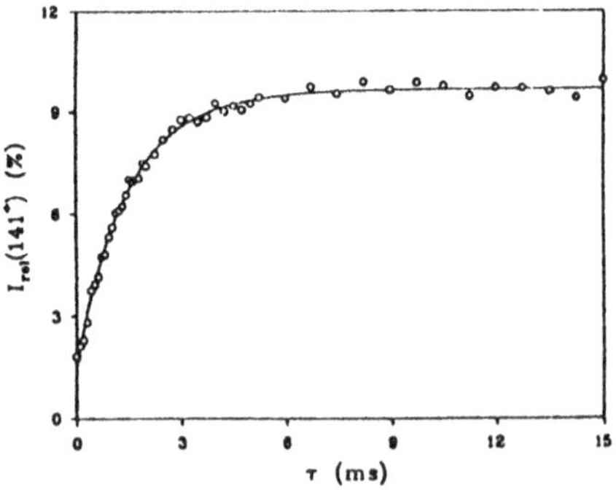

Figure 1

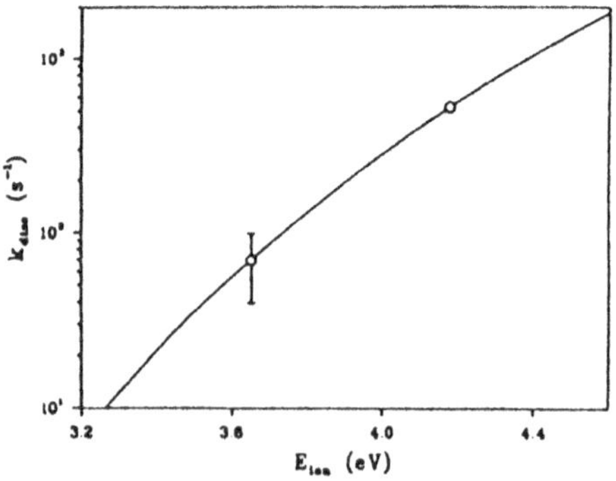

Figure 2

We have been able to measure two points on the rate-energy curve by time-resolved ion cyclotron resonance photodissociation[1], using wavelengths of 355 and 308 nm. Figure 1 shows the time resolved dissociation at 308 nm. The rate-energy plot is shown in Figure 2. The large error bar on the 355 nm point arise from the uncertainty in correcting for the competition between dissociation and radiative relaxation of the photoexcited ion. This radiative relaxation process, which has a rate constant on the order of 100 s^{-1} in this case, in the principal source of difficulty in the interpretation of dissociation rates in the slow regime below 10^3 s^{-1}. The solid line in Figure 2 is the RRKM fit to these rate-energy points, using an activation energy of 2.25 ± 0.2 eV, and a ΔS_{1000K} of -15 cal $K^{-1}mol^{-1}$. The relatively tight transition state suggests a rate-limiting rearrangement preceding dissociation, perhaps to the benzotropylium product structure. E_0 provides a new upper limit of 207 kcal/mol on the heat of formation of $C_{11}H_9{}^+$.

1. R.C. Dunbar, J. Chem. Phys. 1987, 91, 2801.

COLLISIONAL STUDIES OF GASEOUS NEGATIVE IONS

Alex. G. Harrison
Department of Chemistry
University of Toronto
Toronto, Ontario M5S 1A1
Canada

Abstract

The use of variable low energy collision-induced dissociation to probe the energy dependence of the fragmentation of gaseous anions and, thus, to probe the potential energy surfaces involved, is illustrated using $R(CH_3)_2CO^-$ ions as an example. The fragmentation of proton-bound cluster ions $[RO\text{--}H\text{--}OR']^-$ is discussed and it is shown that the relative intensities of RO^- and $R'O^-$ can be related to the relative gas phase acidities of the alcohols ROH and R'OH. The use of both high energy and low energy collisional methods in confirming the structures of gaseous anions and in differentiating isomeric neutral structures is illustrated. It is shown that, for small anions at least, low energy collisional activation provides more certain structural information than high energy collisional activation.

Introduction

The field of negative ion mass spectrometry has undergone considerable development over the past 15 years, particularly in the areas of negative ion chemical ionization [1-4] and negative ion fast atom bombardment [5,6]. These "soft" ionization techniques usually yield ions characteristic of the molecular weight of the compound of interest but often do not produce fragment ions which are necessary to derive structural information. Collision-induced dissociation (CID) of polyatomic ions has proven [7,8] to be a powerful method of providing this structural information through formation of fragment ions from the "quasi-molecular" ions formed in the initial ionization process. In addition to the analytical uses, collisional activation studies can provide much fundamental information

K. R. Jennings (ed.), Fundamentals of Gas Phase Ion Chemistry, 289–313.

concerning the chemistry of gaseous ions [8.9].

Such collisional activation studies can be carried out by examining ion/neutral collisions in the keV collision energy range using either B/E linked scans in conventional double-focussing mass spectrometers [10] or electric sector scans in reversed-geometry instruments [11] to detect the product ions. At the other extreme, the ion/neutral collisions may occur at energies in the range 5-100 eV using quadrupole collision cells and quadrupole mass analyzers [12,13], multi-Fourier transform mass spectrometric techniques [14,15] or ion-trap mass spectrometers [16]. The present paper reviews collisional studies of negative ions carried out at the University of Toronto over the past five years using a hybrid BEqQ mass spectrometer [17]. This configuration permits studies both at keV collision energies using the BE portion of the instrument and at eV collision energies using the r.f.-only quadrupole collision cell q and the quadrupole mass analyzer Q. The first part of the review will explore the utility of variable low energy collisional activation (the ERMS technique [18,19]) to explore the potential energy surfaces for the fragmentation of gaseous anions and will compare the results obtained with the results obtained by multi-photon photodissociation experiments and by kinetic experiments. The second part will explore the use of collision-induced (and unimolecular) dissociation of proton-bound complexes, [RO--H--OR']$^-$, to establish gas-phase acidities of alcohols. The final part of the review illustrates the use of CID studies in ion and neutral structure elucidation and compares the utility of low energy and high energy collisional activation in such studies.

Potential Energy Surfaces for the Fragmentation of Alkoxide Ions

It is well-established [6,20-26] that the fragmentation of

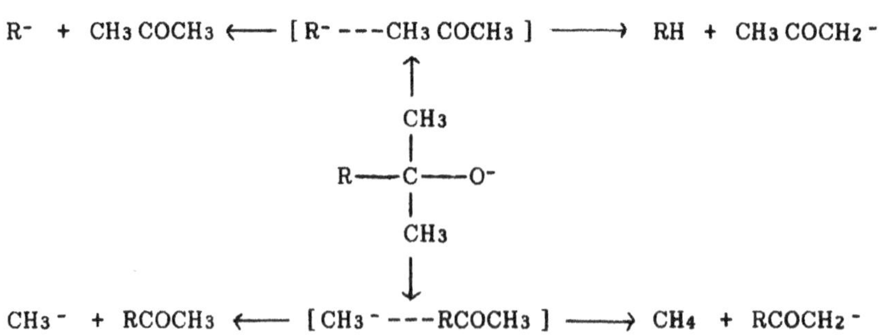

Scheme 1

alkoxide ions occurs by initial formation of an ion/dipole complex, $[R^- {-}{-}{-} CH_3COCH_3]$, as illustrated in Scheme 1 for $R(CH_3)_2CO^-$ ions. This complex may either fall apart to R^- + CH_3COCH_3 or undergo an internal proton transfer reaction to produce RH + $CH_3COCH_2^-$. An alternative fragmentation channel involves formation of a $[CH_3^- {-}{-}{-} RCOCH_3]$ complex which would be expected to yield primarily CH_4 + $RCOCH_2^-$ after an internal proton transfer reaction. Brauman and co-workers [20-23] have examined, in detail, the IR multi-photon photodissociation of a number of $R(CH_3)_2CO^-$ ions and, with one exception, have shown that the single reaction observed involves elimination of RH. In particular they observed no elimination of CH_4. The one exception occurred for $R=CF_3$ where a low yield of CF_3^-, increasing with laser fluence, was observed. Elimination of RH is thermochemically favoured over formation of R^- + CH_3COCH_3 and, for the case of $R=CF_3$, they concluded that there must be a substantial internal energy barrier to proton transfer in the ion/dipole complex.

The photodissociation experiments sample only the lowest energy fragmentation channel since, once the threshold for fragmentation is reached, the rate of fragmentation becomes larger than the rate of photon absorption. By contrast, single-collision CID experiments provide the opportunity of imparting greater energy to the ion and, thus, of exploring higher energy reaction channels. By varying the collision energy it is possible to probe more fully the potential energy surfaces involved in fragmentation of the ion. In addition, reaction of R^- with CH_3COCH_3 proceeds initially to form the same $[R^- {-}{-}{-} CH_3COCH_3]$ ion/dipole complex which may either revert to reactants or proceed to formation of then products RH + $CH_3COCH_2^-$. It is well-established [27-31] that if there is a significant barrier to proton transfer in the ion/dipole complex the rate of the overall proton transfer reaction will be lower than the collision rate. In effect, the transition state for proton transfer is a "tight" transition state whereas reversion to reactants proceeds through a "loose" transition state. In terms of statistical rate theory, the transition state for proton transfer has a lower density of states than the transition state for reversion to reactants. Thus, the measurement of the rate constant for the proton transfer reaction

$$R^- + CH_3COCH_3 \longrightarrow RH + CH_3COCH_2^- \qquad (1)$$

provides a probe of the potential energy surface, since, the higher the internal barrier, the slower will be the reaction rate. In the same vein, multi-photon photodissociation or collisional induced dissociation of $R(CH3)_2CO^-$ should probe the same surface with the implication that the higher the barrier for proton transfer in the ion/dipole complex the greater should be the yield of R^- in the dissociation spectrum.

To compare these methods of probing the potential energy surface we have studied [32] three $R(CH_3)_2CO^-$ systems ($R=CF_3$, CH_2CH and C_6H_5) by variable, low energy collisional activation as well as determining the rate constants for reaction (1) using a flowing afterglow-selected ion flow tube [33]; these three systems have previously been studied by IR multi-photon photodissociation methods by Brauman and co-workers [21,23].

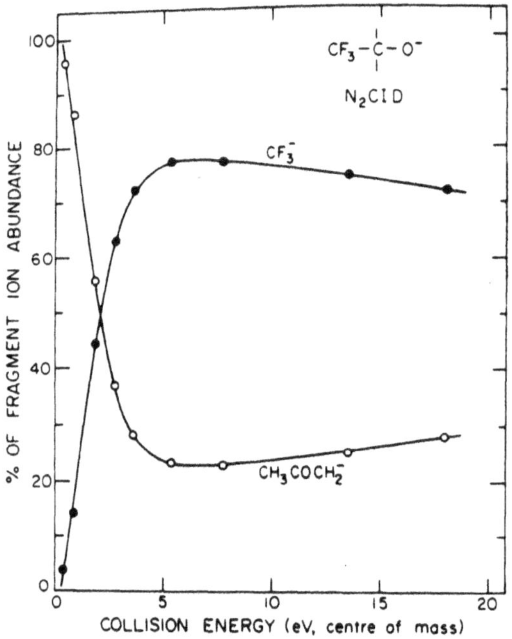

Figure 1 Breakdown graph for $CF_3(CH_3)_2CO^-$, N_2 collision gas. Reproduced, with permission, from Int. J. Mass Spectrom. Ion Processes, 95, 55 (1989).

Figure 1 shows the breakdown graph (% fragment ion abundance versus centre-of-mass collision energy) for the $CF_3(CH_3)_2CO^-$ ion. Clearly, formation of $CH_3COCH_2^-$, by elimination of CF_3H, is the dominant low energy fragmentation channel, in agreement with the photodissociation results [21,23]. However, with increasing collision energy, formation of CF_3^- increases rapidly in importance and accounts for 70-80% of the total ion signal for centre-of-mass collision energies greater than about 5 eV. Brauman and co-workers [21,23] have observed both

$CH_3COCH_2^-$ and CF_3^- in the multi-photon photodissociation experiments; the yield of CF_3^- increased with increasing laser fluence although the fractional yield never exceeded 0.3. The fact that CF_3^- is observed in appreciable yield in both the CID and the photodissociation experiments indicates that there is a substantial barrier for proton abstraction in the [CF_3^----CH_3COCH_3] complex with the result that the more endothermic reaction channel, simple bond cleavage to form CF_3^- + CH_3COCH_3, becomes competitive with proton abstraction. In agreement with these observations, measurement of the rate constant for the proton abstraction from acetone by CF_3^- yielded [32] a value of 0.43×10^{-9} cm^3 $molecule^{-1}$ s^{-1}, approximately 20% of the calculated collision rate constant. All three approaches are in agreement that there is a substantial barrier for proton transfer in the ion/dipole complex.

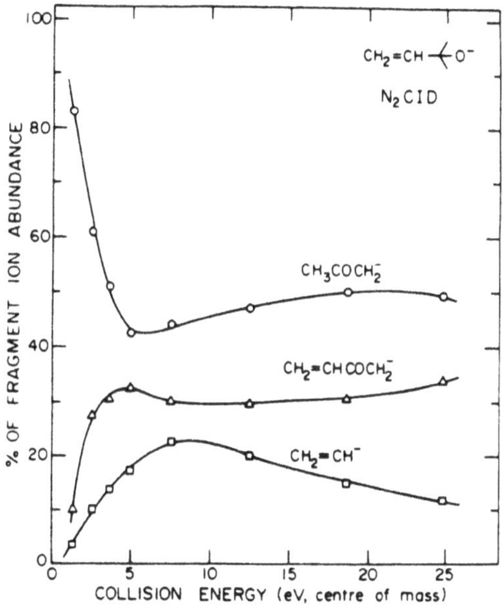

Figure 2. Breakdown graph for $CH_2CH(CH_3)_2CO^-$, N_2 collision gas. Reproduced, with permission from Int. J. Mass Spectrom. Ion Processes, 95, 55 (1989).

Figure 2 shows the breakdown graph for fragmentation of the $CH_2CH(CH_3)_2CO^-$ ion. In contrast to the $CF_3(CH_3)_2CO^-$ system, the yield of R^- (CH_2CH^-) never exceeds the yield of $CH_3COCH_2^-$, indicating a relatively low barrier for proton abstraction in the [CH_2CH^----CH_3COCH_3] ion/dipole complex.

This observation is in agreement with the photodissociation experiments in which only formation of $CH_3COCH_2^-$ was observed [21,23] and with the kinetic measurements [32] which showed that CH_2CH^- abstracted a proton from acetone with 96% efficiency. An interesting observation in the CID of this system is the formation of $CH_2CHCOCH_2^-$ by elimination of CH_4 at higher centre-of-mass collision energies. This channel, which is, thermochemically, the most favourable, was not observed in the multi-photon photodissociation experiments. Thermochemical calculations [32] indicate that the [CH_2CH^----CH_3COCH_3] complex and the [CH_3^----$CH_2CHCOCH_3$] complex should be similar in energy, so fragmentation by elimination of CH_4 is not surprising at higher internal energies. No formation of CH_3^- is observed, indicating a facile internal proton transfer reaction in the latter ion/dipole complex. In contrast to the R=CH_2CH system, in the R=CF_3 system thermochemical calculations [32] indicate that the [CH_3^----CH_3COCF_3] complex is much higher in energy than the [CF_3^----CH_3COCH_3] complex, with the result that elimination of CH_4 from $CF_3(CH_3)_2CO^-$ is not observed.

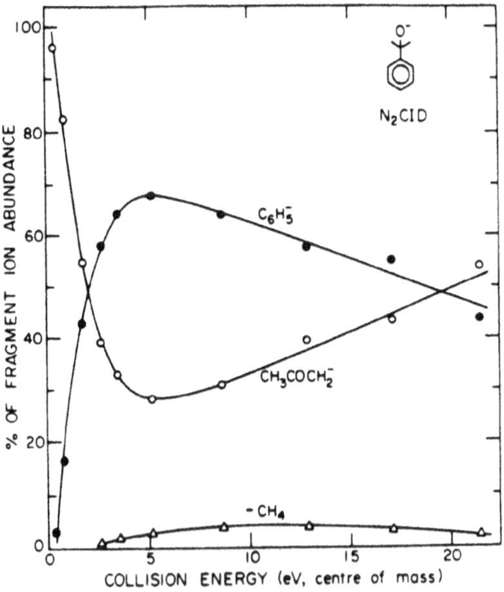

Figure 3. Breakdown graph for $C_6H_5(CH_3)_2CO^-$, N_2 collision gas. Reproduced, with permission, from Int. J. Mass Spectrom. Ion Processes, 95, 55 (1989).

Figure 3 shows the breakdown graph obtained for the $C_6H_5(CH_3)_2CO^-$ system. Clearly, benzene elimination to form $CH_3COCH_2^-$ is the lowest energy fragmentation reaction. However, the yield of $C_6H_5^-$ increases rapidly with increasing collision energy, in a fashion similar to the CF_3^- yield in Figure 1. This behaviour suggests that there is a significant internal barrier for proton transfer in the $[C_6H_5^- \text{---} CH_3COCH_3]$ complex. However, Brauman and co-workers [21] did not observe formation of $C_6H_5^-$ in the multi-photon photodissociation experiments, suggesting a relatively low internal barrier. Our measurements [32] of the rate constant for proton abstraction from acetone by the phenyl anion yielded a rate constant of 1.8×10^{-9} cm^3 molecule $^{-1}$ s^{-1}, about 86% of the collision rate. This reaction is about 32 kcal $mole^{-1}$ exothermic and in the absence of an internal barrier for proton transfer it is surprising that it is a slow as it is. It appears probable that the kinetic and CID probes of the potential energy surface are more sensitive to the presence of such barriers than the multi-photon photodissociation method which probes only the lowest energy reaction and that there is a barrier for proton abstraction in the phenyl anion/acetone ion/dipole complex, although this barrier probably is smaller than that in the CF_3^-/acetone complex.

Gas Phase Acidities from Dissociation Studies

There is a great interest in determining the gas phase acidity of a compound since, from this quantity, one can make many predictions about the reactivity of the compound and its conjugate base. Many gas phase acidities have been measured by measurement of equilibrium constants for the bimolecular proton transfer reaction

$$B'^- + BH \rightleftharpoons B'H + B^- \qquad (2)$$

either directly, by measurement of equilibrium concentrations, or by determining the rate constants for the forward and reverse reactions [34,35]. Accurate measurements of the equilibrium constant may be complicated by a slow approach to equilibrium (when equilibrium concentrations are measured) or by slow forward and/or reverse reactions when rate constants are measured. In addition, impurities present in the reagents and side-reactions of the species of interest also may inhibit accurate determination of equilibrium constants. Further, the equilibrium approach is applicable only for pure species with an appreciable vapour pressure.

Consequently, an alternative approach to gas phase thermodynamic properties involving studies of the competitive fragmentation reactions of weakly bound clusters has been developed. This was applied initially to the determination of gas phase proton affinities [36] but has since been used for the determination of relative gas phase acidities [37,38] as well as the thermochemical properties of other classes of compounds [39-41]. Using gaseous alkoxide ions as an example, the basis of this kinetic method involves the formation of a loosely bound complex of two anions. Such proton-bound cluster ions fragment by the reaction channels

$$[RO{-}{-}H{-}{-}OR']^- \rightarrow RO^- + R'OH \quad (3)$$

$$[RO{-}{-}H{-}{-}OR']^- \rightarrow R'O^- + ROH \quad (4)$$

If RO^- and $R'O^-$ are similar species the two reactions should have similar frequency factors. In addition, the fragmentation reactions should have very small reverse critical energies with the result that the relative rates of the two reactions are determined by the relative critical reaction energies, the difference in which is equivalent to the difference in the gas phase acidities of the two alcohols ROH and R'OH. Since the relative abundances of RO^- and $R'O^-$ are determined by the relative rate constants one can derive

$$\ln(RO^-/R'O^-) = \ln(k_3/k_4) = \Delta E/RT = \delta\Delta G_{ac}/RT \approx \delta\Delta H_{ac}/RT \quad (5)$$

Absolute differences cannot be measured by this method since the effective temperature is not known. However if $R'O^-$ is kept constant and $G_{ac}(R'OH)$ or $H_{ac}(R'OH)$ is known one can derive

$$\ln(RO^-/R'O^-) = m\Delta G_{ac}(ROH) + b \quad (6)$$

$$\ln(RO^-/R'O^-) = m\Delta H_{ac}(ROH) + b' \quad (7)$$

where the slopes m and m' are equal to 1/RT and the intercepts b and b' are equal respectively to $\Delta G_{ac}(R'OH)/RT$ and $\Delta H_{ac}(R'OH)/RT$. By studying a series of complexes where the thermodynamic properties of both ROH and R'OH are known the straight line relationships can be established enabling one to determine ΔG_{ac} and ΔH_{ac} for unknown alcohols.

In the present study [42] we have examined in the r.f.-only quadrupole both the unimolecular metastable ion fragmentation and the 50 eV (lab scale) collision-induced fragmentation of $[RO{-}{-}H{-}{-}OR']^-$ clusters. The clusters were prepared by ionizing a mixture of two alcohols or by reacting the appropriate alkoxide ion with the appropriate formate ester according to the reaction [43]

$$R'O^- + HCOOR \longrightarrow [R'O\text{--}H\text{--}OR]^- + CO \quad (8)$$

More intense cluster ion signals were obtained by the latter method. By suitable overlapping measurements all alkoxide ion ratios were referenced to i-PrO⁻. Figure 4 shows a plot of ln(RO⁻/i-PrO⁻) versus ΔH_{ac} obtained from the 50 eV data.

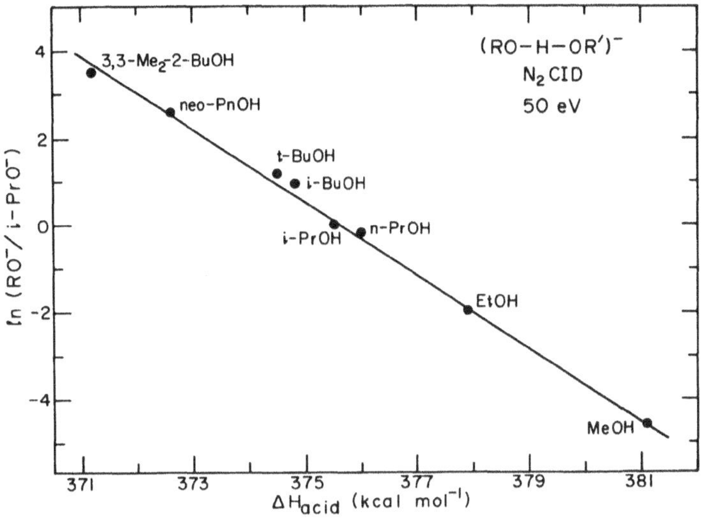

Figure 4. The plot of ln(RO⁻/i-PrO⁻) versus ΔH_{ac} for the 50 eV CID data

A good straight line is obtained leading to the relationship

$$\ln(RO^-/i\text{-}PrO^-) = -0.832_7 \Delta H_{ac}(ROH) + 312.8_1 \quad (9)$$

The study of the unimolecular fragmentation of the cluster ions was limited to the alcohols with acidities between n-PrOH and 3,3-dimethyl-2-butanol and yielded

$$\ln(RO^-/i\text{-}PrO^-) = -1.94_8 \Delta H_{ac}(ROH) + 731.7_3 \quad (10)$$

The greater slope and intercept of equation (10) relative to equation (9) reflects the lower effective temperature of the metastable ions relative to those activated in the 50 eV collisions. A greater sensitivity to differences in acidity is found in the metastable ion studies but this is counterbalanced by larger uncertainties in measuring the weaker ion signals.

Table 1 presents a selection of these results and compares

the values obtained with literature values [35]; the agreement is generally within 0.5 kcal $mole^{-1}$ for these relatively short chain alcohols. Of particular interest are the results for C_2D_5OH and CD_3OH presented in the Table, which show that the deutero alcohols are less acidic than there protio counterparts by approximately 0.6 kcal $mole^{-1}$. DeFrees et al. [44] have derived ΔG^o = -0.50 kcal $mole^{-1}$ for the reaction

$$CD_3O^- + CH_3OH \longrightarrow CH_3O^- + CD_3OH \qquad (11)$$

TABLE 1. ΔH_{ac} (kcal $mole^{-1}$) for representative alcohols

Alcohol	this work	literature [35]
2-Me-3-pentanol	371.6	371.4
3-Me-3-pentanol	372.0	371.9
2-Me-2-pentanol	372.6	372.1
3-Me-2-butanol	372.7	373.1
3,3-Me_2-1-butanol	373.1	372.6
t-Pentanol	373.2	373.1
i-Butanol	374.5	374.8
n-Butanol	375.5	375.2
n-Propanol	376.2	376.0
Ethanol	378.0	377.8
Ethanol-d_5	378.6	
Methanol	381.2	381.1
Methanol-d_3	381.8	

while Grabowski et al. [45] have reported a ΔG^o of -0.47 kcal $mole^{-1}$; our acidity difference is in reasonable agreement with these values. Engelking et al. [46] have reported $EA(CH_3O) - EA(CD3O)$ = 0.42 kcal $mole^{-1}$ while Ellison et al. [47] have reported $EA(C_2H_5O) - EA(C_2D_5O)$ = 0.55 kcal $mole^{-1}$. Since

$$\Delta H_{ac} = D(RO{-}{-}H) + IE(H) - EA(RO) \qquad (12)$$

it is apparent that the difference in the acidities of the deutero and protio alcohols arises mainly from the differences in the electron affinities of the alkoxide radicals.

While the present study appears to provide reliable values for gas phase acidities of branched alcohols containing up to about seven carbons, there appear to be some difficulties in using this approach to determine the acidities of long chain n-alkanols. Figure 5 compares the results we have obtained for n-alkanols in metastable ion and CID studies in the quadrupole cell with the results obtained by Boand et al. [37] examining metastable ion fragmentation reactions by the MIKES technique. Up to n-butanol the results are in reasonable agreement, but with increasing chain length there

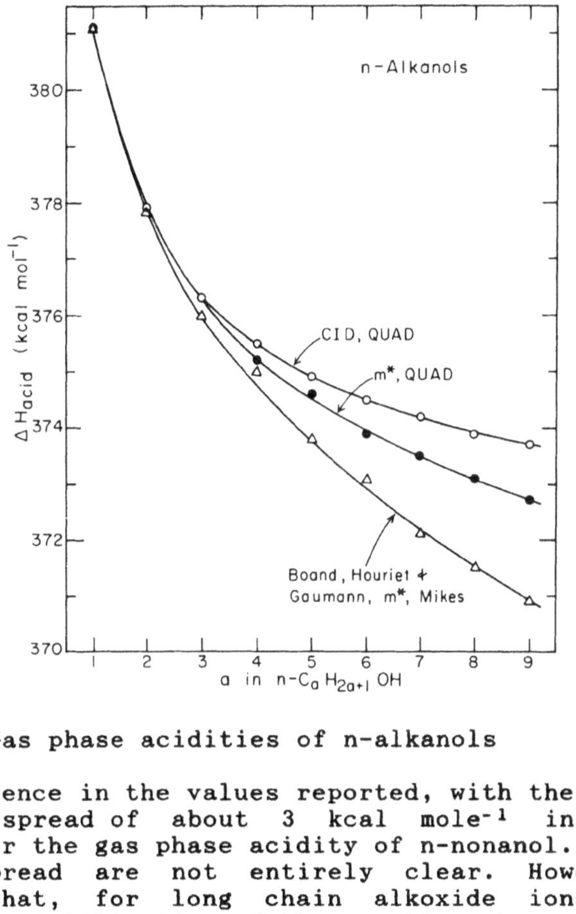

Figure 5. Gas phase acidities of n-alkanols

is a divergence in the values reported, with the result that there is a spread of about 3 kcal $mole^{-1}$ in the values reported for the gas phase acidity of n-nonanol. The origins of this spread are not entirely clear. However, it is probable that, for long chain alkoxide ions, there is substantial stabilization of the negative charge by through-space interactions with polarizable alkyl groups in a coiled conformation. It appears that the kinetic method used in the studies represented in Figure 5 probes this stabilization to different extents in the various approaches. Clearly, there is a need for equilibrium studies of the gas phase acidities of longer chain n-alkanols.

Collision-Induced Dissociation in Structure Elucidation

A major use of collisional activation studies is to elucidate the structure of gaseous ions or, indirectly, the neutral species from which the ion was derived [7,8]. In this section we will illustrate the former in addressing the structure of an unexpected product ion observed in the $O^{-\cdot}$ CI of 2-hexanone. Finally, a comparison will be made between high energy and low energy collisional activation of gaseous anions in providing structural information and, thus, in distinguishing among isomeric neutral species.

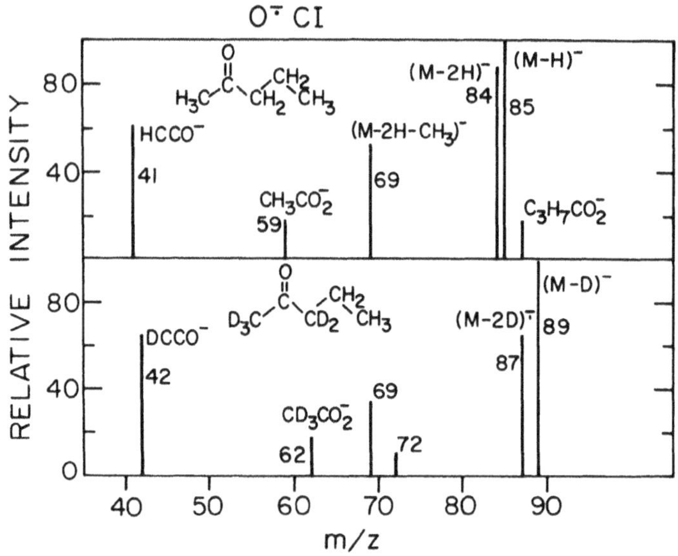

Figure 6. The $O^{-\cdot}$ CI mass spectra of 2-pentanone and 2-pentanone-1,1,1,3,3-d_5

The oxide radical anion, $O^{-\cdot}$, is unusual among negative ion chemical ionization reagent ions in that not only is it a Bronsted base but also it is a radical. Thus, it reacts with organic substrates not only by proton abstraction, typical of Bronsted bases, but also by H-atom abstraction, by $H_2^{+\cdot}$-abstraction and by H-atom and alkyl group displacement [48-54]. The reactions of $O^{-\cdot}$ with carbonyl compounds [50,52] illustrate the majority of these reactions, viz,

$$O^{-\cdot} + RCH_2COCH_2R' \longrightarrow OH^- + [M-H]^\cdot \qquad (13)$$

$$\longrightarrow [M-H]^- + OH\cdot \qquad (14)$$

$$\longrightarrow H_2O + RC^{-\cdot}COCH_2R' \qquad (15)$$
$$\downarrow$$
$$RCCO^- + R'CH_2\cdot$$

$$\longrightarrow H_2O + RCH_2COC^{-\cdot}R' \qquad (16)$$
$$\downarrow$$
$$R'CCO^- + RCH_2\cdot$$

$$\longrightarrow RCH_2COO^- + R'CH_2\cdot \qquad (17)$$

$$\longrightarrow R'CH_2COO^- + RCH_2\cdot \qquad (18)$$

In the earlier study [50,52] it was concluded that both hydrogens in the $H_2^{+\cdot}$-abstraction were removed from the same alpha carbon as illustrated in reactions (15) and (16). However, we have observed [55], as illustrated in Figure 6, that the m/z 69 fragment in the $O^{-\cdot}$ CI of 2-pentanone, originating, nominally, by the reaction sequence (15) (R'=H), shifts in part to m/z 72 in the $O^{-\cdot}$ CI mass spectrum of 2-pentanone-1,1,1,3,3-d_5. The product at m/z 72 can most readily be rationalized by reaction sequence (19), involving

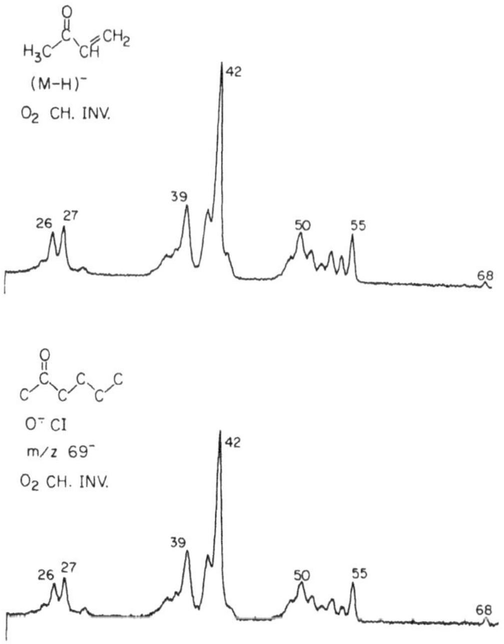

Figure 7. Charge inversion mass spectra of m/z 69 ions from 2-hexanone and methyl vinyl ketone

abstraction of one H/D from each alpha carbon. A similar reaction sequence, (20), appears to be responsible for formation of a product ion at m/z 69 in the $O^{-\cdot}$ CI mass spectrum of 2-hexanone. To provide evidence that the un-

$$O^{-\cdot} + CD_3COCD_2CH_2CH_3 \longrightarrow D_2O + {}^-CD_2COCD^{\cdot}CH_2CH_3 \quad (19)$$
$$\downarrow$$
$${}^-CD_2COCDCH_2 + CH_3^{\cdot}$$

$$O^{-\cdot} + CH_3COCH_2C_2H_5 \longrightarrow H_2O + {}^-CH_2COCH^{\cdot}CH_2C_2H_5 \quad (20)$$
$$\downarrow$$
$${}^-CH_2COCHCH_2 + C_2H_5^{\cdot}$$

expected product at m/z 69 in the 2-hexanone spectrum is, in fact, the enolate ion of methyl vinyl ketone, we have

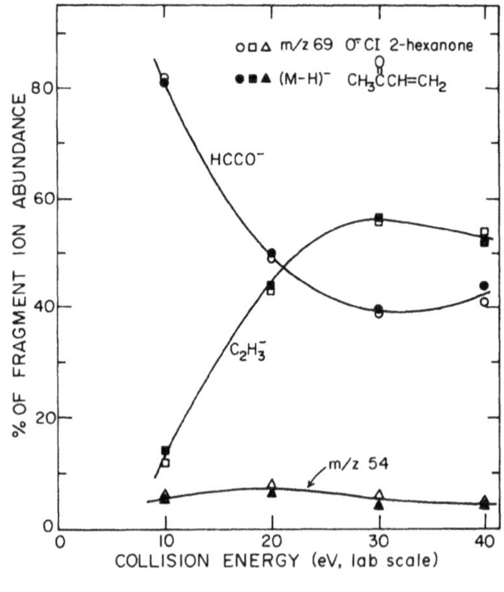

Figure 8 Breakdown graphs for the m/z 69 ions derived from 2-hexanone and methyl vinyl ketone

compared the charge inversion mass spectrum [56] of the m/z 69 ion from 2-hexanone with the charge inversion mass spectrum of the $[M-H]^-$ ion derived by reaction of OH^- with methyl vinyl ketone. The excellent agreement of the two spectra (Figure 7) provides strong support for the reaction sequence (20). Further supporting evidence comes from a study of the low energy CID of the two ions as a function of collision energy (Figure 8). The excellent agreement of the two breakdown graphs leaves little doubt that the two ions have same structure and that 1,3-$H_2^{+\cdot}$-abstraction does occur. It should be noted that Dawson et al. [57] reported, some years ago, that $O^{-\cdot}$ reacted with acetone in part by 1,3-$H_2^{+\cdot}$-abstraction. However, the present work is the first that has shown that 1,3-$H_2^{+\cdot}$-abstraction is a general reaction and that has elucidated the fragmentation reactions that occur following such an abstraction reaction.

The example presented above is one illustration of the use of collisional mass spectrometry to provide confirmation of the structure of the anion. A major use of collisional mass spectrometry of anions is to provide structural information about neutral species and, in particular, to distinguish among isomeric neutral molecules through collisional studies

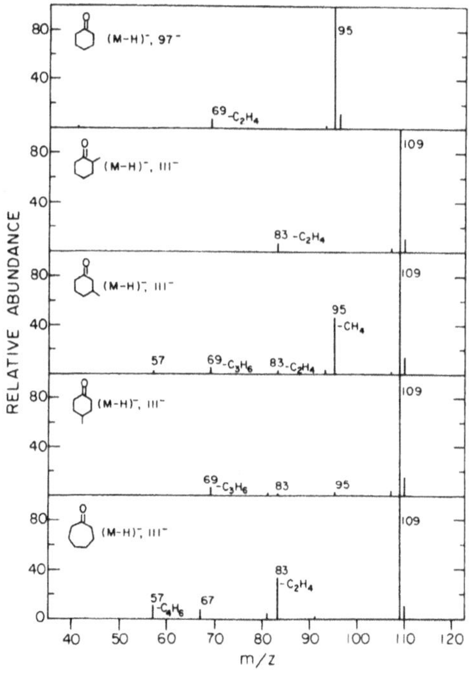

Figure 9. 8 keV CID mass spectra of cycloalkanone enolates using He collision gas. Reproduced, with permission, from J. Am. Chem. Soc.,110, 7345 (1989).

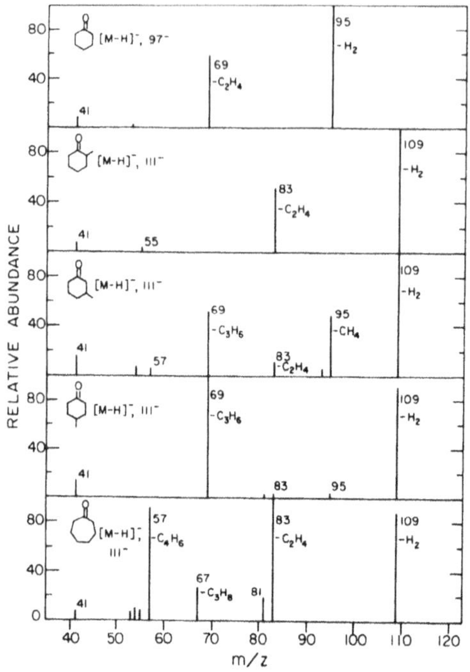

Figure 10. 50 eV CID mass spectra of cycloalkanone enolates obtained using N_2 collision gas. Reproduced, with permission, from J. Am. Chem. Soc., 110, 7345 (1989)

of representative ions. For example, Raftery and Bowie [58] have shown that the isomeric methylcyclohexanones can be distinguished by CID studies of the enolate ions derived therefrom. Figure 9 shows the 8 keV CID mass spectra of the cycloalkanone enolates obtained on our instrument [59]. In agreement with Raftery and Bowie, we observe the major fragmentation to be loss of H_2 for all enolate ions. Isomer identification arises from consideration of the relatively weak signals resulting from olefin elimination from the enolate ions. As Figure 10 shows, the 50 eV CID mass spectra show much more intense ion signals for these olefin-loss processes making isomer identification much easier. We have observed that this is a common phenomenon for relatively small anions at least. Figures 11 and 12 compare the 8 keV and 50 eV CID mass spectra of $C_5H_9O^-$ ions derived from isomeric $C_5H_{10}O$ alcohols. In all cases identification of the alcohol is possible from the CID mass spectrum, although, in several cases, the intensities of the distinctive ions are greater in the 50 eV CID mass spectra. Thus, for example, 3-penten-2-ol and 1-cyclopropylethanol are much more readily distinguished from each other by their 50 eV CID mass spectra than by their 8 keV CID mass spectra.

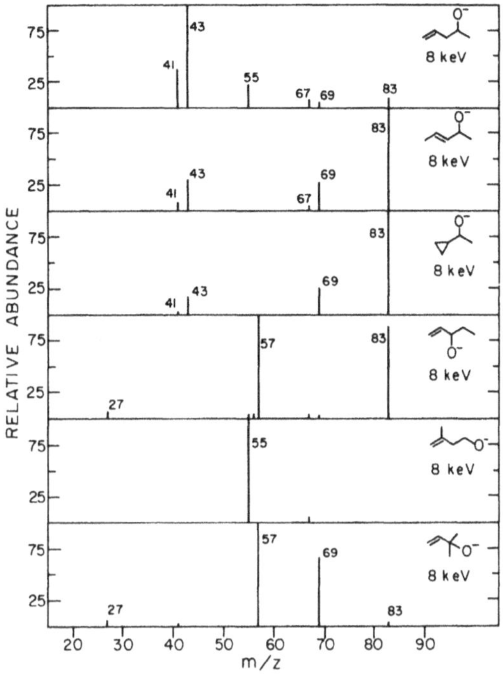

Figure 11. 8 keV CID mass spectra of isomeric $C_5H_9O^-$ ions. He collision gas.

The differences in the low energy and high energy CID mass spectra of anions appear to arise from differences in the mean internal energies of the fragmenting ions after activation by 8 keV or 50 eV collisions. Figure 13 shows the breakdown graph obtained for the enolate ion of 4-methylcyclohexanone. Clearly, elimination of H_2 is the lowest energy fragmentation reaction with the elimination of alkene increasing in importance with increasing centre-of-mass collision energy. The 50 eV CID data of Figure 10 correspond to 10 eV centre-of-mass collision energy. It is apparent that the 8 keV CID data of Figure 9 fit the breakdown graph best for a centre-of-mass collision energy of approximately 2-3 eV. Accepting, as has been clearly shown in positive ion studies [60-65], that the mean internal energy of the activated ions increases with increasing collision energy, it is clear that the ions fragmenting to yield identifiable products have a lower mean internal energy following activation by 8 keV collisions than they do after 50 eV collisions. Similar conclusions are derived from a comparison of the results of Figures 11 and 12 with the breakdown graphs obtained for the $C_5H_9O^-$ ions from the low energy CID mass spectra.

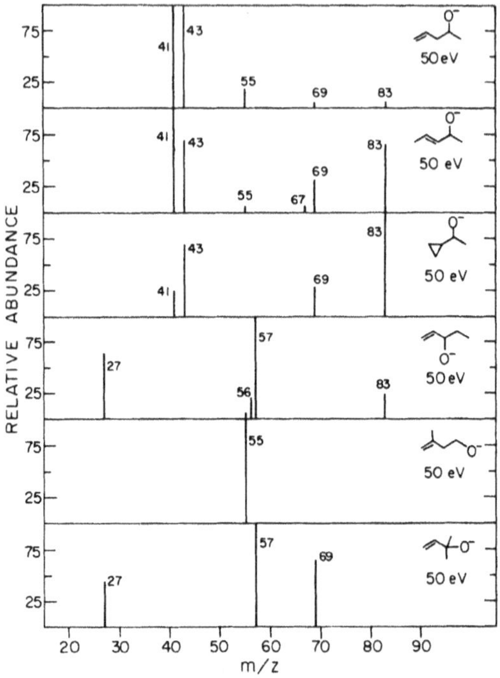

Figure 12. 50 eV CID mass spectra of isomeric $C_5H_9O^-$ ions. N_2 collision gas

It is likely that this difference in mean internal energy arises from difference in the mechanisms of excitation in low and high energy collisions. Low energy collisional activation is considered [66,67] to involve translational to vibrational energy transfer. On the other hand, high energy collisional activation is thought [68] normally to involve electronic excitation. Since few negative ions have stable excited electronic states, electronic excitation will lead primarily to electron detachment. It appears that vibrational excitation of anions is inefficient in high energy collisional activation and that considerably more energy can be deposited in vibrational modes in low energy collisional activation.

Conclusions

The survey presented has shown that a large amount of information can be obtained from collisional studies of negative ions. Energy-resolved CID studies can be used to probe the energy dependence of fragmentation reactions and,

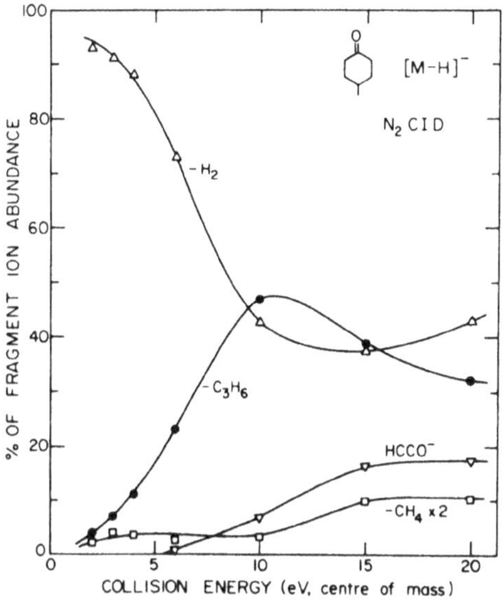

Figure 13. Breakdown graph for the enolate of 4-methylcyclohexanone. Reproduced, with permission, from J. Am. Chem. Soc., 110, 7345 (1989)

thus, to provide information on the potential energy surfaces for ion fragmentation. Systematic studies of the fragmentation of proton-bound cluster ions also can be used to derive relative gas phase acidities, which can be put on an absolute basis by the use of suitable standards. In addition, collisional studies of anions readily provides information concerning ion structures and neutrals from which the ions were derived.

Acknowledgements

I am indebted to the Natural Sciences and Engineering Research Council of Canada for financial support and to my co-workers, listed in the references, for their diligent work. The rate constant measurements reported herein were made during the course of a sabbatical leave at the University of Colorado and I am grateful to Professor C.H. DePuy for his hospitality.

References

1. Jennings,K.R. (1977) 'Negative chemical ionization mass spectrometry', in R.A.W. Johnstone (Ed.), Mass Spectrometry, Vol. 4, The Chemical Society, London, pp. 203-16.
2. Harrison, A.G. (1983) Chemical Ionization Mass Spectrometry, CRC Press Inc, Boca Raton.
3. Budzikiewicz,H.(1986) 'Negative chemical ionization (NCI) of organic compounds, Mass Spectrom. Rev., 5, 345-80.
4. Harrison, A.G. (1989) 'Chemical ionization mass spectrometry', Adv. Mass Spectrom.,11, 582-96.
5. Bowie, J.H. (1984) 'The formation and fragmentation of negative ions derived from organic molecules', Mass Spectrom. Rev., 3, 161-208.
6. O'Hair, R.A.J. and Bowie, J.H. (1989) 'Organic negative ions: structure, reactivity and mechanisms' in M.E. Rose (Ed.) Mass Spectrometry, Vol. 10, Royal Society of Chemistry, London, pp.145-80.
7. McLafferty, F.W. (1983), (Ed.) Tandem Mass Spectrometry, Wiley, New York.
8. Busch, K.L., Glish, G.W. and McLuckey, S.A. (1988) Mass Spectrometry/Mass Spectrometry. Techniques and Applications of Tandem Mass Spectrometry, VCH Publishers, New York
9. Levsen, K. (1983) 'Structural information from tandem mass spectrometry', in F.W. McLafferty (Ed.) Tandem Mass Spectrometry, Wiley, New York, pp.41-66.
10. Jennings, K.R. (1984) 'Scanning methods for double-focussing mass spectrometers', in M.A. Almoster Ferreira (Ed.), Ionic Processes in the Gas Phase, Reidel, Dordrecht, pp.7-21.
11. Cooks, R.G., Beynon, J.H., Caprioli, R.M. and Lester, G.R. (1973) Metastable Ions, Elsevier, New York
12. Dawson, P.H. and Douglas, D.J. (1983) 'Collisionally activated dissociation of low kinetic ions', in F.W. McLafferty (Ed.), Tandem Mass Spectrometry, Wiley, New York, pp.125-48.
13. McLuckey, S.A. and Cooks, R.G (1983) 'Angle- and energy-resolved fragmentation spectra from tandem mass spectrometry', in F.W. McLafferty (Ed.), Tandem Mass Spectrometry, Wiley, New York, pp.303-320.
14. Cody, R.B. and Freiser, B.S. (1982) 'Collision-induced dissociation in a Fourier-transform mass spectrometer', Int. J. Mass Spectrom. Ion Phys., 41,199-204.
15. Cody, R.B., Burnier, R.C. and Freiser, B.S. (1982) 'Collision-induced dissociation with Fourier-

transform mass spectrometry', Anal. Chem., 54, 96-101.
16. Louris, J.N., Cooks, R.G., Syka, J.E.P., Kelley, P.E., Stafford, G.G. Jr. and Todd, J.F.J. (1987) 'Instrumentation, applications and energy deposition in quadrupole ion trap tandem mass spectrometry', Anal. Chem., 59, 1677-85.
17. Harrison, A.G., Mercer, R.S., Reiner, E.J., Boyd, R.K., March, R.E. and Porter, C.J. (1986) 'A hybrid BEQQ mass spectrometer for studies in gaseous ion chemistry', Int. J. Mass Spectrom. Ion Processes, 74, 13-32.
18. McLuckey, S.A., Glish, G.L. and Cooks, R.G. (1981) 'Kinetic energy effects in mass spectrometry/mass spectrometry using a sector/quadrupole instrument', Int. J. Mass Spectrom. Ion Phys., 39, 219-30.
19. Fetterolf, D.D. and Yost, R.A. (1982) 'Energy-resolved collision-induced dissociation tandem mass spectrometry', Int. J. Mass Spectrom. Ion Phys., 44, 37-50.
20. Tumas, W., Foster, R.F., Pellerite, M.J. and Brauman, J.I. (1983) 'Isotope effects as a mechanistic probe of unimolecular ion decomposition. tert-Butoxide ion', J. Am. Chem. Soc., 105, 7464-65.
21. Tumas, W., Foster, R.F., Pellerite, M.J. and Brauman, J.I. (1984) 'Unimolecular decomposition of gas-phase alkoxide ions', J. Am. Chem. Soc., 106, 4053-54.
22. Tumas, W., Foster, R.F., Pellerite, M.J. and Brauman, J.I. (1987) 'A step-wise mechanism for gas-phase unimolecular ion decomposition. Isotope effects in the fragmentation of tert-butoxide anion', J. Am. Chem. Soc.,109, 961-70.
23. Tumas, W., Foster, R.F. and Brauman, J.I. (1988) 'Mechanistic studies of gas-phase negative ion unimolecular decompositions. Alkoxide ions', J. Am. Chem. Soc., 110, 2714-22.
24. Hayes, R.N., Sheldon, J.C., Bowie, J.H. and Lewis, D.E. (1984) 'Elimination of dihydrogen from collision activated alkoxide negative ions in the gas phase. An ab initio and isotope effect study', J. Chem. Soc. Chem. Commun., 1431-32.
25. Hayes, R.N., Sheldon, J.C., Bowie, J.H. and Lewis, D.E. (1985) 'Elimination of molecular hydrogen and methane from collision-activated alkoxide ions in the gas phase. An ab initio and isotope effect study', Aust. J. Chem., 38, 1197-1208.
26. Sulzle, D. and Schwarz, H. (1989) 'Stereochemical criteria providing unambiguous experimental evidence for the intermediacy of ion/molecule complexes in the generation of enolates from alkoxides', Helv. Chim. Acta, 72, 320-22.

27. Farneth, W.E. and Brauman, J.I. (1976) 'Dynamics of proton transfer involving delocalized negative ions in the gas phase', J. Am. Chem. Soc., 98, 7891-98.
28. Brauman, J.I. (1979) 'Factors influencing thermal ion-molecule rate constants' in P. Ausloos (Ed.), Kinetics of Ion Molecule Reactions, Plenum Press, New York, pp. 153-64.
29. Pellerite, M.J. and Brauman, J.I. (1980) 'Intrinsic barriers in nucleophilic displacements', J. Am. Chem. Soc., 102, 5993-99.
30. DePuy, C.H. and Bierbaum, V.M. (1981) 'Gas-phase reactions of organic anions as studied by the flowing afterglow technique', Accts. Chem. Res., 14, 146-53.
31. Magnera, T.F. and Kebarle, P. (1984) 'Exothermic bimolecular ion-molecule reactions with negative temperature dependence' in M.A. Almoster Ferreira (Ed.) Ionic Processes in the Gas Phase, Reidel, Dordrecht, pp. 135-57.
32. Chowdhury, S.K., Mercer, R.S., Harrison, A.G., Barlow, S.E. and DePuy, C.H. (1989) 'An energy-resolved study of the fragmentation of some 2-substituted 2-propoxide ions and the rates of proton abstraction from acetone' Int. J. Mass Spectrom. Ion Processes, 95, 55-69.
33. Van Doren, J.M., Barlow, S.E., DePuy, C.H. and Bierbaum, V.M. (1987) 'The tandem flowing afterglow-sift-drift', Int. J. Mass Spectrom. Ion Processes, 81, 85-100.
34. Bartmess, J.E. and McIver, R.T. Jr. (1979) 'The gas-phase acidity scale', in M.T. Bowers (Ed.) Gas Phase Ion Chemistry, Vol. 2, Academic Press, New York, pp.87-121.
35. Lias, S.G.,Bartmess, J.E., Liebman, J.F., Holmes, J.L., Levin, R.D. and Mallard, W.G. (1988) Gas Phase Ion and Neutral Thermochemistry, J. Phys. Chem. Ref. Data, 17, Suppl. 1.
36. McLuckey, S.A., Cameron, D. and Cooks, R.G. (1981) 'Proton affinities from dissociations of proton-bound dimers', J. Am. Chem. Soc., 103, 1313-17.
37. Boand, G., Houriet, R. and Gaumann, T. (1983) 'Gas-phase acidity of aliphatic alcohols', J. Am. Chem. Soc., 105, 2203-06.
38. Graul, S.T., Schnute, M.E. and Squires, R.R. (1990) 'Gas-phase acidities of carboxylic acids and alcohols from collision-induced dissociation of dimer cluster ions', Int. J. Mass Spectrom. Ion Processes, 96, 181-98.
39. McLuckey, S.A., Schoen, A.E. and Cooks, R.G. (1982) 'Silver ion affinities of alcohols as ordered by mass spectrometry/ mass spectrometry', J. Am. Chem. Soc., 104, 848-50.

40. Burinsky, D.J., Fukuda, E.K. and Campana, J.E. (1984) 'Electron affinities from dissociation of mixed negative ion dimers', J. Am. Chem. Soc., 106, 2770-71.
41. Bursey, M.M., Hass, J.R., Becker, E.I. and Arison, B.H. (1984) 'A correlation between fragment intensity and critical energy in collisional activation of some ions produced by electron impact', Org. Mass Spectrom., 19, 160-64.
42. Haas, M.J. and Harrison, A.G. (1991) 'Fragmentation of [RO--H--OR']$^-$ cluster ions and the gas-phase acidities of alcohols', Int. J. Mass Spectrom. Ion Processes, submitted; M.J. Haas, 'The dissociation of proton-bound adduct ions and the gas-phase acidities of alcohols', M.Sc. Thesis, University of Toronto, 1990.
43. Faigle, J.F.G., Isolani, P.C. and Riveros, J.M. (1976) 'The gas phase reaction of F^- and OH^- with alkyl formates', J. Am. Chem. Soc., 98, 2049-52.
44. DeFrees, D.J., Bartmess, J.E., Kim, J.K. and McIver, R.T. Jr. (1977) 'In support of anion hyper-conjugation: the versatile methyl group', J. Am. Chem. Soc., 99, 6451-52.
45. Grabowski, J.J., DePuy, C.H., Van Doren, J.M. and Bierbaum, V.M. (1985) 'Gas-phase hydrogen-deuterium exchange reactions of anions: kinetics and detailed mechanisms', J. Am. Chem. Soc., 107, 7384-89.
46. Engelking, P.C., Ellison, G.B. and Lineburger, W.C. (1978) 'Laser photodetachment electron spectrometry of methoxide, deuteromethoxide and thiomethoxide: electron affinities and vibrational structure of CH_3O, CD_3O and CH_3S', J. Chem. Phys., 69, 1826-32.
47. Ellison, G.B., Engelking, P.C. and Lineburger, W.C. (1982) 'Photoelectron spectroscopy of alkoxide and enolate negative ions', J. Phys. Chem., 86, 4873-78.
48. Goode, G.C. and Jennings, K.R. (1974) 'Reaction of $O^{-\cdot}$ with some unsaturated hydrocarbons', Adv. Mass Spectrom., 6, 797-802.
49. Dawson, J.H.J. and Jennings, K.R. (1976) 'Production of gas-phase radical anions by reaction of $O^{-\cdot}$ ions with organic substrates', J. Chem. Soc. Faraday Trans II, 72, 700-706.
50. Harrison, A.G. and Jennings. K.R. (1976) ' Reactions of $O^{-\cdot}$ with carbonyl compounds', J. Chem. Soc. Faraday Trans I, 72, 1601-08.
51. Bruins, A.P., Ferrer-Correia, A.J., Harrison, A.G., Jennings, K.R. and Mitchum, R.K. (1978) 'Negative ion chemical ionization mass spectra of some aromatic compounds using $O^{-\cdot}$ as a reagent ion', Adv. Mass Spectrom., 7A, 355-58.
52. Jennings, K.R. (1978) 'Investigation of selective

reagent ions in chemical ionization mass spectrometry', in M.L. Gross (Ed.) High Performance Mass Spectrometry, Am. Chem. Soc., Washington, pp. 179-87.

53. Harrison, A.G. and Tong, H.-Y. (1988)'Characterization of C_8H_{10} alkyl benzenes by negative ion mass spectrometry', Org. Mass Spectrom. ,23, 135-39.

54. Grabowski, J.J. and Melly, S.J. (1987) 'Formation of carbene radical anions: gas-phase reactions of the atomic oxygen anion with organic molecules', Int. J. Mass Spectrom. Ion Processes, 81, 143-64.

55. Marshall, A., Tkaczyk, M. and Harrison, A.G. (1991) '$O^{-\cdot}$ chemical ionization of carbonyl compounds', J. Am. Soc. Mass Spectrom., in press.

56. Bursey, M.M. (1990) 'Charge inversion of negative ions in tandem instruments', Mass Spectrom. Rev., 9, 555-74.

57. Dawson, J.H.J., Noest, A.J. and Nibbering, N.M.M. (1979) '1,1 and 1,3 elimination of water from the reaction complex of $O^{-\cdot}$ with 1,1,1-trideuteroacetone', Int. J. Mass Spectrom. Ion Phys., 30, 189-91.

58. Raftery, M.J. and Bowie, J.H. (1987) 'Collision-induced dissociation of enolate negative ions. Deprotonated cyclohexanones', Int. J. Mass Specrom. Ion Processes, 79, 267-85.

59. Chowdhury, S. and Harrison, A.G. (1988)'Fragmentation of the enolate ions of some cycloalkanones. A comparison of high-energy and low-energy collisional activation', J. Am. Chem. Soc.,110, 7345-49.

60. Douglas, D.J. (1982) 'Mechanism of the collision-induced dissociation of polyatomic ions studied by triple quadrupole mass spectrometry', J. Phys. Chem., 86, 185-91.

61. Dawson, P.H. and Sun, W.-F. (1982) 'Comparison of low-energy collisionally induced dissociation of n-butylbenzene ions with photodissociation', Int. J. Mass Spectrom. Ion Phys, 44,51-59.

62. Dawson, P.H. (1983) 'A study of the collision-induced dissociation of $C_2H_5OH_2^+$ using various target gases', Int. J. Mass Spectrom. Ion Phys., 50, 287-97.

63. Nacson, S. and Harrison, A.G. (1985) 'Energy transfer in collisional activation. Energy dependence of the fragmentation of n-alkylbenzene molecular ions', Int. J. Mass Spectrom. Ion Processes, 63, 325-37.

64. Zwinselman, J.J., Nacson, S. and Harrison, A.G. (1985) 'A comparison of charge exchange, energy resolved and angle resolved mass spectrometric methods for studying the energy dependence of ionic fragmentation reactions', Int. J. Mass Spectrom. Ion

Processes, 67, 93-108.

65. Wysocki, V.H., Kentammaa, H.I. and Cooks, R.G. (1987) 'Internal energy distributions of isolated ions after activation by various methods', Int. J. Mass Spectrom. Ion Processes, 75, 181-208.

66. Yamaoka, H., Ding, P. and Durup,J. (1969) 'Energetics of the collision-induced dissociations $C_2H_2^+ \longrightarrow C_2H^+$ + H and $C_2H_2^+ \rightarrow H^+ + C_2H$', J. Chem. Phys., 51, 3465-76.

67. Durup, J.(1970), 'Mechanism of collision-induced dissociation of fast ions', in K. Ogata and T. Hayakawa (Eds.), Recent Developments in Mass Spectrometry, University Park Press, Baltimore, pp.921-34.

68. Todd, P.J. and McLafferty, F.W. (1983) 'Collisionally activated dissociation of high kinetic energy ions' in F.W. McLafferty (Ed.), Tandem Mass Spectrometry, Wiley, New York, pp.149-74.

FUNDAMENTALS OF COLLISIONAL ACTIVATION

Summary of a workshop organized by

Alex. G. Harrison
Department of Chemistry
University of Toronto
Toronto, Ontario M5S 1A1
Canada

Collision-induced dissociation has become a well-established method to probe the structure and chemistry of gas-phase ions [1]. A second workshop at this meeting was concerned with the applications of collisional activation in these areas. The present workshop addressed three more fundamental areas, the mechanism of collisional activation, the use of energy-controlled collisional activation to measure threshold energies for ion dissociation and the use of collisional techniques (NRMS) to probe the stability of novel neutral species in the gas phase. The contributions to this workshop are summarized in the following.

THEORY OF IMPULSIVE ENERGY TRANSFER IN COLLISIONAL ACTIVATION OF IONS. Einar Uggerud (University of Oslo) and Peter J. Derrick (University of Warwick)

Knowledge of the mechanism of collisional activation is important for many applications in mass spectrometry. A consideration of the time it takes an ion to traverse the diameter of the collision target (usually a noble gas atom) indicates that vibration energy transfer is important for a range of ion masses and energies. In a series of recent papers [2-4] we have discussed the consequences of an impulsive energy transfer mechanism. Originally, the theory was developed for macromolecular ions with kiloelectronvolt energies, but it should be generally valid for treatment of any impulsive collision between a molecular projectile and a target at rest.

The first stage of our theoretical treatment is conceptually simple, although mathematically somewhat cumbersome. The idea is that the process of impulsive

K. R. Jennings (ed.), Fundamentals of Gas Phase Ion Chemistry, 315–332.

transfer of translational energy into internal degrees of freedom can be isolated to involve only the target gas atom and the atom of the ion which is hit by the target. The derivation involves no other approximations, and leads to several important results.

- The average translational energy lost by the ion, $\langle E \rangle$, is proportional to the initial energy of the ion, E.
- The average energy taken up by the ion, $\langle Q \rangle$, increases linearly with the initial energy of the ion, E.
- There is an approximate inverse relationship between the average energy taken up by the ion, $\langle Q \rangle$, and the mass of the ion, m_{ion}.
- When all other factors are equal, increasing the target atom mass, m_g, leads to a larger $\langle Q \rangle$.
- Scattering increases when m_g increases and when m_{ion} decreases. For heavy targets this leads to considerable ion losses because mass spectrometers in general only detect ions scattered below a certain angle (e.g. 1°).

In order to gain some more insight on the effect of a more detailed molecular model on the collision dynamics the next step in the theoretical development involved computer simulations (trajectory calculations). A simple intra and inter molecular potential was applied. In essence, the results from the computer simulations are qualitatively similar to those of the simple theory, although deviations can be observed at higher collision energies. This and other aspects of the calculations will be investigated further using a more elaborate molecular model.

ROLE OF EXCITED ELECTRONIC STATES IN COLLISION-INDUCED DISSOCIATION OF POLYATOMIC IONS. Anil Shukla, Kuangnan Qian, Stephen Anderson and Jean Futrell (University of Delaware)

Utilizing a unique tandem high resolution mass spectrometer/supersonic jet crossed beam apparatus recently constructed in our laboratory, we have just completed a study of the collision-induced dissociation (CID) reaction dynamics of the molecular ions of acetone, nitromethane and propane. The results for these three examples are quite different, Propane is an archetypical example of RRKM-type ions for which the rapid intramolecular vibrational relaxation of internal energy precedes dissociation of the collisionally-excited ion. A statistical model is the appropriate description for this ion. Our dynamics study of nitromethane and acetone ions provides clear evidence for isolated electronic states playing a major role in the CID of non-RRKM ions. The acetone ion is the most striking example; at low energy (below 10 eV LAB) a long-lived excited state dominates the CID of this species.

Our recent publication [5] on the threshold energy CID of acetone molecular ion with helium

$$CH_3COCH_3^+ + He \longrightarrow CH_3CO^+ + CH_3 + He \quad (1)$$

provided conclusive evidence that a significant population of acetone ions formed by electron impact are in an electronically excited state which survives more than 38 μs prior to dissociation. We reported that low energy collisions of ions in this excited state convert their internal energy (electronic and possibly vibrational) largely into recoil kinetic energy of the acetone ion which then rapidly dissociates into CH_3CO^+ and CH_3 on the ground state potential surface. Fragment ions from the decomposition of superelastically scattered molecular ions are back-scattered in the center-of-mass (CM) frame. Reducing the energy of the ionizing electrons in the ion source to 12 eV completely suppressed the superelastic scattering mechanism, demonstrating the excited state(s) origin of this mechanism. The superelastic scattering mechanism also is suppressed at high collision energies [6] where forward-scattered highly inelastic (endothermic) processes dominate acetone ion CID.

We have now extended our study of acetone ion CID to cover the energy range between threshold and a few eV using both He and Ar collision gases. Our specific objective was to understand details of the transition between the threshold region dominated by the back-scattered exothermic mechanism and the high energy region dominated by the forward-scattered endothermic reaction mechanism. The issue of specific target gas effects was queried by conducting experiments with He and Ar at similar center-of-mass (CM) collision energies.

Figure 1a illustrates schematically both the generation of experimental data and the transformation from laboratory to center-of-mass coordinates. Since our experiment involves a supersonic beam of neutral particles, the neutral beam- shown vertically in the figure- is not at rest as in a collision chamber experiment. Both the ion and the neutral beams are sufficiently narrow and monoenergetic that they can be represented as vectors in Figure 1a. The measured ion beam energies, laboratory intersection angle, and masses allow us to calculate the CM velocity and the CM relative

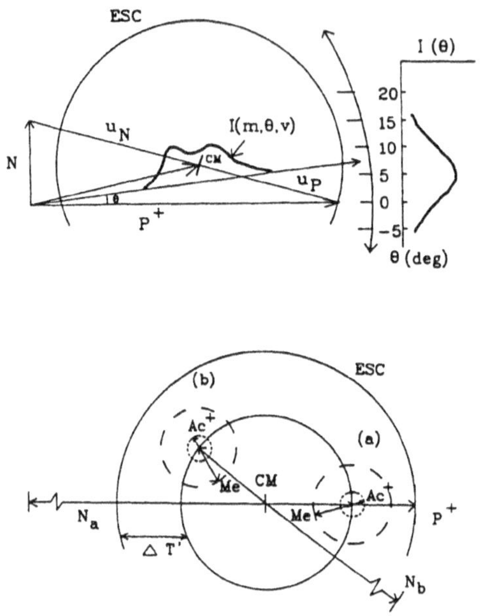

Figure 1 (a) Schematic illustration of the mapping of reactive ion-neutral scattering data into the CM frame. The vectors N and P^+ are the measured laboratory vectors for the supersonic jet molecular beam (N) and primary ion (P^+). The detector mass spectrometer is set to observe the product ion mass at a given laboratory angle θ, the energy distribution and total intensity per unit time are recorded. Energy is converted to velocity and the data are plotted as I(m,θ,v). When referenced to the CM, the resulting plot is the CM contour diagram for the reaction investigated. The CM velocities of the reactant ion, u_P, and neutral collider, u_N, are also shown.

(b) Hypothetical CM scattering diagram illustrating the reduction of the three-body CAD process to a two-body scattering problem. The reactant acetone ion, P^+, and neutral collider, N, collide at the CM, and the post-collision vectors of elastically scattered P^+ (ESC) and dissociation products are shown for two dynamical mechanisms. Excitation of the projectile ion by ΔE through transfer of translational to internal energy exchange, $\Delta T'$, is illustrated for a large- impact-parameter glancing collision (forward-scattered, zero scattering angle) and a strongly impulsive, small-impact-parameter collision (back-scattered, large scattering angle). The excited acetone ion dissociates isotropically with a small kinetic energy release. Momentum conservation is illustrated by the CM vectors for the recoiling collider and methyl radical fragments, neither of which is detected in these experiments.

velocities shown schematically in this figure. The detector (comprising a hemispherical energy analyzer, quadrupole mass filter and pulse-counting multiplier) is pivoted about the intersection point of the ion and neutral beams. At each angle, the kinetic energy of the detected ion is scanned as shown schematically by the $I(m,\theta,v)$ insert. The reactant ion and neutral and product ion energies are converted to velocities, $I(v)$, for a series of laboratory angles. The laboratory velocity of the CM is calculated from the conservation of linear momentum constraint, and product ion velocities with respect to the CM are calculated. Mathematically this involves vector subtraction of the CM velocity from the observed laboratory velocities of the detected product. Figure 1a illustrates schematically the geometric construction for this vector subtraction for a selected scattering angle. The CM velocity vectors of the primary ion, u_P, and the neutral collider, u_N, are shown explicitly to conform to this vector subtraction. The elastic circle (ESC) that we use as a starting point in our discussion of energy transfer is also shown.

Figure 1b illustrates two assumed dynamics models for acetone CID. The baseline of Figure 1b is the hypotenuse of the CM scattering triangle that we have just discussed in Figure 1a. In this postcollision diagram the reactant acetone ion, Ac^+, retreats from the CM towards the right and the neutral atom retreats from the CM towards the left. These two directions in the CM frame define the forward scattering (direction of primary ion beam) and back-scattering (direction of primary neutral beam) hemispheres, respectively. The endpoints of the CM ion and neutral vectors match their laboratory counterparts, as shown in Figure 1a. After the collision, the excited ion and neutral collider recoil from the CM. Model (a) illustrates forward scattering of the excited ion, while model (b) is back-scattered. The neutral atom velocity vectors for models (a) and (b) represent recoil from the CM with momenta equal and opposite to those of the excited acetone molecular ions. The subsequent dissociation of the activated acetone ions into acetyl ions (and methyl radicals) gives us the observable flux of CID products for constructing the scattering diagrams.

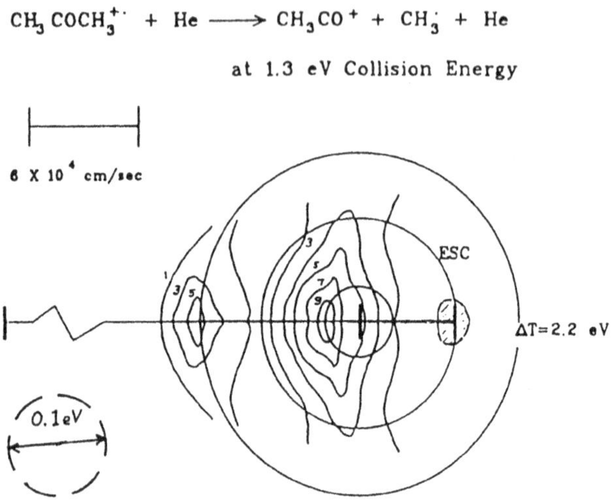

Figure 2 Scattering contour diagram for the CID of acetone ion on collision with He neutrals at 1.3 eV collision energy. The small dashed line oval at the right of the figure shows the full FWHM profile of the primary ion beam and the dashed circle below the figure shows the broadening of the peak that would be observed on the release of 0.1 eV kinetic energy on dissociation.

Before discussing these models further we must confront the unfortunate fact illustrated in Figure 1b that there is a fundamental indeterminacy in discussing the reaction dynamics of CAD processes--namely, that two particles collide and a minimum of three leave the scattering center. Only the product ion velocity vector is measured, and the scattering angles and velocities of the two neutral species can have all values allowed by conservation of energy and angular momentum. Consequently, the scattering diagram does not provide a unique determination of the energetics and angular properties of CID reactions. This dilemma is resolved by assuming that there is a time delay between the excitation process and the decomposition of the excited ion. We think that this assumption is very well founded, especially for polyatomic ions, as the time delay need only be long enough for the neutral collider to escape the potential field of the ion. For typical collisions this will be less than 10^{-13} s. Since the decomposition step is preceded by at least one and typically thousands to millions of vibrational oscillations, we postulate that acetone ion CID is a two-step process in our analysis.

Our two models (a) and (b) assume identical energy depositions and dissociation kinetics. The inner circle is offset from the ESC reference circle by a constant velocity shift corresponding to the translational to internal energy excitation, T'. The zero-scattering angle excitation model (a) corresponds dynamically to a large-impact-parameter glancing collision. The back-scattered mechanism shown as model (b) assumes that a small-impact-parameter impulsive collision is involved in the excitation step. For each mechanism, the centroid for the acetyl ion product velocities can be used to deduce T', which defines the energy deposited in the acetone ion by the collision. The width of the distribution is a measure of the kinetic energy release in the dissociation step.

Figure 2 illustrates the dissociation pattern observed for the acetone molecular ion CID by helium at 1.3 eV collision energy. The data are presented in the CM frame as contour plots identifying lines of equal intensity depicting the relative probability that products are formed with a particular CM velocity and CM scattering angle. The inner circle marked ESC on this diagram is the elastic scattering circle described in Figure 1a. The outer circle, marked T= 2.2 eV, is drawn through the center of the maximum contour of acetyl product ion intensity for a mechanism which corresponds to the exothermic conversion of 2.2 eV of internal energy of the ion into translational energy. The fact that this peak maximum corresponds to a kinetic energy gain by the system of 2.2 eV demonstrates that a large excess of internal energy is converted into translation of the recoiling acetone ion and helium atom before the dissociation occurs. The acetone ion then dissociates to generate acetyl ions with a centroid located on this circle, which is drawn to correspond to a recoil energy of 3.5 eV (2.2 + 1.3) with respect to the CM.

A "normal" endothermic product peak located within the ESC near the CM is also observed in Figure 2. The maxima of both the superelastic (2.2 eV exothermic) and inelastic (1.1 eV endothermic) scattering mechanisms lie directly on the relative velocity vector. Consequently, we may identify these collisions as occurring with near zero impact parameters such that the excited acetone ions and neutral reverse their initial directions and recoil from the collision center. These "hard core" collisions are responsible both for triggering the decay of the long-lived (lifetime greater than 38 μs) electronically excited state acetone molecular ion into acetyl ion and methyl radical and for an inelastic scattering mechanism giving the same products.

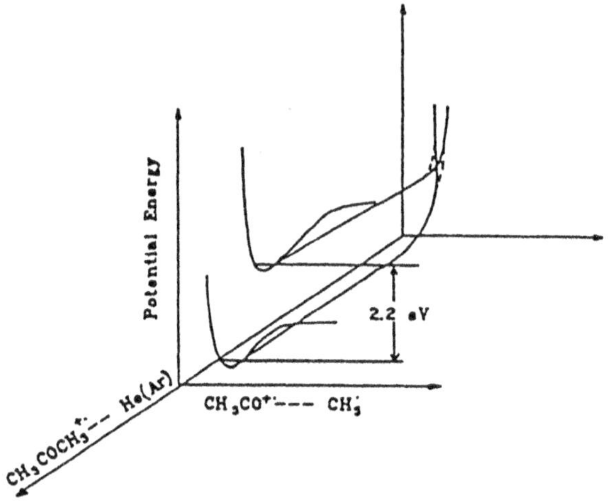

Figure 3 Schematic three-dimensional pseudo-triatomic potential energy curves representing the collision path of an excited acetone ion with the neutral crossing into ground state acetone ion with excess energy released into translation via a curve crossing mechanism

The fact that the scattering contours are relatively tight also demonstrates that the dissociation of the excited acetone into acetyl ion and methyl radical occurs with a relatively small release of kinetic energy in the dissociation step. In fact, the breadth of the distribution for the exothermic channel is only slightly greater than can be accounted for by the velocity distribution of our reactant ion beam. Contours for the ion beam FWHM velocity and angular distribution are shown in Figure 2 as dashed lines; convolution with the supersonic jet neutral broadens this only slightly in defining the CM distribution of reactants. Since about half of the actual distribution for the superelastic mechanism in Figure 2 is accounted for by the distribution of the primary ion and neutral velocities, the combination of two dynamics peak broadening effects-- dispersion in E->T energy exchange and kinetic energy release-- taken together is estimated to be less than 0.1 eV. Clearly energy exchange and angular scattering characteristics are extraordinarily well-defined for this polyatomic ion CID mechanism.

The fundamental reasons for this extraordinary behaviour

are found in the photoelectron spectroscopy of acetone. The adiabatic ionization potential for forming the A state of acetone is located 2.2 eV above the ground state [8]. The vertical ionization is 2.9 eV above the X state. Therefore, the most probable ionization process generates ions with around 0.7 eV of vibrational excitation. This internal vibrational energy is sufficient for C-C bond scission on the ground state surface but apparently not enough on the electronic state surface. Since the electronic transition between the A state and the ground state is optically forbidden, and since internal conversion also is slow, this electronic energy can be stored for a relatively long period of time under collision-free conditions

We infer the following mechanism for CID of excited A-state acetone ions in He and Ar collisions. Collision induces a curve crossing of the ion-neutral potential surface of excited acetone ions with that of the ground state acetone ion as suggested qualitatively in Figure 3. Here the acetone ion is represented as a quasi-diatomic potential curve and the ion-neutral potential is shown on an orthogonal axis, the z-axis in this figure. The excited and ground states are shown as isolated systems which are coupled by collision with the rare gas ion, These collisions lead to de-excitation of the ion to the ground state releasing the excess energy to translational recoil of the acetone ion and neutral with release of the adiabatic energy difference. Vibrational excitation present in the ion is preserved as internal energy in the ground state and provides the energy required to drive the dissociation to acetyl ion and methyl radical. The nearly exact energy balance is responsible for the very narrow peak in Figure 2.

We have further tested this mechanism by examining the microscopic inverse--excitation of ground state ions to the excited state surface. At and just above 2.2 eV collision energy we find a pronounced inelastic scattering peak for stable acetone molecular ions, demonstrating a bound-bound transition (X->A). Figure 3 presents these data for non-dissociative, inelastic scattering of acetone ions at 2.3 eV collision energy; the peak near the CM demonstrates a large cross-section for a transition absorbing 2.2 0.1 eV. At this energy we also find two dissociative channels, one which is endothermic by 2.2 eV and one which is exothermic by 2.2 eV. The endothermic channel represents the excitation of vibrationally excited ground state acetone ions above the dissociation limit of the excited state hypersurface. The exothermic channel is the process discussed above. Clearly, we are able to trace the reaction pathways of Figure 2 in both directions.

In conclusion, our experiments have clearly shown that

at least three distinct mechanisms are involved in the CID of acetone ions: At very low collision energies close to the thermochemical threshold, acetone ion CID is completely dominated by the decomposition of highly-excited, long-lived acetone ions. Collision of an acetone ion with a neutral triggers the release of most of the internal energy into translational energy of the ion and the neutral collider (superelastic process) and the dissociation later takes place from the ground state surface. These near-zero impact parameter collisions result in strong backward scattering of the dissociating ion. As the collision energy is increased, small impact parameter collisions resulting in inelastic backward scattering of ions takes over. At higher energies, from 6 eV up to keV, highly inelastic collisions with forward scattering dominates the CID process. At all energies, intensity maxima are found at non-zero scattering angles, suggesting that small impact parameter, impulsive collisions dominate the CID of acetone ions at all collision energies. These results strongly support the surprising conclusion that acetone CID takes place predominantly by electronic excitation mechanisms at all energies.

Acknowledgement: We gratefully acknowledge the financial support of this work by the National Science Foundation, Grant No. CHE-8796291.

SYSTEMATIC EFFECTS ON CID THRESHOLDS. Richard H. Schultz and P.B. Armentrout (University of Utah)

Collision-induced dissociation(CID) has long been used as a structural probe of polyatomic gas-phase ions [9]. In addition, CID has seen some use as a means of determining the thermochemistry of these ions [10]. CID should be an ideal method of obtaining binding energies in polyatomic ions, since the thermodynamic threshold for the CID process, eq.2, (shown here for a positive ion) is simply

$$ML_x^+ + X \longrightarrow ML_{x-1}^+ + L + X \qquad (2)$$

the binding energy of interest, assuming no reverse activation energy.

This statement obviously only holds true in the case where there are no systematic effects that would obscure the true CID threshold and give inaccurate thermochemistry. We present here results from two systems, N_4^+ and $Fe(CO)_x^+$ (x=1-5), that clearly show that systematic effects can have a major influence on observed CID thresholds. These two systems were chosen because the binding energies have been determined by methods other than CID and thus provide a check of the accuracy of the CID threshold measurements. The net result is that if the systematic effects discussed below are taken into account, then accurate thermochemistry can be obtained, but if not, the observed CID thresholds will not necessarily lead to correct binding energies.

.***Collision partner.*** The choice of collision partner can affect the observed CID threshold in two ways. For a system such as N_4^+, if the ionization energy of the neutral partner is lower than the electron affinity of the ion, then efficient charge transfer can compete with CID. In such cases, e.g. N_4^+ + Xe, the CID threshold is obscured and the CID product rises quite slowly from threshold, making accurate determination of the threshold difficult if not impossible. In cases where charge transfer is more endothermic than CID, the collision partner can still have an effect on the CID threshold. For example, in the case of $Fe(CO)_x^+$ + Rg (Rg = Ar or Xe), while the CID product rises from its thermodynamic threshold in both cases, it rises much more slowly from threshold when Ar is used instead of Xe. Again, this slow rise can make the true threshold difficult to identify. Thus, for metal clusters and ligated metal ions, Xe is the collision partner of choice.

.***Internal energy.*** Clearly, if the ions have an excess of internal energy, then the observed threshold for CID will be lower than the true thermodynamic threshold. Unless great care is taken to make sure that the ions are thermalized, spuriously low binding energies can result. This effect is shown graphically in Figure 4. $Fe(CO)_5^+$ was produced in three ways: by 30 eV electron impact ionization (EI) on $Fe(CO)_5$, by 10 eV EI of $Fe(CO)_5$, and by EI in a flow tube followed by 10^5 thermalizing collisions with He. The latter method obviously is the only one that can yield an accurate binding energy. More significantly, the ions produced by 10 eV EI, only 2 eV above the ion's appearance energy, clearly have a great deal of internal energy. Furthermore, the two EI cross sections are exactly the same to within experimental error. Thus. lowering the EI energy until no change is observed will lead one to the incorrect conclusion that the low-energy EI results represent thermal binding energies.

.***Pressure effects.*** Even under "single collision" conditions, there always is the possibility of multiple collisions in the gas cell. Especially for larger ions with more degrees of freedom, multiple collisions can lead to a spuriously low value for the CID threshold, as multiple collisions below the thermodynamic threshold can lead to CID if the total amount of energy deposited in the ion is greater than the binding energy. For instance, $Fe(CO)_4^+$ shows a marked lowering of the threshold even at single-collision probabilities below 5%. Especially in these cases, an accurate CID threshold can only be obtained by taking data at several gas cell pressures and extrapolating the results to zero pressure.

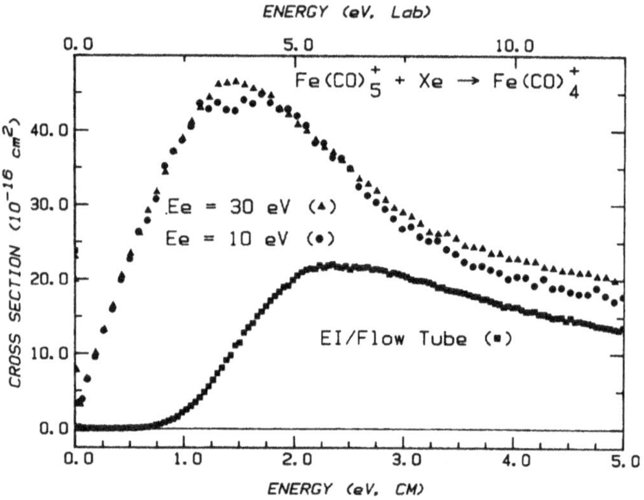

Figure 4 Threshold curves for collision-induced fragmentation of $Fe(CO)_5^+$

DETERMINATION OF BOND DISSOCIATION ENERGIES via ENERGY-RESOLVED COLLISION-INDUCED DISSOCIATION IN A FT-ICR SPECTROMETER. C.E.C.A. Hop, T.B. McMahon and G.D. Willett (University of Waterloo)

The use of FT-ICR for collision-induced dissociation (CID) has been well-established. However, to date, ion beam instruments and triple quadrupoles have been the instruments of choice to determine bond dissociation energies via translational energy controlled collisions between mass-selected ions and stationary target molecules. ICR was not considered for these experiments because it was assumed that r.f. excitation of the ions prior to CID results in a large and/or ill-defined translational energy distribution [11]. In order to test the validity of the latter statement we determined threshold energies for fragmentation with our FT-ICR spectrometer.

All experiments were carried out on a Spectrospin CMS 47 FT-ICR spectrometer equipped with a 6 cm x 6 cm cylindrical cell. The ions of interest were generated in an external high pressure ion source with a pulsed beam of 2 keV electrons [12]. The CI product ions were accelerated to 2 keV and, after passing the flight tube, were decelerated to the ion source potential (ca 0.7 - 0.8 V). A subsequent delay allowed multiple ion-molecule collisions in the ICR cell to relax the ions ensuring that the ions had an essentially thermal distribution of velocities after this delay. The ion of interest was then isolated by ejection of all other ions in the ICR cell. A short r.f. pulse at the exact cyclotron frequency of the isolated parent ion increased the translational energy, allowing dissociation to occur during the subsequent CID delay. The translational energy of the ion was changed by changing the duration of the r.f. pulse while keeping its amplitude constant. To obtain the most accurate results it was necessary to measure the ratio of the intensities (peak areas) of the fragment and parent ion.

A mixture of $CH_3CH_2OCH_2CH_3$ and CH_4 was introduced into the external ion source to generate the proton bound dimer of diethyl ether. $(H_2O)_4H^+$ was produced from a mixture of H_2O and CH_4. Ar at a pressure of 3.3 x 10^{-8} mbar was used as the target gas.

The 2 seconds relaxation period for the ions ensured that the ions were in the centre of the ICR cell. If it is assumed that the parent ion of mass m_p were stationary prior to acceleration, their translational energy (in eV) in the centre-of-mass frame after an r.f. pulse of length t_{rf} can be calculated with equation (3) [13}, where m_t is the mass of the target molecule, V_{p-p} is the peak-to-peak of the r.f. pulse, S^1_{E1} is a geometry factor (0.814) [13b], e is the electronic charge and d is the diameter of the cell. This equation provides the energy scale for the threshold curves.

$$E_{cm} = E_{lab} \cdot \frac{m_t}{m_t + m_p} = \frac{(V_{p-p} \cdot S^1_{E1})^2 \cdot t_{rf} \cdot e}{8 \cdot d^2 \cdot m_p} \cdot \frac{m_t}{m_t + m_p} \qquad (3)$$

The random thermal motion of the target molecules and the parent ions prior to acceleration introduces a spread in the centre-of-mass interaction energy. The resultant distribution of centre-of-mass interaction energy causes Doppler broadening of the threshold curve. Using the model developed by Chantry [14], threshold curves including Doppler broadening were computed. At energies above the threshold the computed curves become linear and extrapolate to $E_{cm,T} - 3\gamma k_BT$, instead of the true threshold energy in the centre-of-mass frame, $E_{cm,T}$ ($\gamma = m_p/(m_p + m_t)$). This method was used here to determine bond dissociation energies.

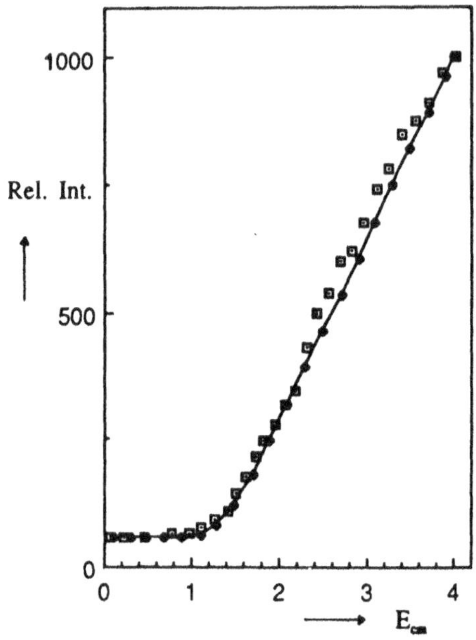

Figure 5 Threshold curve for collision-induced $CH_3CH_2OCH_2CH_3$ loss from $(CH_3CH_2OCH_2CH_3)_2H^+$ with Ar as target gas and a CID delay time of 5 ms. The y-axis represents the ratio of intensities of the fragment and parent ions normalized to 1000 for the largest ratio. The solid line represents the theoretical, Doppler broadened threshold curve using the experimentally obtained threshold energy, 1.43 eV, as input.

The experimental threshold curve obtained for diethyl ether loss from the proton bound dimer by collision with Ar (t_{CID} = 5 ms) is shown in Figures 5 and 6. Extrapolation of the linear part of the threshold curve to the baseline and addition of 0.06 eV ($3\ \gamma\ k_BT$) to the thus obtained value gave a bond dissociation energy of 1.43 ± 0.12 eV, which is only marginally higher than the literature values, 1.36 ± 0.10 eV [15a] and 1.29 ± 0.10 eV [15b]. Using the experimental value for the bond dissociation energy, 1.43 eV, the shape of the theoretical, Doppler broadened threshold curve was computed; see Figure 5. Because the experimental and theoretical curves almost coincide close to the onset, 1.43 eV, it was concluded that <u>the translational energy distribution of the ions after r.f. activation must have</u>

been very narrow. Increasing the ion source potential to 1.5 eV reduced the threshold energy obtained by only 0.07 eV. Note that the degree of fragmentation ion our experiments was very small; at a centre-of-mass energy of 7.8 eV the intensity of m/z 75 was only 1.6% of that of m/z 149.

To obtain realistic bond dissociation energies great care was taken to ensure that single collision conditions were maintained. The kinetic theory of gases was used to approximate the average time between ion-molecule collisions in the ICR cell, t_c. This revealed an advantageous feature of FT-ICR for the present collision experiments: the time between collisions is longer at the threshold energy for dissociation than at higher impact energies. (The opposite applies to ion beam instruments and triple quadrupoles.) For an Ar pressure of 3.0 x 10^{-8} mbar the average time between collisions for

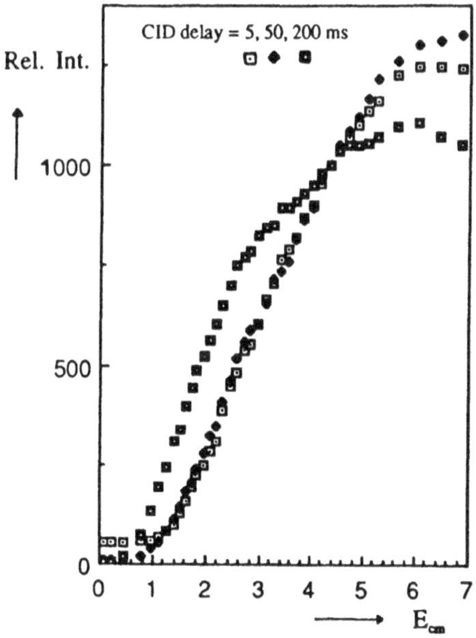

Figure 6 Threshold curve for collision-induced $CH_3CH_2OCH_2CH_3$ loss from $(CH_3CH_2OCH_2CH_3)_2H^+$ with Ar as target gas and CID delay times of 5,50 and 200 ms. The y-axis represents tho ratio of intensities of the fragment and parent ions normalized to 1000 for the ratio at E_{cm} = 4.4 eV

$(CH_3CH_2OCH_2CH_3)_2H^+$ ions was calculated to be 187 ms, which is much longer than t_{CID}, 5 ms. More importantly, using a Poisson distribution the ratio of double to single collisions (Q_2/Q_1) was calculated to be only 1.3 x 10^{-2} at a centre-of-mass energy of 2.0 eV. Thus, a significant effect on the bond dissociation energy obtained due to double collisions is unlikely.

The contribution from double collisions also can be assessed by observing the threshold as a function of delay time, see Figure 6. For CID delay times of 5, 10, 50 and 200 ms, the threshold energies for dissociation were 1.43 ± 0.12, 1.42 ± 0.12, 1.27 ± 0.10 and 0.79 ± 0.07 eV, respectively. This indicates that for CID delay times of 10 ms or less the conditions must have been essentially single collision, whereas at longer CID delay times contributions from multiple collisions lowered the observed threshold. (Note that Q_2/Q_1 is 0.53 for E_{cm} = 2.0 eV and t_{CID} = 200 ms.)

The threshold energy for loss of H_2O from $(H_2O)_4H^+$ (t_{CID} = 10 ms) was determined to be 0.70 ± 0.06 eV, which is very close to the literature value, 0.73 ± 0.03 eV [16].

Thus, a comparison between bond dissociation energies obtained with FT-ICR and literature values showed that r.f. excitation of ions in the ICR cell produced ions with a well-defined translational energy. Moreover, the translational energy distribution of the ions was narrow, because the experimentally obtained threshold curves for dissociation were close to computed curves corrected for Doppler broadening.

OBSERVATION OF HYPERVALENT RADICALS BY NRMS. <u>J.L. Holmes</u> (University of Ottawa)

Hypervalent radicals have provided considerable interest in neutralization-reionization mass spectrometry (NRMS) and related experiments [17]. The most thoroughly studied system is $H_3\cdot$, which has a good pedigree- see references in the work of Selgren and Gellene [18] and Helm [19], who, respectively, observed the photo-assisted decomposition of $H_3\cdot$ and measured its ionization energy (IE). The only other species free from difficulty (e.g, inability of various groups to reproduce results) are the N_xD_{2x+3} series of radicals [20]. $CD_5\cdot$ has proved hard to identify, but Selgren and Gellene [21] have reported its generation as a high Rydberg radical when Zn is used as a neutralization target for CD_5^+ ions.

More recently, Wesdemiotis et al [22] have identified the $CH_3CH_2OH_2\cdot$ radical as the only stable hypervalent form resulting from the neutralization of protonated ethanol with Hg. The writer's group have identified the hypervalent isomeric $[C_2H_7O]\cdot$ radical formed in the Xe, Cd or Hg neutralization of $CH_3O^+(D)CH_3$ [23]. Surprisingly, stable radicals are only formed from excited energy rich ions. The ground state species, produced by protonation of dimethyl ether with isobutene, yield no stable hypervalent radical. On the other hand, exothermic ($\leq$ 110 kJ mol^{-1}) protonation with D_3O^+ yields stable $CH_3O{\cdot}(D)CH_3$ radicals. The internal energy content of the protonated (deuteronated) ether can be monitored via the fraction of ions which can fragment metastably to yield CH_4 + $[CH_2OD]^+$, a reaction which is endothermic by ca 100 kJ mol^{-1} (including a reverse energy barrier). The energy content of the $CH_3O^+(D)CH_3$ ions can be controlled via the partial pressure of D_2O in the ion source, and which at high pressures acts as a thermal sink for the deuteronated ether. The dissociations of the hypervalent species can be observed via the kinetic energy released when the reactions

$$CH_3O{\cdot}(D)CH_3 \longrightarrow CH_3OCH_2{\cdot} + HD \qquad (4)$$

and

$$CH_3O{\cdot}(D)CH_3 \longrightarrow CH_3OCH_3 + D{\cdot} \qquad (5)$$

take place. The activation energy of the related reaction

$$CH_3OCH_3 + H \longrightarrow CH_3OCH_2{\cdot} + H_2 \qquad (6)$$

(6) is known to be 19.7 kJ mol^{-1}. A complete energy surface for the system can be proposed [24], but a measure of the binding energy in the hypervalent radical awaits a measurement of its I.E. or a theoretical calculation.

REFERENCES

1. K.L. Busch, G.L. Glish and S.A. McLuckey, Mass Spectrometry/Mass Spectrometry: Techniques and Applications of Tandem Mass Spectrometry, VCH, New York, **1988**
2. E. Uggerud and P.J. Derrick, Z. Naturforsch.,**1989**, 44a, 245
3. M.M. Sheil, E. Uggerud and P.J. Derrick, Adv. Mass Spectrom., **1989**, 11, 1012
4. E. Uggerud and P.J. Derrick, J. Phys. Chem., submitted
5. K. Qian, A. Shukla, S. Howard, S. Anderson and J. Futrell, J. Phys. Chem., **1989**, 93, 2889
6. A.K. Shukla, S.G. Anderson, S.L. Howard, K.W. Sohlberg and J.H. Futrell, Int. J. Mass Spectrom. Ion Processes, **1988**, 86, 61
7. A.K. Shukla, K. Qian, S.L. Anderson and J.H. Futrell, J. Am. Soc. Mass Spectrom., **1990**, 1, 6
8. R. Bombah, J.P. Stadelman and J. Vogt, J. Chem. Phys., **1982**, 72, 259

9. K. Levsen and H. Schwarz, Mass Spectrom. Rev. **1983**, 2, 77; D.B. Jacobson and B.S. Freiser, J. Am. Chem. Soc., **1983**, 105, 736; B.S. Larsen and D.P. Ridge, J. Am. Chem. Soc., **1984**, 106, 1712; M.F. Jarrold and J.E. Bower, J. Phys. Chem., **1988**, 92, 5702; S.K. Loh, L.Lian, D.A. Hales and P.B. Armentrout, J. Chem. Phys., **1988**, 89, 610.
10. P.B. Armentrout and J.L. Beauchamp, J. Chem. Phys., **1980**, 50, 21; K. Ervin, S.K. Loh, N. Aristov and P.B. Armentrout, J. Phys. Chem., **1983**, 87, 3593; E.K. Parks and S. Wexler, J. Phys. Chem., **1984**, 88, 4492; N.M. Seno and W.S. Koski, J. Phys. Chem., **1984**, 88, 5320; N. Aristov and P.B. Armentrout, J. Phys. Chem., **1986**, 90, 5135; L. Hanley, J.L. Whitten and S.L. Anderson, J. Phys. Chem., **1988**, 92, 5803; S.W. Buckner and B.S. Freiser, Polyhedron, **1988**, 7, 1583; P.J. Marinelli and R.R. Squires, J. Am. Chem. Soc., **1989**, 111, 4101.
11. K.M. Ervin and P.B. Armentrout, J. Chem. Phys., **1985**, 83, 166
12. P. Kofel and T.B. McMahon, Int. J. Mass Spectrom. Ion Processes, **1990**, 98, 1
13. (a) A.G. Marshall and D.C. Roe, J. Chem. Phys., **1980**, 73, 1581. (b) P. Kofel, M. Allemann, H.-P. Kellerhals and K.P. Wanczek, Int. J. Mass Spectrom. Ion Processes, **1986**, 74,1
14. P.J. Chantry, J. Chem. Phys., **1971**, 55, 2746
15. (a) J.W. Larson and T.B. McMahon, J. Am. Chem. Soc.,**1982**, 104, 6255. (b) M. Meot-Ner, J. Am. Chem. Soc., **1984**, 106, 1257
16. (a) A.J. Cunningham, J.D. Payzant and P. Kebarle, J. Am. Chem. Soc., **1972**, 94, 7627. (b) M. Meot-Ner and F.H. Field, J. Am. Chem. Soc., **1977**, 99, 998. (c) Y.K. Lau, S. Ikuta and P. Kebarle, J. Am. Chem. Soc., **1982**, 104, 1462
17. For a review see, J.L. Holmes, Mass Spectrom. Rev. **1989**, 8, 513
18. S.F. Selgren and G.I. Gellene, Chem. Phys. Lett. **1988**, 146, 485
19. H. Helm, Phys. Rev. A, **1988**, 38, 3425
20. S.J. Jeon, A.B. Raksit, G.I. Gellene and R.F. Porter, J. Am. Chem. Soc., **1985**, 107, 4125
21. S.F. Selgren and G.I. Gellene, J. Chem. Phys., **1987**, 87, 5804
22. C. Wesdemiotis, A. Fura and F.W. McLafferty, 38th Annual ASMS Conference on Mass Spectrometry and Allied Topics, Tucson, **1990**
23. J.L. Holmes and M. Sirois, Org. Mass Spectrom., **1990**, 24, 481
24. J.L. Holmes and M. Sirois, manuscript in preparation

TIME-RESOLVED STUDIES OF UNIMOLECULAR GAS PHASE ION DECOMPOSITIONS BY FIELD IONIZATION KINETICS

Nico M.M. NIBBERING
Institute of Mass Spectrometry
University of Amsterdam
Nieuwe Achtergracht 129
1018 WS Amsterdam
The Netherlands

ABSTRACT. Time-resolved studies of unimolecular gas phase ion decompositions, performed over the last five years by the use of the method of field ionization kinetics, are reviewed.

It is shown that in combination with this method the application of stereochemical model compounds, stable isotopic labelling and maxima in normalized rate curves revealed by measurements under relatively high kinetic energy resolution conditions is very successful in providing detailed information on the dynamics of and intermediates in the unimolecular dissociation reactions studied.

Results of collision-induced dissociation experiments performed on fragment ions generated at different molecular ion lifetimes with the aim to determine their structures, that is time-resolved MS/MS, are reported.

1. Introduction

One of the interesting research areas in the field of gas phase ion chemistry is the study of the evolution of the chemistry of unimolecularly decomposing molecular ions as a function of their lifetimes. To this end the method of field ionization kinetics (FIK) has been applied successfully by various groups [1], although many of them unfortunately have left this research area in the past few years.

The purpose of this article is to describe some of the results obtained in the author's group since the appearance of a comprehensive review on mechanistic studies by FIK [1] with the hope for a revival of FIK as one of the various forms of time-resolved mass spectrometry.

Subsequent to a brief description of the principles of the FIK method [1], mechanisms of unimolecular decompositions of molecular ions as a function of their lifetimes will be discussed as probed by the utilization of stable isotopic labelling, stereochemistry and maxima in normalized rate curves.

2. The FIK method

FIK experiments can be performed on double-focusing mass spectrometers having either the conventional (EB) or the reversed (BE) geometry, in which E and B represent the electric and magnetic sectors, respectively. A schematic diagram of a field ionization

K. R. Jennings (ed.), Fundamentals of Gas Phase Ion Chemistry, 333–349.

double-focusing mass spectrometer with the conventional geometry is given in Figure 1.

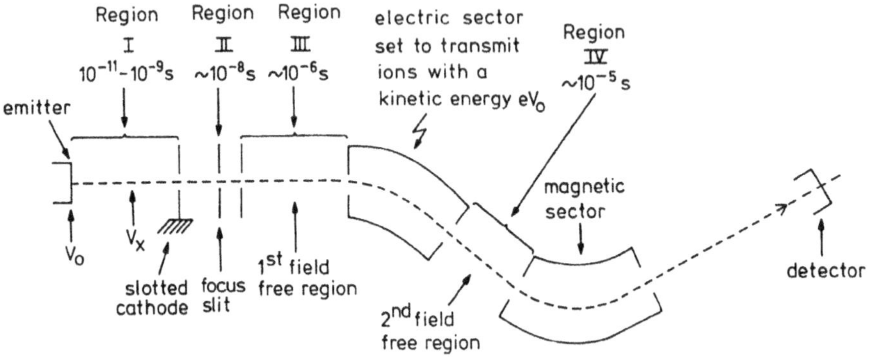

Figure 1. Schematic diagram of a double-focusing field ionization mass spectrometer of conventional geometry (not drawn to scale) showing the four regions in which field ionization kinetic measurements can be made.

Ions are generated from gas phase molecules via electron tunneling in the strong electric field and therefore in a very narrow region close to a 10 µm activated tungsten wire or emitter at a potential V_o (~ 8 kV) which is positioned approximately 1.5 mm from a slotted cathode at ground potential. If these ions are stable for $\geq 10^{-5}$ s, they will acquire a kinetic energy eV_o during acceleration from the ionization region (where to a good approximation the potential is equal to the emitter potential V_o) to the slotted cathode and will pass the electric sector set to transmit ions with a kinetic energy eV_o. Consequently, they will be mass analyzed at their correct m/z values if the magnetic field is scanned. In this way, a FI spectrum is obtained which contains peaks corresponding to the ions generated very near to the emitter, i.e. within approximately 10^{-11} s following ionization. Fragment ions m^+, generated according to equation (1) between the emitter and cathode (region I in Figure 1) at a potential V_x, will not be transmitted through the electric sector because they have insufficient kinetic energy $[e(m/M)(V_o-V_x) + eV_x]$ to be focused by this sector.

$$M^{+\cdot} \rightarrow m^+ + (M\text{-}m)^{\cdot} \qquad (1)$$

However, these ions can be transmitted through the electric sector by application of the most widely used method of increasing the emitter potential at a constant electric sector potential set to transmit ions with a kinetic energy eV_o.

The determination of the amount ΔV with which the emitter potential V_o has to be increased to satisfy equation (2) permits a calculation of the potential V_x at which reaction (1) has occurred.

$$eV_o = e(m/M)(V_o + \Delta V - V_x) + eV_x \qquad (2)$$

It is then possible to calculate the ion lifetime, i.e. the time elapsed between the formation of the molecular ion $M^{+\cdot}$ and its subsequent fragmentation at potential V_x, if the potential distribution between the emitter and cathode is known. Thus, a scan of the emitter potential achieved by increasing ΔV allows one to monitor reaction (1) following FI of M over a continuous time range of approximately 10^{-11} - 10^{-9} s (region I in Figure 1).

Decompositions of $M^{+\cdot}$ ions at times longer than 10^{-9} s can be investigated in the focusing region II and in the first and second field free regions III and IV (see Figure 1) by using metastable ion methods.

The experimental results take the form of curves of product ion current I_m (t) versus kinetic energy, which are transformed according to the principles outlined above to curves of product ion current I_m (t) versus reaction time t. The product ion current I_m (t) is related to the rate of reaction dI_m (t) /dt by equation (3), where Δt is a small interval within which decomposing $M^{+\cdot}$ ions contribute to the product ion current I_m (t) at time t:

$$\frac{dI_m(t)}{dt} = \frac{I_m(t)}{\Delta t} \qquad (3)$$

This interval is determined by the energy resolution of the electric sector and depends on the molecular and product ion masses and on the reaction time t, i.e. this time window interval will decrease with increasing mass difference between the molecular and product ions (not unlimited because of the difficulties in increasing the emitter potential due to arcing problems) and at shorter decomposition times of the molecular ions because of the increase of the electric field gradient with decreasing distance from the emitter.

Division of equation (3) by the measured mass-resolved molecular ion current I_M at an emitter potential V_o, i.e. in the normal FI spectrum, yields the so-called normalized rate k (t):

$$k(t) = I_m(t)/I_M\Delta t \qquad (4)$$

The maximum, usually observed in k (t) versus t plots, has been suggested to represent the k_{max} (E) in the distribution of the time-independent rate constants k (E)'s (see [1] for the relationship between k (t) and k (E)). This maximum generally occurs at longer times for rearrangement reactions than for simple cleavage reactions, which usually compete with each other continuously over the time range of 10^{-11} - 10^{-5} s.

The strength of the FIK method in mechanistic studies is shown to full advantage when it is used in combination with stable isotopic labelling. In that case, the fragment ions from competing reactions are nearly equal in mass, so that the ratios of their ion currents measured at any time t are practically equal to the ratios of the normalized rates k (t) (see equation (4)). Mechanistic studies then may and usually are based upon changes in relative rates as a function of the ion lifetimes.

3. Mechanistic Studies

3.1. STEREOCHEMISTRY

An interesting approach in mass spectrometry to study details of unimolecular fragmentations is to make use of stereochemical model compounds, often in conjunction with isotopic labelling. This may reveal, for example, in the case of substituted cycloalkanes whether the fragmentation reactions take place from either still ring-closed or ring-opened/isomerized molecular ion structures. However, unless the fragmentations occur highly site specific or stereospecific and in the absence of closely related competing channels, such as 1,3- and 1,4-eliminations, details of such reactions are largely lost in electron impact (EI) mass spectrometry, which integrates all the events up to 10^{-6} s following ionization.Moreover, further fragmentation of the daughter ions, such as by loss of a hydrogen atom, is frequently observed upon EI, thereby complicating the mechanistic picture even more. Of course, metastable ion studies in which the relationships between molecular and fragment ions are known, can alleviate the problems mentioned, but refer to ions decomposing after a long lifetime and in a relatively small time window of 10^{-6} - 10^{-5} s following ionization. However, for studies as described above the time-resolved FIK method combined with the fact that upon FI practically only primary fragmentations of molecular ions take place, appears to be very suited as will be exemplified in the two following sections.

3.1.1. *Elimination of Methanol from the Molecular Ion of Methoxycyclohexane.* FIK and specific D-labelling have shown earlier [2] that the molecular ion of cyclohexanol loses a water molecule via a 1,2-elimination within 10^{-10} - 5.10^{-10} s following FI and at longer times via a 1,4-elimination. For the 1,4-elimination reaction, about 90% of the C-4 hydrogen atom in cis-position is retained in the water molecule, the remaining 10% coming from the C-4 hydrogen atom in trans-position. These obervations are in favour of the view that the loss of water from ionized cyclohexanol proceeds via the boat form of the still intact cyclohexane ring. The FIK experiments have been interpreted to show that the hydrogen atoms from positions 3 and 5 do not participate at all in the water elimination, which is in sharp contrast with the observations made upon EI where the hydrogen atoms from the positions 3 and 5 have been found to take part in this water loss. A ring-opened molecular ion is probably involved. Appearance energy measurements under EI conditions have been used to confirm that indeed the 1,3-elimination requires a higher critical energy than the 1,2- and 1,4-eliminations [2].

We have addressed ourselves to the question whether or not a simple replacement of the hydoxy group in cyclohexanol by a methoxy group leads to similar obervations for the loss of methanol from the corresponding molecular ion [3]. To this end the analogues of methoxycyclohexane, site specifically deuterated in the 1, 2 and 6, 3 and 5, and 4 positions and stereospecifically deuterated in the 3 and 5, and 4 positions, have been subjected to FIK. The measurements have indicated that similar to the 1,2-elimination of water from ionized cyclohexanol [2] a 1,2-elimination of methanol occurs to a minor extent (< 20%) at $10^{-10.5}$ s which gradually decreases at longer times and becomes practically negligible at molecular ion lifetimes of > 10^{-9} s [3]. The major part of methanol is lost at molecular ion lifetimes of $\leq 10^{-10.1}$ s via a 1,4-elimination and from the stereospecific D labelling it has been found that ~ 19% of this reaction channel corresponds to a stereospecific cis-elimination, whereas the remaining 81% is only site-specific. This observation indicates that in the 1,4-elimination the major part of methanol is lost from a ring-opened structure of methoxycyclohexane [3] which is in sharp contrast with the highly stereospecific (> 90%) cis-1,4-elimination of a molecule of water from the consequently ring-closed molecular ion of cyclohexanol following FI [2]. An obvious reason for this difference is that the critical energy for rupture of the C_1-C_2 bond by α-cleavage is ~ 0.3 eV lower for the ether than for the alcohol [3]. Another interesting difference compared to the case of cyclohexanol is that at molecular ion lifetimes of between 10^{-10} and 10^{-9} s a very sudden increase of a 1,3-elimination of methanol is observed at the expense of the 1,4-elimination. A stereospecific loss of methanol, however, is not observed at all for the 1,3-elimination indicating that also this reaction occurs from a ring-opened structure of methoxycyclohexane.

Apart from the minor stereospecific 1,4-cis-elimination occurring from the ring-closed structure, the structures of the ring-opened intermediates in the site-specific 1,4- and

1,3-eliminations must be different. Based upon D-labelling and FIK experiments on some methoxycyclohexenes, these intermediates have been shown to be most probably 6-methoxyhexene-1 and 1-methoxyhexene-1, respectively [3]. These methoxyhexenes can account not only for the site-specific 1,4- and 1,3-eliminations of methanol as visualized in Schemes 1 and 2, respectively, but also for the sudden change in competition between these two channels at molecular ion lifetimes of between 10^{-10} and 10^{-9} s (vide supra): the heat of formation of ionized 6-methoxyhexene-1 is ~ 150 kJ/mol higher than that of 1-methoxyhexene-1, so that on energetic grounds the former structure may well not be accessible anymore for longer-lived, i.e. lower energized methoxycyclohexane molecular ions, their heat of formation being comparable to that of ionized 1-methoxyhexene-1. Of course, this energetic argument neglects possible barriers in the potential energy surfaces when going from ionized methoxycyclohexane to either 6-methoxy- or 1-methoxyhexene-1, which seems not unreasonable as they are expected to be similar in view of the similar hydrogen shifts required to effect the isomerizations (see Schemes 1 and 2).

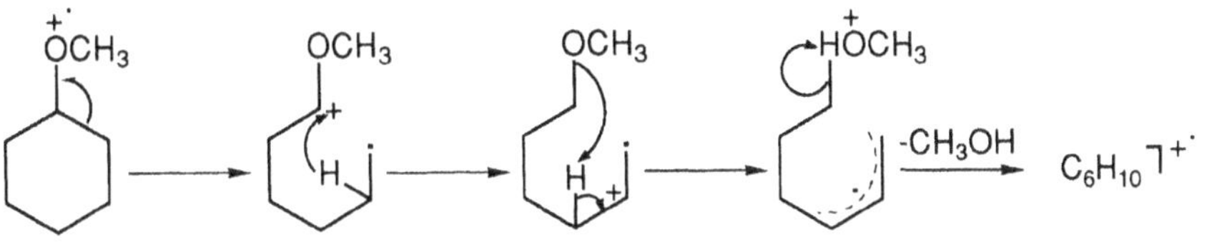

Scheme 1. Proposed mechanism of the site-specific 1,4-elimination of methanol from the molecular ion of methoxycyclohexane within $10^{-10.1}$ s following FI.

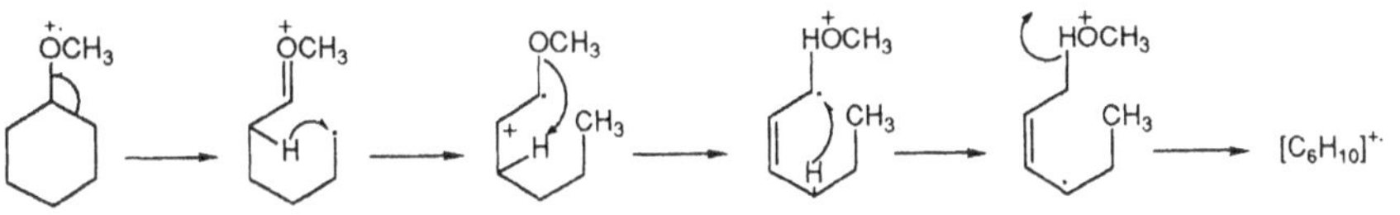

Scheme 2. Proposed mechanism of the site-specific 1,3-elimination of methanol from the molecular ion of methoxycyclohexane within the time interval of $10^{-10.1}$ - 10^{-9} s following FI.

3.1.2.*Dyotropic Hydrogen Rearrangement in the Molecular Ions of 8,9-disubstituted triyclo [5.2.1.0^{2,6}] Decenes*. In an EI study it has been shown, that the exo and endo isomers of the 8,9-disubstituted tricyclo [5.2.1.0^{2,6}] decenes 1 and 2, respectively, fragment essentially different [4]. The ionized endo isomers 2 eliminate C_5H_8 via a retro-Diels-Alder reaction after migration of the hydrogen atoms from positions 8 and 9 to the double bond. This reaction is not observed for the exo isomers 1 where the distance between the hydrogen atoms of positions 8 and 9 and the double bond is too large for the hydrogen migration required to effect the retro-Diels-Alder reaction (Scheme 3).
From that study it is not known, however, whether the double hydrogen migration in ionized 2 occurs in a concerted or stepwise fashion. If the hydrogen atoms would migrate in a concerted fashion, then it is an example of a dyotropic reaction being defined as an uncatalyzed process in which two σ-bonds simultaneously, but not necessarily via a fully symmetrical mode, migrate intramolecularly [5]. Such a suprafacial dyotropic transfer of

hydrogen is a thermally "allowed" $[\sigma 2_s + \sigma 2_s + \pi 2_s]$ pericyclic reaction [6].

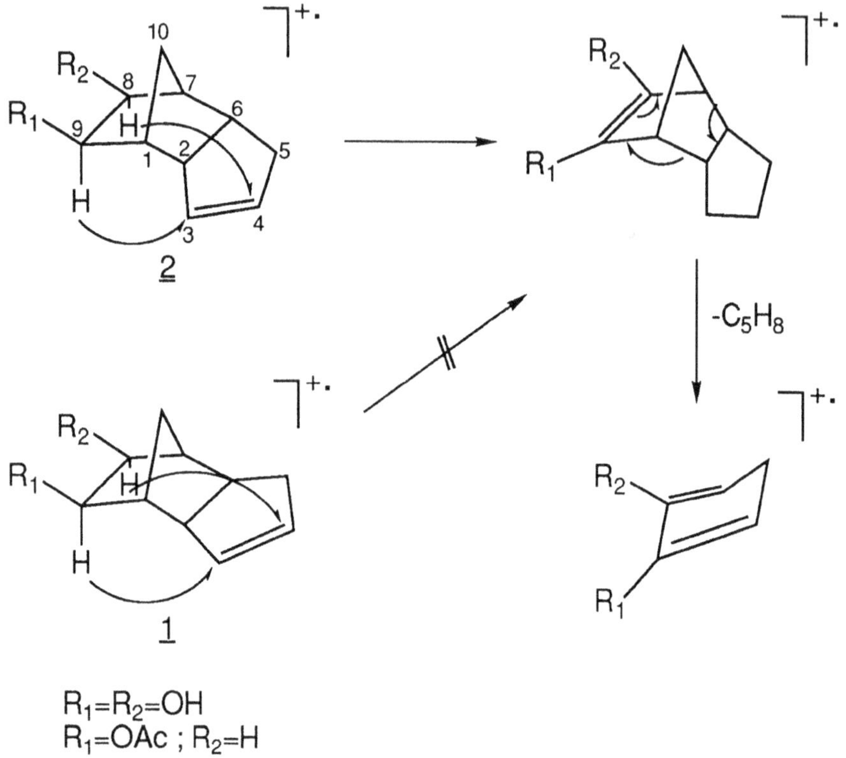

Scheme 3. Double hydrogen transfer in the ionized endo isomer 2 (not possible in the ionized exo isomer 1) followed by elimination of C_5H_8 via a retro-Diels-Alder reaction.

In order to obtain experimental support for either a concerted or a stepwise double hydrogen migration in ionized 2 prior to elimination of C_5H_8 via a retro-Diels-Alder reaction (Scheme 3) FIK has been applied to both exo and endo isomers 1 and 2, respectively [7].

The exo isomer 1 with R_1 and R_2 (Scheme 3) both being OCH_3 appears to eliminate exclusively C_5H_6 (see Scheme 4 for mechanism) over all molecular ion lifetimes studied, while the corresponding endo isomer 2 eliminates both C_5H_6 and C_5H_8, but not $C_5H_7\cdot$. As expected, the C_5H_8 loss is a delayed process with respect to the C_5H_6 loss because of the required double hydrogen rearrangement, i.e. this channel starts to compete with the C_5H_6 loss at molecular ion lifetimes of ~ 10^{-10} s and becomes dominant beyond molecular ion lifetimes of ~$10^{-9.25}$ s [7]. Important, however, is the absence of $C_5H_7\cdot$ loss, which might have been expected to occur for a stepwise, but not concerted double hydrogen rearrangement. In that respect it is relevant to note that the molecular ion of the endo isomer 2 with R_1 = OH and R_2 = OCH_3 eliminates $C_5H_7\cdot$ in addition to C_5H_6 and C_5H_8. This loss of $C_5H_7\cdot$, however, behaves similar to that of C_5H_6 in that it decreases in going from short to longer ion lifetimes. Moreover, the normalized rate k (t) curves for these two channels are similar in shape and have their maximum at the same position, being in both aspects different from the curve for loss of C_5H_8 [7]. In other words, the losses of $C_5H_7\cdot$ and C_5H_6 in this case most probably occur via a common intermediate as rationalized in Scheme 4, but via a different route than the loss of C_5H_8. For the last reaction the experimental facts seem to point indeed to the occurrence of a dyotropic hydrogen rearrangement in the molecular ions of the endo isomers 2 prior to the eventual retro-Diels-Alder reaction (Scheme 3).

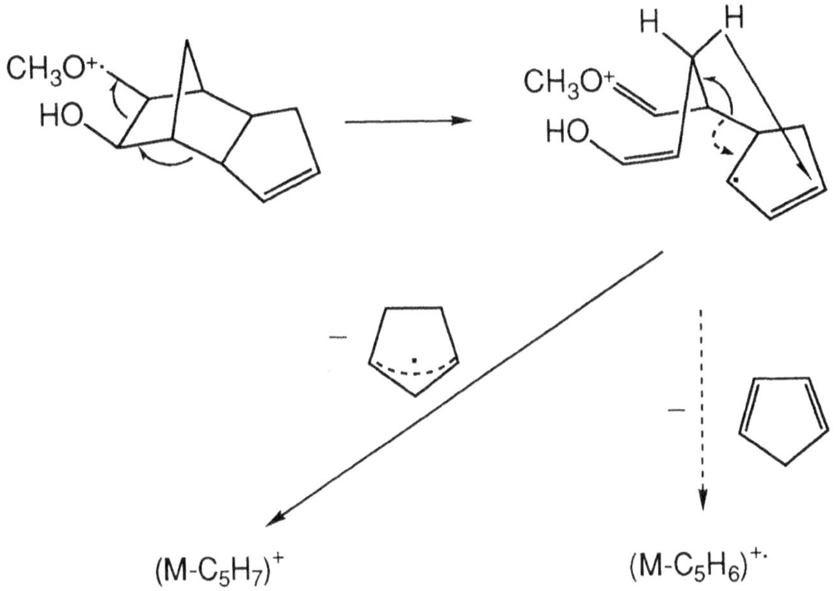

Scheme 4. Rationalization of the losses of C_5H_6 and $C_5H_7\cdot$ from the molecular ion of the endo isomer 2 with $R_1 = OH$ and $R_2 = OCH_3$ via a common intermediate (see further text).

3.2. STABLE ISOTOPIC LABELLING

As noted at the end of section 2, FIK in combination with stable isotopic labelling is very powerful to obtain insight into mechanistic details of reactions of unimolecularly decomposing field ionized molecules.

In addition to the isotopic labelling applied as described in section 3.1.1., four other papers dealing with the combined use of FIK and isotopic labelling [8-11] have been published since the appearance of the review on mechanistic studies by FIK [1]. The main results as described in two of these papers will be presented and discussed below.

3.2.1. *Methyl Elimination from ionized Homo-adamantane.* The interest to study the methyl elimination from ionized homoadamantane 3 has found its origin in a previously performed EI study on this compound and its 1-methyladamantane and 2-methyladamantane isomers 4 and 5, respectively [12]. In that EI study it has been shown that the radical cation of homoadamantane exclusively rearranges to that of 1-methyladamantane and that their metastable decomposing ions do interconvert to a considerable extent prior to methyl loss. Such an interconversion is, of course, a time-dependent process and therefore very suited to be followed by FIK.

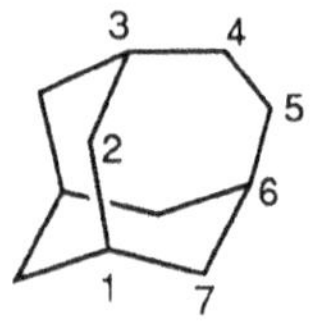

3

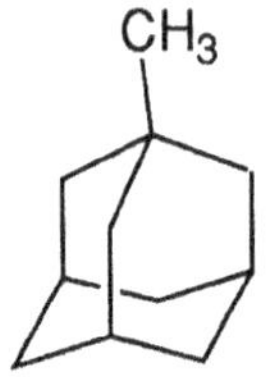

4

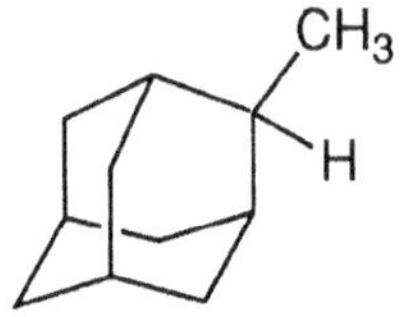

5

First, the methyl group eliminated from field ionized 4 over ion lifetimes ranging from ~ 10^{-11} s to 10-5 s exclusively is the original methyl group as shown by ^{13}C- and D-labelling [10]. In contrast to the previous EI report [12] it must therefore be concluded that, within the limits of detection, the molecular ions of 4 do not rearrange or interconvert at all to the molecular ions of 3 prior to methyl loss.

Concerning the methyl loss from field ionized 3 itself, the following observations have been made [10]. Molecular ions of 4 in which the methylene group at position 4 has been replaced by either $^{13}CH_2$ or CD_2 show up to ion lifetimes of ~ $10^{-10.3}$ s a 1:1 loss of $^{13}CH_3$ versus $^{12}CH_3$ and CD_2H versus CH_3, respectively. At ion lifetimes longer than $10^{-10.3}$ s a gradual increase of unlabelled versus labelled methyl loss is observed for both cases which goes through a maximum between ion lifetimes of 10^{-9} s and 10^{-6} - 10^{-5} s. For metastable decomposing molecular ions the ratios of $^{13}CH_3$ versus $^{12}CH_3$ and CD_2H versus CH_3 losses have the tendency to become similar to those observed at ion lifetimes shorter than $10^{-9.5}$ s. Similar observations are made for molecular ions of 4 in which the hydrogen atoms at both bridgehead positions 3 and 6 have been substituted by deuterium: the initial 4:1 ratio for loss of CDH_2 versus CH_3 gradually decreases from ion lifetimes of ~ $10^{-10.3}$ s, but is practically recovered again for the metastable decomposing molecular ions.

These observations point to the occurrence of at least three channels for the methyl elimination.

Two of them, i.e. those taking place at ion lifetimes shorter than $10^{-10.3}$ s and longer than 10^{-5} s, correspond to a rather specific loss of methyl which contains the methylene group of either position 4 or position 5 and one of the bridgehead hydrogen atoms of positions 3 and 6. These channels have been rationalized via the routes given in Scheme 5, where for entropic reasons pathway a → b → d has been assigned to occur within $10^{-10.3}$ s and pathway a → c → d at ion lifetimes of 10^{-5} s.

The less specific methyl loss at the intermediate ion lifetimes has been rationalized as being due to an initial cleavage of the C_3-C_4 (or C_5-C_6) bond in field ionized 4, for which the reader is referred to the original publication [10].

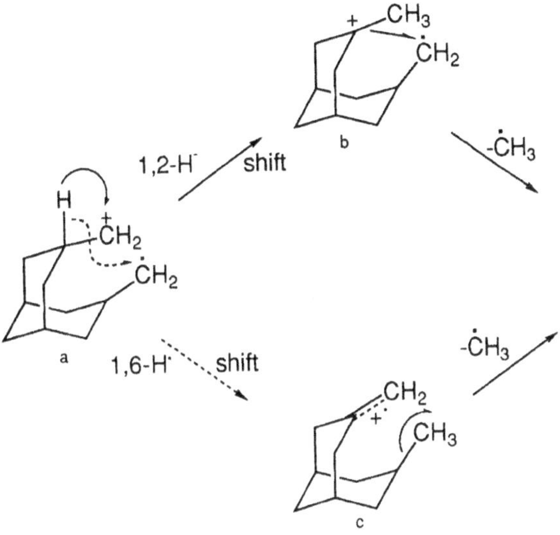

Scheme 5. Proposed pathways for the elimination of a methyl radical from the molecular ion of homoadamantane at ion lifetimes ≤ $10^{-10.3}$ s and ~ 10^{-5} s as sustained by isotopic labelling (see text).

3.2.2. *Ethene and Ethyl Eliminations from the Molecular Ion of 2-ethylbutanoic acid.* The molecule of 2-ethylbutanoic acid has been chosen for a FIK study [11] with the aims (i) to see wehther the hydrogen atoms at positions C-4 and C'-4 interchange positionally following a 1,5-hydrogen shift from C-4 (or C'-4) to the carbonyl oxygen prior to ethene elimination, a process which would provide another argument in favour of the stepwise character of the McLafferty rearrangement and (ii) to obtain further support for the generality of the mechanism of alkyl radical loss from ionized alkanoic acids as advanced and summarized for 2-methylbutanoic acid [13] in Scheme 6.

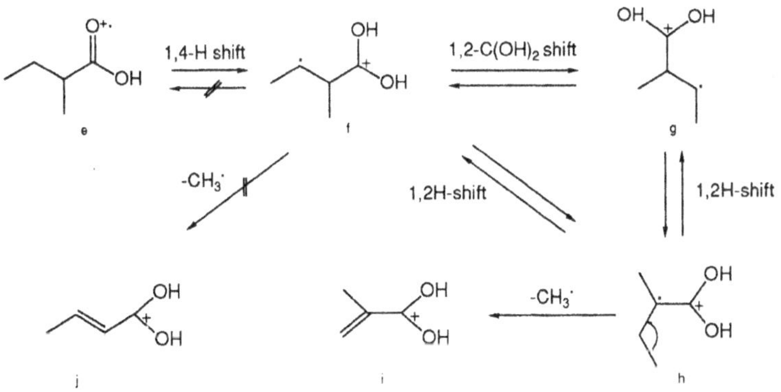

Scheme 6. Route for methyl loss from ionized 2-methylbutanoic acid according to [13].

FIK measurements on unlabelled 2-ethylbutanoic acid have shown that both ethene and ethyl are eliminated indeed from its molecular ion: up to molecular ion lifetimes of 10^{-10} s exclusively ethene is lost, while at longer ion lifetimes ethyl loss starts to compete and becomes the dominant channel at ion lifetimes longer than $10^{-9.1}$ s [11].

FIK measurements on 2-ethylbutanoic acids labelled with deuterium and ^{13}C in the various positions of both ethyl groups have demonstrated that the ethene and ethyl eliminations occur without any detectable exchange between the hydrogen or carbon atoms of the two original ethyl groups [11]. The same observations have been made for molecular ions decomposing in the metastable time frame of 10^{-6} to 10^{-5} s.
These data suggest that the ethene elimination occurs as visualized in equation (5).

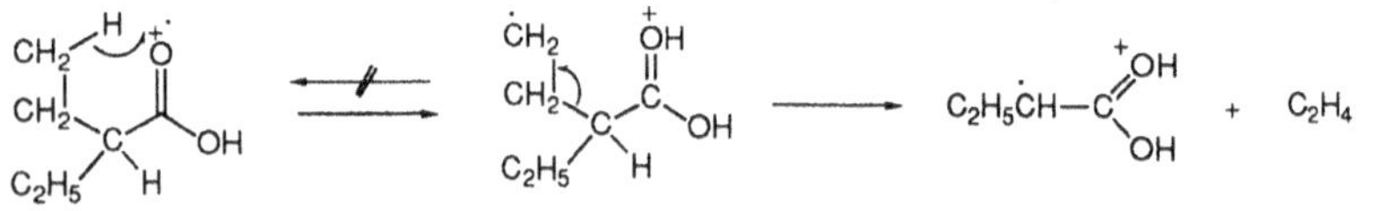

(5)

k l m

Intermediate l in equation (5) in that respect might also account for the ethyl elimination at longer ion lifetimes via the ring closure reaction as pictured in equation (6).

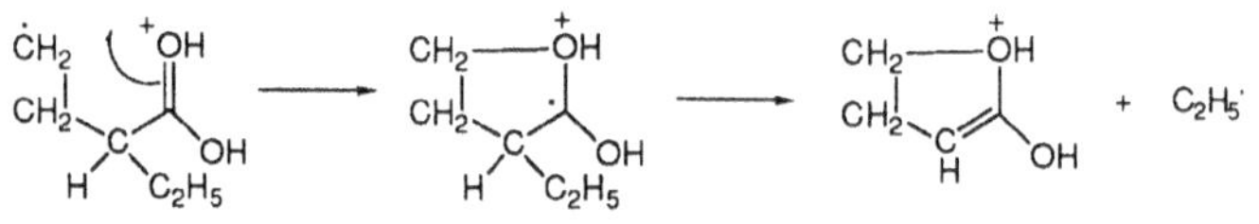

(6)

l n o

This has been checked by collision-induced dissociation experiments on the (M-ethyl)$^+$ ions to obtain information on their structure(s). Comparison with the spectra obtained by collision-induced dissociation experiments on various reference ions has shown that 95% of the (M-ethyl)$^+$ ions have the structure of carbonyl oxygen-protonated crotonic acid and the remaining 5% that of carbonyl oxygen-protonated methacrylic acid.
The following conclusions may therefore be drawn:
(i) The ethene elimination following a 1,5-hydrogen shift from C-4 (or C'-4) to the carbonyl oxygen (equation (5)) is occurring such fast that it is not possible to make clear statements in favour of or against a stepwise character of this process.
(ii) Only 5% of the ethyl loss proceeds via a route similar to the sequence e → f → g → h → i in Scheme 6, while the remaining 95% of the ethyl loss is due to a direct C-2 - C-3 (or C'-3) bond cleavage following a 1,4-hydrogen shift from C'-3 (or C-3) to the carbonyl oxygen as visualized in equation (7).

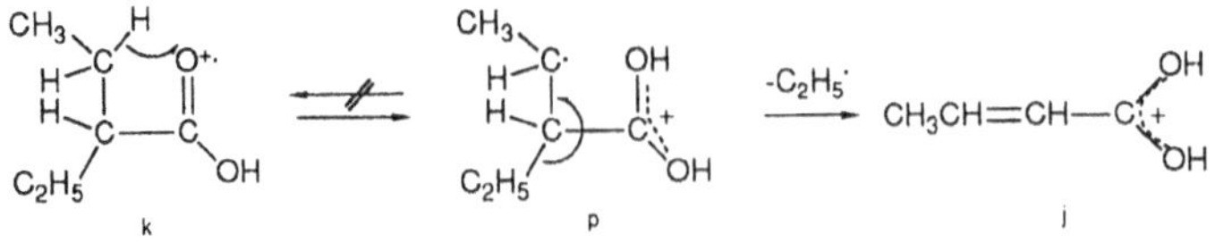

(7)

k p j

This result is in sharp contrast with that for 2-methylbutanoic acid where the corresponding direct cleavage step f → j (Scheme 6) is not operative and shows that it is sometimes difficult to generalize mechanisms for ionic decompositions.

3.3. NORMALIZED RATE CURVES

As shown in the preceding sections, the combined application of stereochemistry, stable isotopic labelling and FIK is very powerful to obtain insight in reaction mechanistic details. In some cases relative rate measurements of competing channels without isotopic labelling can provide also sufficiently detailed mechanistic information. Examples are the competing eliminations of C_5H_6 and C_5H_8 from ionized endo 8,9-disubstituted tricyclo[5.2.1.0^{2,6}] decenes (see section 3.1.2) and CN· and HCN from ionized β-thiocyanatovinyl aldehydes, which both occur via rearrangement [14]. Of course, such relaive rate measurements cannot be made, if no neutral species with different masses are lost from the molecular ions. Yet, the exclusive loss of a neutral species with a well-defined elemental composition from the molecular ion may still proceed via more than one reaction channel. In that case either stable isotopic labelling should be applied or normalized rates should be measured under relatively high kinetic energy resolution conditions to reveal whether more than one reaction channel contributes to the fragmentation. The last approach will be exemplified below by two recent studies [15,16] and is useful especially for those cases where the neutral species cannot change its composition by isotopic exchange within the molecular ion prior to elimination, such as a halogen atom.

3.3.1 *Mechanism of Alkyl Ion Formation from Ionized Alkyl Halides.* For various isomeric and homologous alkyl chlorides and bromides the position of the maximum in the normalized rate curves for alkyl ion formation has been measured with the aim to possibly correlate this position with (i) the stability of the alkyl ion formed, (ii) the strength of the carbon-halogen bond which is broken and (iii) with the size of the alkyl group [15]. The results of these measurements, obtained at a relatively high kinetic energy resolving power of ~ 150000, have been listed in Table 1.
First the data in the second column of Table 1, designated by log $t_{k\ 1st\ max}$, will be considered. They suggest that the following conclusions may be drawn:
(i) The more stable the carbenium ion formed is, the faster the rate of its formation is. This can be derived from a combined inspection of the values for the isomeric butyl bromides and chlorides showing that in decreasing order the rates of formation of the isomeric butyl ions are t-butyl 〉 s-butyl 〉 i-butyl 〉 n-butyl. It also agrees with the Hammond postulate [17] which states that the stability order of the product ions will be reflected in the magnitude of the corresponding critical energy to generate them from the molecular ions. In this explanation it is assumed that the energy deposition upon ionization is equal for the different isomeric butyl bromides and chlorides, which seems not unreasonable as described in the original publication [15].
(ii) The rate of the alkyl ion formation is inversely proportional to the strength of the bond being broken. This can be derived from comparison of the values of the bromides and chlorides which shows that the C-Br bond cleavage has a maximum in the normalized rate curve at shorter molecular ion lifetimes than has the C-Cl bond cleavage. Note, for example, that the i-butyl cation from i-butyl bromide is formed even faster than the t-butyl cation from t-butyl chloride.
(iii) The rate of C-X bond cleavage is inversely proportional to the number of degrees of freedom in the molecular ion. This can be seen from the positions of the maxima in the normalized rate curves of the primary propyl, butyl and pentyl bromides, indicating that

the corresponding carbenium ions are formed increasingly later in time.
It should be noted that these three experimentally supported conclusions are in full agreement with the well-known rate equation for unimolecularly dissociating isolated ionic species as formulated in QET-RRKM theory [18,19].

TABLE 1. Maxima observed in the normalized rate curves for halogen loss from alkyl halides following field ionization.

Compound	log t_k 1st mass	log t_k 2nd mass
C-C-C-C-Cl	–[a]	
C C-C-Cl C	–[a]	
C C-C-C-Cl	-10.3	
C C-C-Cl C	-10.6	
C-C-C-C-Br	-10.6	-10.25
C C-C-Br C	-10.75	-10.32
C C-C-C-Br	<<-11.2[b]	
C C-C-Br C	<<-11.2[b]	
C-C-C-C-C-Br	-10.45	-10.15
C C-C-C-C-Br	-10.55	-10.25
C-C-C-Br	<-11.0[b]	
C-C-C Br	<<-11.2[b]	

a No loss of the chlorine atom observed.
b Loss of the bromine atom is too fast to resolve a maximum in the normalized rate curve.

Interesting are the observations made which have been summarized in the third column of Table 1 being designated as log $t_{k\ 2nd.max.}$ They refer to a second maximum in the normalized rate curve, an example being given in Figure 2.

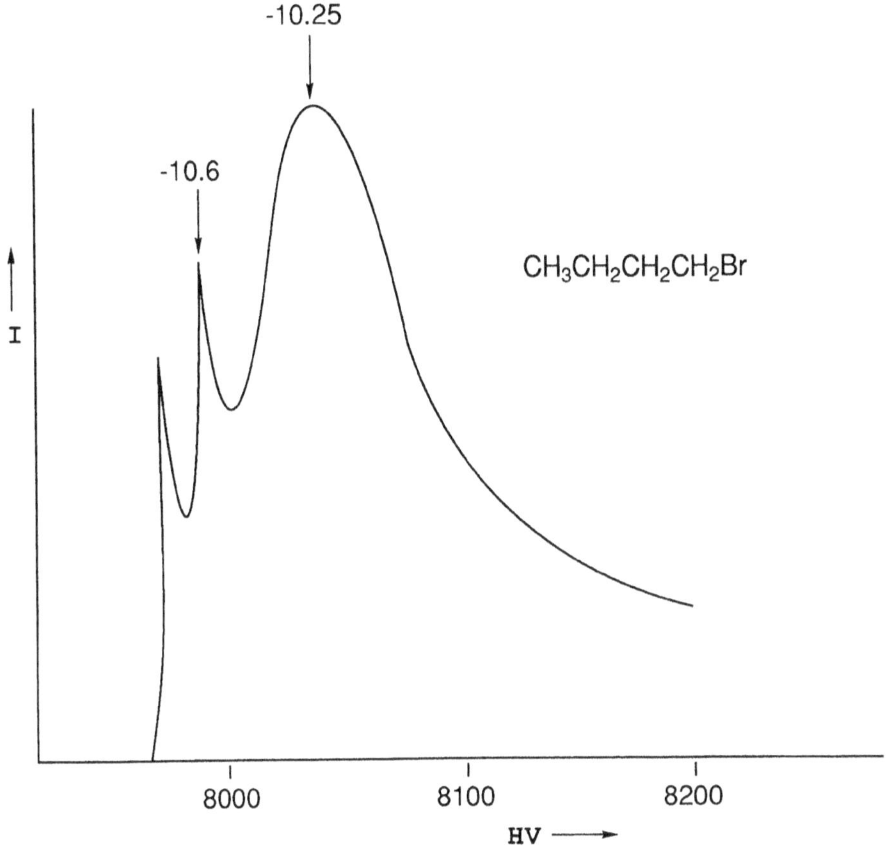

Figure 2. The intensity of the signal due to m/z 57, which corresponds to the loss of the bromine atom from the molecular ion of n-butyl bromide, as a function of the emitter potential The numbers of -10.6 and -10.25 s given at the maxima correspond to the logarithm of the molecular ion lifetimes after field ionization.

Note from the data in Table 1 that these second maxima are only observed for those alkyl halides, which formally would generate a primary carbenium ion upon cleavage of the C-X bond in their molecular ions. They have been interpreted as being due to an additional C-X bond cleavage, i.e. a 1,2-hydride shift assisted C-X cleavage mechanism which becomes operative at longer molecular ion lifetimes (equation (8)) [15].

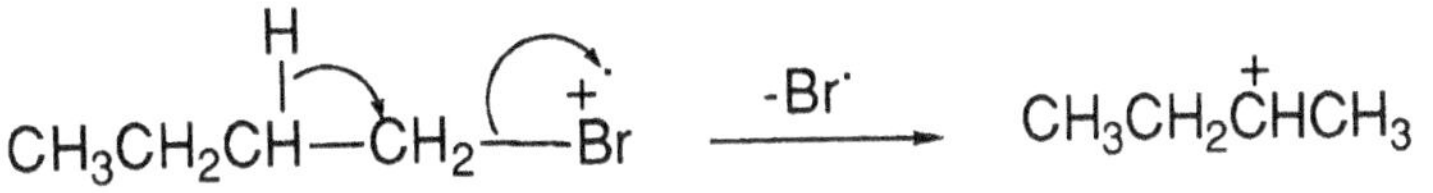

(8)

In other words, for the relevant alkyl halides there is a direct C-X bond cleavage at short ion lifetimes and a pseudo direct C-X bond cleavage at longer ion lifetimes. For the loss of X· during the direct cleavage mechanism no rearrangement of the alkyl group is required, but for the loss of X· during the pseudo direct cleavage mechanism apparently the energy released during the rearrangement from a nascent primary carbenium ion to a secondary or tertiary carbenium ion fuels the cleavage reaction. Of course, the primary carbenium ions formed via the direct C-X bond cleavage at short ion lifetimes most probably rearrange also to more stable secondary or tertiary carbenium ions after the loss of X·.

Finally, it should be noted that the pseudo direct cleavage mechanism, described above, has been suggested before in fragmentations of similar compounds studied by photoionization [20] and photoelectron-photoion coincidence techniques [21,22].

3.3.2. *Methyl Loss from Ionized 3-penten-2-ol.* A previous deuterium labelling study [23] has revealed three pathways for the loss of methyl from ionized 3-penten-2-ol as summarized in Scheme 7.

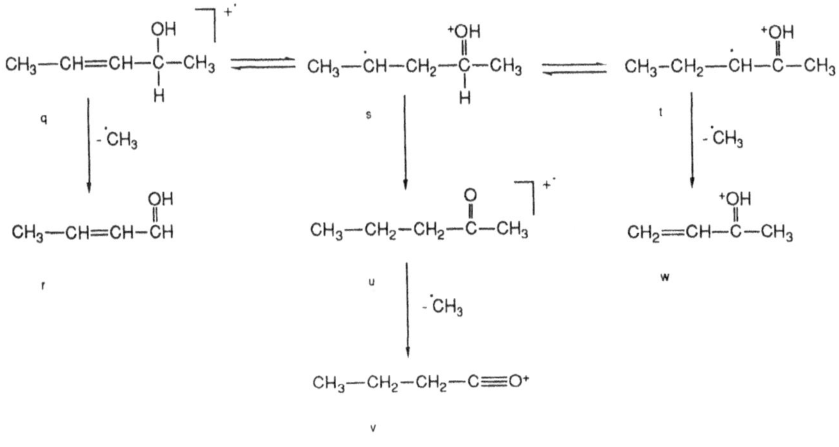

Scheme 7. Routes for methyl loss from ionized 3-penten-2-ol according to [23].

Route q → r is operative at ion lifetimes < $10^{-10.5}$ s and has been suggested to lead to oxygen protonated crotonaldehyde, route q → s → t → w is dominating at ion lifetimes between $10^{-10.5}$ and 10^{-9} s and has been suggested to lead to oxygen protonated methyl vinyl ketone, while route q→ s → u → v is coming up at longer ion lifetimes and has been suggested to produce the butyryl cation (Scheme 7).

In view of the maxima observed in the normalized rate curves for loss of halogen from ionized alkyl halides as described in section 3.3.1, the molecule of 3-penten-2-ol seemed therefore ideally suited to check whether the diferent channels for methyl loss from its molecular ion are reflected as different maxima in the corresponding normalized rate curve.

This has been shown to be indeed the case in the sense that the normalized rate curve for methyl loss from the molecular ion of unlabelled 3-penten-2-ol contains two distinct maxima at ion lifetimes of $10^{-10.5}$ s and $10^{-10.1}$ s after ionization [16].

In the normalized rate curve for CD_3 loss from the molecular ion of $CH_3CH{=}CHCH(OH)CD_3$ similar maxima are observed, but in that for CH_3 loss only the maximum at $10^{-10.1}$ s is present.
This means that the maximum at $10^{-10.5}$ s corresponds to the route $\underline{q} \rightarrow \underline{r}$ of Scheme 7, as expected for such a direct cleavage reaction, and that the maximum at $10^{-10.1}$ s is due to the in time overlapping routes $\underline{q} \rightarrow \underline{s} \rightarrow \underline{t} \rightarrow \underline{w}$ and $\underline{q} \rightarrow \underline{s} \rightarrow \underline{u} \rightarrow \underline{v}$ of Scheme 7.

Experimental support for the structures of the eventually generated $(M\text{-methyl})^+$ ions has been obtained from collision-induced dissociation experiments on these ions formed at different molecular ion lifetimes, i.e. at times < $10^{-10.6}$, $10^{-10.4}$ and $10^{-10.0}$ s after ionization. Such so-called time-resolved MS/MS spectra, in which the peaks are rather noisy due to their low intensities, have appeared to match well with the MS/MS spectra of the independently made reference ions obtained by protonation of crotonaldehyde and methyl vinyl ketone and by loss of a chlorine atom from the molecular ion of butyryl chloride [16]. For further details the reader is referred to the original publication [16].

4. Final Remarks.

This review of work performed over the last five years has sought to show that the FIK method is still unique in providing a detailed view on the dynamics of unimolecular dissociation reactions of gas phase ions. Since the previous review [1] further progress has been made in the sense that it has become possible to obtain also time-resolved MS/MS spectra of daughter ions generated at different molecular ion lifetimes and to detect various maxima in normalized rate curves because of the higher kinetic energy resolving power of the presently available instrumentation. A limitation of the FIK method is the frequently encountered minor amount of primary fragmentation of the molecular ions because of the small amount of energy deposited during the FI process. This inhibits to study a large variety of molecules. It is hoped, however, that the planned application of lasers in the FIK research of the Institute of Mass Spectrometry at the University of Amsterdam, so as to excite the neutral molecules prior to field ionization, will assist to overcome successfully the problem mentioned.

5. Acknowledgements.

The author is greatly indebted to his co-workers and colleagues, listed in the references, for their enthusiastic and expert contribution to this work and to the Netherlands Organization for Scientific Research (SON/NWO) for the continuous financial support.

References.

1. Nibbering, N.M.M. (1984) 'Mechanistic studies by field ionization kinetics', Mass Spectrom. Rev. 3, 445-477.
2. Derrick, P.J., Holmes, J.L. and Morgan, R.P. (1975) 'Kinetics and mechanisms of the loss of water from the cyclohexanol radical ion at times from 50 picoseconds to 10 microseconds following field ionization', J. Am. Chem. Soc. 97, 4936-4941.
3. Molenaar-Langeveld, T.A. and Nibbering, N.M.M. (1986) 'A combined time-resolved and stereospecific deuterium labelling study of the elimination of methanol from the molecular ion of methoxycyclohexane', Org. Mass Spectrom. 21, 637-644.

4. Herzschuh, R., Rabbi, R.M., Kühn, H. and Mühlstädt, M. (1983) 'Stereochemical effects on electron impact induced rearrangement reaction of tricyclo [5.2.1.$0^{2,6}$] decane and -decene derivatives', Int. J. Mass Spectrom. Ion Phys. 47, 387-390.
5. Reetz, M.T. (1973) 'Primary and secondary orbital effects in dyotropic rear-rearrangements', Tetrahedron 29, 2189-2194.
6. Woodward, R.B. and Hoffmann, R. (1970) The Conservation of Orbital Symmetry, Verlag Chemie, Weinheim.
7. Kluft, E., Nibbering, N.M.M., Kühn,H. and Herzschuh, R. (1986) 'Dyotropic hydrogen rearrangement in the molecular ions of 8,9-disubstituted tricyclo [5.2.1.$0^{2,6}$] decenes as supported by field ionization kinetics', J. Am. Chem. Soc. 108, 7201-7203.
8. McAdoo, D.J., Hudson, C.E., Zwinselman, J.J. and Nibbering, N.M.M. (1985) "Carbocation-like isomerisations at radical sites precede methyl losses from ionised butanoic acid', J. Chem. Soc. Perkin Trans. II, 1703-1709.
9. Kluft, E. and Nibbering, N.M.M. (1986) "A combined time-resolved and $^{2}H/^{13}C$ labelling study of the elimination of hydrogen cyanide from the molecular ion of pyridine', Int. J. Mass Spectrom. Ion Processes 69, 339-342.
10. Kluft, E., Nibbering, N.M.M., Schwarz, H., Müller, R. and Zeller, K.-P. (1986) 'A time-resolved study of the methyl elimination from ionized homoadamantane in combination with ^{2}H- and ^{13}C-labelling', Int. J. Mass Spectrom. Ion Processes 73, 167-174.
11. Kluft, E., Nibbering, N.M.M., Stringer, M.B., Eichinger, P.C.H. and Bowie, J.H. (1988) 'A combined field ionization kinetics, collision-induced, and stable isotopic labelling study of the ethene and ethyl (and subsequent) eliminations from the molecular ion of 2-ethylbutanoic acid', Int. J. Mass Spectrom. Ion Processes 85, 215-228.
12. Schwarz, H., Wesdemiotis, C., Weiske, T., Zeller, K.-P. and Müller, R. (1980) 'Methyl loss from metastable homoadamantane cation radical', Z. Naturforsch. B, 35, 207-211.
13. Weiske, T., Halim, H. and Schwarz, H. (1985) 'Zur Rolle von Enol-Radikalkationen als reaktive Zwischenstufen bei Isomerisierungs-/Dissoziationsprozessen in der Gasphase' , Chem. Ber. 118, 495-509.
14. Kluft, E., Nibbering, N.M.M., Schulze, B. and Herzschuh, R. (1987) 'The time-resolved CN· and HCN loss from ionized β-thiocyanatovinyl aldehydes', Org. Mass Spectrom. 22, 119-120.
15. Kluft, E. and Nibbering, N.M.M. (1989) 'Rate and mechanism of alkyl ion formation from alkyl halides as studied by field ionization kinetics', Int. J. Mass Spectrom. Ion Processes 92, 171-184.
16. Zappey, H.W., Ingemann, S. and Nibbering, N.M.M. (1990) 'A study of the rate of methyl loss and the structure of the product ion as a function of the lifetime of the molecular ion of 3-penten-2-ol', Org. Mass Spectrom. submitted.
17. Hammond, G.S. (1955) 'A correlation of reaction rates', J. Am. Chem. Soc. 77, 334-338.
18. Rosenstock, H.M. and Krauss, M. (1963) 'Quasi-equilibrium theory of mass spectra', in F.W. McLafferty (ed.), Mass Spectrometry of Organic Ions, Academic Press, New York, pp. 1-64.
19. Robinson, P.J. and Holbrook, K.A. (1972) Unimolecular Reactions, Wiley-Interscience, London.
20. Traeger, J.C. (1980) 'Photoionization mass spectrometry of the propyl halides', Int. J. Mass Spectrom. Ion Phys. 32, 309-319.

21. Rosenstock, H.M., Buff, R., Ferreira, M.A.A., Lias, S.G., Parr, A.C., Stockbauer, R.L. and Holmes, J.L. (1982) 'Fragmentation mechanism and energetics of some alkyl halide ions', J. Am. Chem. Soc. 104, 2337-2345.
22. Oliveira, M.C., Baer, T., Olesik, S. and Almoster Ferreira, M.A. (1988) 'The dissociation energies and mechanism of energy-selected bromo- and iodobutanes', Int. J. Mass Spectrom. Ion Processes 82, 299-318.
23. Zwinselman, J.J., Nibbering, N.M.M., Midlemiss, N.E., Hegedüs Vajda, J. and Harrison, A.G. (1981) 'A field ionization kinetics and metastable ion study of the fragmentation of some pentenols', Int. J. Mass Spectrom. Ion Phys. 38, 163-179.

ION-MOLECULE REACTIONS AND STEREOCHEMISTRY IN TANDEM MASS SPECTROMETRY

JEAN-CLAUDE TABET
Laboratoire de Chimie Organique Structurale
Université P. et M. Curie, 5 Place Jussieu,
Paris Cedex 05, FRANCE

1. Introduction

Fundamental considerations, such as kinetics, thermodynamics and various computational methods, are not well suited to describe the reactivity of organic molecules containing a large number of carbon atoms, such as natural products and similar synthetic compounds [1]. In fact, of the different, complementary experimental methods used in the determination of a mechanism (e.g. labelling, measurements of rate and equilibrium constants, etc.), the analysis of stereochemical effects constitutes a suitable, alternative approach [2] and appears to be more appropriate for organic chemists concerned with the elucidation of gas-phase reaction mechanisms. Indeed, the analysis of stereochemical effects can be of assistance in (i) the determination of the molecularity of gas phase processes (e.g. unimolecular or bimolecular pathways), and (ii) distinguishing the relative importance of thermochemical and kinetic control of ion-molecule reactions [3] without extensive calculations. The instrumentation employed for such investigations may be a low pressure system [4] such as ion cyclotron resonance or an ion trap, or a high pressure source of a more conventional mass spectrometer, including tandem instruments [5].

In this review, several studies of stereochemical effects in reactions of positive and negative even electron ions are described. These studies were carried out in (i) a double focussing mass spectrometer of reverse geometry, (ii) a triple quadrupole instrument or (iii) a tetraquadrupole instrument built in this laboratory operated in the MS/MS or MS/MS/MS modes [6]. Collision spectra are investigated either to determine structures of ions produced in ion-molecule reactions in a high pressure source or to compare the behaviour of selected species, such as MH^+ and $[M + NH_4]^+$ towards reactive collisions in the collision cell of a tandem mass spectrometer as a function of the method of preparation of the selected ions, e.g. fast atom bombardment or chemical ionisation. The observed reactions are mainly those induced by the NH_3/NH_4^+ system [7]: solvation, protonation (and/or nucleophilic substitution) under positive chemical ionisation conditions and deprotonation reactions under negative chemical ionisation conditions. Under high pressure source conditions, "survivor ion" species, stabilised by multiple collisions may, when selected, have the original structure or an isomeric structure, e.g. an ion-dipole form. The reactivity of the ion is investigated as a function of the reagent gas (introduced either in the source or in the collision cell) and depends on factors such as the active site stereochemistry and the rigidity of the ring conformation. The relative influence of various energetic factors (such as proton affinity, ΔH^o_{acid} and ΔH^o_f of ions) and activation energies (e.g. steric effects, anchimeric assistance, etc) on the orientation of the pathway followed are compared qualitatively with various stereochemical experimental approaches.

K. R. Jennings (ed.), Fundamentals of Gas Phase Ion Chemistry, 351–372.

2. Positive Ion-molecule Reactions

2.1. STEREOSPECIFIC ION-MOLECULE REACTIONS IN A HIGH PRESSURE SOURCE.

2.1.1. *The Dependence of Proton Transfer Processes on Stereochemical Factors.*

It is known that under chemical ionisation conditions (e.g. CI-NH_3), protonation of molecules occurs via exothermic or slightly endothermic pathways (PA(NH_3) < PA(M)) [8]. This process takes place via the formation of an excited adduct ion, $[M + NH_4]^{+*}$, which can decompose (equation 1) or react with ammonia [9]. The relative rate constants k_{-1} and k_2 for the decomposition of the adduct ion to give $[M + NH_4]^+$ or $[MH^+ + NH_3]$ are related to the enthalpies of formation, $\Delta H^o{}_f$, of the products if one considers that the energy of the transition state is close to that of the products. In addition, the proton affinity of the substrate M is an important factor in determining the relative importance of the two decomposition reactions:

$$M + NH_4^+ \underset{k_{-1}}{\overset{k_1}{\rightleftharpoons}} [M + NH_4]^{+*} \xrightarrow{k_2} MH^+ + NH_3 \qquad (1)$$

This clearly describes the behaviour of monofunctional molecules, RX, in the ion source, in which the nucleophilic function, X, is specifically protonated (Scheme 1a). For molecules containing two nucleophilic functions, i.e. X-R-Y, if both reactive sites can solvate the NH_4^+ ion, competitive decompositions giving isomeric protonated species $XRYH^+$ and $YRXH^+$ can take place, depending on the relative proton affinities of R'X (R' = -RY) and R"Y (R" = -RX) (Scheme 1b). Decompositions of the MH^+ ions depend on their structures, e.g. the spatial proximity of groups present in the diastereoisomer under study. Hence the analysis of stereochemical effects can be very informative, for example, in the investigation of intramolecular proton transfer.

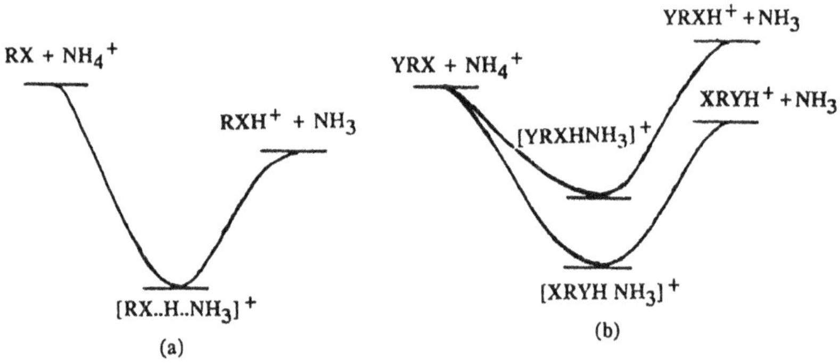

Scheme 1. Hypothetical and simplified reaction coordinate energy diagrams for (a) monofunctional RX compounds (PA(RX) > PA(NH_3)) and (b) bifunctional YRX compounds (PA(R'X) < PA(NH_3) < PA(R"Y)) in the case where $[YRX..H..NH_3]^+$ $[XRY..H..NH_3]^+$ is not sterically possible.

(a) Site of protonation on diastereoisomeric protonated species as indicated by CAD processes [10].

The following cis- and trans- hydroxybenzoates 1a and 1b provide an interesting example of such a situation. Solvation of NH_4^+ can take place competitively at -OH and COO- sites yielding $[M + NH_4]^+$. However, *a priori*, protonation should occur at the benzoate site since this reaction is slightly less endothermic (PA ($C_6H_5COOCH_3$) = 852 kJ mol^{-1}, PA (NH_3 = 853.5 kJ mol^{-1} [11]) and so gives rise to a very low contribution to solvation at the hydroxylic site since PA (ROH) < PA (NH_3) by more than 20-30 kJ mol^{-1}.

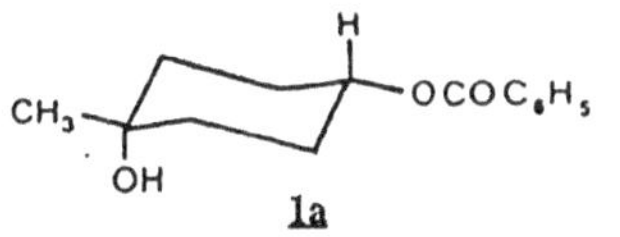

1a

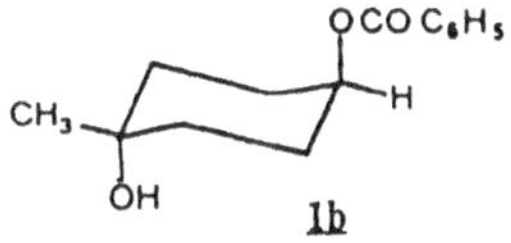

1b

The MIKE/CAD spectra of epimeric MH^+ ions [12] show that the protonated 1a (cis-isomer), $1aH^+$, loses H_2O more readily than the protonated 1b, (trans-isomer), $1bH^+$, giving peaks of relative intensity 0.46 and 0.05 of the total ion current arising from the ions $[1aH - H_2O]^+$ and $[1bH - H_2O]^+$. The favourable formation of the former ion is expected since the proton is solvated by both nucleophilic sites. In the decomposition of $1aH^+$, the group having the lower PA (i.e. OH) must be lost since the ion $[MH - C_6H_5COOH]^+$ is not observed [10]. On the other hand, the loss of benzoic acid is stereospecific from the trans $1bH^+$ ion since the ester site specifically is protonated, reducing the probability of H_2O loss. The formation of $[1bH - C_6H_5COOH]^+$ is likely to be favoured by anchimeric assistance by the hydroxyl group (equation 2).

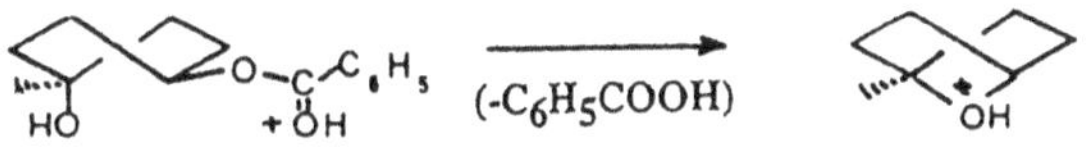

$1bH^+$ $[1bH\text{-}C_6H_5COOH]^+$ (2)

(b) Stereospecific consecutive decompositions of $[M + NH_4]^+$ ions via formation, under CAD conditions, of unusual MH^+ ions (not favoured under high pressure source conditions), observed by means of sequential MS/MS/MS experiments [14].

Low energy CAD spectra of the epimeric adduct $[M + NH_4]^+$ ions generated under ammonia CI conditions from the exo 2a and endo 2b hydroxyketal isomers illustrated consist mainly of the daughter ions MH^+ and NH_4^+. The $[MH^+]/[NH_4^+]$ ratio given by the $[2b + NH_4]^+$ ion is much greater than that given by the $[2a + NH_4]^+$ ion. The higher PA of 2b arises from the double solvation of the proton in the $2bH^+$ ion. Furthermore, of the other fragment ions observed, $[MH - C_2H_4O]^+$, $[MH - C_2H_6O_2]^+$ and $[MH - C_2H_6O_2 - H_2O]^+$, the first is formed specifically from $[2a + NH_4]^+$ since $I([2aH - C_2H_4O]^+)/I(2aH^+) = 0.65$ and $I([2aH - C_2H_4O]^+)/I([2aH - C_2H_6O_2]^+) = 4.6$ whereas the ion $[2bH - C_2H_4O]^+$ is not observed in the CAD spectrum of the endo $[2b + NH_4]^+$ ion.

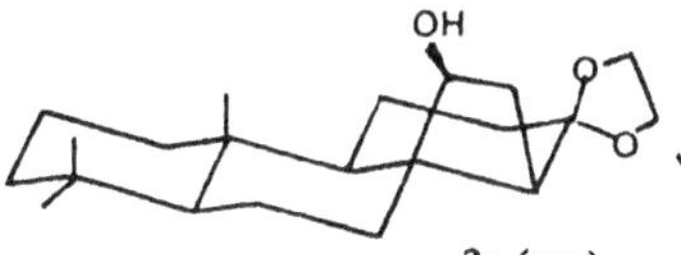

2a (exo)

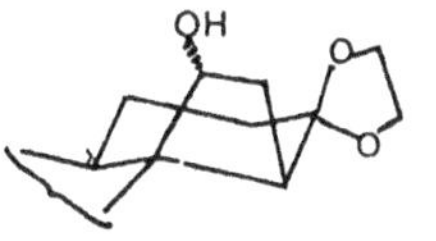

2b (endo)

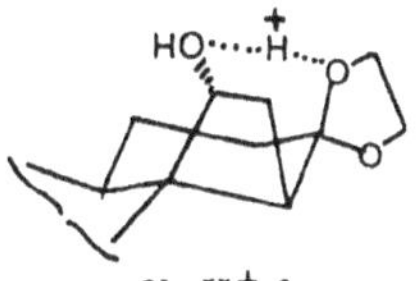

$2b_1H^+$ form

This behaviour of the $[2a + NH_4]^+$ ion under CAD conditions is in marked contrast to that of the $2aH^+$ ion for which the corresponding intensity ratios are respectively 2.5 x 10^{-3} and 10^{-2}. This difference in behaviour of these ions under CAD conditions is somewhat contradictory [16] and in order to rationalise this contradiction, sequential MS/MS/MS experiments [6] were carried out as follows:

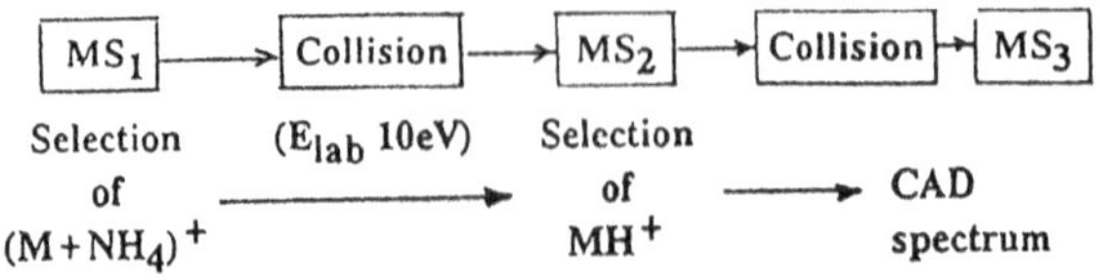

The ion $[\underline{2a} + NH_4]^+$ is selected by MS1 and $\underline{2a}H^+$ is formed by CAD in the first collision region and selected by MS2. Products formed by the CAD of this ion in the second collision region are recorded by scanning MS3. The base peak is $[\underline{2a}H - C_2H_4O]^+$ and the intensity ratio $I([\underline{2a}H - C_2H_4O]^+)/ I(\underline{2a}H^+)$ increases to 1.2. Conversely, loss of C_2H_4O does not occur from selected $\underline{2b}H^+$ under the same conditions, confirming the stereochemical effect already observed in the CAD spectra of isomeric $[M + NH_4]^+$ ions. Labelling studies on $\underline{2a}$ indicate that the $[\underline{2a}H - C_2H_4O]^+$ is formed by the decomposition of the cyclobutanol ring, (equation 3).

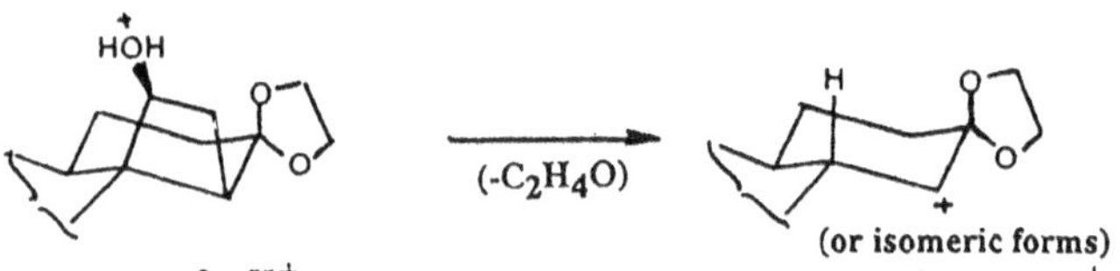

$\underline{2a}_1H^+$ $[\underline{2a}_1H\text{-}C_2H_4O]^+$ (3)

Such an interpretation requires the proton to be located at the -OH site when $\underline{2a}H^+$ is formed from $[\underline{2a}H + NH_4]^+$ in the first collision region. Since this is not the favoured site for protonation in the source, the structure of the selected $[\underline{2a}H + NH_4]^+$ ions must be considered. In the source, the NH_4^+ ion is solvated by both nucleophilic sites, i.e. by the hydroxyl and the ketyl groups, independently, yielding the three isomeric forms shown below:

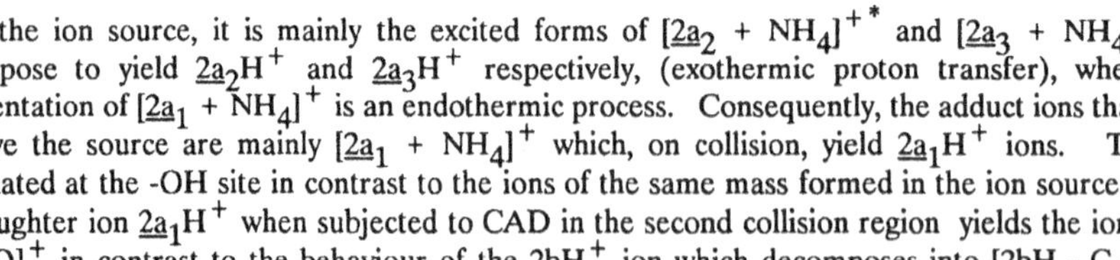

$[\underline{2a}_1 + NH_4]^+$ $[\underline{2a}_2 + NH_4]^+$ $[\underline{2a}_3 + NH_4]^+$

In the ion source, it is mainly the excited forms of $[\underline{2a}_2 + NH_4]^{+*}$ and $[\underline{2a}_3 + NH_4]^{+*}$ that decompose to yield $\underline{2a}_2H^+$ and $\underline{2a}_3H^+$ respectively, (exothermic proton transfer), whereas the fragmentation of $[\underline{2a}_1 + NH_4]^+$ is an endothermic process. Consequently, the adduct ions that survive to leave the source are mainly $[\underline{2a}_1 + NH_4]^+$ which, on collision, yield $\underline{2a}_1H^+$ ions. These are protonated at the -OH site in contrast to the ions of the same mass formed in the ion source. Hence the daughter ion $\underline{2a}_1H^+$ when subjected to CAD in the second collision region yields the ion $[\underline{2a}_1H - C_2H_4O]^+$ in contrast to the behaviour of the $\underline{2b}H^+$ ion which decomposes into $[\underline{2b}H - C_2H_6O_2]^+$ rather than $[\underline{2b} - C_2H_4O]^+$ which is not observed.

Finally, these results suggest that the cross section for NH_4^+ solvation in the ion source must be comparable for the different nucleophilic sites since the formation of the $[\underline{2a}_1 + NH_4]^+$ ion, protonated at the -OH site, is supported by the specific decomposition via formation of $\underline{2a}H^+$ ions.

2.1.2. *Different nucleophilic substitution pathways determined by active site stereochemistry.*

An illustration of such a reaction is endothermic proton transfer (equation 4) exemplified by an alcohol for which the main product is $[M + NH_4 - H_2O]^+$, designated M_sH^+ [7-10,17].

$$R\text{-}OH + NH_4^+ \longrightarrow [ROH{+}NH_4]^+ \longrightarrow H_2O + R\text{-}NH_3^+ \qquad (4)$$

This reaction proceeds either *via* a unimolecular (S_Ni mechanism) or a bimolecular (S_N2 mechanism). Since the former pathway is accompanied by retention of configuration and the latter occurs with inversion of configuration, the investigation of stereochemical effects may allow one to determine the stereochemistry of the NH_3^+ group introduced [17]. (Note that the third possible mechanism, S_N1 *via* carbonium ion formation, appears to be less efficient). Consequently, the analysis of such experimental effects should allow one to distinguish which of the SN mechanisms is followed.

(a) Steric effects on the S_Ni/S_N2 orientation [18].

In the case of monoalcohols, steric hindrance does not lead to spectacular stereochemical effects so that (i) the influence on the SN yield may be weak, and (ii) it may be difficult to attribute a configuration to the NH_3^+ group introduced [19]. Nevertheless, stereochemical effects have been used for the study of the ammonia SN process occurring from the exo (3a) and endo (3b) norborneols in the ion source [20].

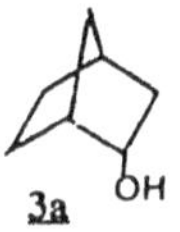

3a

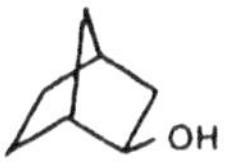

3b

Indeed, under these conditions, MH^+ ions are not observed when $[M + NH_4]^+$ and $[M + NH_4 - H_2O]^+$ ions are very abundant. This is consistent with the proton affinities of these diastereoisomers relative to $PA(NH_3)$, as shown in Scheme 2a, $(PA(C_7H_{11}OH) = 816\ kJ\ mol^{-1}$ approximately).

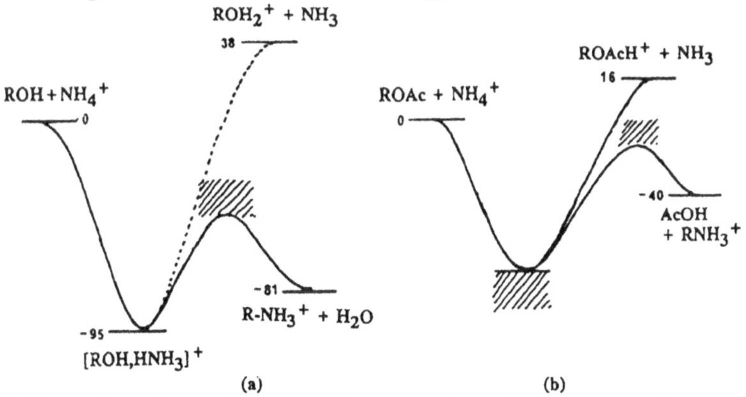

(a) (b)

Scheme 2. Energetic pathways for competitive protonation and SN processes for norbornyl systems (a) alcohol and (b) acetate derivatives. Energies ($kJ\ mol^{-1}$) relative to initial state.

The problem is to determine by which mechanism this reaction proceeds. An indication is provided by comparing the low energy (E_{Lab} 10 eV) CAD spectra of $C_7H_{11}NH_3^+$ (M_sH^+) with those of corresponding ions generated in the source by direct protonation of the exo and endo isobornylamines (4a and 4b respectively), see Table 1. In particular, the direct competitive fragmentation processes are scrutinised (Equations 5a and 5b) to compare M_sH^+ structures.

$$C_7H_{11}NH_3^+ \longrightarrow C_7H_{11}^+ + NH_3 \qquad (5a)$$

$$C_7H_{11}NH_3^+ \longrightarrow C_7H_{10} + NH_4^+ \qquad (5b)$$

Table 1 : Daughter ion $I(NH_4^+)/I(C_7H_{11}^+)$ ratios measured in CAD spectra of $C_7H_{11}NH_3^+$ formed from isomeric exo and endo norborneols and norbornylamines.

	3a exo	3b endo	4a exo	4b endo
$I(NH_4^+)/I(C_7H_{11}^+)$	1.6	0.78	1.71	0.66

The similarity of the ratios measured for the pairs 3a/4a and 3b/4b indicates that the initial stereochemistry is preserved during nucleophilic substitution. This reaction therefore proceeds with retention of configuration and the $C_7H_{11}NH_3^+$ formation from both isomeric norborneols occurs *via* intramolecular nucleophilic substitution, i.e. S_Ni, equation 6.

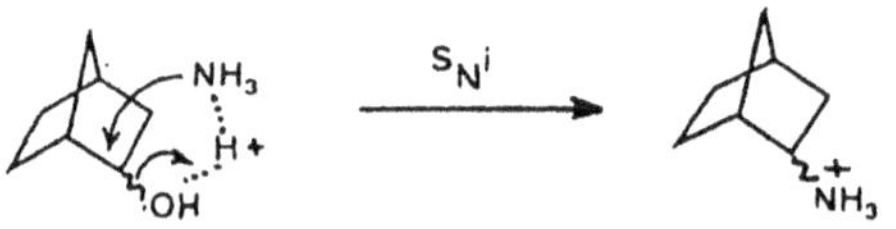

(6)

$[3a + NH_4]^+$, exo ⟶ $4aH^+$, exo
$[3b + NH_4]^+$, endo ⟶ $4bH^+$, endo.

For the exo 3a isomer, however, it is difficult to rule out the classic antiplanar C-C bond migration assisting the SN process since the exo configuration is conserved during formation of the exo protonated $4aH^+$ amine.

Alternatively, the reaction of the norbornyl acetate derivatives (5a and 5b, exo and endo epimers, respectively) with NH_3 differs from that of the norborneols since there is competition between protonation and nucleophilic substitution within the ion source. Protonation indicates that the PA(acetate) is greater than PA(alcohol), see Scheme 2b. The fact that protonation of the acetates is favoured despite (i) its endothermicity and (ii) the SN pathway which is exothermic can be explained in terms of the role played by the back activation energy of the latter process. It is probable that this arises from the requirement that the transition state should be planar, (Scheme 3), [21]. Stereochemical effects may be observed because of steric hindrance arising from neighbouring groups and atoms which modify the enthalpy of formation of the transition state.

R=H and $COCH_3$

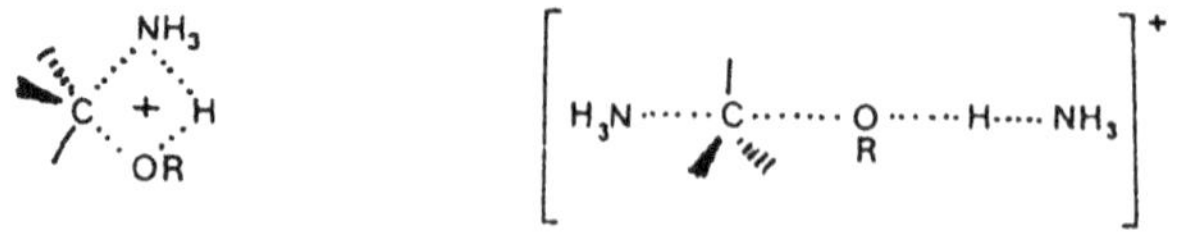

(a) (b)

Scheme 3 : Assumed transition states for (a) S_Ni and (b) S_N2 pathways.

The stereochemistry of the NH_3^+ group introduced is determined from the CAD spectra of the $C_7H_{11}NH_3^+$ ion given by 3a and 3b. These spectra closely resemble that given by the exo protonated 4aH$^+$ norbornylamine. These results may be rationalised by suggesting that different SN pathways are followed which reflect the acetate stereochemistry, i.e. S_Ni and S_N2 from $[5a + NH_4]^+$ and $[5b + NH_4]^+$ respectively, (equation 7).

endo $[5b+NH_4]^+$

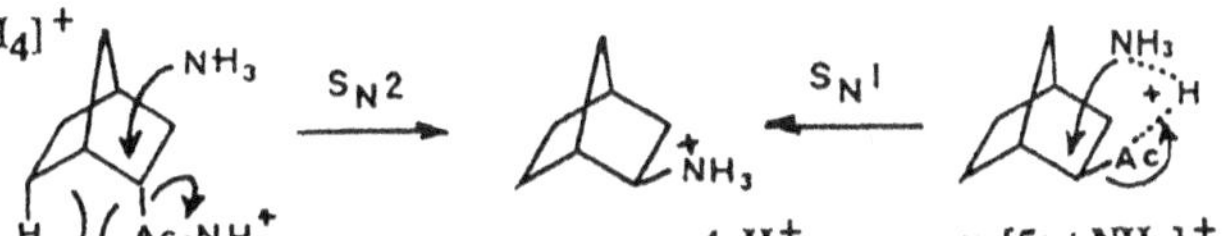

exo 4aH$^+$ exo $[5a+NH_4]^+$ (7)

It is probable that the endo hydrogen atom at C_6 sterically hinders the S_Ni process from $[5b + NH_4]^+$ so that the S_N2 pathway is favoured. The eventual formation of a common 4aH$^+$ ion from each of the isomeric adduct ions could take place by adduct isomerisation [22] into a three-body ion-dipole intermediate (equations 8a, 8b and 8c). The daughter ion, $AcOH_2^+$, was not, however, observed since its formation is too endothermic a process.

$$[C_7H_{11}\text{-OAc},NH_4]^+ \quad [C_7H_{10}\cdot\cdot H\cdot\cdot AcOH,NH_3]^+ \longrightarrow C_7H_{11}NH_3^+ + AcOH \quad (8a)$$

$$\longrightarrow C_7H_{10} + NH_4^+ + AcOH \quad (8b)$$

$$\longrightarrow C_7H_{10} + NH_3 + AcOH_2^+ \quad (8c)$$

(b) Regioselective and Stereospecific S_N2 Reactions evidenced by MS/MS [23].

The diastereoisomeric diols 6a (cis) and 6b (trans) yield M_sH^+ ions without the formation of MH^+ ions. The structure of these M_sH^+ ions is clearly indicated by their CAD spectrum since it fragments by the competitive processes 9a and 9b:

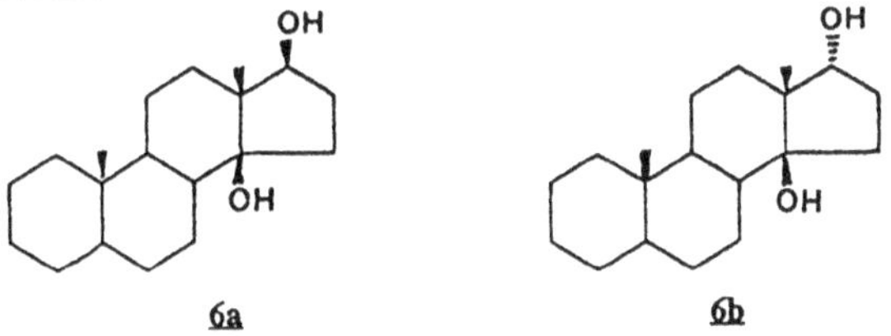

6a 6b

$$M_sH^+ \longrightarrow [M_sH\text{-}H_2O]^+ + H_2O \quad (9a)$$

$$M_sH^+ \longrightarrow [M_sH\text{-}NH_3]^+ + NH_3 \quad (9b)$$

The ratio $I([M_sH - H_2O]^+)/I([M_sH - NH_3]^+)$ is strongly influenced by the initial stereochemistry of the diols since it has a value of 0.05 for the 6a (cis) isomer and 1.0 for the 6b (trans) isomer. The loss of water occurs despite (i) the charge being located at the nitrogen function and (ii) the endothermicity of proton transfer from NH_3^+ to OH, and is observed under CAD conditions.

If the two groups (e.g. $-NH_3^+$ and -OH) are in close spatial proximity, loss of H_2O is favoured. This elimination is therefore indicative of the cis stereochemistry of the hydroxy- ammonium M_sH^+ ion. Because this decomposition is stereospecific for the trans 6b diol, the M_sH^+ structure is that of the cis protonated amino-alcohol generated *via* the Walden inversion (i.e. an S_N2 mechanism). On the other hand, the trans protonated aminoalcohol ion is produced from the cis 6a isomer which cannot lead to the elimination of water.

Furthermore, the CAD spectra of the fragment ions $[M_sH - NH_3]^+$ formed in the source from 6a and 6b can be superimposed which suggests that NH_3 is lost from the C_{17} position rather than from the C_{14} position of M_sH^+, yielding a carbonium form with a common structure. Indeed, the M_sH^+ ion having NH_3^+ in the C_{14} position should yield $[M_sH - NH_3]^+$ ions characterised by a different CAD spectrum because of the different OH stereochemistry at the C_{17} position. Finally, as shown in equations 10a and 10b, nucleophilic substitution occurs *via* an S_N2 pathway from the -OH site at the C_{17} position.

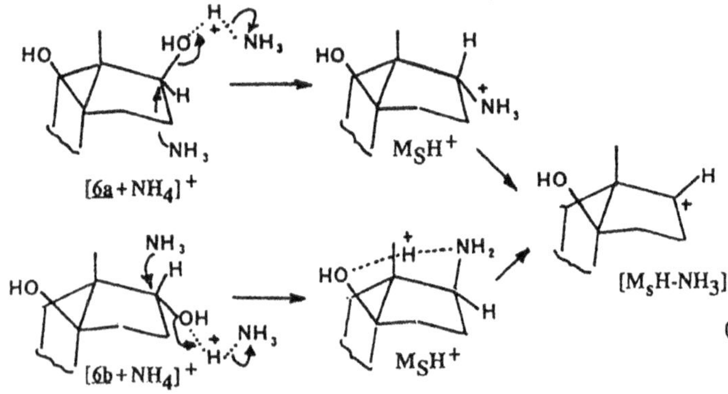

(10a)

(10b)

Regioselective nucleophilic attack by NH_3 at the C_{17} site can be rationalized by considering the following two possible reasons: (i) The NH_4^+ solvation by the hydroxylic site at the C_{14} position may be difficult because of steric hindrance due to substituents located at the C_{13} and C_{17} positions (i.e. either CH_3 and OH for 6a or both methyl groups for 6b). (ii) Cis ring junction at the C_{14} position may hinder the approach of the NH_3 group by the back side of the $[M + NH_4]^+$ skeleton (ζσον υθε α side), the S_N2 process disfavoured and the ion-molecule reaction at the C_{17} position is regioselective. Such assumptions have their limitations, however, since in the cases of 3,17-androstane diols [24], S_Ni and S_N2 processes compete and the reactivity is not regioselective, behaviour which is difficult to explain.

(c) Regioselectivity dependent on polarisability of nucleophilic sites. Hindrance of S_Ni by hydrogen bonding [25].

Cis and trans indane-1-2-diols 7a and 7b react with the NH_3/NH_4^+ system in a high pressure source to yield mainly $[M + NH_4]^+$ and M_sH^+ in relating abundances that depend on the stereochemistry of the diols. The intensity ratio $I(M_sH^+)/I([M + NH_4]^+)$ is 0.34 for the cis isomer, 7a, and 1.84 for the trans isomer, 7b.

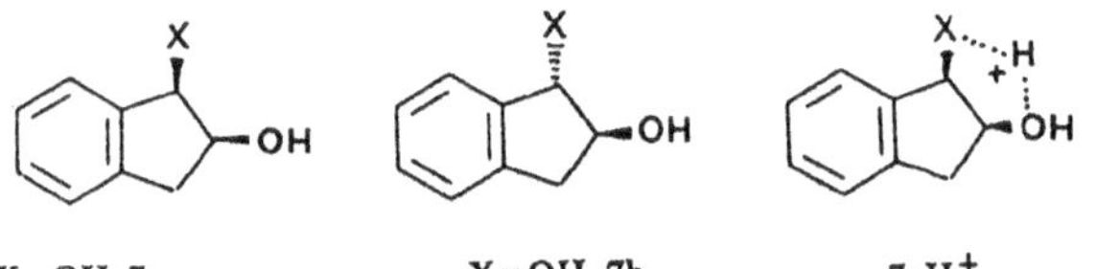

X = OH, 7a
X = NH_2, 8a

X = OH, 7b
X = NH_2, 8b

$7aH^+$
$8aH^+$

The plot of the intensity ratio $I(M_sH^+ cis)/I(M_sH^+ trans)$, ion abundances being relative to those of the respective adduct ions, against $P(NH_3)$ in the ion source is linear with a positive slope. This suggests that the ion $[7a + NH_4 - H_2O]^+$ is formed by a process that is characterised by a higher molecularity than the reaction yielding the ion $[7b + NH_4 - H_2O]^+$, i.e. $[7a + NH_4]^+$ and $[7b + NH_4]^+$ respectively follow S_N2 and S_Ni pathways. On the other hand, the CAD spectra of both M_sH^+ ions closely resemble that of the $8bH^+$ ion (trans chemistry). All these results can be explained by an inversion of configuration from 7a (cis) and configuration retention from 7b (trans) as shown in equation (11).

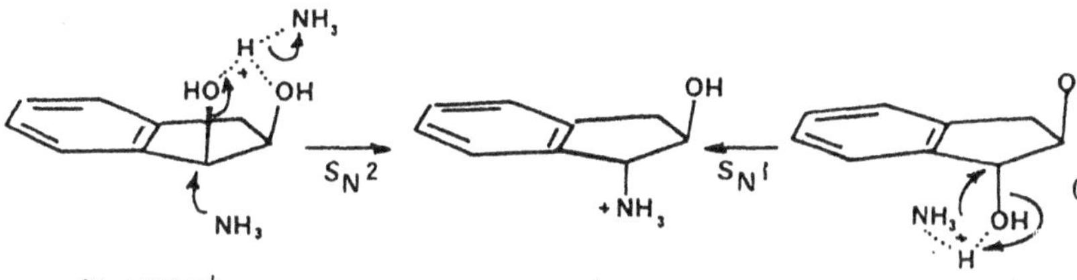

(11)

[7a+NH4]+ 8bH+ [7b+NH4]+

This suggestion is consistent with the lower intensity ratio $I(M_sH^+)/I([M + NH_4]^+)$, measured from the spectrum given by the cis 7a isomer, since the M_sH^+ structure is common and the cis $[7a + NH_4]^+$ ion is more stable than the trans $[7b + NH_4]^+$ ion, as expected since hydrogen bonding is possible only in the former adduct ion.

The fact that the S_Ni pathway from the cis $[7a + NH_4]^+$ ion is not favoured must also be due to the hydrogen bonding which hinders the formation of the planar transition state of the S_Ni reaction (scheme 3) and favours the S_N2 pathway. From the trans $[7b + NH_4]^+$ ion, there is no steric hindrance so that the rate constant for the formation of the planar transition state of the S_Ni pathway is higher.

In addition, since the CAD spectra of the isomeric M_sH^+ ions can be superimposed on that given by the $8bH^+$ ion (trans stereochemistry and containing NH_3^+ in the benzylic position), the nucleophilic substitution is regioselective at the benzylic position. This behaviour can be explained by the higher polarisability of the benzylic hydroxyl site relative to the homobenzylic -OH group. It is probable that both hydroxylic groups compete in the solvation of the NH_4^+ ion but during the second step (i.e. the NH_3 attack), the rate constant for the nucleophilic attack on the activated site (*via* an intramolecular or bimolecular pathway), at the benzylic position is higher than at the homobenzylic position, which must be characterised by a higher enthalpy of formation of the transition state. Note that similar variations of specificity for these reactions were observed for similar compounds having larger saturated rings (e.g. cycloheptane) [25b].

(d) Applications of specific ion molecule reactions for the differentiation of enantiomeric compounds.

By the analysis of a racemic mixture of deuterated di-isopropyl-d_{14}-L- and di-isopropyl-D-tartrates, Fales and Wright [26] were able to distinguish the L and D enantiomeric tartrates under CI conditions. This enantiomeric discrimination was attributable to geometric and steric factors [27]. More recently, a chiral reagent gas was used to distinguish enantiomeric R and S amino acids [28]. The diagnostic ions, however, are usually adduct ion species, associated by hydrogen bonding, whose stability is strongly dependent on the CI source conditions, which may be a disadvantage for analytical purposes.

The use of a specific SN process, induced by a chiral amine (R_sNH_2) reacting with enantiomeric substrates, R'_sOH and R'_rOH, (where s and r indicate sinister and rectus configurations of the alcohol), can be an efficient alternative [29]. Indeed, such reactions yield diastereomeric ions (equations 12 and 13) that contain only covalent bond linkages between the two parts of the diagnostic ions. These ions, $R'_sNH_2R_s^+$ and $R'_rNH_2R_s^+$, can be differentiated by CAD after selection by the first analyser.

$$R'_sOH + R_sNH_3^+ \rightleftharpoons [R'_sOH+R_sNH_3]^+ \longrightarrow R'_sNH_2R_s^+ + H_2O \qquad (12)$$

$$R'_rOH + R_sNH_3^+ \rightleftharpoons [R'_rOH+R_sNH_3]^+ \longrightarrow R'_rNH_2R_s^+ + H_2O \qquad (13)$$

The following chiral reagents have been used to differentiate menthol 12r and 12s enantiomers [29]:

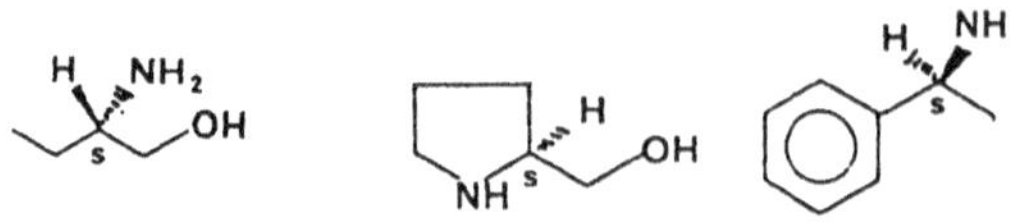

9s 10s 11s

12r 12s

It has been demonstrated that the ion-molecule reaction involved is (i) a bimolecular nucleophilic reaction (e.g. S_N2) and (ii) a regioselective process since the OH/NH_2 substitution occurs only between the -OH site of menthol and the $-NH_2$ group of the chiral reagent (equation 14), rather than between the two hydroxyl groups.

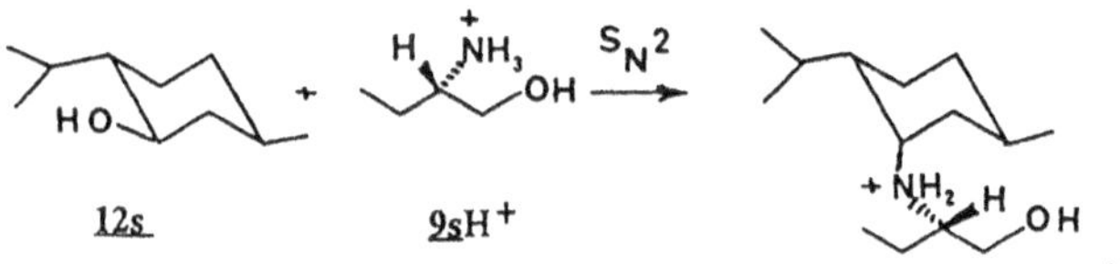

12s 9sH+ $[12r + 9sH-H_2O]^+$ (14)

The MIKE/CAD spectra of the substituted $[12r + 9sH - H_2O]^+$ and $[12s + 9sH - H_2O]^+$ ions (m/z 228) display characteristic differences, particularly in the daughter ion abundances (e.g. m/z 210, 139, 90 and 73, scheme 4). For example, by changing the R configuration of the hydroxylic site into an S configuration, the relative abundance of the daughter ion $[12s + 9sH - 2H_2O]^+$ is increased by a factor of two and the intensity ratio of the two daughter ions I(m/z 210)/I(m/z 73] is increased from 0.66 to 3.85.

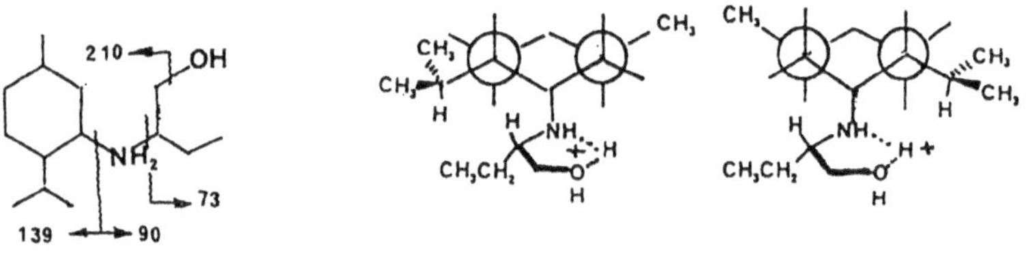

Scheme 4.

Steric hindrance is a possible reason for the enhancement of the loss of water from the 12s enantiomer. Indeed, from the examination of a Dreiding model, (see the Newman representation in scheme 4), it appears that the proximity of neighbouring groups such as (i) isopropyl, methyl (equatorial configuration on the menthol skeleton) and ethyl (aminoalcohol skeleton), and (ii) the quasiaxial hydrogen atom at the isopropyl substituent in $[12r + 9sH - H_2O]^+$, hinder the proton transfer that occurs via a five-centre transition state which is needed for the loss of H_2O. Such an interaction is not observed for the $[12s + 9sH - H_2O]^+$ ion. It should therefore be possible, *a priori*, to assign the initial chirality of unknown alcohols.

The other chiral reagent gas, (10s and 11s) gives less spectacular differences. Other enantiomeric substrates such as mandelic acid [30] and pyrolidin pheromones [31] can be distinguished by means of this method which is sufficiently sensitive and reproducible for analytical purposes.

2.2. STEREOSPECIFIC ION-MOLECULE REACTIONS INDUCED FROM SELECTED IONIC SPECIES IN THE COLLISION CELL.

Several studies [32 - 37] have shown that a tandem mass spectrometer in which the collision cell is based on an RFonly quadrupole, e.g. a triple quadrupole or hybrid instrument, can be used for the analysis of ion-molecule reactions. Mass-selected ions of low kinetic energy (E_{lab} ≈ 3eV) are introduced into the collision cell and collide with the target reagent gas introduced into the RF-only quadrupole. This method therefore allows one to control the composition and energies of selected ions [38]. A number of different substrates have been analysed by means of this method. These low collision-energy spectra have been called collision-activated-reaction (CAR) spectra.

2.2.1. *Stereospecific Proton Transfer to Nucleophilic Reagents.*

Previous publications of Bursey [39] provide examples of the differentiation of stereoisomers (e.g. 9a cis and 9b trans 1,2-cyclopentane diols) by means of their CAR spectra obtained with different reagent gases. In particular, it was shown that protonated diol isomers react with NH_3 by proton transfer leading to NH_4^+ in an abundance which depends upon the diol stereo-chemistry. By varying E_{Lab} he was able to demonstrate that for the cis 1,2-diol 9aH$^+$, the reaction is 32 kJ mol^{-1} endothermic (measurements obtained from TIKE experiments [40]), whereas the proton transfer reaction from the trans 1,2-diol 9bH$^+$ is exothermic.

(a) CAR spectra of protonated molecules as a general method for the comparison of proton affinities of diastereoisomeric compounds.

The protonated hydroxy-ketals 2aH$^+$ and 2bH$^+$ (see section 2.1.1.b) formed in the ion source and submitted to low energy collisions (E_{Lab} ≈ 3eV) with trimethylamine as target reagent gas yield spectacularly different CAR spectra [41]. This is because reaction (15) takes place much more rapidly with the cis 2aH$^+$ than with the trans 2bH$^+$ ions leading to intensity ratios of I(HN$(CH_3)_3{}^+$)/I(2aH$^+$) = 3.9 and I([HN$(CH_3)_3]^+$)/I(2bH$^+$) = 0.03. Furthermore, at higher values of E_{Lab}, the abundance of the [HN$(CH_3)_3]^+$ formed from 2aH$^+$ decreases somewhat whereas it increases if formed from 2bH$^+$. This particular behaviour therefore indicates that proton transfer is endothermic for the endo-isomer (hydrogen bonding leads to an increase in PA) in contrast to the reaction with 2aH$^+$, so that PA(2a) < PA($(CH_3)_3$N) < PA([2b]).

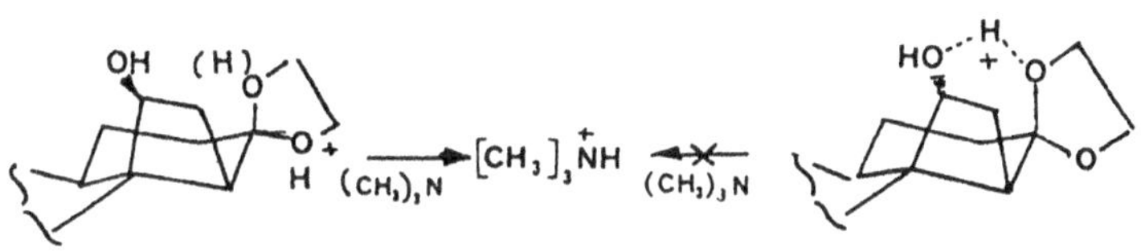

2aH$^+$ 2bH$^+$

If one uses a nucleophilic reagent gas, such as acetone or methanol, which have lower proton affinities, proton transfer occurs at higher collision energies but in much lower yields than was observed for NH_3 and $(CH_3)_3N$.

(b) Stereochemical dependence of the respective production of NH_4^+ and $[M + NH_4]^+$ from MH^+ [10,42].

Although 2aH$^+$ and 2bH$^+$ ions do not form adduct ions, such ions are formed as stable species by the isomeric hydroxy-benzoates 1a and 1b, together with NH_4^+ and $N_2H_7^+$ ions (see Table 2). These ions are produced in reactions (16a) and (16b). Other processes, such as SN reactions, also occur and will be discussed later.

$$\underline{1a}H^+ + NH_3 \underset{k_{2a}}{\overset{k_{-2a}}{\rightleftharpoons}} [\underline{2a}H + NH_3]^+ \xrightarrow{k_{-1a}} \underline{2a} + NH_4^+ \qquad (16a)$$

$$\underline{2b}H^+ + NH_3 \underset{k_{2b}}{\overset{k_{-2b}}{\rightleftharpoons}} [\underline{2b}H + NH_3]^+ \xrightarrow{k_{-1b}} \underline{2b} + NH_4^+ \qquad (16b)$$

Table 2. Observed ion abundance relative to MH^+ measured in CAR spectra of $\underline{1a}H^+$ and $\underline{1b}H^+$ (multiple collision conditions)*

	$[NH_4]^+ + [N_2H_7]^+$	$[MH + NH_4]^+$
$\underline{1a}H^+$	0.3	0.6
$\underline{1b}H^+$	2.2	1.3

* M_sH^+ ions and fragment ions produced by low energy CAD are observed. Contribution of CAD process is included in MH^+.

The relative abundances of product ions of ion molecule reactions occurring in the collision cell are presented in Table 2 and allow one to make two stereochemical observations: (i) $\underline{1a}H^+$ is more reactive than $\underline{1b}H^+$ towards NH_3; (ii) the intensity ratio $I(NH_4^+ + N_2H_7^+)/I([M + NH_4]^+)$ observed for the cis-isomer $\underline{1a}H^+$ is lower than that obtained for the trans-isomer, $\underline{1b}H^+$. This latter finding demonstrates that PA($\underline{1a}$) > PA($\underline{1b}$) > PA(NH_3); a similar trend was observed from CAD spectra of MH^+ ions (see section 2.1.1.a). Based upon this trend in PA values, it is possible to produce approximate potential energy diagrams (Scheme 5) that illustrate the respective role played by the cis/trans stereochemistry of the protonated species on the ion-molecule reactions occurring with NH_3.

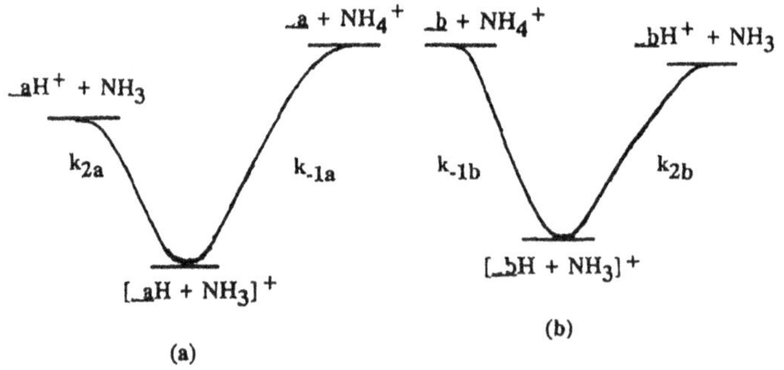

Scheme 5 : Simplified reaction-coordinate-energy diagram for (a) cis $\underline{1a}H^+$ and (b) trans $\underline{1b}H^+$.

The rate constant, k_{-2}, for the exothermic solvation of both $\underline{1a}H^+$ and $\underline{1b}H^+$ by NH_3 must be much greater than (i) that for the reverse reaction (k_2) and (ii) that for NH_4^+ formation (k_{-1}) and only the electrostatic and steric effects associated with the approach of NH_3 are limiting factors. Since CAR experiments were performed under low energy multiple collision conditions, the observation of adduct ions was to be expected. Furthermore, since the reactive collisions occurred at $E_{Lab} > 0$, it is possible that endothermic reactions may be observed (e.g. formation of NH_4^+). If one assumes that, in general, for competitive fragmentations, the process having the higher rate constant orients the decompositions, (i.e. k_{2a} and k_{2b}), then the results of table 2 may be explained. Since (i) k_{2a} is greater than k_{2b}, as a result of the effect of proton solvation multiplicity, it is expected that the intensity ratio I([1aH + $NH_3]^+$)/I([1bH + $NH_3]^+$) would be less than unity, and (ii) PA(1a) > PA(1b), then $k_{-1a} < k_{-1b}$ so that the formation of NH4+ from the decomposition of [1bH + $NH_3]^+$ is favoured.

The analysis of this example demonstrates that the combination of CAR spectra [43] and investigations of stereochemical effects can be very instructive in the study of the mechanisms of ion-molecule reactions.

2.2.2. *Stereospecific nucleophilic substitution in the collision cell.*

(a) Stereospecific and regioselective S_N2 reactions of selected MH^+ substrates; S_N2 enhancement by intramolecular anchimeric assistance.

In previous CAR spectra of the diastereoisomeric 1aH$^+$ and 1bH$^+$ hydroxybenzoates, in addition to the ions NH_4^+, $N_2H_7^+$ and $[MH + NH_3]^+$ (and other fragment ions produced by the CAD process), the substituted ions $[MH + NH_3 - H_2O]^+$ are observed in abundances that depend upon the cis/trans stereochemistry [42]. Note that the nucleophilic substitution occurs at the C-OH site rather than on the benzoate group since formation of $[MH + NH_3 - C_6H_5COOH]^+$ is not observed.

The stereochemical effects observed indicate that S_N2 or S_Ni reactions may occur from MH^+ either directly (S_N2, equation 17) or *via* formation of $[M + NH_3]^+$ which reacts by S_N2 and/or S_Ni pathways, as shown in equations 18a and 18b respectively.

$$MH^+ + NH_3 \xrightarrow[k_4]{S_N2} [MH{+}NH_3{-}H_2O]^+ \qquad (M_sH^+) \tag{17}$$

$$MH^+ + NH_3 \rightleftharpoons [MH{+}NH_3]^+ \xrightarrow{SNi} [MH{+}NH_3{-}H_2O]^+ + H_2O \ \text{(retention)} \tag{18a}$$

$$[MH{+}NH_3]^+ \xrightarrow[S_N2]{+NH_3} [MH{+}NH_3{-}H_2O]^+ + H_2O + NH_3 \ \text{(inversion)} \tag{18b}$$

From the trans 1bH$^+$ isomer, substituted M_sH^+ ions are formed in low relative abundances (i.e. $I(M_sH^+)/I(MH^+) = 1.6$) in comparison with those generated from the cis 1aH$^+$ ion for which the corresponding intensity ratio is over 4. Two complementary reasons can provide a possible interpretation of these results: (i) in the trans 1bH$^+$ ion, protonation occurs at the benzoate site with a low contribution from the OH site (see section 2.1.1.a) and the SN process occurring at the OH position remains improbable; and (ii) anchimeric assistance favoured the S_N2 process from 1aH$^+$ as shown in equation 19:

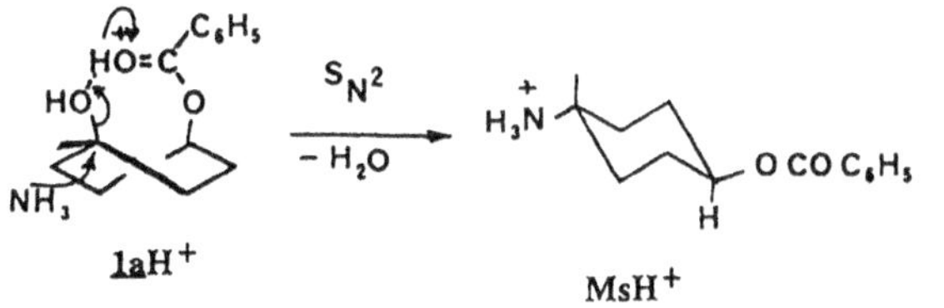

1aH$^+$ MsH$^+$ (19)

An S_Ni process is ruled out since it has been shown previously that hydrogen bonding hinders the S_Ni process (see section 2.1.2.c). Although one may explain the effects of stereochemistry in terms of MH^+ reacting with NH_3 to give mainly M_sH^+ (equation 17), one cannot rule out a significant contribution from $[M + NH_4]^+$ in the nucleophilic substitution.

(b) The role of the transition state in the competition between the formation of MH^+ and M_sH^+ in the $[M + NH_4]^+/NH_3$ system [42].

For this type of approach, because $\underline{1a}H^+$ reacts rapidly with NH_3, the $\underline{1a}H^+$ and $[\underline{1a} + NH_4]^+$ ions were selected for comparison of their reactivity under identical CAR conditions. A particularly striking difference in ion abundances is the intensity ratio $I([\underline{1a} + NH_4 - H_2O]^+)/I(\underline{1a}H^+)$ which decreases from 4.1 for the precursor ion $[\underline{1a}H + NH_4]^+$ to 0.3 for $\underline{1a}H^+$, indicating a marked increase in the formation of $\underline{1a}H^+$ rather than $[\underline{1a} + NH_4 - H_2O]^+$ from the adduct ion. This inversion of behaviour is difficult to explain only in terms of the relative values of $\Delta H^o{}_f$ of the competing final states, $[\underline{1a}H^+ + NH_3]$ and $([\underline{1a} + NH_4 - H_2O]^+ + H_2O)$. If one assumes that (i) an activation energy barrier exists for the S_N2 transition state, and (ii) the barrier for $\underline{1a}H^+$ is lower than that from $[\underline{1a} + NH_4]^+$, i.e. $k_{4a} > k_{5a}$, the results may be explained. The value of $\Delta H^o{}_f$ of the transition state for the S_N2 reaction of $\underline{1a}H^+$, however, is likely to be higher than that for the corresponding reaction of the ion $[\underline{1a} + NH_4]^+$ since the charge is more dispersed in the latter transition state (scheme 3).

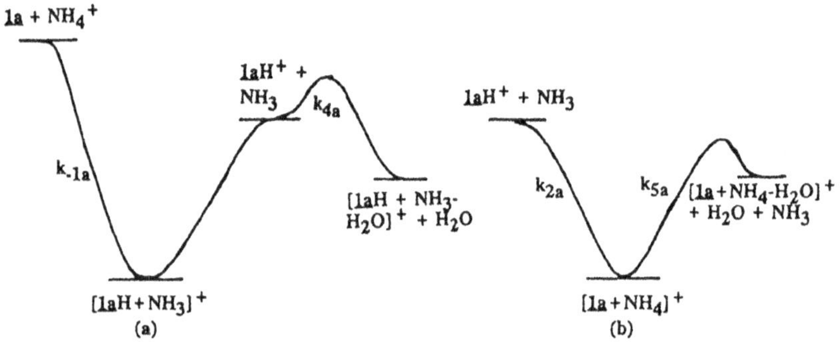

Scheme 6 : Comparison of approximate reaction profiles for competitive reactions occuring with NH_3 in the collision cell for (a) $\underline{1a}H^+$ and (b) $[\underline{1a} + NH_4]^+$.

For a reactive collision energy of $E_{Lab} \approx 3eV$, the formation of $\underline{1a}H^+$ rather than $[\underline{1a}H + NH_3 - H_2O]^+$ from the ion $[\underline{1a} + NH_4]^+$ is favoured since despite its formation being more endothermic, the reaction forming it has a higher frequency factor than that for M_sH^+ production which, since it is formed in a rearrangement process, will have a lower frequency factor. Reactions of $\underline{1a}H^+$ under similar conditions are different since $k_{4a} >> k_{5a}$ and the reactions leading to the formation of $[\underline{1a} + NH_4]^+$ and $[\underline{1a}H + NH_3 - H_2O]^+$ are each exothermic and in competition.

3. Deprotonation reactions induced by negative ion reagents.

3.1. STEREOSPECIFIC DECOMPOSITIONS OF $[M - H]^-$ IONS, AND ION-DIPOLE FORMATIONS.

In a high pressure source, some ion-molecule reactions have been observed that give rise to various adduct ions the structures of which have not been clearly established [44]. Structures of $[M - H]^-$ ions have, however, been widely studied, particularly by means of CAD, and the discussion here will be limited to this type of ion.

The $[M - H]^-$ ions are prepared using negative reagent ions A-, (e.g. NH_2-, OH-, CH_3O-, etc.) in an exothermic process (equation 20):

$$M + A^- \longrightarrow (M\text{-}H)^- + AH \quad \text{with} \quad \Delta G^\circ{}_{Acid(M)} < \Delta G^\circ{}_{Acid(AH)} \tag{20}$$

For polyfunctional compounds, deprotonation can occur competitively from different acidic sites, depending on their relative acidities [46]. The decompositions of $[M - H]^-$ ions under CAD conditions depend, as expected, upon the charge localisation. The analysis of stereochemical effects should consequently be a useful tool for the study of decomposition processes of negative ions.

The extent of stereochemical effects during ion decomposition depends on the mechanism of the reaction. Indeed, one-step pathways,(quasi-concerted processes) give rise to the more important effects. On the other hand, step-wise pathways (e.g. formation of ion-dipole intermediates) can hinder these effects; in certain cases, however, the stereochemical effects can be preserved and they may even enhance the contribution of stereospecific decomposition pathways.

3.1.1 *Stereospecific decompositions of selected bifunctional $[M - H]^-$ ions under CAD conditions.*

Stereochemical effects have been observed for fragmentations of diastereoisomeric deprotonated cyclohexane diols (cis 13a and trans 13b) that occur either in the ion source, or in the second field-free region (FFR) in a MIKES experiment. In contrast to the cis $[13a - H]^-$ ion which is stabilised by hydrogen bonding [48], the trans $[13b - H]^-$ ion loses H_2 specifically, both in a metastable transition or under CAD conditions [47]. The stereospecific loss of H_2 yields a keto-alkoxide ion by hydride transfer from the initially-formed alkoxide ion (equation 21).

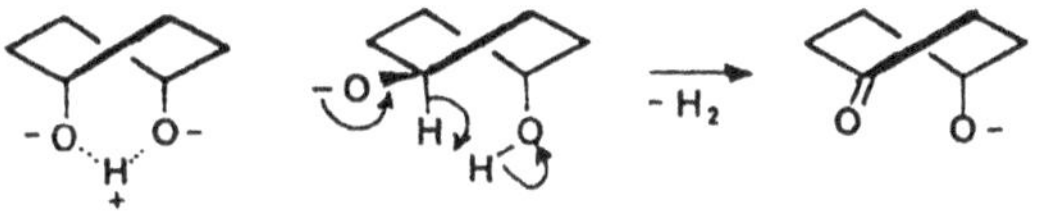

(21)

$[13a\text{-}H]^-$ $[3b\text{-}H]^-$

Such a hydride transfer also occurs specifically from more complex molecules (e.g. 14-17 androstanediols 6b with a trans relative stereochemistry, equation (22)). If H- can migrate, the R- can also be transferred giving rise to the loss of RH only from the trans stereoisomers [49].

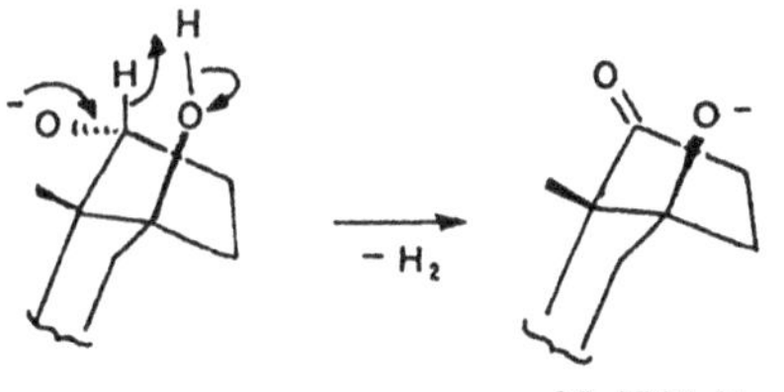

(22)

$[6b\text{-}H]^-$ $[6b\text{-}H\text{-}H_2]^-$

In addition, hydride transfer is able to lead to the specific reduction of a ketal group in the gas phase [50]. Indeed, the ketal-alkoxide with the exo-stereochemistry $[14a - H]^-$ leads to loss of C_2H_4O (equation 23) by the endo hydride attack on the ketal function which is adjacent to the CH-O- site. This results in the formation of the keto-alkoxide the stereochemistry of which is specifically exo [50]. It is probable that the H/ketal proximity strongly increases the reactivity of the H- ion and discriminates against the formation of an ion-dipole complex. Recently, this intramolecular ketal reduction (equation 23) has been shown to occur in solution [51].

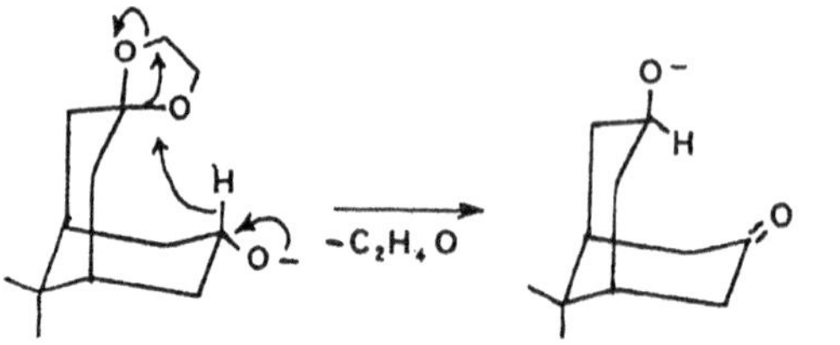

(23)

[14a-H]$^-$ [14a-H-C_2H_4O]$^-$

In the case of the endo isomer [14b - H]$^-$ ion, the formation of a hemiketal form is sterically preferred (ketal/alkoxide as neighbouring groups) and is produced by the intramolecular hydrolysis of the ketal from the endo side rather than by the ketal reduction pathway (equation 24). The isomeric ions [14a - H - C_2H_4O]$^-$ and [14b - H - C_2H_4O]$^-$ decompose in different ways, the former ion specifically losing water.

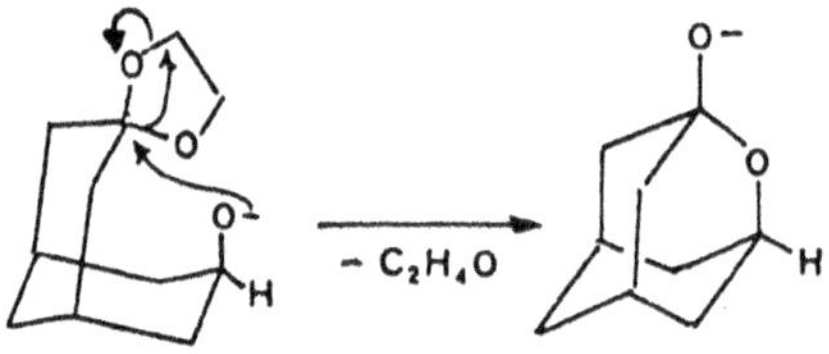

(24)

[14b-H]$^-$ [14b-H-C_2H_4O]$^-$

Many examples are found in the literature [45] which demonstrate the role played by stereochemical effects in the decomposition of [M - H]$^-$ ions.

3.1.2. *Ion dipole complex formation as a limiting factor in stereochemical effects in [M - H]$^-$ decompositions.*

As discussed in the previous section, stereochemical effects are important in negative chemical ionisation and depend mainly upon the particular spatial arrangement of atoms at reactive sites. Therefore, ΔH^o_f of the transition state rather than that of the initial state, [M - H]$^-$, plays the major role. Indeed, the stereochemical factor has little effect on the acidities of diastereoisomeric molecules as may be seen from the fact that the acidity values differ only by 4-5 kJ mol^{-1} for cis and trans cyclohexane diols [53] and for cis and trans methyl-2-cyclohexanols [54,55] (Table 3).

Table 3. Acidity values of cis/trans diastereoisomers.

1,4-cyclo hexane diols	$£G^o_{Acid}$ /kJ mol^{-1}	2-methyl cyclo hexanols	$£G^o_{Acid}$ /kJ mol^{-1}
cis	1516	cis	1523
trans	1521	trans	1527

Another study [56] shows that stereochemical evidence can provide an experimental criterion for the formation of ion-molecule complexes from [M - H]$^-$ ions. *A priori*, metastable decompositions and CAD processes arising from the exo-labelled borneols (15a and 15b) should show loss of H_2 from [15a - H]$^-$ and loss of HD from [15b - H]$^-$ (equation 25) since endo H/H (for 15a) and endo H/D (for 15b) are in *syn* alignment.

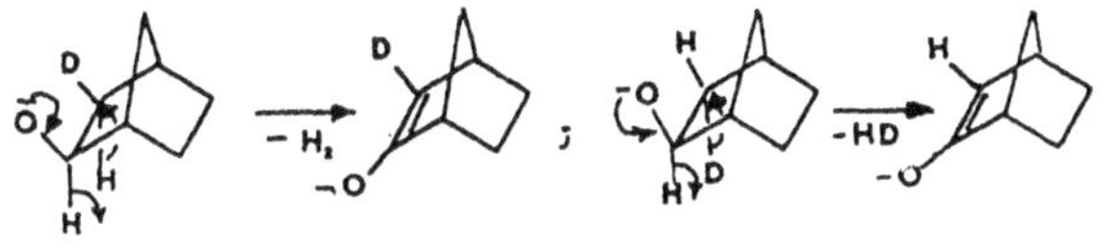

(25)

[15a-H]$^-$ [15a-H-H_2]$^-$ [15b-H]$^-$ [15b-H-HD]$^-$

The MIKE/CAD spectra of [15a - H]$^-$ and [15b - H]$^-$, however, show HD and H_2 losses in the ratio 0.9 and 1.1 respectively which differ from the ratios predicted for a one-step process. This observation rules out the loss by a concerted pathway and suggests that the pathway actually followed is a stepwise process *via* the formation of an ion-dipole complex [$C_{10}H_{15}DO.H$]$^-$ (equation 26).

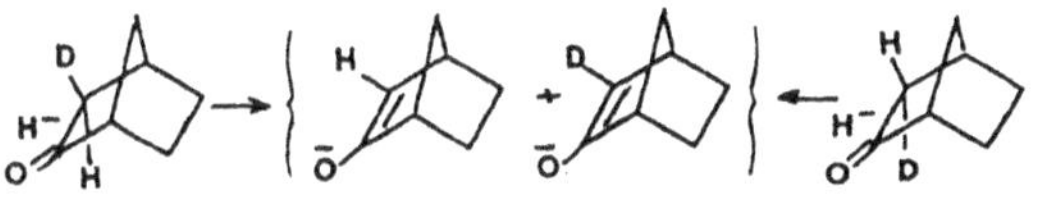

(26)

[15a-H]$^-$ [15b-H]$^-$

Finally, one should note that the competitive losses of H_2 and HD from this species occur with no observable kinetic isotope effect. If a similar intermediate were to be formed from the endo borneol isomer, it is to be expected that the stereochemical effects will be very small since the -OH stereochemistry is lost.

3.2. INFLUENCE OF THE NEGATIVE CHARGE SITE ON THE STRUCTURE OF THE ION-DIPOLE COMPLEX.

3.2.1. *The dependence of regioselective deprotonation on the mode of preparation of [M - H]$^-$.*

Recently, the deprotonated 17α- and 17β-oestradiol-17-stearate esters [16a - H]$^-$ and [16b - H]$^-$, generated under FAB and NICI conditions, have been investigated under low energy CAD conditions [57]. In order to simplify the discussion, the CAD spectrum of [16a - H]$^-$ is considered first as a function of the method of preparation of the ion. In particular, the following fragmentations are observed under low energy CAD conditions, the daughter ion abundance depending on the mode of ionisation (FAB or NICI-NH_3) as shown in Table 4.

R = -H_3C-$(CH_2)_{13}$-
16a = 17α Isomer
16b = 17β Isomer

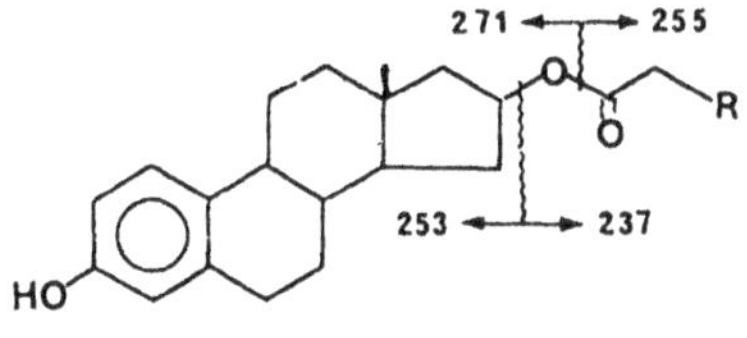

Table 4. Daughter ion abundances (relative to base peak) formed under CAD conditions from the [16b - H]$^-$ ion (m/z 509)

	Fragment ions (m/z)			
	237	253	255	271
FAB (glycerol)	8.1	45	100	6.9
NICI-NH_3	32	14	29.7	100

This table shows that when the [16b - H]$^-$ ion is generated by FAB, the major products are the ions m/z 253 and 255 whereas under NICI-NH_3 conditions, the major products are m/z 237 and 271. This difference signifies the formation of two isomeric forms of the ion such as [$16b_1$ - H]$^-$ and [$16b_2$ - H]$^-$ which can coexist without any possible isomerisation, e.g. *via* a long distance proton transfer, which is sterically unlikely.

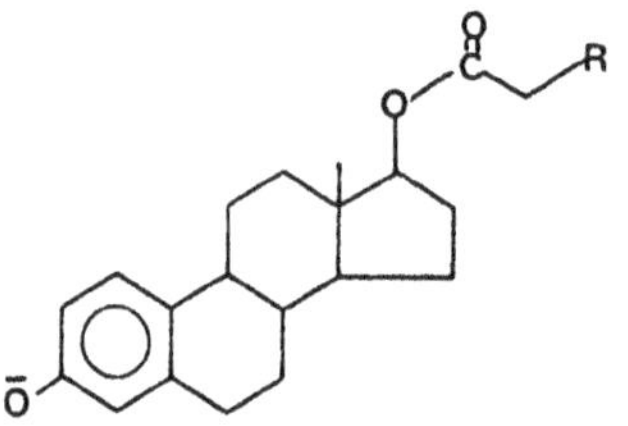

[$16b_1$-H]$^-$

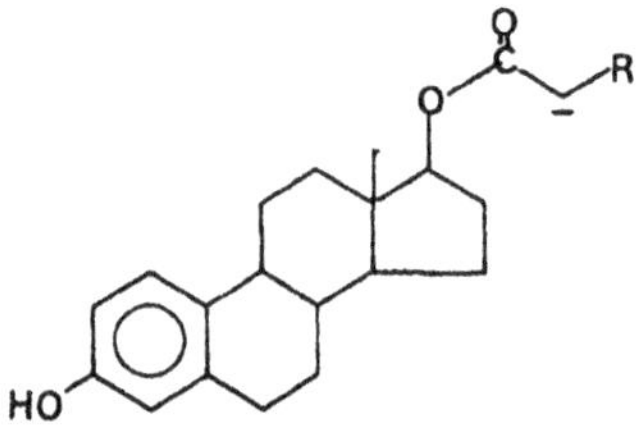

[$16b_2$-H]$^-$

One can prepare 16b(d_1), labelled at the phenol site, under FAB conditions but the peak at m/z 509 is not shifted to m/z 510. This indicates that the labelling is lost so that the charge must be located at the phenol site ([16b(d_1) - D]$^-$ which corresponds to the [$16b_1$ - H]$^-$ form). Thus deprotonation in the matrix is regioselective in accordance with the relative pKa's of the acidic sites. The production of the isomeric [$16b_1$ - H]$^-$ form under FAB conditions makes it difficult to rationalise the formation of the m/z 255 ion, $CH_3(CH_2)_{14}COO^-$, since the charge is located at the C_3 position. The formation of the carboxylate ion can be explained by a stepwise process based upon the Longevialle-Botter model [58]. In this, an ion-dipole [16b' - H]$^-$ intermediate must be produced by an isomerisation of [16b - H]$^-$ that is not promoted by the negative charge (equation 27).

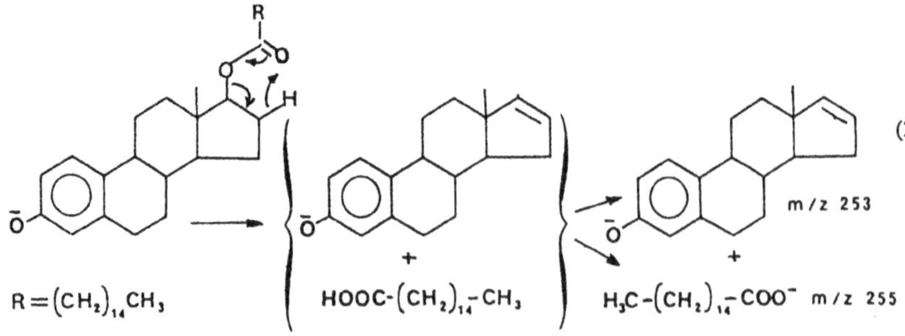

(27)

[$16b_1$-H]$^-$ [$16b'_1$-H]$^-$

On the other hand, under NICI-ND_3 conditions, both m/z 509 and m/z 510 are observed in similar abundances. This suggests that deprotonation is not regioselective and takes place at both acidic sites (phenol and enolic positions). The CAD spectra of the two ions show marked differences, the former yielding mainly the pair of complementary daughter ions m/z 253 and 255, whereas the latter is characterised by the pair of ions at m/z 237 and 271 (not shifted to m/z 272 despite D-labelling at C_3) together with the ions m/z 253 and 255 in low abundance. The formation of the unlabelled daughter ion m/z 271 from the labelled precursor [16b(d_1) - H]$^-$ can be explained by assuming that the latter isomerises to a second ion-dipole complex [16b'$_2$ - H]$^-$ induced by the negative charge at the enolic site (equation 28).

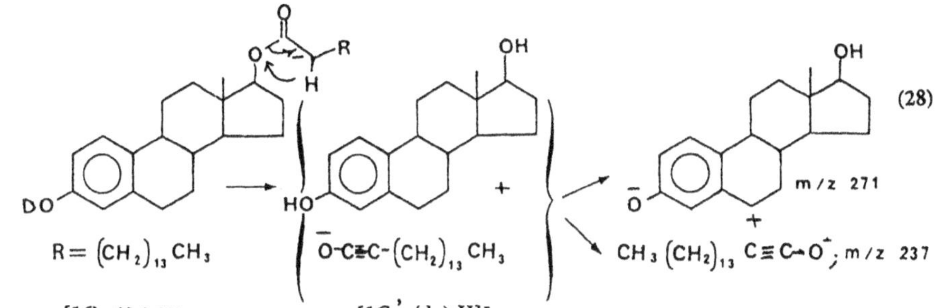

(28)

[16b$_2$(d_1)-H]$^-$ [16b'$_2$(d_1)-H]$^-$

Dissociations of the ion dipole complexes [16b'$_1$ - H]$^-$ and [16b$_2$'(d_1) - H]$^-$ occur by competitive proton or deuteron transfer since each component of the complex must be characterised by closely similar acidities.

Such step-wise decompositions are confirmed by labelling of the skeleton and appear to be general processes since analogous behaviour is observed for other long chain esters of oestradiol.

3.2.2. *Influence of stereochemistry on the formation of the ion-dipole complex.*

An unexpected stereochemical effect was observed for the deprotonated 17α-oestradiol-17 ester, [16a - H]$^-$, generated by FAB ionisation (deprotonation at the phenol site). Indeed, the daughter ion intensity ratio, I([m/z 253])/I([m/z 255]), measured under CAD conditions for this ion is close to 10 rather than the value of 0.5 observed for the same ratio from [16b - H]$^-$. This difference is behaviour may be explained by assuming that the rate constant for the formation of the ion-dipole complex from the 17α isomer is much lower than the corresponding rate constant for the 17β isomer. The ion-dipole complex formed from the latter isomer must be produced mainly prior to collision whereas that formed from the former isomer is generated mainly under CAD conditions and its dissociation is faster than any proton transfer process yielding the m/z 255 carboxylate ion.

3.3 THE FORMATION OF ION-DIPOLE INTERMEDIATES AS A POSSIBLE ORIGIN OF STRUCTURAL MEMORY EFFECTS OBSERVED FOR NATURAL ISOMERIC HETEROSIDES.

Various studies demonstrate that NICI is a suitable ionisation method for the analysis of polyhydroxy compounds [59], and is particularly suitable for the structural differentiation of heterosides containing isomeric sugar [60]. Recently, NICI combined with low energy CAD has been used [61] to distinguish the monoterpene glucosides 17a (stereochemistry Z) and 17b (stereochemistry E), isomeric heterosides with isomeric aglycone moieties, by means of consecutive decompositions of [M - H]$^-$ ions under CAD conditions.

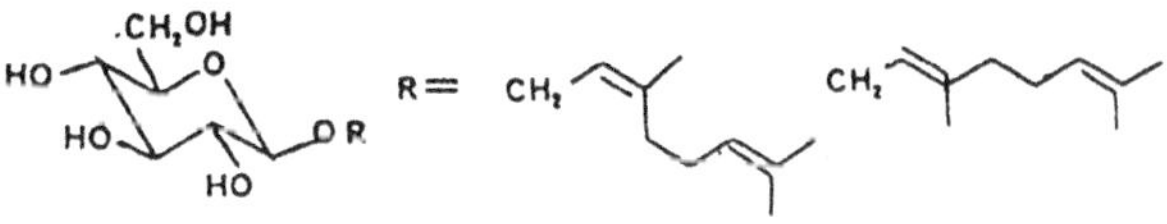

isomer Z (17a) isomer E (17b)

The deprotonated [17a - H]$^-$ and [17b - H]$^-$ ions fragment into the $C_6H_{11}O_6^-$ alkoxide ion (equation 29) which has been shown to have a common structure when formed from each of the precursors [62].

$$[C_6H_{10}O_5\text{-}O\text{-}C_{10}H_{17}]^- \longrightarrow C_6H_{11}O_6^- + C_{10}H_{16} \quad (29)$$

A fraction of the daughter ions $C_6H_{11}O_6^-$ readily decomposes into grand-daughter ions such as m/z 149, 119 and 89. The abundances of these ions, relative to the base peak at m/z 119 are very similar when formed from either precursor which confirms that the $C_6H_{11}O_6^-$ intermediate ions have a common structure. The total ionic current of the grand-daughter ions relative to their direct precursors (mainly m/z 179, $C_6H_{11}O_6^-$), however, depends on the initial aglycone stereochemistry. The values of the total ion current (arbitrary units) are 0.35 and 2.3 for [17a - H]$^-$ and [17b - H]$^-$ respectively. This suggests that the fraction of the intermediate $C_6H_{11}O_6^-$ ions that decomposes depends on the Z and E double bonds of the aglycone. This effect was called the "structural memory effect" since it appears that grand-daughter ion abundances depend on grand-parent ion structures and the structure of the intermediate ion is common for both heterosides. A study of collision energy effects confirms this suggestion since similar grand-daughter ion abundances are found for each of the deprotonated isomers whereas the abundance of the $C_6H_{11}O_6^-$ ions varies. The mechanism of formation of the unstable $C_6H_{11}O_6^-$ ions must be stepwise, *via* an ion-dipole intermediate where the charge is retained (equation 30) at the carbonyl site of the alkoxide position. Such a model has been suggested recently by Promé [63] for less complex glucosides.

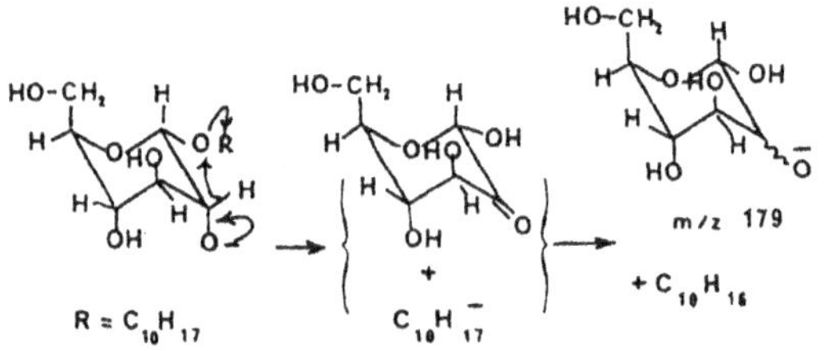

(30)

This possible model is consistent with the CAD spectra obtained from the deprotonated molecules labelled at different positions of the glucose and terpene moieties. It has been applied to the analysis of the natural mixture of heterosides contained in fruit juices [64].

References

1. (a) J.C. Lorquet, Org. Mass Spectrom. 16, 469, 1981; (b) L. Radom, M.W. Wong, and P.W.M. Gill, Advances in Mass Spectrometry, Vol. 11A, Ed. P. Longevialle, Heyden and Son, London, p 702, 1989; (c) C. Lifshitz. Advances in Mass Spectrometry, Vol. 11A, Ed. P. Longevialle, Heyden and Son, London, p 713, 1989.

2. (a) A. Mandelbaum, Mass Spectrometry Reviews 2, 223, 1983; (b) F. Turecek, Coll. Czeck. Chem. Comm. 52, 1921, 1987.

3. J.L. Holmes, Advances in Mass Spectrometry, Vol. 11A, Ed. P. Longevialle, Heyden and Son, London, p 53, 1989.

4. R. Houriet, Advances in Mass Spectrometry, Vol. 11A, Ed. P. Longevialle, Heyden and Son, London, p 626, 1989.

5. (a) P.H. Dawson, Int. J. Mass Spectrom. Ion Phys. 50, 287, 1983; (b) D.J. Douglas, J. Phys. Chem. 86, 185, 1982.

6. S. Le Meillour, R. Cole, F. Clairet, F. Fournier, J.C. Tabet, T. Blasco, C. Beaugrand and G. Devant, Advances in Mass Spectrometry, Vol. 11A, Ed. P. Longevialle, Heyden and Son, Lodnon, p 236, 1989.
7. (a) J.B. Westmore and M.M. Alauddin, Mass Spectrometry Reviews, 5, 381, 1986; (b) H. Budzikiewicz, Mass Spectrometry Reviews, 5, 345, 1986.
8. J. Keough and A.J. DeStefano, Org. Mass Spectrom. 16, 527, 1981.
9. Y.Y. Lin and L.L. Smith, Biomed. Mass Spectrom. 5, 604, 1978.
10. S. Le Meillour and J.C. Tabet, Spectroscopy Int. J. 5, 135, 1987.
11. S.G. Lias, J.E. Bartmess, J.F. Liebman, J.L. Holmes, R.D. Levin and W.G. Maillard, "Gas phase ion and neutral thermochemistry", in J. Phys. Chem. Ref. Data 17, supp. 1, 1988.
12. S. Le Meillour, thesis, 1991, Paris.
13. I. Jardine and C. Fenselau, J. Amer. Chem. Soc. 98, 5088, 1976.
14. R.B. Cole, C. Prevost and J.C. Tabet, Advances in Mass Spectrometry, Vol. 11B, Ed. P. Longevialle, Heyden and Son, London, p 1056, 1989.
15. F.J. Winkler and D. Stahl, J. Amer. Chem. Soc. 101, 3685, 1979.
16. R.B. Cole and J.C. Tabet, Org. Mass Spectrom. in Press.
17. (a) J. Bastard, Dokhac Manh, M. Fetizon, J.C. Tabet and D. Fraisse, J. Chem. Soc., Perkin Trans. II, 1591, 1981; (b) J.C. Tabet, A. Bouillot, C. Prevost, J. Bastard, Dokhac Manh, Y. and Tondeur, Helv. Chim. Acta 69, 809, 1986.
18. (a) D. Despeyroux, Thesis, Paris, 1990 and (b) D. Despeyroux, R.B. Cole, M. Ricard and J.C. Tabet, Advances in Mass Spectrometry, Ed. P. Longevialle, Heyden and Son, London, Vol. 11B, p 1056, 1989.
19. J. Jalonen, J. Taskinen and C. Glidewell, Int. J. Mass Spectrom. Ion Phys. 46 , 467, 1983.
20. F.O. Galucar, F. Mermoud, F.J. Winkler and A. Buchs, Helv. Chim. Acta 64, 488, 1984.
21. J.C. Tabet and D. Despeyroux, 37th Annual Conference on Mass Spectrometry and Allied Topics, p 829, 1989.
22. D.J. McAdoo, Mass Spectrometry Reviews, 7, 363, 1988.
23. J.C. Tabet, M. Bertranne, J.C. Beloeil and D. Stahl, Org. Mass Spectrom. 19, 363, 1984.
24. J.C. Promé and D. Stahl, Org. Mass Spectrom. 20, 528, 1985.
25. (a) J.C. Tabet, H.E. Audier and J.P. Denhez, Tetrahedron 43, 873, 1985; (b) K.P. Madhusudanan, M. Prasad, S.N. Rastoyi and D. Fraisse, Org. Mass Spectrom. 20, 630, 1985.
26. H.M. Fales and G.J.J. Wright, J. Amer. Chem. Soc. 99, 2339, 1977.
27. F.J. Winkler, D. Stahl and F. Maquin, Tetrahedron Letters 27, 335 1986.
28. H. Suming, C. Yaozu, J. Longfei and X. Shuman, Org. Mass Spectrom. 21, 7, 1986.
29. J.C. Tabet, Tetrahedron, 43, 3413, 1987.
30. J.C. Tabet, Spectroscopy Int. J. 5, 83, 1987.
31. J.C. Tabet, G. Lhommet, J. Royer and P. Husson, Advances in Mass Spectrometry, Vol. 11B, Ed. P. Longevialle, Heyden and Son, London, p 530, 1989.
32. J. Chakel, H. Gregg, J. Allison and K. Enke, 28th Annual Conference on Mass Spectrometry and Allied Topics, New York, 1980, p 142.
33. J.H. Bakey and J.M. Tedder, J. Chem. Soc. Perkin Trans II, 1263, 1963.
34. (a) G.L. Glish, P.H. Hemberger and R.G. Cooks, Anal. Chim. Acta 119, 137, 1980; (b) R.R. Pachuta, H.I. Kenttämaa, R.G. Cooks, T.M. Zennie, C. Ping, C. Chang and J.M. Cassady, Org. Mass Spectrom, 23, 10, 1988.
35. J.A. Freeman, J.V. Johnson, M.E. Hail, R.A. Yost and D.W. Kuehl, J. Amer. Soc. Mass Spectrom, 1, 110, 1990.
36. J.P. Schmit, P.H. Dawson and N. Beaulieu, Org. Mass Spectrom, 20, 269, 1985.
37. (a) R. Orlando, C. Fenselau and R.J. Cotter, Org. Mass Spectrom. 24, 1033, 1989; (b) R. Orlando, C. Murphy, C. Fenselau, G. Hansen and R.J. Cotter, Anal. Chem. 62, 1990; (c) R. Orlando, C. Fenselau and R.J. Cotter, Rapid Comm. Mass Spectrom. 7, 259, 1990.
38. E. Schröder, Org. Mass Spectrom. 24, 205, 1989.

39. (a) E.L. White and M.M. Bursey, Biomed. and Environm. Mass Spectrom. 18, 413, 1989; (b) W.J. Meyerhoffer and M.M. Bursey, Org. Mass Spectrom. 24, 169, 1989; (c) W.J. Meyerhoffer and M.M. Bursey, Org. Mass Spectrom. 24, 246, 1989.
40. (a) M.T. Kinter and M.M. Bursey, J. Amer. Chem. Soc. 108, 1787, 1986; (b) M.T. Kinter and M.M. Bursey, Org. Mass Spectrom. 22, 775, 1987.
41. R.C. Cole, E. Prevost and J.C. Tabet, Advances in Mass Spectrometry, Ed. P. Longevialle, Heyden and Son, London, p. 1054, 1989.
42. S. Le Meillour and J.C. Tabet, Advances in Mass Spectrometry, Ed. P. Longevialle, Heyden and Son, London, p. 572, 1989.
43. H.I. Kenttämaa and R.G. Cooks, J. Amer. Chem. Soc. 111, 4122, 1989.
44. H. Budzikiewicz, Org. Mass Spectrom. 23, 561, 1988.
45. (a) J.H. Bowie, Mass Spectrometry Reviews, 9, 349, 1990; (b) T.A. Roy, F.H. Field, Y.Y. Lin and L.L. Smith, Anal. Chem. 51, 272, 1979.
46. J.E. Bartmess, Mass Spectrometry Reviews 8, 297, 1989.
47. T. Gäumann, D. Stahl and J.C. Tabet, Org. Mass Spectrom. 18, 263, 1983.
48. R. Houriet, D. Stahl and F.J. Winkler, Environmental. Health Perspective 36, 63, 1980.
49. J.C. Beloeil, M. Bertranne, D. Stahl and J.C. Tabet, J. Amer. Chem. Soc. 105, 1355, 1983.
50. J.C. Tabet, I. Hanna and M. Fetizon, Org. Mass Spectrom. 20, 61, 1985.
51. (a) T. Momose, T. Itooka, T. Nishi, M. Uchimoto, K. Ohnishi and O. Maraoka, Tetrahedron 43, 3713, 1987; (b) I. Watt, S.N. Whittleton and S.M. Withworth, Tetrahedron, 42, 1047, 1986.
52. L.G. Wright, S.A. McLuckey and R.G. Cooks, Int. J. Mass Spectrom. Ion Phys. 42, 115, 1982.
53. R. Houriet, J.C. Tabet and A. Tchapla, Spectroscopy Int. J. 3, 132, 1984.
54. F. Clairet, R.G. Cooks and J.C. Tabet, to be published
55. A.G. Harrison, unpublished results.
56. D. Sutzle and H. Schwarz, Helv. Chem. Acta, 72, 320, 1989.
57. (a) F. Fournier, J.C. Tabet, L. Debrauwer, D. Rao, A. Paris and G. Bories, Rapid Comm. Mass Spectrometry in press; (b) F. Fournier, C. Salles, J.C. Tabet, L. Debrauwer, D. Rao, A. Paris and G. Bories, Anal. Chim. Acta, in press.
58. (a) P. Longevialle and R. Botter, Org. Mass Spectrom. 18, 1, 1983; (b) T.H. Morton, Tetrahedron 38, 3195, 1982; (c) D.J. McAdoo, Mass Spectrometry Reviews 7, 363, 1988.
59. W.C. Brumley, D. Andrzejewski and J.A. Sphon, Org. Mass Spectrom. 23, 204, 1988.
60. (a) J.W. Dalling and W. Heerma, Biomed. and Environm. Mass Spectrom. 18, 363, 1989; (b) J.W. Dallinga, F.D. Rinkema and W. Heerma, Biomed, and Environm. Mass Spectrom. 18, 241, 1984.
61. R.B. Cole, J.C. Tabet, C. Salles, J.C. Jallegeas and G. Grouzet, Rapid Comm. Mass Spectrom, 3, 59, 1989.
62. L.B. Ma, F. Fournier, T. Blasco, J.C. Tabet, C. Salles and J.C. Jallageas, J. Amer. Soc. Mass Spectrom., submitted.
63. J.C. Promé, H. Aurelle, D. Promé and A. Savignac, Org. Mass Spectrom. 22, 6, 1987.
64. F. Fournier, L.M. Ma, J.C. Tabet, C. Salles, J.C. Jallageas and G. Grouzet, Spectroscopy Int. J., in press.

E2/S_N2 AND OTHER ORGANIC ION-MOLECULE REACTIONS

Charles DePuy (Colorado) and Hans Grützmacher (Bielefeld), Organizers

Eleven presentations were made, nine on various aspect of gas phase E2 and/or S_N2 reactions and two on positive ion-molecule processes. Presentations were limited to five minutes each and discussion was delayed until all papers were presented.

Chuck DePuy began by posing two problems in which the Boulder group is especially interested. The first of these is to identify those S_N2 reactions of alkyl halides for which a central barrier, as shown in Fig. I, is unequivocally slowing down the reaction rate. In the literature it is frequently assumed that this is the case for reactions of CH_3Cl with HO^-, CH_3O^-, and F^- but evidence was presented that this may not be the case and that only when much weaker nucleophiles are used ($HO^-(H_2O)$, $CF_3CH_2O^-$, HS^-) can one be sure of the influence of a central barrier on the reaction rate. He pointed out that while methyl chloride reacts slightly slower than the collision rate with the strong bases, neopentyl chloride reacts at essentially the same rate, whereas with the weaker bases not only is the reaction efficiency much below the collision rate but neopentyl chloride reacts much slower than does methyl chloride. He also reported that, based on relative rate constants for the series MeCl, EtCl, i-PrCl and t-ButylCl, oxyanions appear to react mainly or exclusively by E2 processes where possible while thio anions of the same basicity react only by S_N2 .

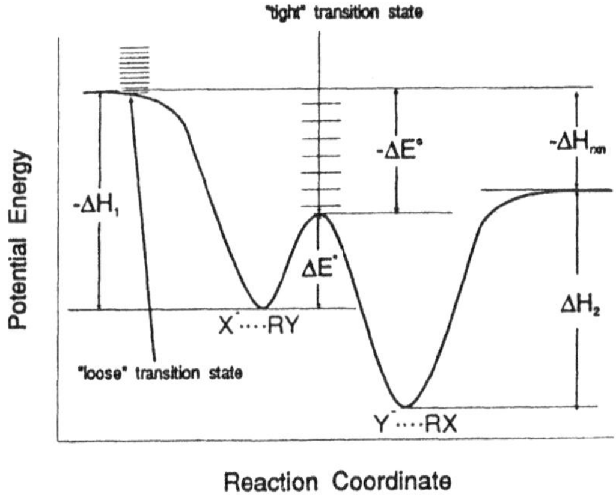

Figure 1. Potential Surface for the Gas-phase S_N2 Reaction

Veronica Bierbaum (Colorado) reported on her results using deuterium isotope effects to probe E2 and S_N2 reactions. She found substantial β-deuterium isotope

K. R. Jennings (ed.), Fundamentals of Gas Phase Ion Chemistry, 373–378.

effects (k_H/k_D as high as 4.7) in those reactions for which E2 reactions are expected to predominate (RO^-, 2° and 3° alkyl bromides and chlorides) and no isotope effect or even a modest inverse isotope effect (k_H/k_D as small as 0.75) for reactions which occur only by S_N2 processes (RS^- with 1° and 2° halides, all nucleophiles with methyl halides).

Bob Morris (Geophysics Laboratory, Mass.) discussed experimental and theoretical work on the kinetic energy and temperature dependencies of the rate constants for the reactions of F^- with methyl chloride, bromide and iodide. These reactions proceed at rates which are less than the collision rate, but this inefficiency may be due, not to the intervention of a central barrier along the low energy path, but rather from improperly oriented collisions which lead to dissociation back to reactants, without necessarily sampling that part of the potential surface which can lead to a successful S_N2 reaction.

Terry McMahon (Toronto) described experiments using a high pressure mass spectrometer (HPMS) in which both well depths and internal energy barriers on the S_N2 potential energy surface were characterized. On such surfaces (Fig I) the first well depth can be characterized by HPMS clustering equilibria as long as the barrier height is near or above the energy of the reactants. Similarly the well depth for the reverse, endothermic S_N2 reaction can always be obtained. If the central barrier is within ± 2 kcal/mol of the reactant energy, application of transition state theory can be used in conjunction with measurement of the temperature dependence of the rate constant for the reaction to yield the energy of the central barrier. Data obtained are summarized in Table I.

Table I. Stationary Point Energies on the Potential Energy Surface of Gas-Phase S_N2 Reaction.

S_N2 Reaction Type	Well I ($\Delta H_1, \Delta S_1$)	T.S (ΔE_o)	ΔE^*_o	Well II ($\Delta H_2, \Delta S_2$)	ΔH_{rxn}
$CN^- + CH_3Cl \rightarrow$ $Cl^- + CH_3CN$	-10.1 -17.8	12[a]	36.8	-15.8 -18.7	-27.2
$Cl^- + CH_3Cl \rightarrow$ $Cl^- + CH_3Cl$	-10.5 -16.3	3[b]	13.5	-10.5 -16.3	0
$CN^- + CH_3Br \rightarrow$ $Br^- + CH_3CN$	-11--14	-0.8	26.9	-14.3 -18.7	-35.4
$Cl^- + CH_3Br \rightarrow$ $Br^- + CH_3Cl$	-11--13	-0.5	14.8	-10.9 -20.3	-8.0
$Cl^- + CH_3CH_2CH_2Br \rightarrow$ $Br^- + CH_3CH_2CH_2Cl$	-14.2 -20.9	0.6	16.7	-12.1 -20.6	-7.0
$Cl^- + CH_3CH(Br)CH_3 \rightarrow$ $Br^- + CH_3CH(Cl)CH_3$	-15.2 -22.4	5.0	21.6	-12.4 -20.6	-6.6
$CN^- +$ t-BuCl $\rightarrow$ $Cl^- +$ t-BuCN	-15.9 -28.0	>10	na	-18.5 -25.0	-29.2

Nico Nibbering (Amsterdam) discussed results on E2 reactions of ethers which Leo de Koning obtained in his laboratory. Diethyl ether reacts with HO^- by an E2 reaction to give ethylene and both $C_2H_5O^-$ and $C_2H_5O^-(H_2O)$. The α- and β-deuterium isotope effects and leaving group effects have been obtained and (combined with data from cyclic systems) show that the unsolvated ion results from anti elimination and the cluster ion results from syn elimination. No S_N2 reactions occur at all. Analogous tracer studies on $C_2H_5SC_2H_5$ are more in agreement with a "merged" E2/S_N2 (E_2C) reaction in which the nucleophile (F^-) attacks the β-hydrogen and α-carbon atoms simultaneously, while for cyclic thioethers the S_N2 pathway can only compete effectively with E2 when the ring contains three or four atoms.

John Bartmess (Tennessee) investigated E2/S_N2 reactions in systems of the type $GCH_2CH_2CH_2X$, where G was varied among the groups Me, H, MeO, MeS, Cl, CN, etc. He showed that for X = Cl reaction with CH_3S^- leads to a Hammett-type plot when log k_{obs} is plotted against σ_I and gives $\rho_I = 6.4$. He estimated $\rho_I = 21$ for a true E_{1cb} reaction with a carbanionic intermediate; therefore the reactions under investigation show a strong leaving group effect and hence display a concerted mechanism rather than a stepwise E_{1cb} process. He also emphasized how gas phase ion molecule reactions can be slowed down by becoming involved in unproductive side reaction channels which return to reactants relatively rapidly. The reaction above with G = OH may be an example of this, where an oxyanion is formed competitively with the complex leading to E2/S_N2 reactions and the overall rate is reduced.

Michael Henchman (Brandeis) communicated work done jointly with Peter Hierl and Sally Meyer (Kansas). He considered two questions: Can central barriers affect the rate constants of highly exothermic reactions? Can accurate barrier heights be derived from measured rate constants?. They measured a negative temperature dependence for the rate constant of the reaction between DO^- and methyl chloride, a reaction which is exothermic by 50 kcal/mol. The reaction efficiency was found to be 0.64 at 213 K falling to 0.47 at 423 K. Because the reaction efficiency is less than unity, there has to be, in operational terms, a barrier. A central barrier provides the simplest mechanism to model the negative temperature dependence, but it is not the only mechanism that can do this. Even for reactions as exothermic as 50 kcal/mol, reaction efficiencies can be measurably less than unity, indicating, operationally, the action of a barrier.

The kinetic data for this reaction have been used to locate the central barrier; the top of the barrier lies 5.5 kcal/mol below the energy of the reactants. For comparison, the same quantity derived from the most accuratge hypersurface is 10 kcal/mol (Jorgensen et al, JACS 1987, 109, 2349). What does this comparison reveal? Can kinetic data yield accurate barrier heights?

This reaction provides a stringent test of this. First, an accurate hypersurface is available and the hypersurface defines the parameters of the transition state (at the top of the barrier) needed to calculate the barrier height. Second, the data cover a temperature range, and fitting these imposes a severe constraint upon the location of the barrier. Accordingly, the "experimental" number of 5.5 kcal/mol is not really adjustable. In contrast, the hypersurface calculation is least certain in the region of the barrier, so that an error of ±5 kcal/mol may not be unreasonable. In any event, a central barrier can account at least qualitatively for the reduced reaction efficiencies observed and for their temperature dependence for a reaction exothermic by 50 kcal/mol.

There is a fundamental reason why a barrier derived from kinetic data many differ from the corresponding barrier on the hypersurface. For the latter the top of the barrier is a single point in configuration space, lying on the minimum energy pathway where the whole system is optimized. In the kinetic measurements with energized reactants, the single points becomes a *hypervolume* in configuration space and the derived barrier height is a phenomenological quantity, averaged over the distribution function. The latter is temperature dependent, the former is not. The two quantities cannot be identical (even at 0 K!). The difference applies in neutral chemistry (Menzinger and Worfgang, Angew. Chem. 1969 81, 483), where the convex hypersurface constrains reactive trajectories to those that just traverse the barrier. For ion - molecule chemistry, where the concave hypersurface does not constrain the trajectories, the effect is accentuated.

Most applications are much less well defined than the reaction between DO^- and methyl chloride studied here. Where nothing is known about the transition state, its parameters have to be guessed. Often barrier heights are estimated from kinetic measurements made at a single temperature. If this is done for the single data points measured here,one obtains a range of barrier heights, which are poorly defined, which vary with temperature and which bear no clear relationship to the barrier height obtained from the full temperature dependence. It is always possible to derive a "barrier height" from a single rate constant. How this relates to the barrier height on the reaction hypersurface remains unclear.

Susan Graul (UC Santa Barbara) reported her preliminary results on the kinetic energy release distributions (KERD) of products from metastable dissociations of ion-dipole complexes which correspond to the intermediate species formed in S_N2 reactions. Ions of the type $X^-(CH_3Y)$ were formed in a cooled (40°) source of a VG ZAB-2F mass spectrometer using electron ionization at low pressures ("EI" conditions) to avoid collisional stabilization. The X^- ions were produced from a precursor other than CH_3X to prevent formation of $Y^-(CH_3X)$. The $X^-(CH_3Y)$ ions were mass-selected and some fraction of them underwent metastable dissociation in the second field-free region (2FFR), corresponding to ion lifetimes of about 10 μsec.

The ionic dissociation products were energy analyzed and KERDs were derived from the peak shapes.

Metastable dissociation of the $Cl^-(CH_3Br)$ ion yields predominantly Br^- and very little Cl^-. The low yield of Cl^- from back dissociation was surprising; under the ion formation conditions used, collisional stabilization is not expected to occur. She suggested that the low yield of Cl^- might be indicative of a back dissociation rate significantly greater than the ca. 10^5 s^{-1} rate required for dissociation in the 2FFR. The average kinetic energy released for the Cl^- product was about 0.05 eV, which probably reflects the thermal energy content of the $Cl^-(CH_3Br)$. The average kinetic energy released for the Br^- product ion, however, was remarkably small -- about 0.025 eV -- from a dissociation that should liberate about 0.3 eV of energy. Similar results were obtained for $Cl^-(CH_3I)$ and $Br^-(CH_3I)$. These results suggest that almost all of the energy liberated in the S_N2 reactions is released as internal excitation of the CH_3X product. The surprisingly small kinetic energy release for the substitution transition state may be indicative of dynamical constraints at the substitution transition state, perhaps on angular momentum. This hypothesis will be tested using phase space theory to model the dissociation. The present results contrast with recent studies by S. W. Buckner and co-workers (1990 ASMS Conference, Tucson) of the reaction of F^- with CH_3Cl, which indicated that almost all of the energy liberated in the exothermic substitution reaction is released as translational excitation of the Cl^- and CH_3F products.

Bob Squires (Purdue) discussed α-elimination reactions. He found, for example, that while it is endothermic to lose X^- from $^-CH_2X$, for many other α-halocarbanions loss of the halide is thermodynamically favorable. For example the reaction of CH_3OCH_2Cl with HO^- is approximately thermoneutral to form CH_3OCH:, H_2O and Cl^- and is exothermic if $Cl^-(H_2O)$ is the ionic product. It should be possible to obtain heats of formation of carbenes from a quantitative study of reactions of this type.

In the second part of the workshop the reaction mechanisms of some organic radical cations were discussed. C.G. deKoster (Gorlaeus Labs, Leiden) reported on work he had done with J. van Thuyl on the reactions of distonic oxenium ions with some nucleophiles. It is known that the α-distonic ion $^+CH_2OCH_2^{\cdot}$, generated from oxirane, transfers $CH_2^{+\cdot}$ to pyridine and acetonitrile. Now the reactions of the distonic ion formed by ring opening of methyl oxirane radical cation with pyridine and pyridine - d_5 were studied. The transfer of $CH_2^{+\cdot}$ and $C_2H_4^{+\cdot}$ was observed without any H exchange between the reactants. However the structure of the product ion $[C_6H_5NC_2H_4]^{+\cdot}$ is not known, and it was not possible to tell from the experiments whether the ion generated from methyl oxirane has an α- or a β-distonic structure. It was mentioned in the discussion that the β-distonic ion, $^+CH_2OCH_2CH_2^{\cdot}$ formed independently from a different source transfers CHO^+ to pyridine making the α-distonic structure likely for the ion derived from methyl oxirane.

In a second contribution from Leiden, W. J. van der Guchte and W.J. van der Hart discussed the isomerization reactions of $C_6H_6^{+\cdot}$ ions formed from 1,5-hexadiyne during charge exchange ionization at low kinetic energy in an ICR instrument. The molecular ion (IE 9.9 eV) is unstable with respect to fragmentation and it has been shown (Lifshitz and Ohmichi, J. Phys. Chem. 1989, 93, 6329) that the appearance energy of $C_6H_5^+$ coincides with the IE. Nevertheless, exchange ionization (CE) of 1,5-hexadiyne with several CE reagents resulted in relatively large abundances of $C_6H_6^{+\cdot}$ which was shown by subsequent ion molecule reactions to be benzene ions. Obviously an isomerization of the diyne takes place in the collision complex with the CE reagent. The heat of formation of the benzene ion is 4.1 eV lower in energy than that of the ion of the diyne and 3.65 eV below the threshold for dissociation to $C_6H_5^+$. This leads to the excess energy E of the charge transfer complex $E = IE_{(CE\ -\ reagent)} - 9.45$ in units of eV. In the case of Hg^+ as the CE reagent E = 4.64 eV, a value very close to the triplet excitation energy (4.67 eV), which may lead to a special isomerization mechanism for 1,5-hexadiyne.

Isomers and Isomerization of Molecular Ions: the Formation of Stable Ion-Neutral Complexes During Unimolecular Dissociation

Steen Hammerum

Department of Chemistry, University of Copenhagen,
The H.C. Ørsted Institute, DK-2100 Copenhagen Ø, Denmark

Abstract

The formation of intermediate ion-neutral complexes in unimolecular dissociation reactions can be confirmed when the life-time of such intermediates is sufficiently long. Evidence to confirm the presence of complexes prior to the elimination of alkane molecules from alcohol and ether molecular ions comes from specific hydrogen atom exchange and from kinetic energy release measurements. The properties of the intermediate complexes formed during dissociation reactions are very similar to those of similar complexes formed by bimolecular reactions. Addition of alkyl radicals to protonated carbonyl groups occurs in [alkyl-radical/oxonium-ion] complexes under both unimolecular and bimolecular conditions. The unimolecular loss of alkane molecules from these complexes exhibit unusually large primary and secondary isotope effects.

Introduction

Some ten years ago it was suggested that cleavage of a single bond in acyclic radical cations would, under certain circumstances, lead to an electrostatically bonded intimate pair, a complex, of the two fragments-to-be rather than directly to dissociation [1-3]. Such a process would bear strong analogy to the formation of ion pairs in condensed phase reactions [4]. The intermediacy of complexes was proposed in order to provide mechanistic interpretation, and the model did indeed yield intuitively pleasing explanations for a number of puzzling observations. However, whether the formation of stable complexes during ionic dissociation reactions under low-pressure conditions would in fact correspond to physical reality was rarely [1] discussed, let alone demonstrated.

$$A\text{-}B^{+\cdot} \rightarrow [\ A \quad B\ ]^{+\cdot} \rightarrow \rightarrow \text{products}$$

$$A\text{-}B^{+\cdot} \rightarrow \rightarrow [\ A' \quad B'\ ]^{+\cdot} \rightarrow \text{products}$$

Explicit suggestions that complexes intervene in gas-phase ionic reactions are becoming increasingly common, but there is a conspicuous absence of consensus with regard to the proper

K. R. Jennings (ed.), Fundamentals of Gas Phase Ion Chemistry, 379–390.

meaning of the term complex. It is being used to describe intermediates, to describe transition states, and at times to describe something else again. The word is a Humpty-Dumpty word [5]. That is perhaps an unavoidable consequence of the different meanings associated in chemistry with "complex", ranging from (in inorganic chemistry) species with in principle infinite life-time (e.g., $Cu[NH_3]_6^{2+}$) to the activated complexes of absolute rate theory that exist only during one vibrational period.

Definition

For the term ion-neutral complex to be useful when describing intermediates in unimolecular dissociation reactions that are capable of existence as independently stable species, the following criteria must be met [6]:

The interactions between the components of a complex must be predominantly electrostatic -- species bonded by covalent bonds (including hydrogen bonds) should not be considered complexes.

The components of a complex must be separated by (at least) the sum of the van der Waals radii -- shorter distances imply the sharing of electrons, i.e., covalent bonding.

The lifetime of the complex must be long enough for us to talk about it -- that is, the complex must exist long enough to have properties of its own.

In other words, the stabilization towards dissociation of the complex is bounded by the van der Waals condition, since the magnitude of the electrostatic interactions depend on the intra-complex distance. This limits the internal energies that complexes may possess, if they are not to dissociate immediately, and constrains the structural reorganization possible within complexes. Electrostatic interactions do not in themselves indicate the presence of an intermediate complex [3]; an ionic dissociation will always involve electrostatic forces between the incipient products. Furthermore, lest it be a metaphysical entity, the complex must possess properties that can be examined, in order that its existence may be verified.

Distinction

It is useful, when considering the properties of intermediate complexes, to distinguish between complexes formed by attempted simple cleavage, complexes formed by reaction, and complexes of the incipient products, even though these categories are not mutually exclusive. This distinction is not always clearly made, which has occasionally caused disagreement on false premises [7].

Prediction

Simple cleavage, loss of an alkyl radical from molecular ions by fission of a C-C bond next to the functional group, is a very common reaction, but it is in many cases observed only in the mass spectrometer ion source. In fact, uncomplicated simple cleavage is, with the exception of the reactions of secondary and tertiary amine molecular ions, almost never observed for metastable ions. Many small aliphatic radical cations instead lose an alkane molecule [8-12] when dissociating just above the ionization threshold [8,12,13].

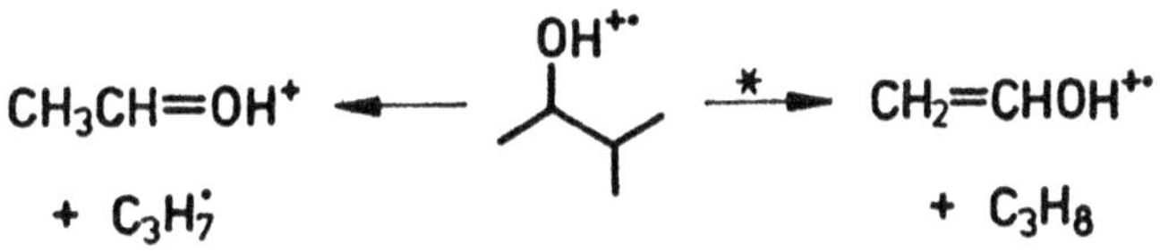

Fig 1 Alkyl radical loss (ion source) and alkane loss (metastable ions)

Such processes may be thought of as proceding via ion-radical complexes (Fig 2) [6,12-15]. In this picture the alkane molecule eliminated from simple alcohol and ether radical cations would be formed by hydrogen atom abstraction within an [oxonium-ion/alkyl-radical] intermediate. On the assumption that such a complex can exist as a stable, well-defined entity for some time it is possible to make definite predictions that may be examined experimentally:

Thermoneutral H-transfer should yield a second, similarly stabilized complex of the incipient products, in which case
(i) the H-transfer will be reversible -- suitably isotope labeled compounds should show specific H/D-exchange, and
(ii) dissociation of either complex will release very little kinetic energy.

Exothermic H-transfer should cause the products to dissociate rapidly, in which case
(i) the H-transfer will not be reversible -- no H/D exchange, and
(ii) the kinetic energy released when the complex dissociates will reflect the exothermicity of the hydrogen atom transfer.

Experimental confirmation of these predictions comes from studies of the unimolecular reactions of the metastable molecular ions in the field-free regions of a double-focusing mass spectrometer (MIKE spectra, recorded on a ZAB-2F mass spectrometer). Studies of the bimolecular reactions of alkanes and alkenes with unsaturated and saturated alcohols provides supporting evidence (FT-ICR studies, performed on a Bruker CMS47-X instrument with an external ion source) [16].

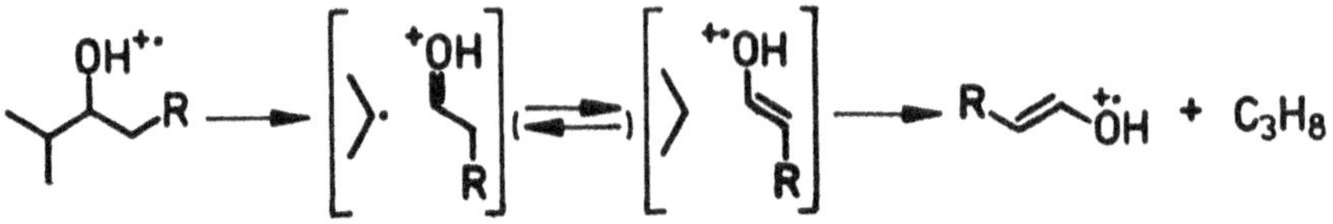

Fig 2 Thermoneutral (R=H) and exothermic ($R=CH_3$) H-atom transfer prior to propane loss.

Interconversion

The loss of alkane molecules from deuterium labeled simple alcohol and ether molecular ions in the microsecond time-frame is preceded by specific intramolecular H/D-exchange when the heat of formation of the [alkyl-radical/oxonium-ion] pair is within some 3 kcal/mol of the heat of formation of the corresponding [alkane/enol$^{+\cdot}$] pair, that is, when the hydrogen transfer step is approximately thermoneutral [17] (Fig 3). One such case is 3-methyl-2-butanol, which dissociates by loss of C_3H_8 after multiple reciprocal hydrogen atom transfers involving specifically the H-atoms at C(1) and C(3); the remaining hydrogen atoms retain their positional identity. Similar behavior is observed for the other radical cations in Fig 3 (except pinacolyl alcohol and its ethers which lack hydrogens at C(3) to exchange). When the hydrogen transfer is exothermic, no such exchange is observed (Fig 4).

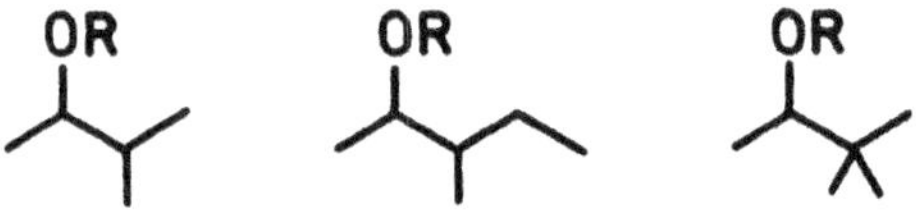

Fig 3 Ions losing alkane molecules after thermoneutral H-transfer. R = H, CH_3, C_2H_5

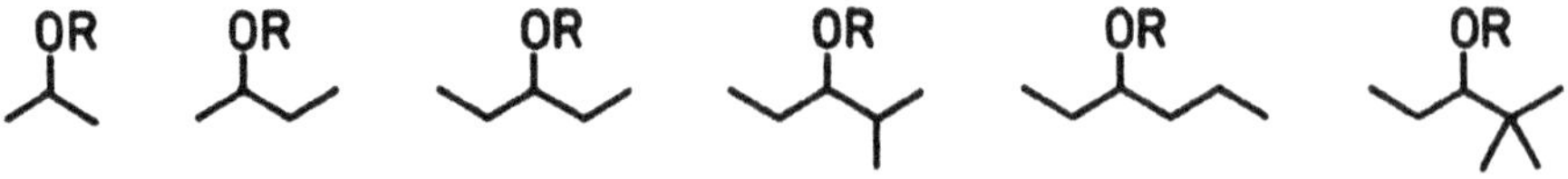

Fig 4 Ions losing alkane molecules after exothermic H-transfer. R = H, CH_3, C_2H_5

The interconversion reactions that take place when the H-transfer is thermoneutral involve not only the two complexes; also the initial simple cleavage of the molecular ion is reversible (Fig 5). Hydrogen atom exchange between C(1) and C(3) precedes the collision induced loss of methyl and isopropyl result of thermoneutral hydrogen transfer (Fig 3). The kinetic energy release (KER) measured ($T_{0,5}$) for loss of the alkane molecule is in each case less than 3 meV.

Conversely, the KER measured for alkane loss is considerably higher, between 9 and 25 meV, when the alkane + enol$^{+\cdot}$ products are formed as the result of exothermic hydrogen abstraction (Fig 4). The kinetic energy released directly reflects the exothermicity of the H-transfer when the energy liberated in this reaction is considerably in excess of the electrostatic interactions in the initial [alkyl-radical/oxonium-ion] complex. The heat of reaction divided by the number of degrees of freedom, $\Delta H_r^o/(3N-6)$, correlates well with the KER for alkane loss (Fig 6), with the exception of loss of CH_4 from isopropyl alcohol and isopropyl ether radical cations (not on the Figure). The aberrant behavior of these ions is not understood, but the loss of methyl and methane frequently exhibits atypical behavior in these and related dissociation reactions of aliphatic radical cations (e.g., the ethers of 2-butanol and 3-methyl-2-butanol do not expell CH_4 at all).

Interaction

The electrostatic forces in complexes with neutral hydrocarbon components arise from ion-induced dipole interactions (polarization interactions). These are not strong, and diminish rapidly with increasing inter-species distance (Fig 7). The internal energy of the [alkyl-radical/oxonium-ion] intermediates

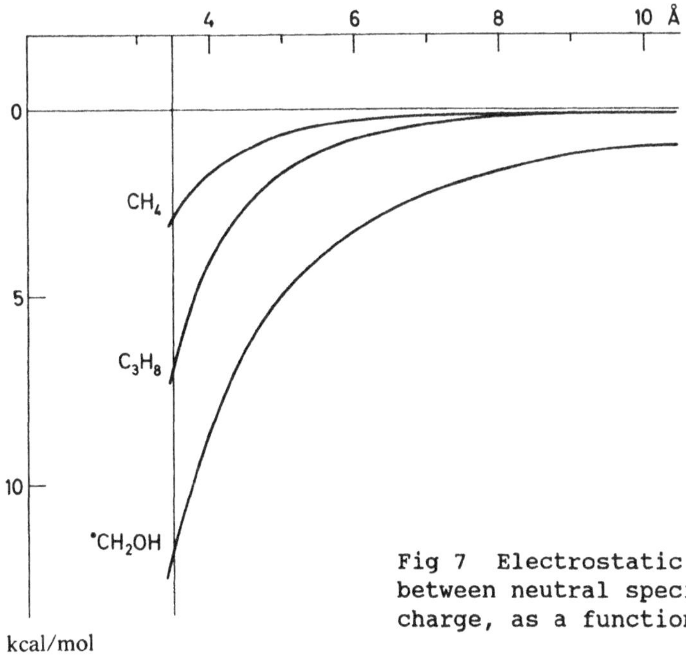

Fig 7 Electrostatic interactions between neutral species and a point charge, as a function of distance.

radicals from the surviving molecular ions, demonstrating that addition of the isopropyl radical to protonated acetaldehyde occurs in the complex. Reversibility of a simple cleavage reaction presumably requires, in the general case, that the critical energy for this process is low. It is probably also in such instances that the formation of long-lived complexes in radical cation dissociation reactions will be particularly favorable.

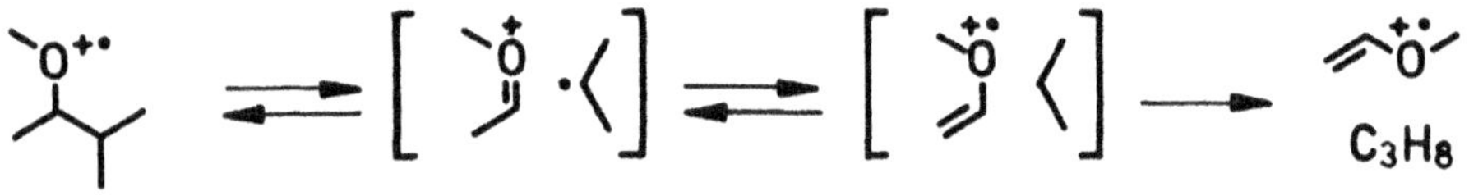

Fig 5 Reversibility of simple cleavage and hydrogen atom transfer

Dissociation

Dissociation of a stable complex of the incipient products requires only the energy necessary to overcome the electrostatic interactions within the complex. The excess energy in the transition state will therefore be low, and only little kinetic energy should be released upon dissociation. The metastable peak should be very narrow. That is indeed the case for those reactions where the alkane molecules are formed as the

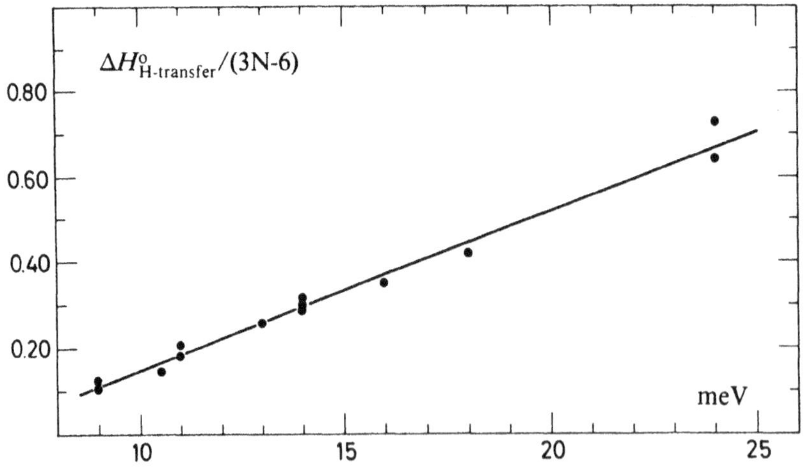

Fig 6 Kinetic energy release vs hydrogen atom transfer exothermicity for alkane loss from simple alcohol and ether molecular ions.

therefore cannot be very high if alkane loss is to compete successfully with simple dissociation, particularly since the inter-species distance is increased at the transition state for hydrogen abstraction (Fig 8). Experimental studies using photoionization or charge exchange ionization [8,12,13] have confirmed this; alkane expulsion takes place only over a narrow energy range just around the threshold for simple cleavage.

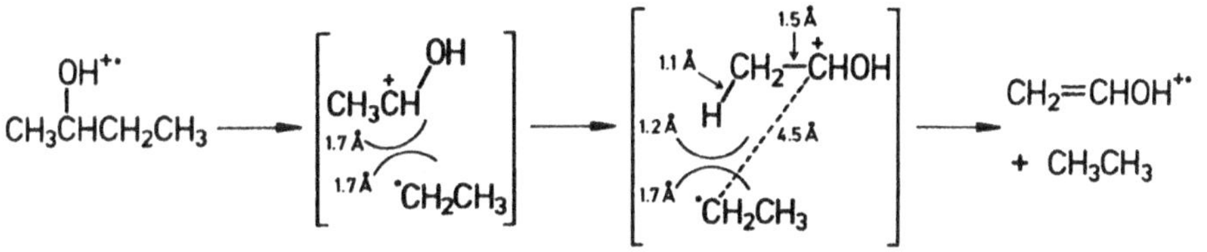

Fig 8 van der Waals radii and approximate inter-complex distances for hydrogen abstraction when C_2H_6 is lost from 2-butanol.

Unusually large primary and secondary kinetic isotope effects accompany the elimination of alkane molecules [18]. They can be measured unambiguously when the reaction is not accompanied by H/D exchange. Considerable variation even between closely related species is observed. The effects on the ethane loss from β-labeled 3-pentanol and its methyl ether are illustrative (obtained from the $M\text{-}C_2H_5D/M\text{-}C_2H_4D_2$ peak ratios for the 2,2-$\underline{d}_2$ and 2,4-$\underline{d}_2$ species):

$$(CH_3CH_2)_2CHOR^{+\cdot} \rightarrow C_2H_6 + CH_3CH{=}CHOR^{+\cdot}$$

R=H	primary: 2.4	secondary: 75
R=CH_3	primary: 20	secondary: 9

Even larger effects are observed for methane loss from labeled isopropyl alcohol and isopropyl ethers. The magnitude of the isotope effects reflects that the reactions take place close to threshold, where the zero-point energy differences will be particularly important, not the mechanism of reaction.

Comparison

The fragmentation of metastable neopentyl alcohol molecular ions produces $C_4H_8^{+\cdot}$ ions by loss of CH_3OH [19]. This reaction may proceed via an intermediate $[C_4H_9^+/HOCH_2\cdot]$ complex. Hydrogen abstraction to produce $[C_4H_8^{+\cdot}/HOCH_3]$ is thermoneutral, and the reactions proceed as expected for a system where two stable complexes interconvert: the metastable peak is very narrow, and considerable (though not statistical) hydrogen exchange is observed [15] prior to methanol expulsion from $(CH_3)_3CCD_2OH^{+\cdot}$. Neopentyl methyl ether shows similar behavior. Conversely, the hydrogen transfer step is exothermic when methanol is lost from the higher homolog, $CH_3CH_2(CH_3)_2CCH_2OH^{+\cdot}$; this reaction proceeds without hydrogen exchange and the KER is not particularly low.

The hydrogen exchange observed for neopentyl alcohol may alternatively be described as occurring via two interconverting distonic [20] ions (Fig 9); the two models differ only by the distance between the oxygen and the <u>tert</u>-butyl group. Note that interconversion via [alcohol/alkene$^{+\cdot}$] complexes (or via distonic ions) lead to exchange of the α-hydrogen atoms, whereas interconversion via [alkane/enol$^{+\cdot}$] complexes involves exchange of the β-hydrogen atoms.

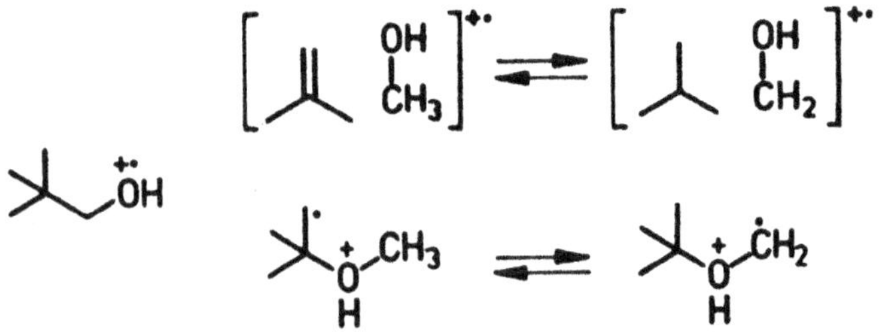

Fig 9 Possible mechanisms for hydrogen exchange when neopentyl alcohol radical cations dissociate.

It has been suggested that complexes with strong ion-dipole interactions, such as the [$C_3H_6^{+\cdot}$ H_2O] species formed when metastable propanol radical cations expell water, will dissociate unusually slowly [21]. It seems not clear if similar arguments apply to the dissociation of complexes where only polarization forces are important.

<u>Confirmation</u>
The suggestion that the low-energy fragmentation reactions of secondary alcohol and ether radical cations proceed via intermediate [alkane/enol$^{+\cdot}$] and [alcohol/alkene$^{+\cdot}$] complexes derives considerable support from the reactions that these complexes exhibit when formed independently, by low-pressure ion-molecule reactions in an ion cyclotron resonance (ICR) spectrometer. When propane and butane react with vinyl alcohol molecular ions (Fig 10) there is specific exchange of carbon-bonded hydrogens between the reactants; extensive CH/CH and CH/OH exchange is observed when the various butene radical cations react with methanol and ethanol.

The exchange of the vinyl alcohol methylene hydrogens with those in propane corresponds to the C(1)-C(3) exchange observed for 3-methyl-2-butanol. Exchange of the alcohol α-hydrogens with the butene hydrogens corresponds to the CH/CH exchange observed for neopentyl alcohol and 3,3-dimethyl-2-butanol radical cations.

Hydrogen exchange in these systems takes place concurrently with alkyl radical addition to the protonated carbonyl in the [alkyl-radical/oxonium-ion] complexes. The resulting saturated alcohol molecular ions react by subsequent α-cleavage, to give quite abundant product ions (Fig 10). This parallels the observation of collision-induced simple cleavage subsequent to exchange in the unimolecular reactions.

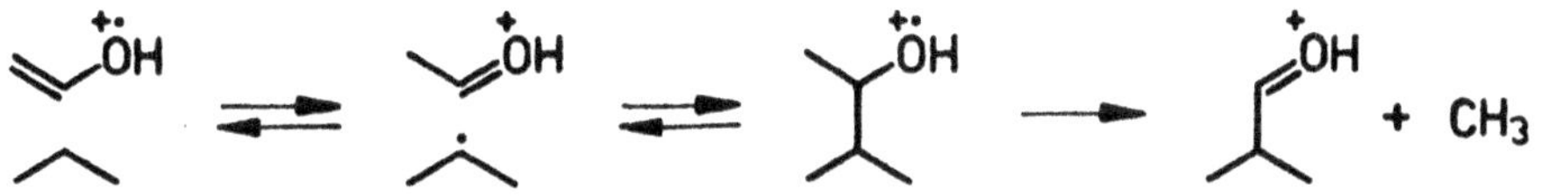

Fig 10 Reactions of propene with vinyl alcohol$^{+\cdot}$ (ICR)

Bimolecular reactions between alkenes and various enol molecular ions also result in addition [19]. The resulting distonic ions are directly accessible (by intramolecular hydrogen abstraction) from the corresponding carbonyl radical cations, and the ICR experiments are not conclusive with regard to whether or not ion-neutral complexes are involved in the McLafferty rearrangement at low energy.

Conclusion

A number of the properties that stable intermediate complexes formed in the course of unimolecular dissociation reactions ought to possess can be predicted on the basis of relatively simple-minded considerations. These predictions have been tested, and confirmed, for the reactions of simple alcohol and ether molecular ions close to threshold. A similar approach could be applied to other systems that may undergo complex-mediated reactions.

The correspondence between the heat of reaction for H-atom transfer and the hydrogen exchange prior to alkane loss as well as the relationship between exothermicity and kinetic energy release indicates that a stable initial complex is formed by attempted simple cleavage. The extensive exchange associated with thermoneutral hydrogen transfer suggests that the initial complex and a complex of the incipient products undergo reversible interconversion, followed either by dissociation or by reversal of the initial cleavage; it also confirms that the ion-neutral complexes involved have an appreciable life-time.

The close similarity between the reactions ascribed to the intermediacy of complexes during unimolecular dissociation and the reactions observed when the same complexes are formed by bimolecular processes provides additional support for the intermediacy of ion-neutral complexes in the unimolecular reactions. The specificity of the hydrogen exchange in both unimolecular and bimolecular processes demonstrates that mechanisms involving intermediate complexes are not inherently any less exacting than conventional mechanisms.

The bimolecular nucleophilic addition of radicals to protonated carbonyl groups in the gas phase is expected to be a quite general reaction; it has a number of intramolecular analogues. The observation that such reactions can occur in intermediate complexes formed during unimolecular fragmentation reaction adds another facet, the reversibility of the simple cleavage, to the concept of isomerization prior to dissociation of radical cations.

References

1 P. Longevialle and R. Botter, J. Chem. Soc. Chem. Commun., 1980, 823; P. Longevialle and R. Botter, Org. Mass Spectrom., 18, 1 (1983)

2 T.H. Morton, J. Am. Chem. Soc., 102, 1596 (1980)

3 R.D. Bowen and D.H. Williams, J. Chem. Soc. Chem. Commun., 1981, 836

4 T.H. Morton, Tetrahedron, 38, 3195 (1982)

5 "When I use a word," Humpty Dumpty said in rather a scornful tone, "it means just what I choose it to mean -- neither more nor less." L. Carroll, "Through the Looking Glass", The Bodley Head, London 1974, p 197.

6 S. Hammerum, J. Chem. Soc. Chem. Commun., 1988, 858

7 T. Weiske, S. Akkök, and H. Schwarz, Int. J. Mass Spectrom. Ion Proc., 76, 117 (1987)

8 K.M.A. Refaey and W.A. Chupka, J. Chem. Phys., 48, 5205 (1968); J. Sunner and I. Szabo, Int. J. Mass Spectrom. Ion Phys., 25, 263 (1977)

9 J.F. Litton, T.L. Kruger, and R.G. Cooks, J. Am. Chem. Soc., 98, 2011 (1976); S. Hammerum, K.F. Donchi, and P.J. Derrick, Int. J. Mass Spectrom. Ion Phys., 47, 347 (1983); C.E. Hudson and D.J. McAdoo, Int. J. Mass Spectrom. Ion Proc., 59, 325 (1984); D.J. McAdoo, Mass Spectrom. Rev, 7, 363 (1988)

10 R.G. Cooks, A. Yeo, and D.H. Williams, Org. Mass Spectrom., 2, 985 (1969); D.J. McAdoo and M.P. Barbalas, Int. J. Mass Spectrom. Ion Phys., 36, 281 (1980); C. Lifshitz and E. Tzidony, Int. J. Mass Spectrom. Ion Phys., 39, 181 (1981); G. Bouchoux and Y. Hoppilliard, Can. J. Chem., 60, 2107 (1982); P.J. Derrick and S. Hammerum, Can. J. Chem., 64, 1957 (1986); J.C. Traeger, Org. Mass Spectrom., 20, 223 (1985); J.A. Abbatt and A.G. Harrison, Org Mass Spectrom., 21, 557 (1986); J.C. Traeger, C.E. Hudson, and D.J. McAdoo, Int. J. Mass Spectrom. Ion Proc., 82, 101 (1988); N. Heinrich, F. Louage, C. Lifshitz, and H. Schwarz, J. Am. Chem. Soc., 110, 8183 (1988)

11 B.G. Keyes and A.G. Harrison, Org. Mass Spectrom., 9, 221 (1974); C.E. Hudson and D.J. McAdoo, Org. Mass Spectrom., 14, 109 (1979); C.E. Hudson, R.D. Lerner, R. Aleman, and D.J. McAdoo, J. Phys. Chem., 84, 155 (1980); H.E. Audier, G. Bouchoux, Y. Hoppilliard, and A. Milliet, Org Mass Spectrom., 17, 382 (1982); J.L. Holmes, P.C. Burgers, and Y.A. Mollah, Org. Mass Spectrom., 17, 127 (1982); R.D. Bowen and A. Maccoll, J. Chem. Soc. Perkin Trans. 2, 1990, 147

12 D.J. McAdoo, J.C. Traeger, C.E. Hudson, and L.L. Griffin, J. Phys. Chem., 92, 1524 (1988)

13 J.C. Traeger, C.E. Hudson, and D.J. McAdoo, J. Phys. Chem., 92, 1519 (1988); J.C. Traeger, C.E. Hudson, and D.J. McAdoo, J. Phys. Chem., 94, 5714 (1990)

14 C.E. Hudson and D.J. McAdoo, Int. J. Mass Spectrom. Ion Proc., 59, 325 (1984)

15 S. Hammerum and H.E. Audier, J. Chem. Soc. Chem. Commun., 1988, 860; S. Hammerum and H.E. Audier, Adv. Mass Spectrom., 11, 894 (1989)

16 Experimental results presented without literature references were obtained in collaboration with H.E. Audier and P. Mourgues, Ecole Polytechnique, France; HEA and PM do not necessarily share (but occasionally disapprove of) the slightly polemic views presented in this paper.

17 Thermochemical data from J.C. Schultz, F.A. Houle, and J.L. Beauchamp, J. Am. Chem. Soc., 106, 3917 (1984); J.L. Holmes and F.P. Lossing, J. Am. Chem. Soc., 102, 1591 (1980); F.P. Lossing, J. Am. Chem. Soc., 99, 7526 (1977); J.C. Traeger, J. Phys. Chem., 90, 4114 (1986); S.G. Lias and P. Ausloos, Int. J. Mass Spectrom. Ion Processes, 81, 165 (1987); J.B. Pedley, R.D. Taylor, and S.P. Kirby, Thermochemical data of organic compounds, Chapman and Hall, 1986

18 S. Hammerum, H.E. Audier, and P.J. Derrick, manuscript in preparation

19 J.L. Holmes, D.C.M. Tong, and R.T.B. Rye, Org. Mass Spectrom., 6, 897 (1972); R.D. Bowen and A. Maccoll, J. Chem. Soc. Perkin Trans. 2, 1984, 1005

20 S. Hammerum, Mass Spectrom. Rev., 7, 123 (1988)

21 J.-D. Shao, T. Baer, J.C. Morrow, and M.L. Fraser-Monteiro, J. Chem. Phys., 87, 5242 (1987)

22 P. Mourgues, C. Monteiro, H.E. Audier, and S. Hammerum, Org. Mass Spectrom., 25, 389 (1990)

INDUSTRIAL APPLICATIONS OF GAS PHASE ION CHEMISTRY

J H SCRIVENS, K ROLLINS

ICI Wilton Materials Research Centre
P O Box 90
Wilton
Cleveland
TS6 8JE

1 INTRODUCTION

Much of industrial interest in gas phase ion chemistry is related to the identification of unknown compounds, often in complex mixtures. In recent years the use of tandem mass spectrometry techniques has enabled previous intractable problems to be solved[1,2]. The power of the approach in structural studies, complex mixture analysis and ultra trace analysis has led to its widespread use in problem solving industrial laboratories.

In order to take full advantage of the techniques of ion selection, ion activation and ion detection provided by tandem mass spectrometry techniques it has been necessary to spend considerable effort studying the areas of reproducibility, sensitivity and general applicability. Out of these more fundamental studies comes an understanding of the limits of applicability of this approach. Tandem mass spectrometry covers a wide range of instrumentation and approaches. Combinations of quadrupoles, magnetic, electrostatic and time of flight analysers have been used together with numerous ion storage devices such as FTICR and ion trap instruments. Ion activation using a collision gas to induce dissociation is the most common approach although collisions with surfaces[3] have been used as well as interactions with electrons[4] and photons[5,6]. These ion activation methods are typically characterised by the amount and distribution of energy deposited, the ease with which this can be varied and how probable is it that the activation will occur.

In this discussion we will limit ourselves to collisional activation which can occur at high (kV) or low (eV) energies. This is thought to be a two step process with an initial activation of the parent ion followed by unimolecular dissociation High energy (kV) collisions proceed via an electronic excitation mechanism whereas low energy (eV) collisions give rise to vibrational excitation.

K. R. Jennings (ed.), Fundamentals of Gas Phase Ion Chemistry, 391–415.

2 PRACTICAL EXAMPLES

(a) Identification of components in a resin extract (Magnetic Sector/ Quadrupole Hybrid Data)

One of the most important uses of tandem mass spectrometry in industry is in the resolution of the spectra of complex mixtures. Figure 1 shows the probe EI spectrum obtained of an uncured epoxy resin extract. In order for tandem mass spectrometry to contribute to this problem it must produce reproducible product ion spectra from selected ions. This is not experimentally trivial since the experimental variables of collision energy, collision gas pressure, gas type, ionisation parameters, and source tuning all need to be kept constant in order for reproducible results to be obtained.

Conventional double focusing EI spectrum of epoxy resin extract.

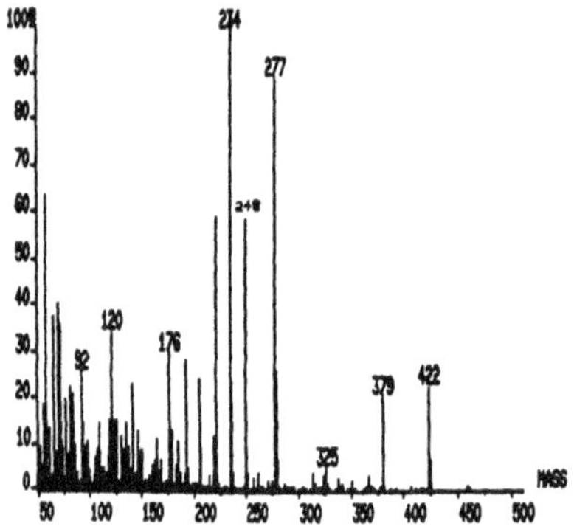

Figure 1

These experiments were carried out on a VG ZAB HSQ hybrid instrument of BEQQ configuration. Low energy collisions with argon were used and the instrument was modified to give a pressure reading, relating to the collision cell region, that was easy to reset to a given value. Collision cell pressures were calculated from conductance considerations. Experiments were carried out at a calculated collision cell pressure of 1.3m Torr using Argon as a collision gas, 20eV collision energy was employed.

Daughter ion spectrum of m/z 422 from extract

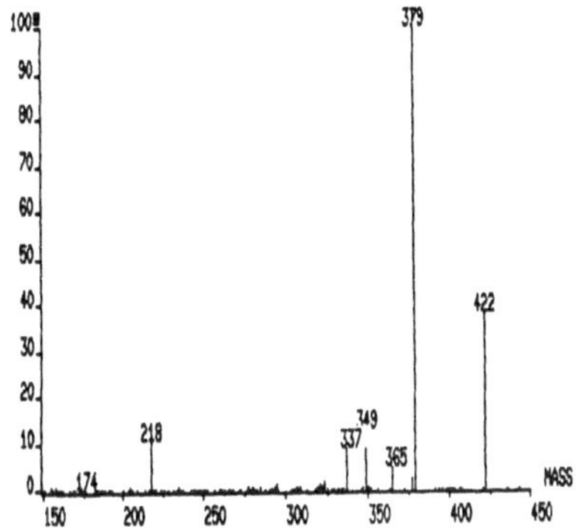

Figure 2a

Daughter ion spectrum of m/z 422 from TGDDM.

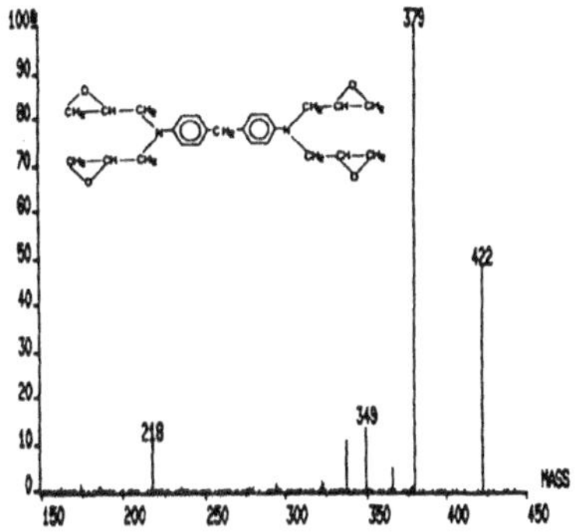

Figure 2b

Daughter ion spectrum of m/z 277 from extract

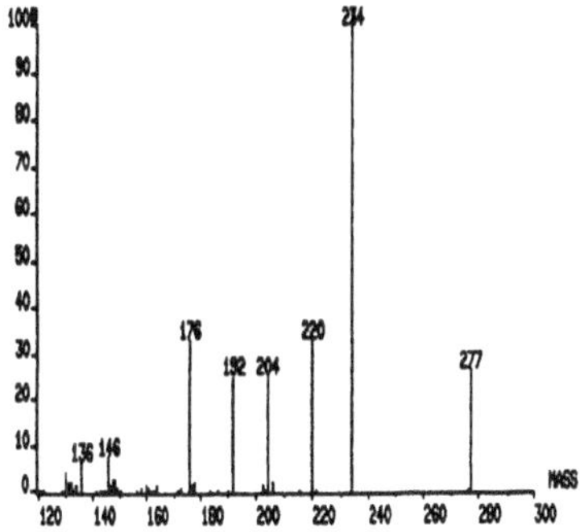

Figure 3a

Daughter ion spectrum of m/z 277 from GDA

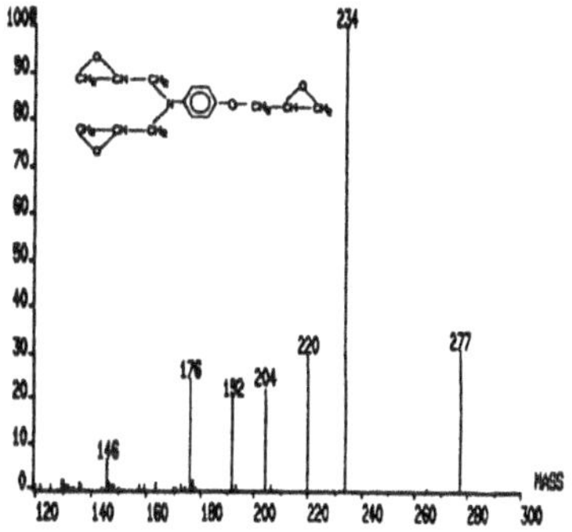

Figure 3b

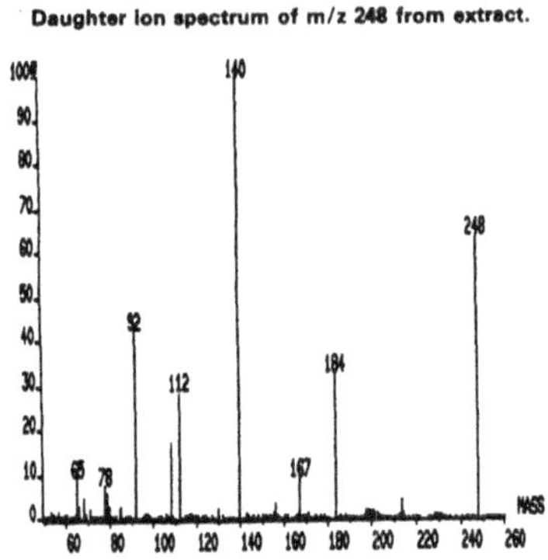

Figure 4a

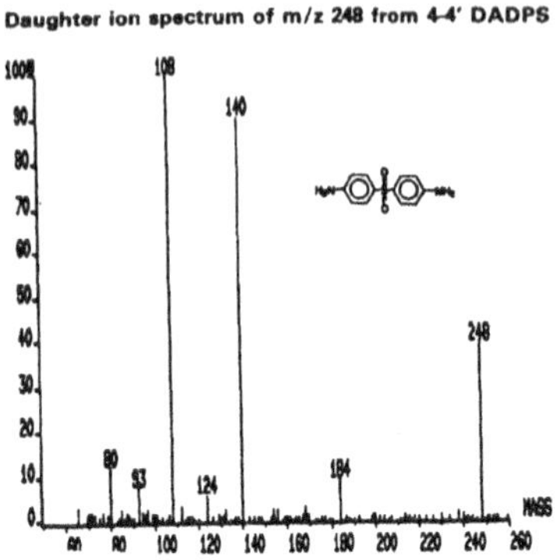

Figure 4b

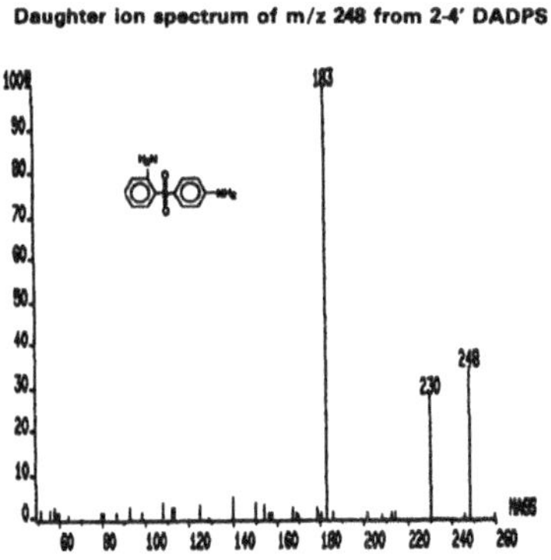

Figure 4c

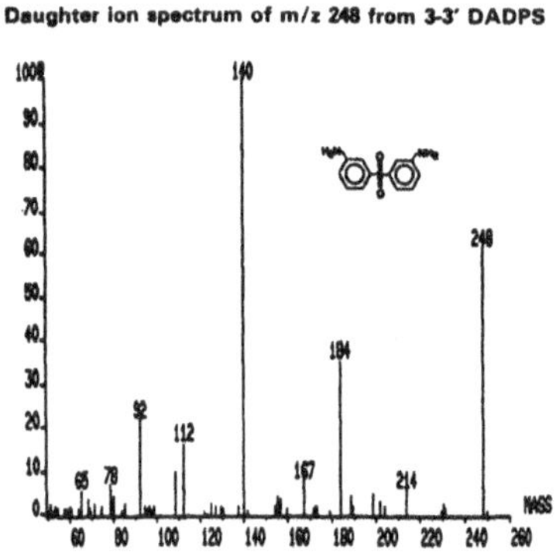

Figure 4d

Figure 2 shows the product ion spectrum of the ion at m/z422 compared with that obtained from the molecular ion of tetraglycidyldiaminodiphenylmethane (TGDDM). Good agreement is obtained giving confidence that this is indeed the identity of the compound. The next highest ion not accounted for by the TGDDM is m/z 277. Figure 3 shows the product ion spectrum of m/z 277 compared with that obtained from the molecular ion of triglycidylaminophenol. Again very good agreement is obtained. The remaining ion at m/z 248 was originally thought to correspond to 4,4′- diaminodiphenylsulphone. Figure 4 shows the product ion spectrum of m/z 248 compared with 44′, 24′ and 33′ diaminodiphenylsulphone. This indicates clearly that the ion at m/z 248 in the extract corresponds to the 33′ isomer. This example shows the power of tandem mass spectrometry techniques to give ion structural information which is both informative and reproducible enabling identifications to be confirmed and, in this case, isomeric differentiation to be carried out.

(b) Identification of components in resin extract (FTMS data)

Recently the above data were also obtained using an FTMS instrument. An EXTREL 2001 Fourier transform mass spectrometer with dual trapped ion cells was used. Within the analyser cell mass selected or precursor ions are accelerated to a chosen kinetic energy and allowed to collide with a collision gas. These MS-MS experiments are performed by using ion ejection techniques such as SWIFT[7]. Figure 5 shows the EI spectrum of the resin mixture together with the product ion spectrum of m/z 422, from the mixture, and that obtained from TGDDM. Figure 6 shows the product ion spectra obtained from m/z 277 in the mixture and from the standard. Figure 7 shows the product ion spectra obtained from the three diaminodiphenylsulphone isomers compared with that obtained from the mixture. In all cases good agreement between reference and sample spectra were obtained but the current level of control over, particularly the collision gas pressure, and the delicate timing of the experiment means that the reproducibility obtained is not as high as that from the more established hybrid experiment.

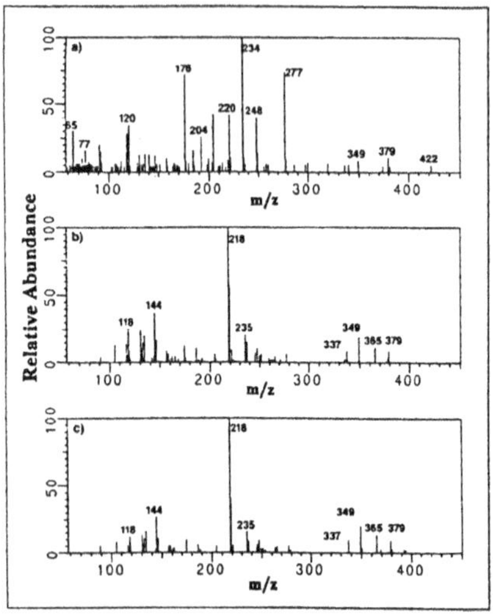

Fig. 5 a) Electron ionization mass spectrum of an uncured epoxy resin mixture. b) Daughter-ion mass spectrum of the m/z 422 species in the resin mixture. c) Daughter-ion mass spectrum of the molecular ion of tetraglycidyldiaminodiphenylmethane (TGDDM).

Figure 5

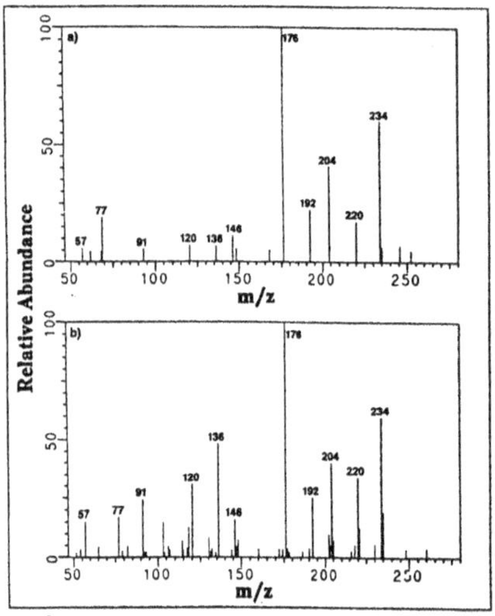

Fig. 6 a) Daughter-ion mass spectrum of the m/z 277 species in the resin mixture. b) Daughter-ion mass spectrum of the molecular ion of triglycidylaminophenol.

Figure 6

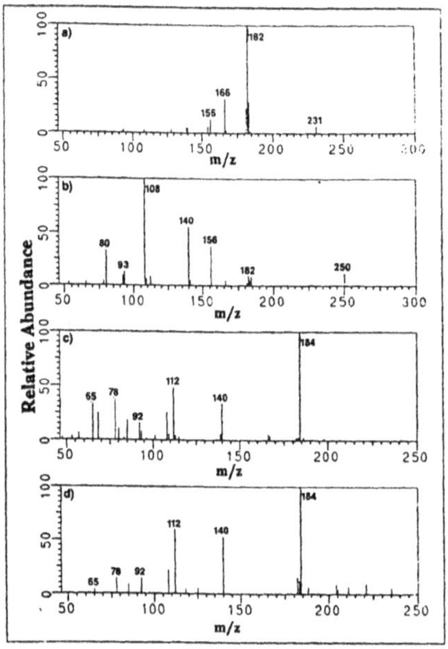

Fig. 7 a) Daughter-ion mass spectrum of the molecular ion of 2,4'-diaminodiphenylsulfone. b) Daughter-ion mass spectrum of the molecular ion of 4,4'-diaminodiphenylsulfone. c) Daughter-ion mass spectrum of the molecular ion of 3,3'-diaminodiphenylsulfone. d) Daughter-ion mass spectrum of the m/z 248 species in the resin mixture.

Figure 7

(c) Characterisation of unknown m/z 59 ion (Magnetic Sector/Quadrupole Hybrid Data)

Another example that shows clearly the ability of hybrid MS-MS instruments to give reproducible data is given by the following study carried out on the ZAB HSQ instrument of BEQQ geometry. The ion of m/z value 59 obtained from the reaction of methanol and acetaldehyde in the gas phase was studied in some detail[8]. Figure 8 shows the CID product ion spectrum of m/z 59 of the unknown compared with that obtained from protonated acetone, protonated methyl vinyl ether, protonated alkyl alcohol, protonated proponaldehyde and an ion of m/z 59 formed by homolytic cleavage in the molecular ion of 2- methylbutane. In all cases the spectra here have been normalised to the largest product ion. The spectra were obtained at 91eV collision energy and at a calculated collision cell pressure of 3 mTorr of Argon. These conditions were used to optimise fragmentation over the area of interest. In order to evaluate the reproducibility of such data the product ion spectrum of protonated acetone was obtained repeatedly over the 36 hour period of the experiments (Figure 9). The peaks at m/z 15, 27, 29 and 41 fluctuated with relative standard deviations in the range 3% to 11% indicating that the reproducibility of the low energy CID data on the hybrid instrument was sufficient to distinguish the set of isomers studied. This enabled the structure of the unknown product ion at m/z 59 to be established as $H_3CH=\overset{+}{O}CH_3$.

Product Ion Spectra of m/z 59 ex Various Precursors

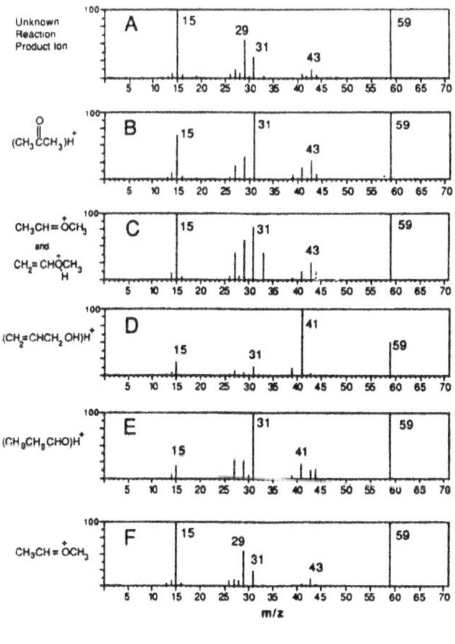

Figure 8

Product Spectra of m/z 59 ex Protonated Acetone obtained over 36 hour Period

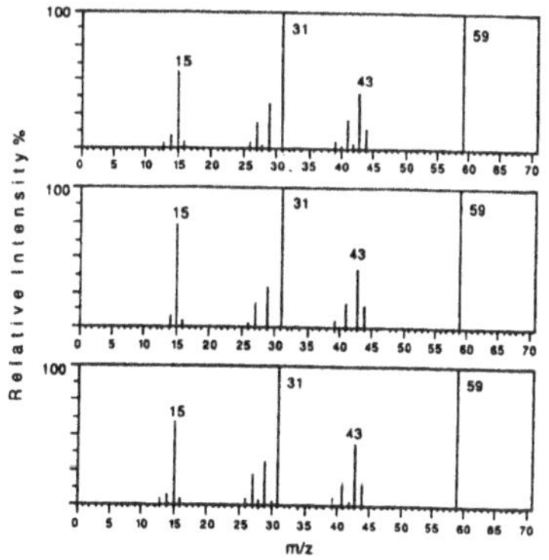

Figure 9

3 IONISATION

Much of the tandem mass spectrometry literature involves the recording of product-ion spectra of precursor ions formed by fast atom bombardment (FAB) or liquid SIMS (LSIMS) techniques. This has had a major impact in the biological area where the combination of the soft ionisation afforded by FAB and LSIMS with the structural detail given by MS-MS has proven powerful. A significant number of experiments have also been carried out using chemical ionisation or thermospray as ionisation methods again giving useful structural information. Many of these ionisation techniques are particularly useful for polar, biological compounds such as peptides but none of them are suitable for the study of non polar, involatile, synthetic polymers. Since these types of compounds are of considerable interest industrially, we have used another ionisation method, field desorption (FD)[9]. FD had been used for biological compounds before the advent of FAB but the experimental simplicity of FAB together with the superior results obtained for peptides led to FD being overtaken as the ionisation technique of choice. Some laboratories, particularly those in Germany and Japan, still maintained FD as a technique although many of the major instrument manufacturers stopped development in favour of FAB. FD suffers from the reputation of being experimentally difficult, having poor sensitivity, suffering from strong matrix effects and poor general applicability. We have used FD in the study of non polar, synthetic molecules in our laboratory for the last two years and have found these difficulties to be only partially true. FD has given us access to experimental data not possible using any other approach.

4 FD RESULTS

Experimental

All FD results were obtained on a VG ZAB-2SE reversed geometry instrument equipped with the standard VG FD/FAB source. The samples were analysed using 13μm activated carbon FD emitters supplied by Linden ChroMasSpec (Bremen FRG) with the sample applied to the emitter using the standard syringe technique. The emitter was typically programmed from 0 - 40 mA or until the signal intensity disappeared. All data were acquired at a resolution of 1000 into the VG Opus datasystem in multichannel analysis (MCA) mode.

(a) Oligomeric comparison

Figure 10 shows the FD spectra of two oligomeric hydrocarbons. These materials had been studied by gel permeation chromatography (GPC) which had indicated that both samples had the same average molecular weight with similar oligomer profiles. However, their properties were significantly different. The oligomers in (a) were made by polymerising dicyclopentadiene. Those in (b) which were a competitive material were obviously made from a mixed (and therefore cheaper) feedstock. This example shows the ability of FD to give stable, reproducible data on non polar ions up to a few thousand daltons. The extra structural detail given by FD over GPC enables the system to be much more fully characterised. This ability of FD to give predominately molecular ions with little fragmentation is of great importance in the study of these, often complex, mixtures.

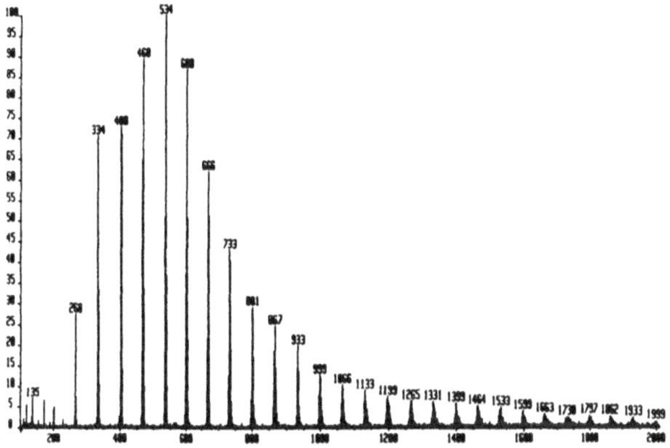

Figure 10a FD Spectra of Oligomers

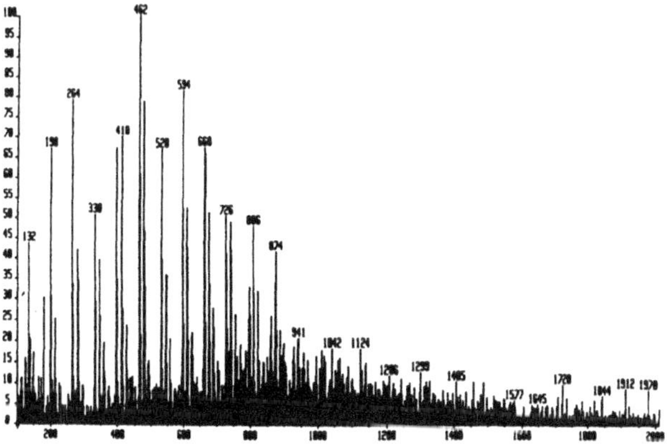

Figure 10b

Structure of 3608 Ester

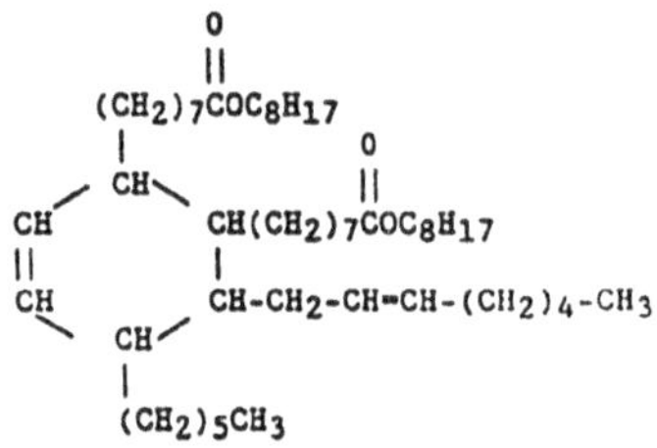

Elemental Composition: $C_{52}H_{98}O_4$

Molecular Weight: 786

Figure 11

Electon Impact and Field Desorption Spectra of 3608 Ester

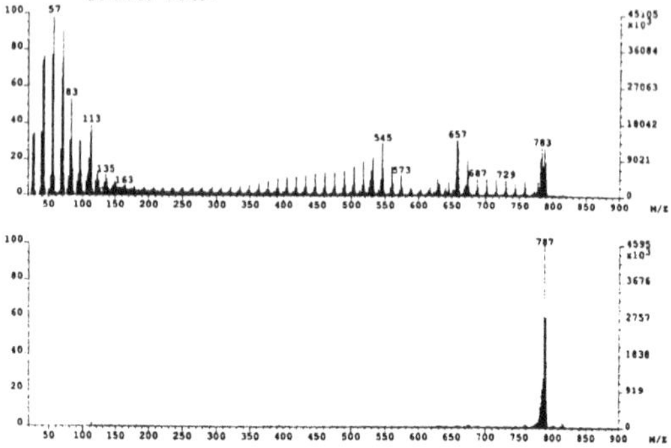

Figure 12

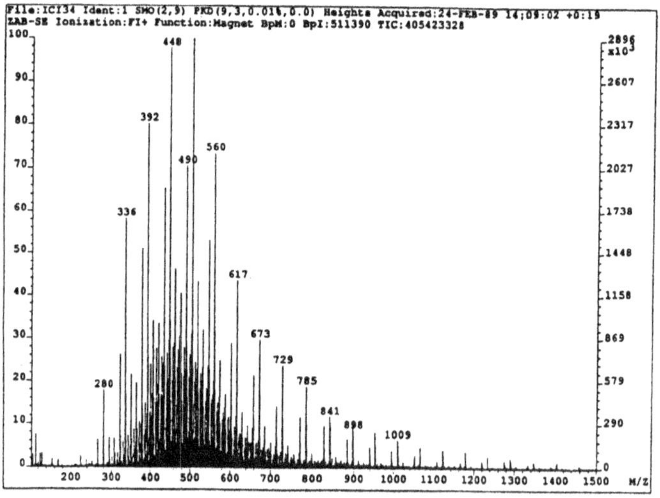

FIGURE 13 : Field Desorption Spectrum - Polyisobutylene Oil

Figure 13.

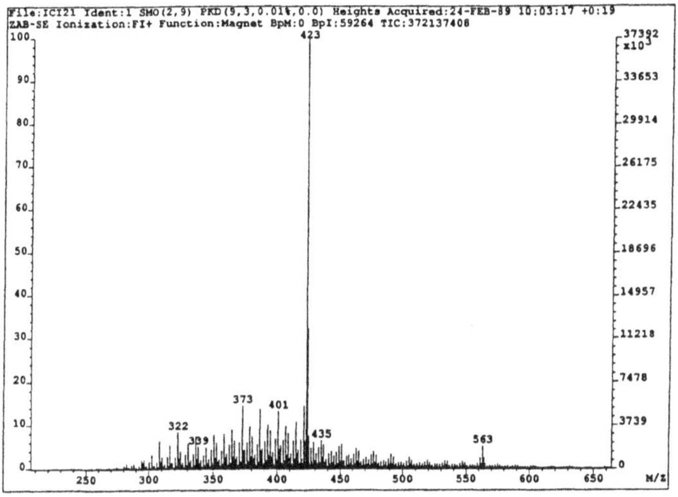

FIGURE 14 : Field Desorption Spectrum - Synthetic Oil/Polyalphaolefin

Figure 14

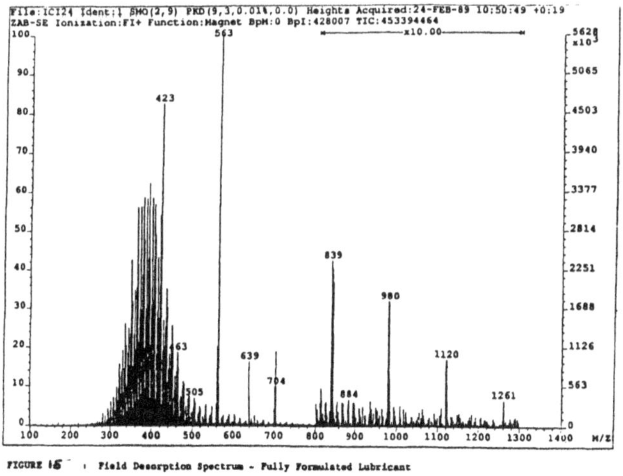

FIGURE 15 : Field Desorption Spectrum - Fully Formulated Lubricant

Figure 15

(b) Lubricant mixtures

The characterisation of the complex formulations that make up commercial lubricants has long proved a difficult analytical problem. The use of FD enables us to save considerable labour when compared with previous methods which rely on chemical separation. The major components of the formulations are a synthetic or mineral oil, polyalphaolefins and complex synthetic esters. These may be all or individually present. A typical ester found in these formulations is oleic acid dimer esterfied with 2-ethylhexanol. The structure of this compound is shown in Figure 11 while Figure 12 shows the marked differences between the EI and FD spectra. As can be seen the FD spectrum is dominated by the molecular ion region enabling the identification of this compound in complex hydrocarbon mixtures to be more easily carried out. Figure 13 shows the FD spectrum of a formulation consisting of an oil. Hydrocarbon type and molecular weight distribution can easily be determined. Figure 14 shows the FD spectrum of a formulation consisting of an oil, together with a polyalphaolefin (decene). Figure 15 shows the fully formulated lubricant. An oil component, polyalphaolefin and esters are all present. This shows the ability of FD to readily give information regarding the components of an extremely complex mixture. These results also indicate that, although the ionisation efficiencies of the different compound classes may be significantly different, making a quantitative interpretation of the results difficult, there is little evidence for significant compound suppression. This observation appears to hold true for a number of systems which we have studied.

(c) Polystyrene

In order to get an understanding of the potential mass range of the technique, together with observing if any significant mass discrimination occurs, we have studied a range of polystyrene standards used to calibrate GPC traces. These well characterised materials can be used as potential calibrants as well as being typical of aromatic polymers that are used in industry. Figure 16 shows the FD spectrum of a polystyrene standard of average molecular weight 5K daltons. The distribution is symmetrical indicating little mass discrimination and multiply charged species with charges = 2-6 are observed as overlapping envelopes at lower mass/charge.

Figures 17 and 18 show polystyrene standards of average molecular weight of 7K and 10K daltons respectively together with their equivalent GPC traces. It can be seen that the shape of the molecular ion (singly charged) region is in good agreement with that obtained by GPC. Figure 19 shows the singly charged region of a polystyrene standard of average molecular weight 12K. A good profile is again obtained with masses to > 15K being easily observable. As can be seen from these results this does not indicate an upper limit for observed masses since there is ample signal to noise. Table 1 gives an indication of agreement between nominal, corrected and observed masses for the 5K sample.

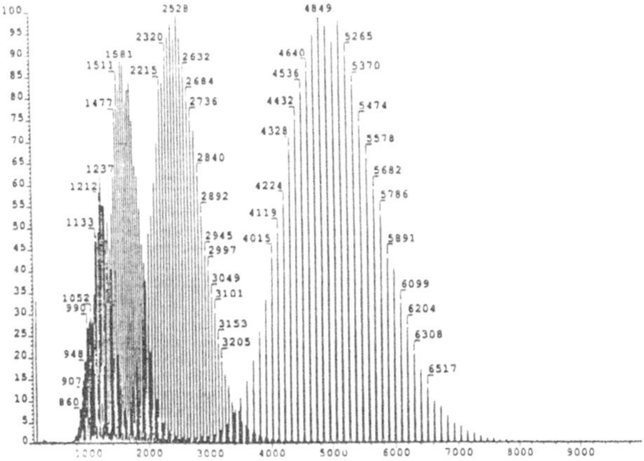

Figure 16. FD Spectrum of 5K Standard Polystyrene.

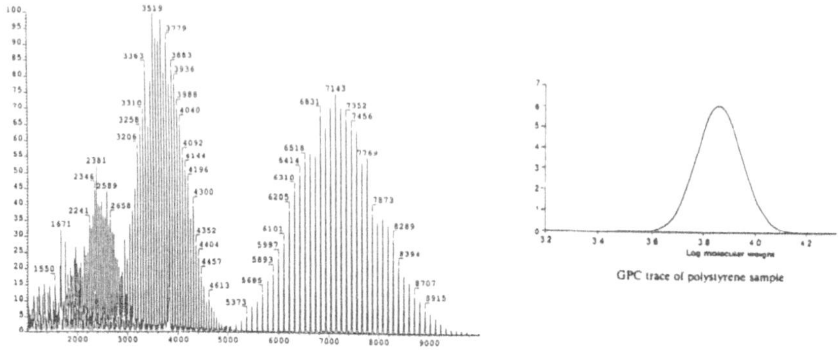

Figure 17. FD Spectrum of 7.5K Polystyrene Standard.

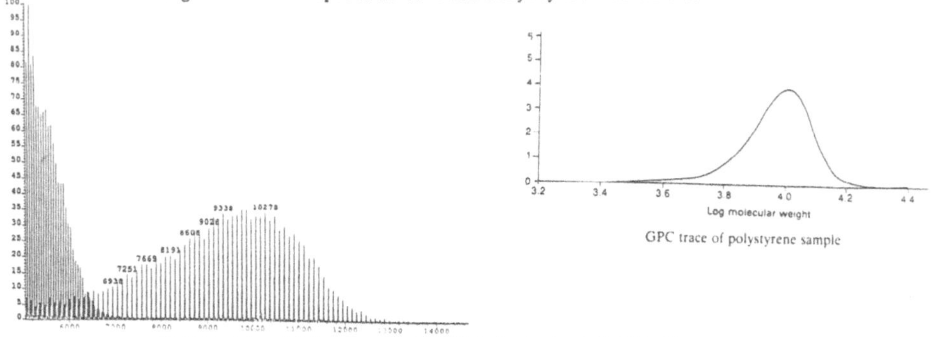

Figure 18. FD Spectrum of 10K Polystyrene Standard.

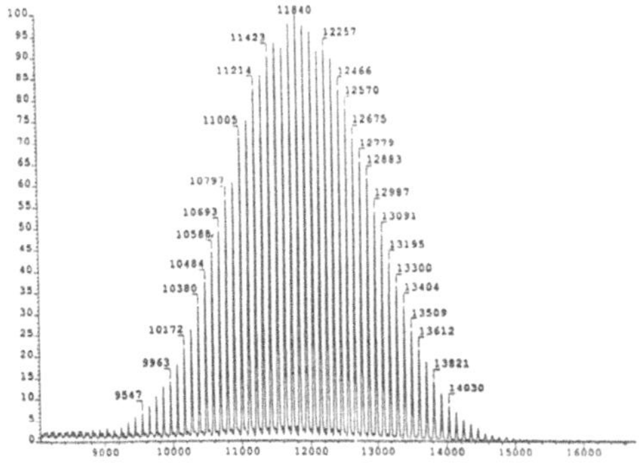

Figure 19. FD Spectrum of 12K Polystyrene Standard.

Table 1

Table 1 Nominal, corrected and observed masses for polystyrene sample A

n	Oligomer[a] formula	Nominal mass	Corrected mass[b]	Observed mass[c]
42	$C_{340}H_{346}$	4426	4432.5	4432
43	$C_{348}H_{354}$	4530	4536.6	4536
44	$C_{356}H_{362}$	4634	4640.7	4640
45	$C_{364}H_{370}$	4738	4744.9	4745
46	$C_{372}H_{378}$	4842	4849.1	4849
47	$C_{380}H_{386}$	4946	4953.2	4953
48	$C_{388}H_{394}$	5050	5057.4	5057
49	$C_{396}H_{402}$	5154	5161.5	5162
50	$C_{404}H_{410}$	5258	5265.7	5266
51	$C_{412}H_{418}$	5362	5369.8	5370

[a] The formula indicated here assumes a polystyrene of structure:

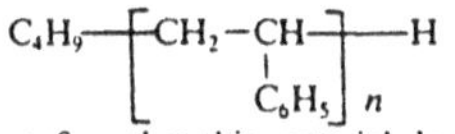

The observed ions confirm that this material does indeed possess a butyl end group.

[b] Corrected using the approach described by Matsuo *et al.*[10]

[c] Observed mass indicates the apex of the unresolved molecular ion region which was detected under the resolution conditions used.

(d) Alkyl phenol-formaldehyde resins

These are typically structures such as;

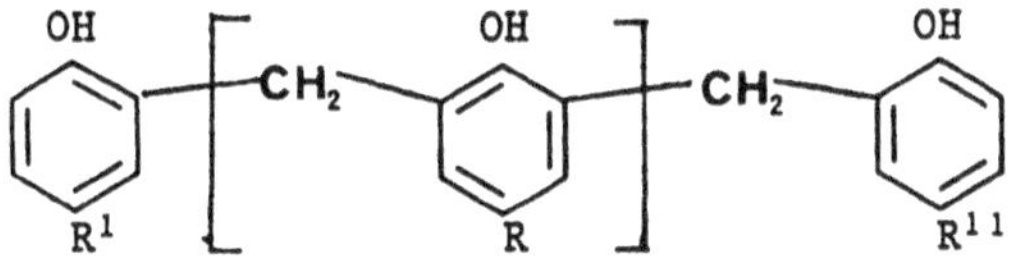

n

R is an alkyl group, typically C_4 to C_{12}. These materials need to be characterised in terms of; the nature of R, the ratio of linear to cyclic material and the identity of the end groups. Average molecular weight and molecular weight distribution are also important. FD is the only ionisation technique capable of giving this information on samples of industrial importance. Figure 20 shows the FD spectrum of a linear resin with R = C_4. Figure 21 also shows a C_4 resin but here the molecular weight is higher and also we see the presence of cyclic material. Figure 22 shows a more complex case where R = C_9 but where there is also cyclic material and material where the end group has been modified to dinonyl. Even in this case the three series of ions can be easily differentiated showing the power of the approach.

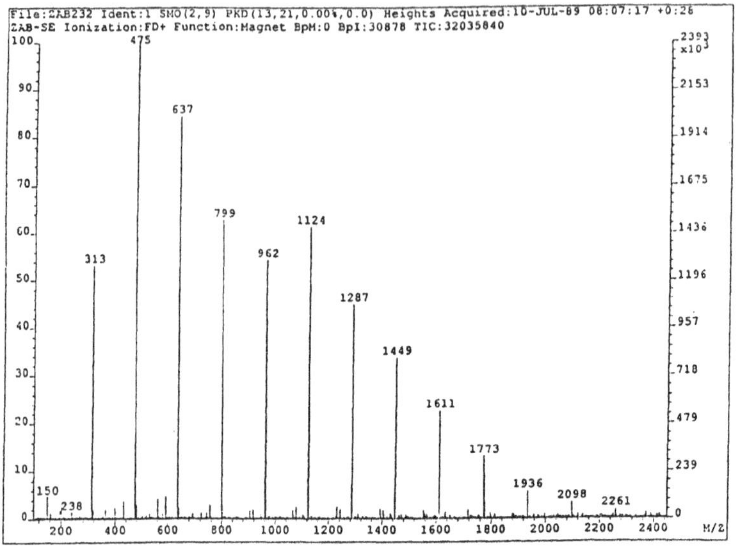

Figure 20. FD Spectrum of C_4 Phenol Formaldehyde Resin (Linear).

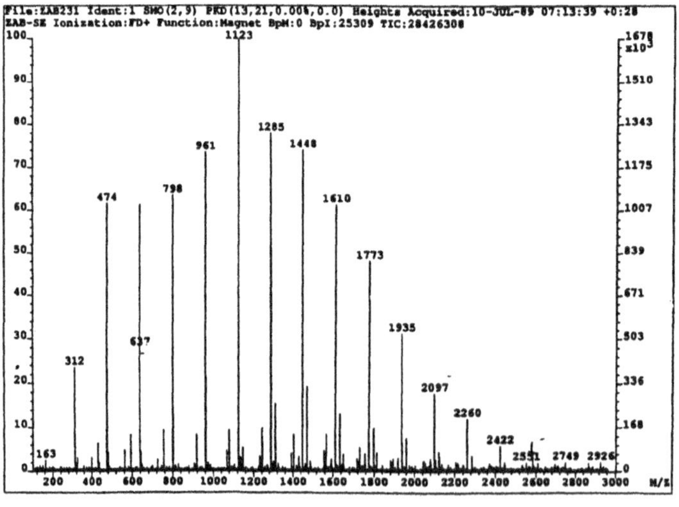

Figure 21. FD Spectrum of C_4 Phenol Formaldehyde Resin (Linear + Cyclic).

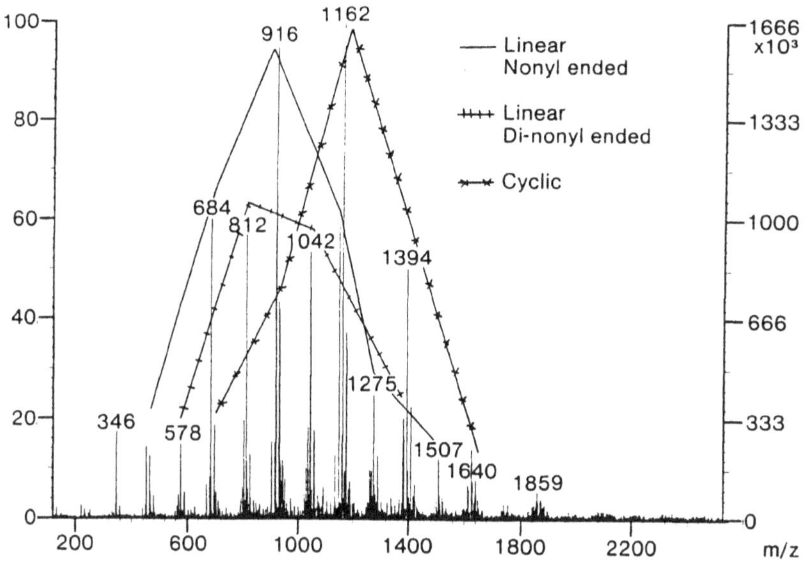

Figure 22. FD Spectrum of C_9 Phenol Formaldehyde Resin

5 SEQUENCING OF AROMATIC OLIGOMERS

Many modern synthetic aromatic polymer systems have structures of the type

$$A \leftarrow\!\left(\emptyset - X - \emptyset - Y - \emptyset \right)\!\rightarrow_n B$$

where $\emptyset$ indicates a para substituted benzene ring, X and Y are typically CO, S, SO_2, O and A and B are F, OH, OCH_3, $C\ell$. These polymers, often only 15 - 30K daltons in size are made up of a relatively small number of complex monomers. Tandem mass spectrometry has been shown to be a powerful technique for the elucidation of the structures of these monomers or early oligomers formed from them. Two typical such molecules are shown in Figure 23 with molecular weights of 942 and 1006 respectively. These molecules give good EI spectra, at probe temperatures of >450°C, up to ~1500 daltons. Since these molecules are often present in complex mixtures tandem mass spectrometry approaches are used to generate structural information. Figures 24 and 25 show the product ion spectra obtained for these molecules indicating that sequencing readily obtained. Figure 26 shows an FD spectrum of a similar molecule of nominal mass 730 but this technique also reveals the presence of two additional high oligomers at 926 and 1122 daltons respectively which had not been observed in the EI spectra. The use of FD for these problems considerably extends the molecular weight range that can be studied but at the price of reduced sensitivity making product ion spectra more difficult to obtain.

FϕCOϕϕCOϕOϕϕOϕCOϕϕCOϕF 942

FϕCOϕϕCOϕOϕSO_2ϕOϕCOϕϕCOϕF 1006

Figure 23. Structures of Aromatic Oligomers.

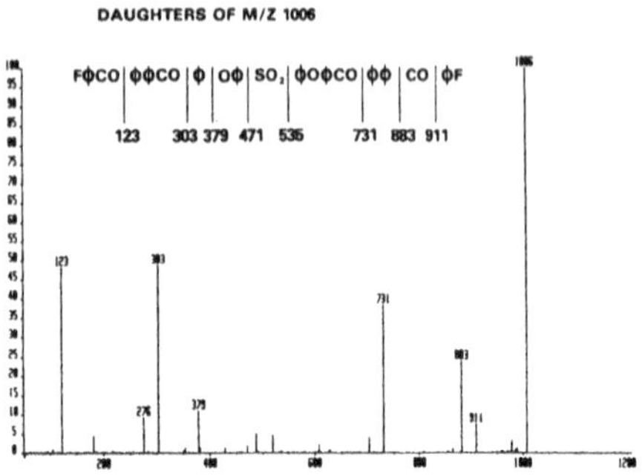

Figure 24.

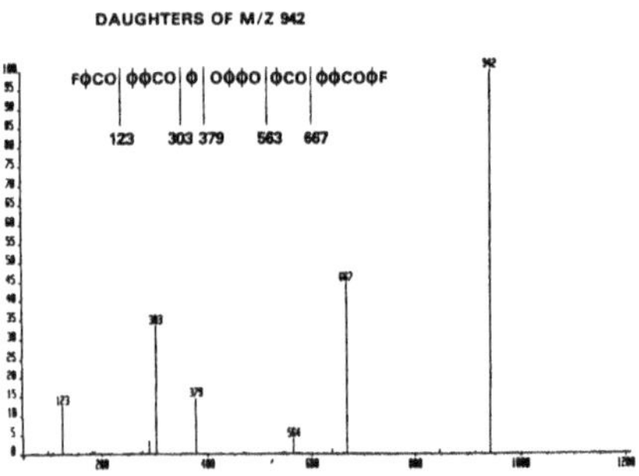

Figure 25.

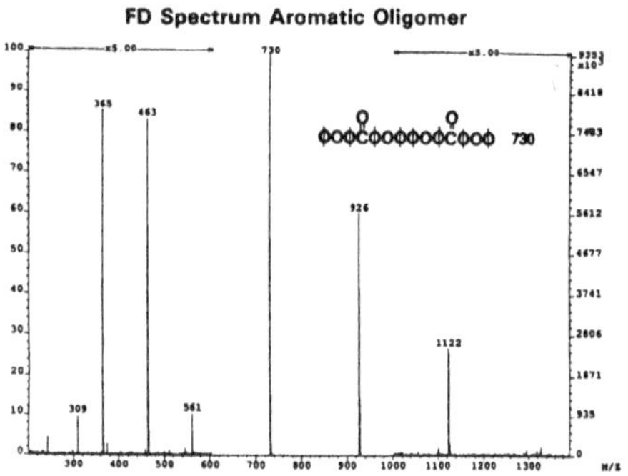

Figure 26.

INSTRUMENTATION

In order to utilise the power of the FD technique in harness with tandem mass spectrometry we have used a four sector instrument, the VG ZAB-T. The ion optics of this instrument, of BEBE geometry, with its combination of two reversed geometry double focusing instruments combined with array detection is shown in Figure 27. This provides a number of unique capabilities previously unavailable for FD-MS-MS operation. The inhomogeneous electric sector E2 employed on the instrument enables the focal length, focal plane angle and focal plane flatness of the ion beam to be varied giving a zoom capability for simultaneous detection from 1.4:1 (at 500 resolution full width at half maximum (FWHM)) to 1.02:1 (5000 resolution FWHM). In addition, the array may be removed giving access to a single point detector with an achievable product ion resolution of 10,000 (10% valley definition). The mid point ion sources on this instrument are fully functional making the second mass spectrometer (MS2), with array detector capability, the primary analyser with the first mass spectrometer (MS1) added to obtain tandem mass spectrometry capability.

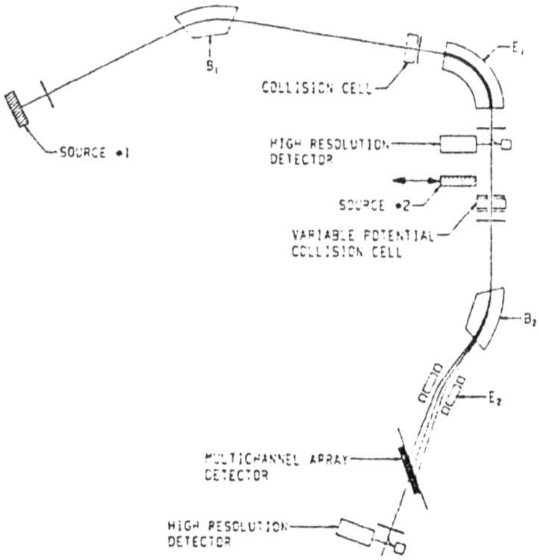

ZAB-T ion-optical arrangement.

Figure 27

The benefits of array detection in techniques with little chemical background such as FD and MS-MS have been reported previously[11]. These detectors are able to provide an increase in sensitivity of up to two orders of magnitude when compared to a conventional point detector. Figure 27 shows the partial product ion spectra obtained from the 262 dalton ion of hexachlorobutadiene obtained by field ionisation. It can clearly be seen that the use of the array detector gives a stable signal which cannot be obtained using the conventional point detector. Figure 29 shows that, for a strong signal, good product ion spectra can be obtained from FD signals using a point detector. The product ion spectrum of m/z 693 ex hexaphenoxy phosphazene is shown. This was obtained using multi-channel acquisition and is actually the sum of a number (>20) spectra. Figure 30 shows a direct comparison of point detector and array using equal acquisition times for single scans. the difference in signal to noise is evident. As we have seen, FD-MS-MS spectra may be obtained using the point detector for pure compounds. In many cases, however, we wish to characterise such compounds when present at much lower concentrations in complex mixtures and it is at this point that the limited sensitivity of FD can present practical problems and require the use of the array detector.

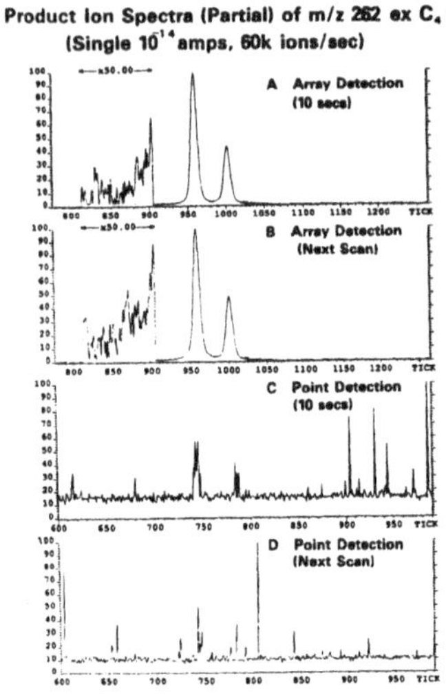

Figure 28.

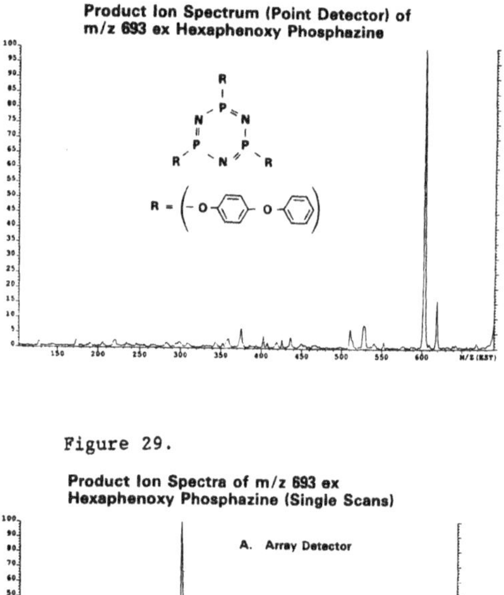

Figure 29.

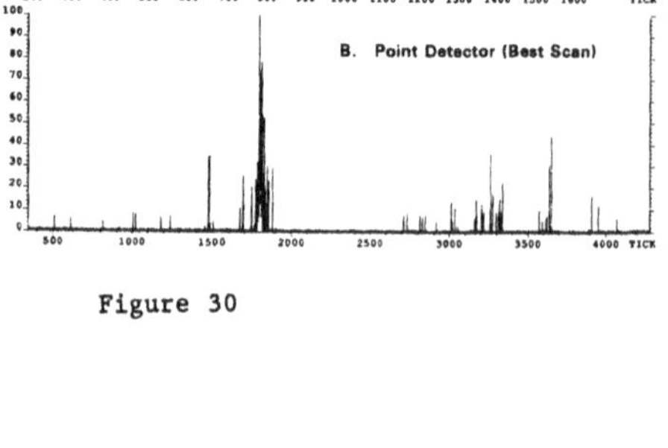

Figure 30

Although the array detector gives access to a 1.4:1 portion of the product ion spectrum in order to observe the whole product ion spectrum a number of separate exposures need to be made and then fitted together. Figure 31 shows the product ion spectrum (point detector) of m/z 546 obtained from the EI spectrum of an aromatic oligomer of structure

```
    O        O
    ||       ||
ø ø C ø O ø C ø O ø.
```

Figure 32 shows the equivalent array spectrum obtained from eight x two second acquisitions. The agreement is excellent. Figure 33 shows the product ion spectrum, obtained on the point detector, for the equivalent ion generated from field desorption. This indicates that, at least for this example, internal energy does not play a large part in defining the product spectrum.

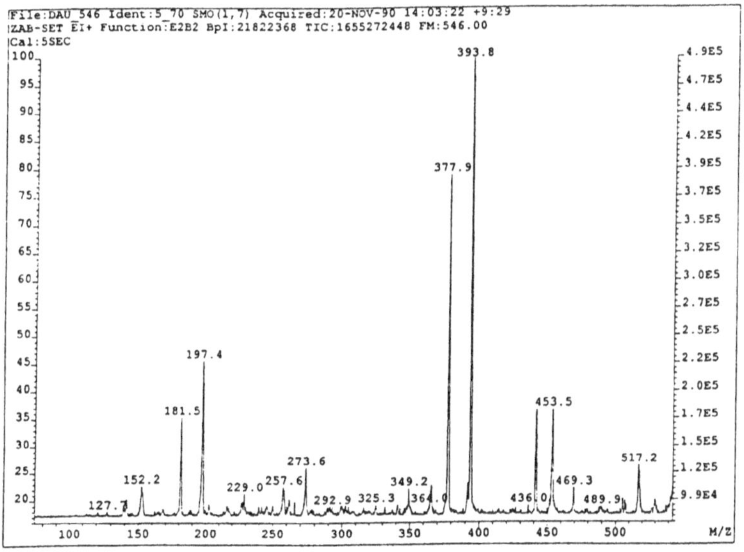

Figure 31. Product Ion Spectrum of m/z 546 (pt Detector).

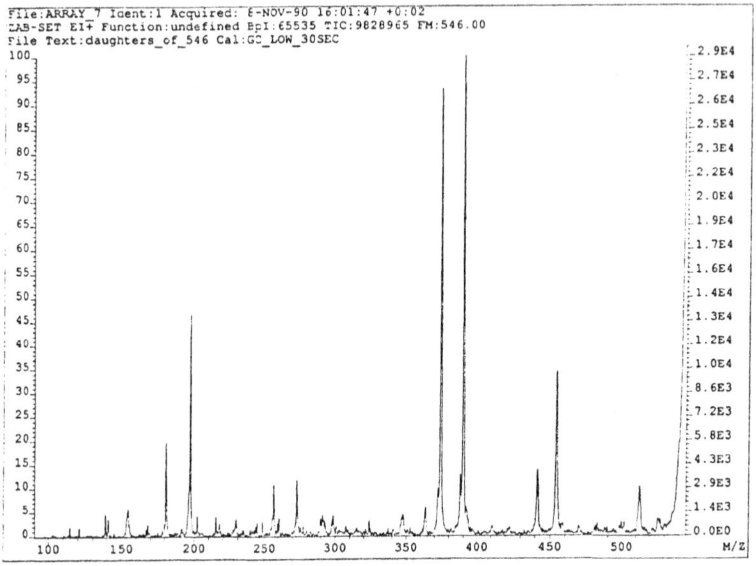

Figure 32. Product Ion Spectrum of M/Z 546 (Array).

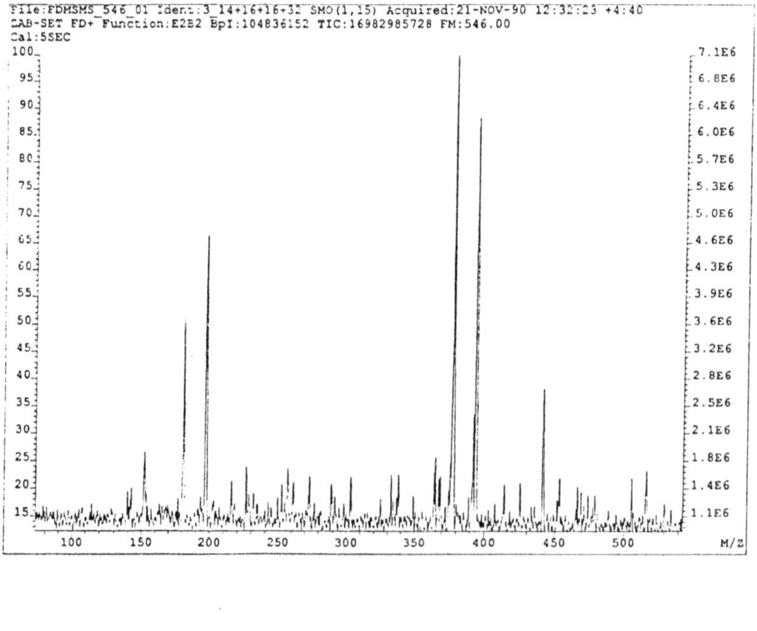

Figure 33. Product Ion Spectrum of M/Z 546 (pt Detector, FD).

7 CONCLUSION

We have seen how the use of tandem mass spectrometry approaches to gas phase ion structural determination plays an important role in industrial problem solving. The use of field desorption techniques to generate the ions of interest has considerably extended the range of compound classes and mass ranges accessible by mass spectrometry. The combination of field desorption with four sector instrumentation equipped with array detection provides a powerful new technique with enormous potential. In the future attention will have to be paid to the area of ion activation in order to extend the mass range available for structural studies with surfaces, electrons or photons being attractive possibilities.

REFERENCES

1 McLafferty, F W (ED) (1983) Tandem mass spectrometry, Wiley, New York.

2 Busch, K L, Glish G L, McLuckey S A, (1988) 'Mass spectrometry/mass spectrometry:Techniques and applications of tandem mass spectrometry, VCH, New York.

3 Mabud, M A ; DeKrey, M J ; Cooks, R G (1985) Int J Mass Spectrom Ion Proc, 67, 285-294.

4 Cody, R B ; Freiser, B S (1987) Anal Chem, 59, 1054-1056.

5 Hill, J LA ; Annan, R S ; Biemann, K (1990) ASMS Proc Tuscon, 495-497.

6 Horning, S R ; Wood, J M ; Gord, J R ; Freiser, B S, Cooks, R G (1990) ASMS Proc Tuscon, 483-484.

7 Chen, L ; Wang, T-C L ; Ricca, T L ; Marshall, A G (1987) Anal Chem, 59, 449-454.

8 Dolnikowski, G G ; Heath, T G ; Throck Watson, J ; Scrivens, J H ; Rolando, C H ; Morden, W (1990) J Am Soc Mass Spectrom 1 481-488.

9 Prokai, L 'Field desorption mass spectrometry' 1990, Marcel Dekker, New York.

10 Matsuo, T ; Matsuda, H ; Katakuse, I (1979) Anal Chem 51 1329.

11 Cottrell, J S ; Evans, S (1987) Anal Chem 59 1990.

ACKNOWLEDGEMENTS

The authors would like to thank the following people for their involvement in the work described in this review.

Prof K R Jennings Dr G Dolwkowski	)	Warwick University
Prof J J Monaghan Mr W Morden	)	ICI Chemicals and Polymers
Dr D Weil		Extrel
Mr M Taylor		ICI Advanced Materials
Mrs H Major Mr R H Bateman	)	VG Analytical

APPLICATIONS OF MASS SPECTROMETRY IN AN INDUSTRIAL RESEARCH LABORATORY

Jan C. Kleingeld*, Fred J. Fels and Wim J.L. Genuit
Koninklijke/Shell-Laboratorium, Amsterdam (Shell Research B.V.)
P.O. Box 3003
1003 AA Amsterdam
The Netherlands

ABSTRACT. Mass spectrometry has many applications in the research laboratory of Shell in Amsterdam, including:

- high resolution studies of oil feedstocks and products,
- identification of components in complex mixtures, and
- accurate mass measurements of a number of compounds.

Of all of these applications examples are given.

1. INTRODUCTION

For quite a long time now, mass spectrometry has been playing an important role in the Research Laboratory of Shell in Amsterdam. In fact, this year (1990) commemorates the fortieth anniversary of the installation of the first mass spectrometer in our Laboratory [1]. Mass spectrometry, which is one of the techniques applied in the analytical department, is continuously used to assist the various research departments in solving their analytical problems. These problems may range from malfunctioning of pilot plants, such as a sudden occurrence of undesired by-products or a decrease in product yield, to the analysis of newly synthesised products. It will be clear that such a variety of problems will require a variety of tools to help solving them. The research that is performed by the mass spectrometry group is, therefore, highly focussed on keeping up with, and where possible keeping ahead of, the analytical needs that exist within the Laboratory. This research is required not only to keep abreast of newly developed mass spectrometric techniques, but also to keep the existing techniques up-to-date, an example being the recent installation of the so-called 'GC/MS Quick Service'. Although genuine routine analyses, in the sense that a large number of samples all have to be tested for a single component or condition, are hardly ever encountered, the rapid growth of GC/MS

---

*Present address:
Koninklijke/Shell Exploratie en Produktie Laboratorium,
P.O. Box 60, 2280 AB Rijswijk, The Netherlands.

K. R. Jennings (ed.), Fundamentals of Gas Phase Ion Chemistry, 417–435.

(combined gas chromatography/mass spectrometry) has called for a more standardised treatment of GC/MS analysis requests. Therefore, the possibility is now offered to analyse samples in an entirely automated way under standard GC/MS conditions, i.e. standard column and temperature program conditions, and then to have the compounds of interest identified by a comparison of their mass spectra with those in a mass spectral library. This will give the mass spectrometrists more time to spend on the more difficult requests, among which might very well be the ones that where not completely answered by the 'GC/MS Quick Service'.

This paper intends to show how mass spectrometry is applied in an industrial research laboratory. It will give a selected number of examples of analyses that have actually been performed in the past few years at the Shell Laboratory in Amsterdam.

2. OIL ANALYSIS

2.1 *Gallegos group-type analysis.* One of the earliest applications of mass spectrometry within the oil industry is, quite naturally, the analysis of oil itself. In particular, high resolution mass spectrometry (HR-MS) is very well suited to the analysis of oils. The main advantage of HR-MS is that it eliminates most of the interferences between saturated and aromatic hydrocarbons. Therefore, no prior separation of the sample is needed. For a very long time, we have applied the group-type analysis method introduced by Gallegos et al. [2]. This method uses some of the characteristic fragment ions of the various types of hydrocarbons present in petroleum fractions to calculate the distribution of these hydrocarbon types. For example, the homologous series m/z 69, 83, 97 etc. is characteristic of the ions formed by fragmentation of the monocycloparaffins (and of alkenes, which,therefore, may not be present in the samples analysed). Of course, these ions are also (minor) fragmentation products of other hydrocarbons, so that a correction for their presence has to be made. This is done by a matrix calculation, covering all relevant fragment ions, which results in a distribution list of 19 compound types [2] (see Table 1).

An example of the use of this method is given in Table 2, which lists the compound types found in several oil products. The difference in nature between the various products is very clear: the absence of aromatic compounds in the naphthenic oil, the difference in paraffins content between the waxy oil and the wax as well as the much cleaner nature, i.e. almost exclusive presence of paraffins, of the synthetic wax show up very nicely.

TABLE 1. The compound types used in the group-type analysis by high resolution mass spectrometry [2].

Compound type	Example of structure
PARAFFINS	C_nH_{2n+2}
MONOCYCLOPARAFFINS	C_nH_{2n}
DICYCLOPARAFFINS	C_nH_{2n-2}
TRICYCLOPARAFFINS	C_nH_{2n-4}
TETRACYCLOPARAFFINS	C_nH_{2n-6}
PENTACYCLOPARAFFINS	C_nH_{2n-8}
HEXACYCLOPARAFFINS	C_nH_{2n-10}
ALKYLBENZENES	
BENZOCYCLOPARAFFINS	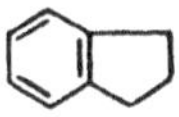
BENZODICYCLOPARAFFINS	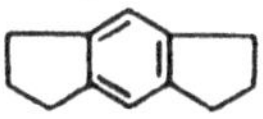
NAPHTHALENES	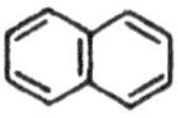
ACENAPHTHENES	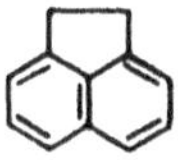
FLUORENES	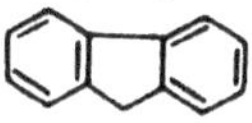
PHENANTHRENES	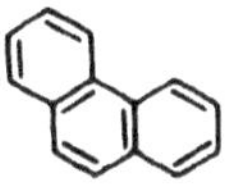
PYRENES	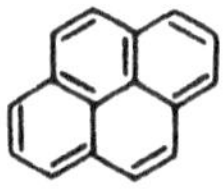
CHRYSENES	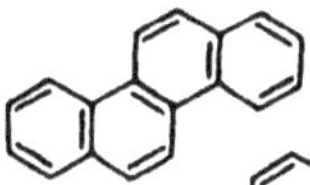
BENZOTHIOPHENES	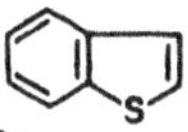
DIBENZOTHIOPHENES	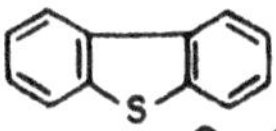
NAPHTHOBENZOTHIOPHENES	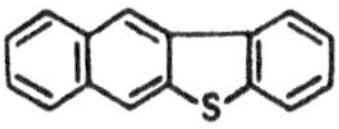

TABLE 2. Compound type distribution of several oil products.

1: naphthenic lubrication oil, 2: waxy oil, 3: wax, and 4: synthetic wax.

Compound type	1	2	3	4
Paraffins	4%	31%	71%	97%
Monocycloparaffins	14%	24%	15%	3%
Dicycloparaffins	16%	16%	3%	0%
Tricycloparaffins	17%	8%	3%	0%
Tetracycloparaffins	31%	8%	1%	0%
Pentacycloparaffins	13%	5%	0%	0%
Hexacycloparaffins	5%	2%	0%	0%
Alkylbenzenes	0%	3%	4%	0%
Benzocycloparaffins	0%	1%	1%	0%
Benzodicycloparaffins	0%	1%	0%	0%
Naphthalenes	0%	0%	0%	0%
Acenaphthenes	0%	0%	0%	0%
Fluorenes	0%	0%	0%	0%
Phenanthrenes	0%	0%	0%	0%
Pyrenes	0%	1%	0%	0%
Chrysenes	0%	0%	0%	0%
Benzothiophenes	0%	0%	0%	0%
Dibenzothiophenes	0%	0%	0%	0%
Naphthobenzothiophenes	0%	0%	0%	0%

2.2 *Field Ionisation Mass Spectrometry.* A disadvantage of the Gallegos method is that it fails to yield information about molecular weight distributions within the various hydrocarbon types. A technique which does provide molecular weight information, and consequently is of increasing importance in the analysis of oils, is Field Ionisation Mass Spectrometry (FI-MS). FI-MS spectra predominantly feature the molecular ions of the oil constituents (see, for example, Figure 1). From these spectra, it is possible to calculate the number of carbon and hydrogen atoms present in the various molecular ions and hence to display the distribution of the hydrocarbons such as shown in Figure 2, which depicts a base oil sample before and after a dewaxing treatment (to avoid confusion, it should be noted, that this is not the same sample as shown in Figure 1). This treatment removes normal paraffins and leaves all other compounds unaffected. The effect of the treatment can clearly be seen in Figure 2b. The concentration of paraffins (z=2) has decreased considerably (although not completely due to the presence of iso-paraffins), whereas the concentration of the other components is not affected.

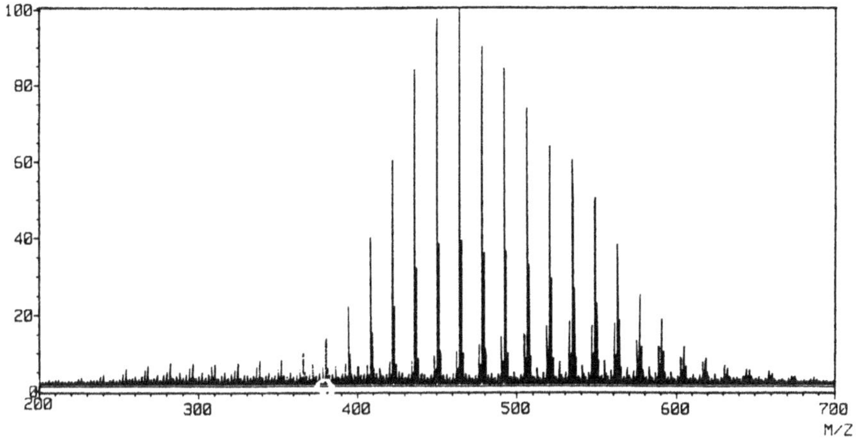

Figure 1. Field ionisation mass spectrum of a wax sample.

The FI-MS method as described here has gradually replaced the old Gallegos method, and is now used as a powerful tool for the analysis of all kinds of petroleum fractions, thereby providing a better insight into the way in which these fractions are made and converted, which is very important in the ongoing research on hydrocarbon processes.

3. IDENTIFICATION OF REACTION (BY-)PRODUCTS

The identification of compounds in complex mixtures and/or in low concentrations is an important task of the mass spectrometrist in an industrial laboratory. Most of these questions are addressed by the combined use of gas chromatography and mass spectrometry (GC/MS). In many cases a compound is identified by comparing its mass spectrum with those in a mass spectral library, as available nowadays with most mass spectral data systems. Since many samples do, however, consist of compounds whose spectra have not been recorded before, a library search fails to produce a satisfactory answer on quite a number of occasions. In such cases the analysis comes down to the interpretation of the mass spectrum, thereby incorporating as many additional information as can possibly be obtained.

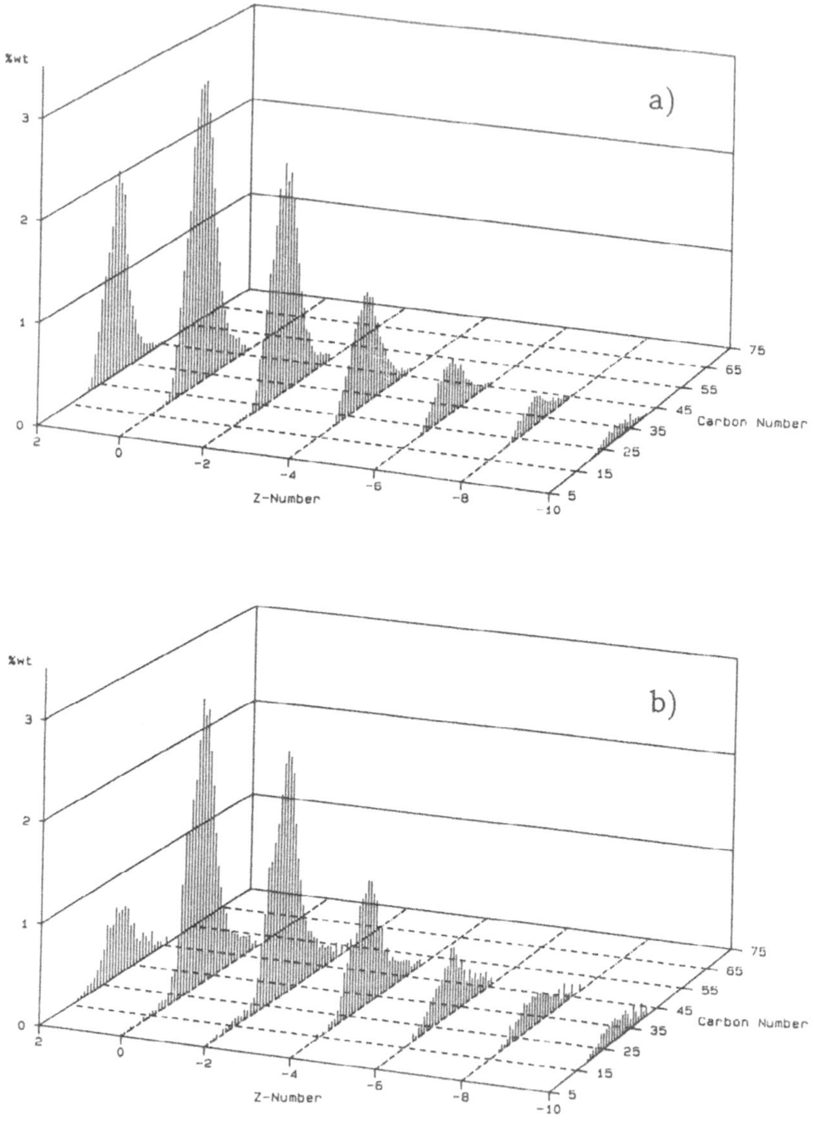

Figure 2. Distribution of hydrocarbon components in a base oil (a) before and (b) after a dewaxing treatment. The components are arranged according to the general formula for a hydrocarbon: C_nH_{2n+z}, where n is the carbon number.

An example of such an analysis is the identification of a by-product in the transesterification reaction of methyl propionate by butanol to butyl propionate:

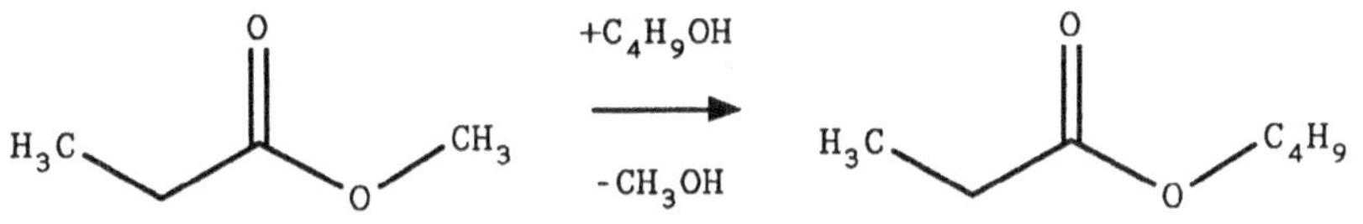

Scheme 1. Transesterification reaction of methyl propionate to butyl propionate.

The Electron Impact (EI) mass spectrum of this by-product is given in Figure 3a. The presence of an abundant ion m/z 57 is indicative of an ethylcarbonyl ($C_2H_5C(O)$-) or of a butyl group (C_4H_9-). The observation of the small peak at m/z 186 might indicate that the molecular weight of the component is 186 amu. This is supported by the occurrence of a peak at m/z 157, which can be explained by the loss of an ethyl radical from the molecular ion, a very common process, in particular in ethylcarbonyl-substituted molecules.

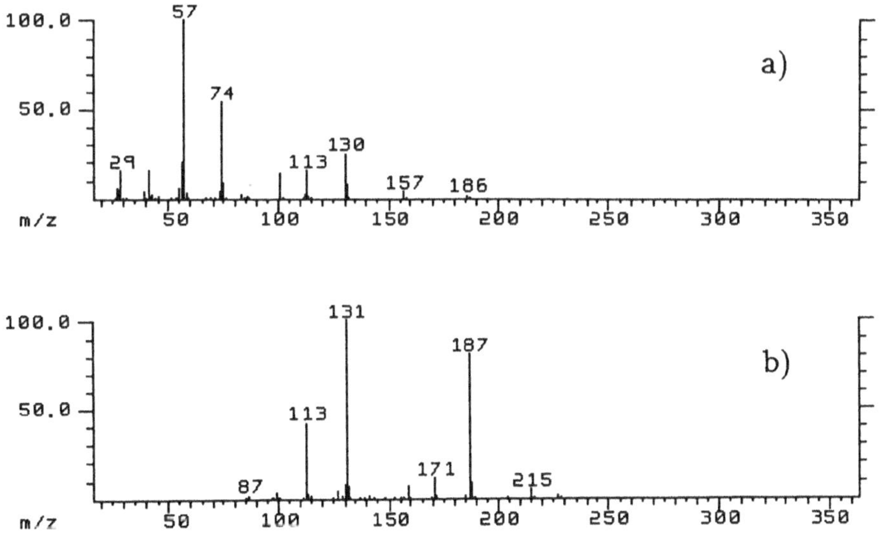

Figure 3. (a) Electron Impact and (b) Chemical Ionisation mass spectra of an unknown product in the transesterification reaction of methyl propionate to butyl propionate.

Further information can be derived from the Chemical Ionisation mass spectrum, which is shown in Figure 3b. The first conclusion that can be drawn from this spectrum is that the molecular weight of the compound is 186 amu indeed, as shown by the occurrence of ions m/z 187 $(M+H)^+$, 215 $(M+C_2H_5)^+$ and 227 $(M+C_3H_5)^+$. Furthermore, conclusions can be drawn from the observation of the abundant ion m/z 131. Depending on the exothermicity of the protonation reaction, the resulting protonated ester ion can fragment via the mechanism shown in Scheme 2 [3].

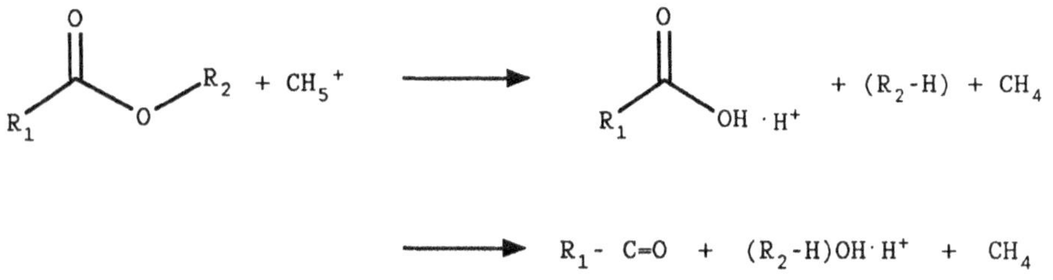

Scheme 2. Fragmentation of alkyl esters.

The resulting protonated acid or alcohol ion can then lose a water molecule (Scheme 3).

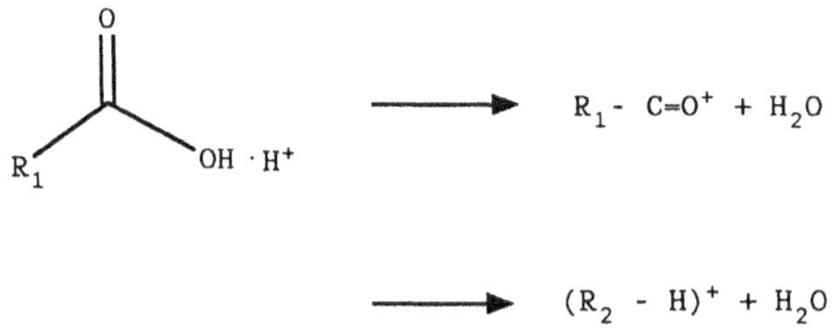

Scheme 3. Loss of water from protonated acid or alcohol ions.

The observation of an ion m/z 131, which corresponds to the loss of a C_4H_8-group from the $(M+H)^+$ ion, therefore warrants the conclusion that the compound contains a butyl-ester function (see Scheme *2*, R_2- is butyl-, and Scheme *3*, the ions m/z 113 are formed via the loss of a water molecule from the ions m/z 131). The observation of the ions m/z 159 and 171, formed by the loss of a C_4H_8-group from the $(M+C_2H_5)^+$ and

$(M+C_3H_5)^+$ ions, respectively, supports this conclusion. When this piece of information is combined with the information obtained from the EI spectrum that part of the molecule consists of an ethylcarbonyl group, it may be concluded that the structure of R_1- is $C_2H_5C(O)C_2H_4$-, although C_6H_{13}- cannot be ruled out entirely.

Conclusive evidence is extracted from the results of accurate mass measurements on the EI spectrum of the compound (see Table 3; the molecular ion m/z 186 is too small to allow accurate mass measurements, whereas ions below m/z 100 cannot at present be mass measured due to calibration restraints in the software).

TABLE 3. Results of accurate mass measurements on the electron impact mass spectrum of an unknown compound in the transesterification of methyl propionate, showing the three 'best' fits and assuming the presence of C, H and O only.

Nominal Mass (amu)	Measured Mass (a) (amu)	Possible Composition	Calculated Mass (b) (amu)	Diff. (a-b) (mmu)
157	157.08552	$C_9H_{17}O_2$	157.12285	-37.3
		$C_8H_{13}O_3$ $(M-C_2H_5)$	157.08646	0.9
		$C_7H_9O_4$	157.05008	35.4
130	130.10046	$C_8H_{18}O$	130.13576	-35.3
		$C_7H_{14}O_2$ $(M-C_2H_4CO)$	130.09938	1.1
		$C_{10}H_{10}$	130.07825	22.2
113	113.06061	$C_7H_{13}O$	113.09663	-36.0
		$C_6H_9O_2$ $(M-C_4H_9O)$	113.06025	0.4
		C_5H_5	113.02387	36.7

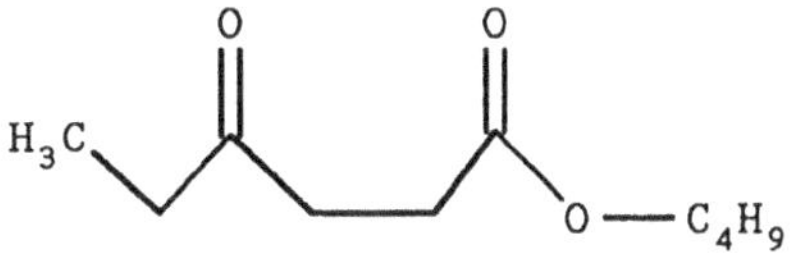

butyl 4-oxo-hexanoate

1

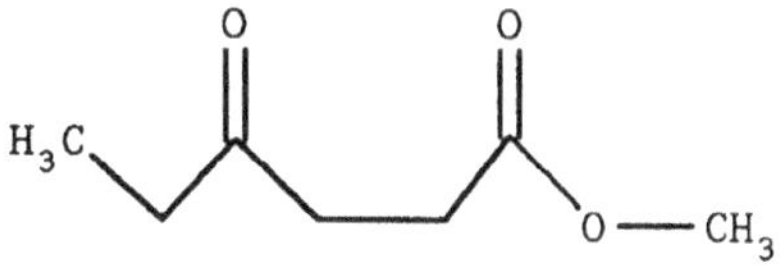

methyl 4-oxo-hexanoate

2

These results clearly show the presence of three oxygen atoms in the molecule and, therefore, justify the conclusion that the compound is butyl 4-oxo-hexanoate (1), which is formed by the transesterification of methyl 4-oxo-hexanoate (2), known to be present in small amounts in the methyl propionate used.

The formation of this compound can be understood when one knows that methyl propionate is synthesised by a reaction of ethene, carbon monoxide and methanol (Scheme 4a): a reaction with additional molecules of ethene and carbon monoxide will then produce methyl 4-oxo-hexanoate (Scheme 4b).

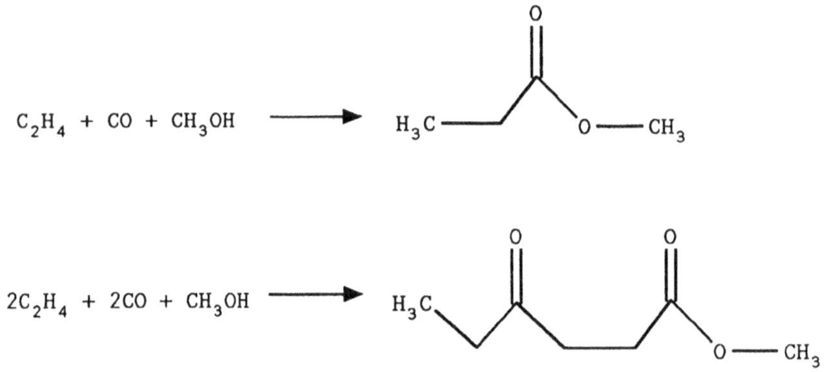

(a)

(b)

Scheme 4. Formation of methyl propionate (a) and methyl 4-oxo-hexanoate (b).

4. ACCURATE MASS MEASUREMENTS

In the previous section, the use of accurate mass measurements by HR-MS was briefly mentioned. This method has in recent years experienced what might be called a revival due to the possibility of performing high resolution mass scans during GC/MS analyses. This possibility circumvents one of the problems encountered when analysing samples introduced via a batch inlet system or a probe, in that in such cases mixtures of compounds such as isomers, which produce common (fragment) ions, are very hard or even impossible to analyse. In the previous section, the application of GC/HR-MS to a practical problem has already been indicated. Here, we will give an example of the structure elucidation of several isomers.

The reaction of diphenylphosphinotrimethoxybenzene (3) with butyllithium in hexane [4] yields several products. The mass spectra of some of those products are shown in Figure 4. The accurate masses of their molecular ions are given in Table 4.

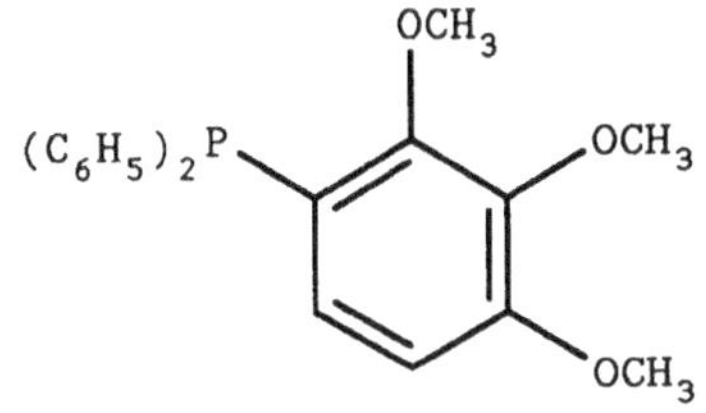

3

Diphenylphosphinotrimethoxybenzene

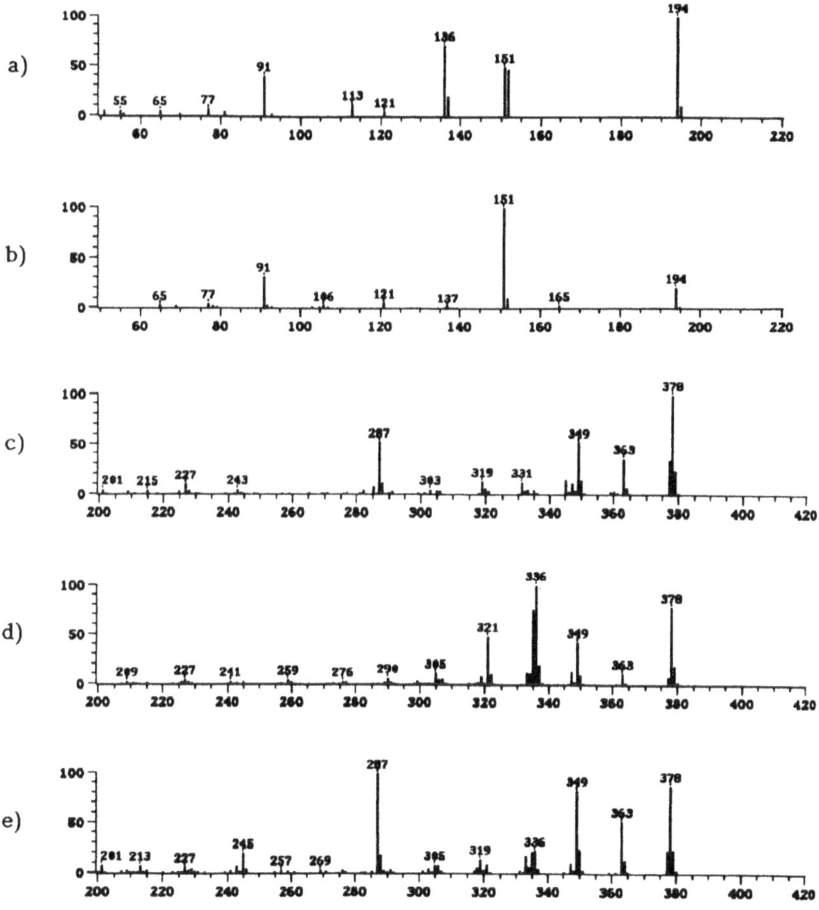

Figure 4. High resolution electron impact mass spectra of several products of the reaction of diphenylphosphinotrimethoxybenzene with butyllithium in hexane [4].

Table 4. Results of accurate mass measurements on the molecular ions of several products from the reaction of diphenylphosphinotrimethoxybenzene with butyllithium in hexane [4].

Compound Number	Proposed Composition	Calculated Mass (a) (amu)	Measured Mass (b) (amu)	Diff. (a-b) (mmu)
4	$C_{12}H_{18}O_2$	194.13068	194.13296	-2.28
5	$C_{12}H_{18}O_2$	194.13068	194.13053	0.15
6	$C_{24}H_{27}PO_3$	378.17487	378.17429	0.57
7	$C_{24}H_{27}PO_3$	378.17487	378.17574	-0.87
8	$C_{24}H_{27}PO_3$	378.17487	378.17463	0.24

Of course, the sole measurement of accurate molecular weights does not prove the structure of any of these compounds. However, the spectra shown in Figure 4 do provide more information than just molecular ions. For example, the spectrum of one of the two compounds with elemental composition $C_{12}H_{18}O_2$ (4, Figure 4a) demonstrates the presence of $(M-C_3H_6)^+$ ions (m/z 152) and $(M-C_4H_{10})^+$ ions (m/z 136). These ions are absent in the spectrum of the other compound with this elemental composition (5, Figure 4b). This difference can be explained from the mechanism of formation of the ions m/z 152 (Scheme 5, the ions m/z 136 are most likely formed by loss of a methane molecule from the ions m/z 152).

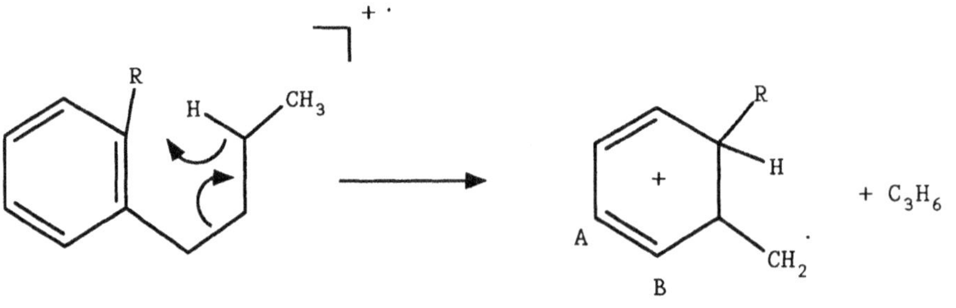

a

Scheme 5. Formation of m/z 152 ions.

The resulting product ion a is stabilised by the presence of a methoxy substituent in the position labelled A, whereas it is not when the methoxy group is substituted in position B [5]. This provides a clear distinction between the possible isomers and supports the identification of these two components as 4 and 5, respectively.

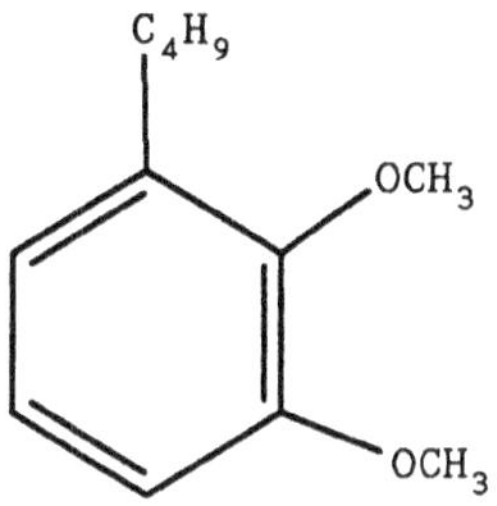

4

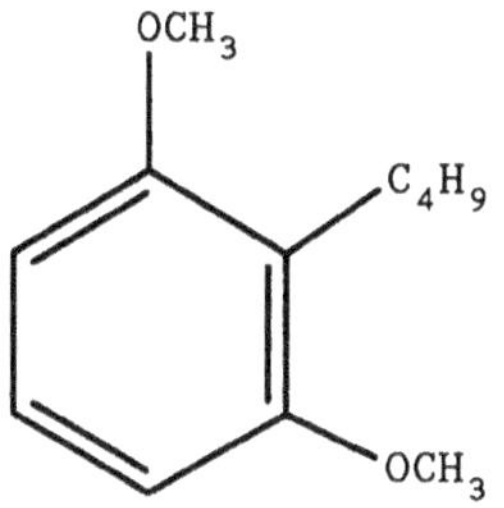

5

The presence of $(M-C_3H_6)^+$ ions does also assist in distinguishing the various isomers with elemental composition $C_{24}H_{27}PO_3$ (6, 7 and 8), where the $(M-C_3H_6)^+$+ ions will show at m/z 336. Another feature in the mass spectra of these compounds (see Figure 4c-e) is the formation of $(M-C_7H_7)^+$ ions (m/z 287) in certain cases. These ions can only be formed when the diphenylphosphino and methoxy groups are substituted at ortho-positions with respect to each other. Both pieces of information support the identification of compounds 6, 7 and 8.

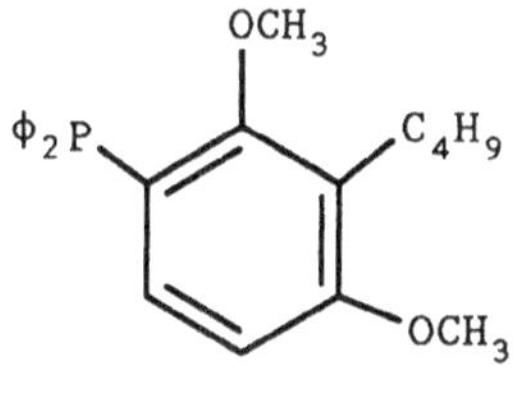

6

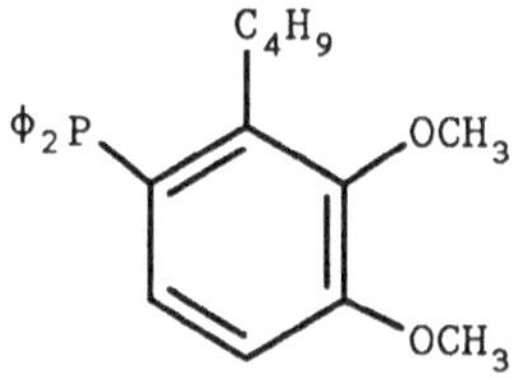

7

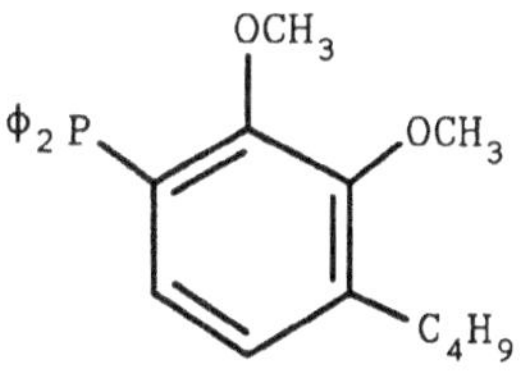

8

As mentioned previously, the measurement of accurate masses is sometimes very difficult. This was the case in analysing a compound with elemental composition $C_{20}H_{14}N_2Cl_2$ (molecular weight 352 amu). The sample appeared to consist of various chlorinated homologues, which made it impossible to analyse it by introduction via a direct insertion probe in electron impact mode, since the molecular ion m/z 352 was largely overshadowed by the various fragment ions of these homologues. Furthermore, the compound appeared to have too high a boiling point to be amenable to a GC/MS analysis. Therefore, it was decided to analyse this compound by FI-MS. The FI mass spectrum (shown in Figure 5) appears to be very clear,

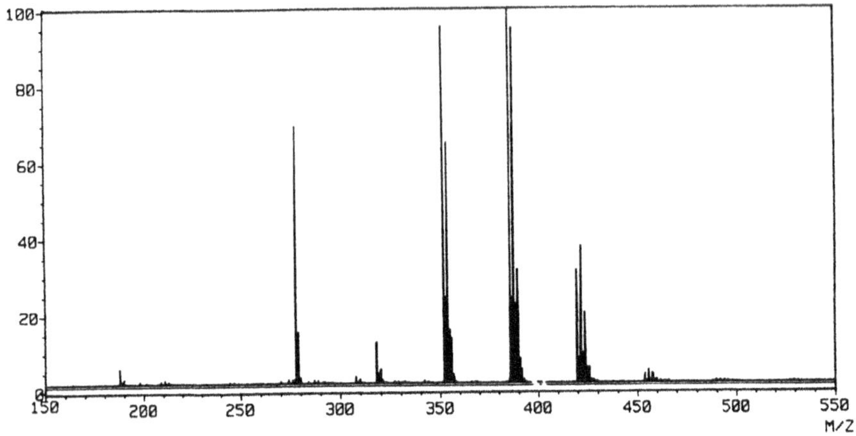

Figure 5. Field ionisation mass spectrum of a compound with elemental composition $C_{20}H_{14}N_2Cl_2$

lacking all fragmentation: the molecular ion at m/z 352 (and, of course, its isotopes) is totally free of interference. The accurate mass of this ion can now be measured by introducing some normal alkanes as reference material (in this case n-$C_{24}H_{50}$, n-$C_{25}H_{52}$ and n-$C_{26}H_{54}$ were used), revealing an accurate mass of 352.0564 amu (calculated mass 352.0533 amu). The accurate mass of the 13C-isotope at m/z 353 is measured to be 353.0560 amu (calculated mass 353.0564).

In conclusion, it can be stated that the power of accurate mass measurements by HR-MS is still unbroken.

5. ANALYSIS OF VERY COMPLEX MIXTURES

In a number of cases it is not feasible to identify every single component in a mixture. One type of example has already been given in the paragraph on oil analysis, where the oils are analysed according to their hydrocarbon component types rather than according to every single constituent. Another example can be found in the analysis of compounds believed to be responsible for the instabilities of oil products during storage. Because oil products often show colouring and gum and sludge formation upon prolonged storage, it is important to know what components in these products are responsible for this behaviour so that countermeasures can be taken. Already for quite a long time the presence of nitrogen-containing compounds was believed to be a key factor in this respect. Therefore, several oils have been analysed by GC/MS to identify the various types of nitrogen components present.

GC/MS analysis of nitrogen compounds in oils is nowadays reasonably straightforward due to the high calculating power of modern mass spectral data systems. The analysis basically comes down to recording the consecutive mass spectra during the GC run and then having the data system search through all these spectra (the number of which may easily amount to around 3000) for molecular and/or fragment ions that are characteristic of the compounds of interest. With present day processors this can be done in a very short time. Of course, looking for all the characteristic ions of (in the present example) all possibly occurring nitrogen-containing compounds can still be a laborious task, in particular because it is still necessary to inspect the complete mass spectra to avoid incorrect identifications, but the total analysis time is considerably reduced. It should also be noted, that, in order to increase the detection limit, the analyses are performed on fractions in which the nitrogen compounds have been concentrated [6].

An example of a successful analysis is given in Figure 6, which shows the identification of dibenzocarbazole (9, molecular weight 167 amu) and its homologues in a light cycle oil.

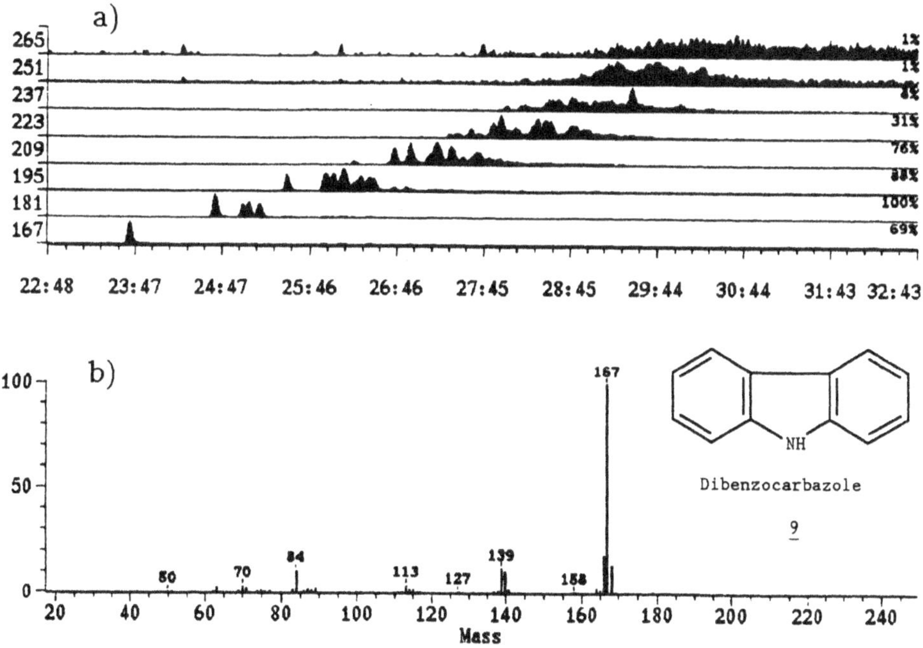

Figure 6. (a) Selected traces of the molecular ions of dibenzocarbazole (9, m/z 167) and its homologues (m/z 181 through 265). (b) Mass spectrum of the compound eluting at 23:44, identified as dibenzocarbazole (9).

6. LIQUID CHROMATOGRAPHY / MASS SPECTROMETRY

If one technique should be mentioned that is rapidly proving itself to be a useful extension to the mass spectrometric toolbox within Shell, it is the combination of liquid chromatography and mass spectrometry (LC/MS). This technique is very well suited to the analysis of compounds which are not amenable to GC/MS, for instance compounds that have too high a molecular weight or that are thermally labile. Classes of compounds for which this holds in particular are e.g. polymer and oil additives, homogeneous catalysts, and epoxy resins.

In Amsterdam use is made of the thermospray LC/MS technique [7]. An example of its use is the analysis of the reaction products of diphenylolpropane (DPP, 10) and allyl chloride. As shown in Scheme 6 this reaction was expected to give two products: the major reaction product in which both hydroxide groups of the DPP had reacted (12) and a minor one in which only a single hydroxide group had reacted (11). A consecutive work-up step would then remove all molecules with free hydroxide groups, hence leaving the major reaction product.

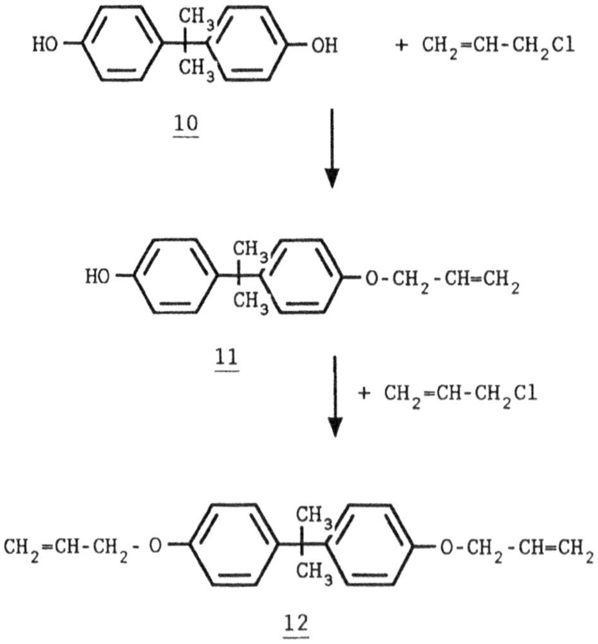

Scheme 6. Reaction of diphenylolpropane and allyl chloride

In Figure 7 it can be seen that the initial reaction mixture consists of four products that are detectable by LC/MS (see Figure 7a, the products are labelled A, B, C and D) and that after the work-up step two products still remain (see Figure 7b). The mass spectra of all four products are given in Figure 8.

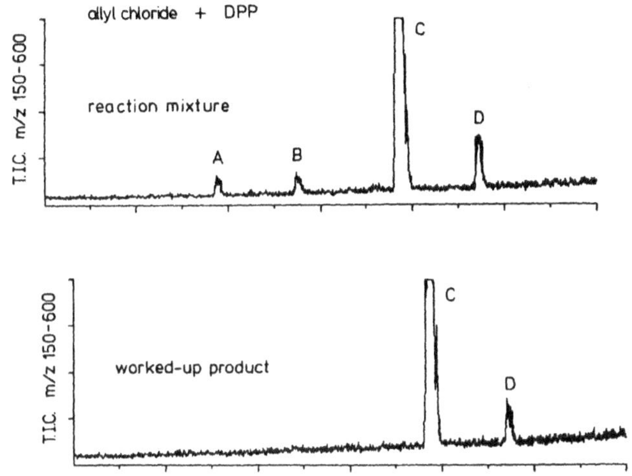

Figure 7. LC/MS chromatogram of the products of the reaction of diphenylolpropane (10) and allyl chloride (a) before and (b) after a work-up step to remove molecules with free hydroxide groups.

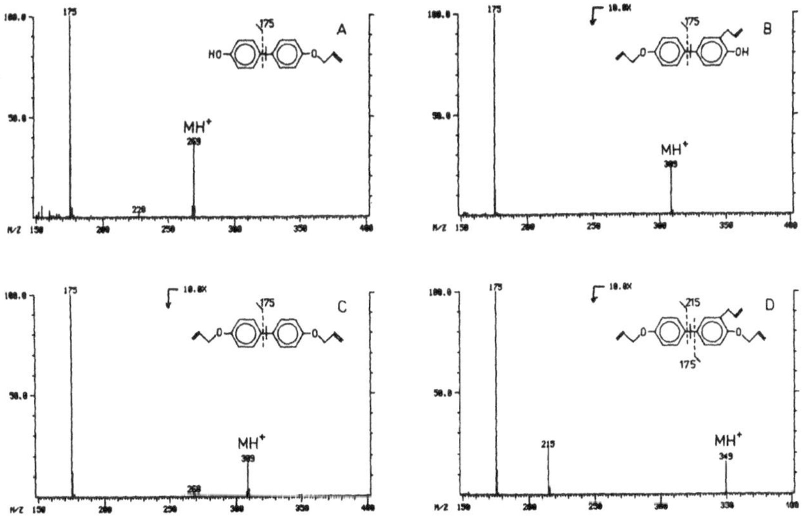

Figure 8. Thermospray mass spectra of the products of the reaction of diphenylolpropane (10) and allyl chloride.

It is not very difficult to identify the two expected reaction products (structures 11 and 12) as products A and C, respectively (see the spectra shown in Figure 8a and 8c, the ions m/z 175 are formed by cleavage of the molecular ion in between the phenyl ring and the propane backbone). The spectrum of product B (Figure 8b) closely resembles that of the major reaction product (Figure 8c). There is, however, one major difference between the two compounds: product B does disappear in the work-up step and is thus known to possess a free hydroxide group. Its formation is explained by the "Claisen rearrangement reaction" shown in Scheme 7 [8].

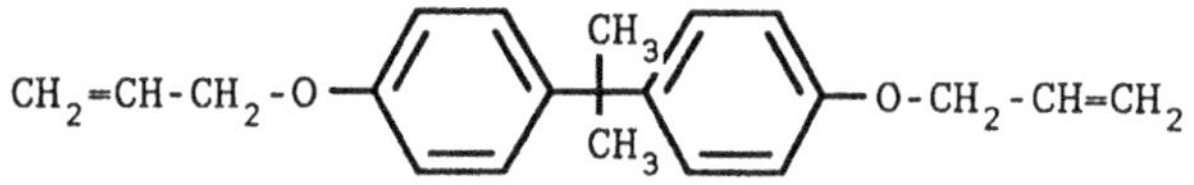

12

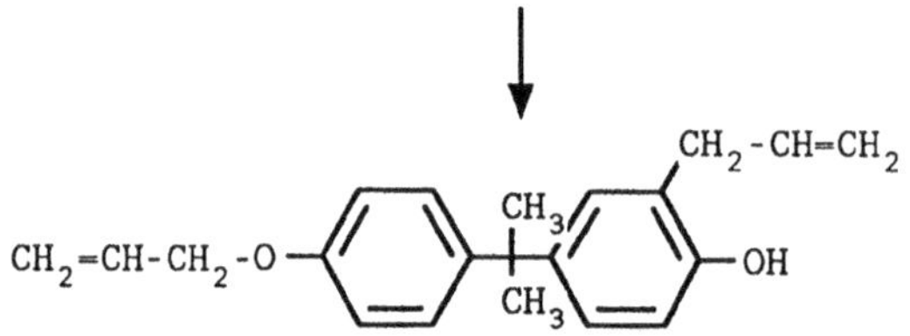

13

Scheme 7. Claisen rearrangement reaction.

Now that the structure of compound B (13) is known, the formation of compound D can easily be explained: the free hydroxide group of B reacts with another allyl chloride to form a compound with structure 14, which explains the formation of ions m/z 175 and 215 (see Figure 8d) due to fragmentation of the molecular ion on either side of the propane backbone).

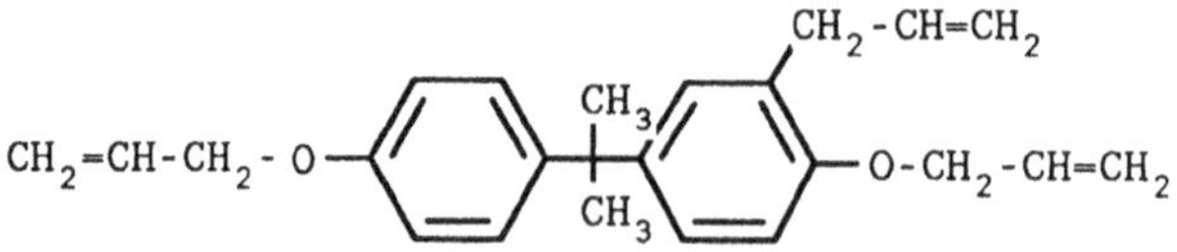

14

7. CONCLUDING REMARKS

In this paper we have tried to give some idea of the type of analyses requested from the mass spectrometry group at the Shell Laboratory in Amsterdam as well as the way in which these analyses are performed. It can in no way give a complete picture of the day-to-day life of a mass spectrometrist in the laboratory, but, on the other hand, the examples given were no exceptions to the ordinary analytical requests.

8. ACKNOWLEDGEMENTS

The authors kindly acknowledge the cooperation of their colleagues Jaap de Boer, Martin van Grondelle, Martin Nekkers, Herman van Binsbergen and Arjan van Dijk in preparing this manuscript.

9. REFERENCES

1. Van Katwijk, J. and Kunst, E.D., CEC Recordings, Fall Issue (1959).

2. Gallegos, E.J., Green, J.W., Lindeman, L.P., LeTourneau, R.L. and Teeter, R.M., Anal. Chem. 39, 1833 (1967).

3. Harrison, A.G., Chemical Ionization Mass Spectrometry, CRC Press, Florida (1983).

4. Budzelaar, P.H.M. and Van Doorn, J.A., Recl. Trav. Chim. Pays-Bas 109, 443-445 (1990).

5. Williams, D.H. and Howe, I., Principles of Organic Mass Spectrometry, McGraw-Hill, London (1972), p. 125.

6. Dorbon, M. and Bernasconi, C., Fuel 68, 1067 (1989).

7. Blakley, C.R. and Vestal, M.L., Anal. Chem. 55, 750-754 (1983).

8. Fessenden R.J. and Fessenden, J.S., Organic Chemistry, 3rd ed., Brooks/Cole Publishing Company, Monterey USA (1986), p. 829.

Index

Batch number: 10809439

Printed by Printforce, the Netherlands